Nuclear Receptors
in Development

T0348752

NUCLEAR RECEPTORS IN DEVELOPMENT

Editor

Reshma Taneja

Mount Sinai School of Medicine
Brookdale Department of Molecular
Cell and Developmental Biology
New York

2006

ELSEVIER

AMSTERDAM • BOSTON • HEIDELBERG • LONDON • NEW YORK • OXFORD
PARIS • SAN DIEGO • SAN FRANCISCO • SINGAPORE • SYDNEY • TOKYO

Elsevier
525 B Street, Suite 1900, San Diego, California 92101-4495, USA
84 Theobald's Road, London WC1X 8RR, UK

This book is printed on acid-free paper.

For information on all Academic Press publications
visit our Web site at www.books.elsevier.com

ISBN-13: 978-0-444-52873-5
ISBN-10: 0-444-52873-3

Printed and bound by CPI Group (UK) Ltd, Croydon, CR0 4YY

Transferred to Digital Print 2011

Contents

List of Contributors

W. Ray Anderson   Department of Cell Biology and Molecular Genetics, University of Maryland, College Park, Maryland

Gerard M. J. Beaudoin, III   Kennedy Krieger Research Institute, Baltimore, Maryland; Department of Neuroscience, Johns Hopkins University School of Medicine, Baltimore, Maryland

Gaétan Bour   Institut de Génétique et de Biologie Moléculaire et Cellulaire, CNRS/INSERM/Université Louis Pasteur, UMR 7104, BP 10142, 67404 Illkirch Cedex, France

Don Cameron   Department of Biochemistry, Division of Cancer Biology and Genetics, Cancer Research Institute, Queen's University, Kingston, Ontario, Canada K7L 3N6

Béatrice Desvergne   Center for Integrative Genomics, National Research Center "Frontiers in Genetics," University of Lausanne, Lausanne, Switzerland

Pascal Dollé   Institut de Génétique et de Biologie Moléculaire et Cellulaire, UMR 7104 du CNRS, U. 596 de l'INSERM, Université Louis Pasteur, BP 10142, 67404 Illkirch Cedex, CU de Strasbourg, France

Douglas Forrest   Clinical Endocrinology Branch, National Institute of Diabetes and Digestive and Kidney Diseases, National Institutes of Health, Bethesda, Maryland

Joel C. Glover   Department of Physiology, University of Oslo, Norway

Dongsheng Guo   Department of Pathology, Feinberg School of Medicine, Northwestern University, Chicago, Illinois

Joung Hyuck Joo   Cell Biology Section, Division of Intramural Research, National Institute of Environmental Health Sciences, National Institutes of Health, Research Triangle Park, North Carolina

Anton M. Jetten   Cell Biology Section, Division of Intramural Research, National Institute of Environmental Health Sciences, National Institutes of Health, Research Triangle Park, North Carolina

Yuzhi Jia   Department of Pathology, Feinberg School of Medicine, Northwestern University, Chicago, Illinois

Xavier Lampe   Institut de Génétique et de Biologie Moléculaire et Cellulaire, CNRS/INSERM/ULP, Illkirch Cedex, France

Fu-Jung Lin   Department of Molecular and Cellular Biology, Baylor College of Medicine, Houston, Texas

David Lohnes   Department of Cellular and Molecular Medicine, University of Ottawa, Ottawa, Canada K1H 8M5

Liliane Michalik   Center for Integrative Genomics, National Research Center "Frontiers in Genetics," University of Lausanne, Lausanne, Switzerland

Lily Ng   Clinical Endocrinology Branch, National Institute of Diabetes and Digestive and Kidney Diseases, National Institutes of Health, Bethesda, Maryland

Karen Niederreither   Department of Medicine, Center for Cardiovascular Development, Baylor College of Medicine, Houston, Texas; Department of Molecular and Cellular Biology, Center for Cardiovascular Development, Baylor College of Medicine, Houston, Texas

Tracie Pennimpede   Department of Biochemistry, Division of Cancer Biology and Genetics, Cancer Research Institute, Queen's University, Kingston, Ontario, Canada K7L 3N6

Martin Petkovich   Department of Biochemistry, Division of Cancer Biology and Genetics, Cancer Research Institute, Queen's University, Kingston, Ontario, Canada K7L 3N6; Department of Pathology and Molecular Medicine, Division of Cancer Biology and Genetics, Cancer Research Institute, Queen's University, Kingston, Ontario, Canada K7L 3N6

Leslie Pick   Department of Entomology, 4112 Plant Sciences, University of Maryland, College Park, Maryland; Department of Cell Biology and Molecular Genetics, University of Maryland, College Park, Maryland

M. Sambasiva Rao   Department of Pathology, Feinberg School of Medicine, Northwestern University, Chicago, Illinois

Janardan K. Reddy   Department of Pathology, Feinberg School of Medicine, Northwestern University, Chicago, Illinois

Jean-Sébastien Renaud   Department of Physiology, University of Oslo, Norway

Charlotte Rhodes   Department of Cellular and Molecular Medicine, University of Ottawa, Ottawa, Canada K1H 8M5

Filippo M. Rijli   Institut de Génétique et de Biologie Moléculaire et Cellulaire, CNRS/INSERM/ULP, Illkirch Cedex, France

Cécile Rochette-Egly   Institut de Génétique et de Biologie Moléculaire et Cellulaire, CNRS/INSERM/Université Louis Pasteur, UMR 7104, BP 10142, 67404 Illkirch Cedex, France

Nicolas Rotman   Center for Integrative Genomics, National Research Center "Frontiers in Genetics," University of Lausanne, Lausanne, Switzerland

Jeffrey Shultz   Department of Entomology, 4112 Plant Sciences, University of Maryland, College Park, Maryland

Reshma Taneja   Brookdale Department of Molecular, Cell and Developmental Biology, Mount Sinai School of Medicine, New York

Ke Tang   Department of Molecular and Cellular Biology, Baylor College of Medicine, Houston, Texas

Catherine C. Thompson   Kennedy Krieger Research Institute, Baltimore, Maryland; Department of Neuroscience, Johns Hopkins University School of Medicine, Baltimore, Maryland; Department of Molecular Biology and Genetics, Johns Hopkins University School of Medicine, Baltimore, Maryland

Ming-Jer Tsai   Department of Molecular and Cellular Biology, Baylor College of Medicine, Houston, Texas; Program of Development, Baylor College of Medicine, Houston, Texas

Sophia Y. Tsai   Department of Molecular and Cellular Biology, Baylor College of Medicine, Houston, Texas; Program of Development, Baylor College of Medicine, Houston, Texas

Walter Wahli   Center for Integrative Genomics, National Research Center "Frontiers in Genetics," University of Lausanne, Lausanne, Switzerland

Craig T. Woodard   Department of Biological Sciences, Mount Holyoke College, South Hadley, Massachusetts

Songtao Yu   Department of Pathology, Feinberg School of Medicine, Northwestern University, Chicago, Illinois

Preface

While the effect of steroid hormones on various physiological processes has been recognized since the early twentieth century, it was only in the 1960s that the notion of specific hormone-binding proteins that transmit this signal in target tissues emerged. It is now over two decades since the first nuclear receptor—the glucocorticoid receptor—was cloned in the mid 1980s, which was quickly followed by the cloning of a number of additional receptors including those for thyroid hormone, estrogen, retinoids, and progesterone among others. The nuclear receptor superfamily now comprises a large family of ligand-activated transcription factors that include receptors for steroid and thyroid hormones, retinoids, vitamin D3, peroxisome proliferators, as well as "orphan" receptors that are essential for many physiological functions including cellular differentiation, growth, morphogenesis, apoptosis, and homeostasis. Nuclear receptors are modular proteins and share structural similarities in several regions. The most conserved domains include the DNA-binding domain (DBD) and the ligand-binding domain (LBD). Two separable transactivation domains have been identified: a ligand-dependent activation function (AF-2) located in the LBD, and a ligand-independent activation domain AF-1 located in the divergent A/B region.

Over the past two decades, the molecular mechanisms by which nuclear receptors are activated and initiate a transcriptional response have been defined. Steroid hormone receptors, which reside in the cytoplasm in complex with heat shock proteins, are transported to the nucleus on hormone binding, where they bind DNA as homodimers and transcriptionally regulate expression of target genes. Other receptors, such as retinoic acid receptors (RARs), thyroid hormone receptors (TRs), and peroxisome proliferator-activated receptor (PPAR), reside in the nucleus even in absence of ligand and bind DNA as heterodimers with their partner retinoid X receptor (RXR). In the absence of ligand, these receptors interact with corepressor molecules and actively repress transcription. Ligand binding results in a conformational change, which permits dissociation of corepressors and allows recruitment of coactivators that enhance the transcriptional activity of receptors. In addition to ligand-dependent transactivation, the transcriptional activity of nuclear receptors can be modulated by phosphorylation through cross talk with various signaling pathways. Nuclear receptors for which ligands have not been identified are classified as "orphan" receptors. This subfamily includes the chicken ovalbumin transcription factors (COUP-TFs), retinoid-related orphan receptors (RORs), and Ftz-F1 in *Drosophila*.

The advent of gene-disruption technologies have helped elucidate the functions of nuclear receptors *in vivo*. As our understanding of the biology of nuclear receptors increases, so does our appreciation of the complexity of their actions. Since it is beyond the scope of any one book to cover all aspects of nuclear receptor biology, I have aimed to focus this book on the developmental roles of several nuclear receptors, as well as those of nuclear receptor coactivators and corepressors.

Thyroid hormone is critical for growth, development, and metabolism. Ng and Forrest provide a comprehensive review on the functions of TR with a specific focus on the developing auditory and visual systems. The chapter by Rotman et al. reviews the function of PPARα, PPARβ, and PPARγ in the formation of placenta, adipocytes, skin, and gut, and discusses potential roles in the nervous system.

Retinoic acid (RA), the active derivative of vitamin A, is critical for numerous aspects of development. Since both an excess and deficiency of RA can lead to severe developmental defects, embryonic RA levels are regulated by controlled synthesis and catabolism mediated by retinaldehyde dehydrogenases (RALDHs) and the cytochrome oxidase P450 subfamily 26 (CYP26) enzymes, respectively. The chapter by Pennimpede et al. covers our overall understanding of RA signaling and the functions of RARs and RXRs in embryonic development. In a related chapter, Niederreither and Dollé describe embryonic phenotypes related to loss of RALDH2, which is critical for RA synthesis *in vivo*.

Anterior–posterior (A–P) patterning of the hindbrain is dependent on retinoids. Hox genes, which are directly regulated by RA signaling, play a vital role in this process. The chapter by Glover et al. focuses on the role of endogenous retinoids and Hox genes in vertebrate hindbrain development. Similar to the hindbrain, RA and Hox genes are essential for vertebral A–P patterning. However, mesodermal Hox genes require an intermediary homeoprotein Cdx1, which is directly regulated by RA and Wnt signaling. Rhodes and Lohnes review the role of RA and Wnt signaling in regulation of Cdx1 and Hox genes in axial patterning.

The RA-responsive F9 embryonal carcinoma (EC) cell line has been used as a cell-autonomous model system to determine the role of RA signaling in endodermal differentiation. The chapter by Bour et al. covers our understanding of RAR function in primitive, parietal, and visceral endoderm formation, and how phosphorylation of specific residues in RARs is essential for the formation of these distinct cell types.

The chapter by Tang et al. discusses the functions of the orphan receptor COUP-TFI in the development of the nervous system. Jetten and Joo provide an extensive overview of RORα, RORβ, and RORγ in brain development, circadian rhythms, and development of secondary lymphoid tissues. They also discuss the role of DHR3, the ROR homologue in *Drosophila*, which regulates the nuclear receptor Ftz-F1. In a related chapter, Pick et al. review the role of Ftz-F1, an orthologue of the mammalian nuclear receptor SF-1, in metamorphosis and embryonic development. The protein partners that interact with and modulate Ftz-F1 activity are discussed in relation to its function.

The last section of this book covers the role of nuclear receptor coregulators. Thompson and Beaudoin review the role of the corepressor Hairless, which interacts with TR, ROR, and vitamin D receptor (VDR) and is essential for skin function. Many coactivators including CBP/p300, the SRC-p160 family, the TRAP/DRIP/ARC complex, RIP140, and PGC-1 enhance the transcriptional activity of nuclear receptors. Reddy et al. provide a detailed review on the role of coactivators in modulating nuclear receptor function in differentiation and in metabolism.

This compilation of reviews represents a continuum of progress toward understanding the biological functions of the nuclear receptor superfamily. Recent developments include gene disruption studies of nuclear receptors and their cofactors, which has advanced our understanding of their roles in differentiation and development. In cases where embryonic or perinatal lethality has precluded full understanding of the biological functions, efforts are being directed to generate conditional mutants. Many additional issues are presently being addressed which include identification of molecular targets of nuclear receptors, defining the mechanisms by which cross talk with other signaling pathways modulates their function, and determining the role of various cofactors that positively or negatively regulate nuclear receptor activity in different cellular contexts. These ongoing studies will help define and understand the cellular and molecular basis of nuclear receptor function.

RESHMA TANEJA

Developmental roles of the thyroid hormone receptor α and β genes

Lily Ng and Douglas Forrest

Clinical Endocrinology Branch, National Institute of Diabetes and Digestive and Kidney Diseases, National Institutes of Health, Bethesda, Maryland

Contents

Advances in Developmental Biology
Volume 16 ISSN 1574-3349
DOI: 10.1016/S1574-3349(06)16001-9

Thyroid hormone (T3) has numerous actions in development in vertebrate species. Thyroid hormone receptors (TRs) that act as T3-regulated transcription factors are central to these actions. Several TRα and TRβ isoforms encoded by two genes, *Thra* and *Thrb*, provide biological versatility. Gene targeting has indicated specific roles for each gene. For example, *Thra* determines heart rate, thermogenesis, and bone and intestinal differentiation, whereas *Thrb* controls the pituitary–thyroid axis and liver function. *Thrb* is also critical for hearing and color vision. Together, *Thra* and *Thrb* coregulate a wider spectrum of actions in brain function, growth, and fertility. The deletion of all α and β receptors results in viable mice, but these exhibit immaturity in multiple systems, indicating the importance of T3 signaling at the later, maturational stages of development. The tissue specificity and timing of TR functions *in vivo* are likely to be determined in cooperation with other factors, including deiodinase enzymes that regulate hormone availability in a given tissue.

1. Introduction

The thyroid has long been associated with developmental functions. Preceding even the isolation of the hormonal product, thyroxine, it had been shown in 1891 that injection of sheep thyroid extracts could reverse many of the symptoms of adult myxedema, which nowadays is known to be caused by too little thyroid hormone (Murray, 1891). Thyroid extracts also gave remarkable improvements in children suffering from cretinism, the severe physical and mental retardation that is now known to result from a developmental lack of thyroid hormone (McCarrison, 1908). The wider importance of the thyroid agent in development was indicated by Gudernatsch (1912) who found that extracts of thyroid, but not of other tissues, promoted premature metamorphosis in amphibian species.

The mechanisms by which the diverse actions of thyroid hormone are enacted remained a puzzle for much of the twentieth century. Moreover, within a given tissue, the actions can change with the developmental stage indicating a time dependency in many responses. This is exemplified in the need to provide timely thyroid hormone replacement to hypothyroid infants soon after birth: if hypothyroidism is not diagnosed and treated promptly, permanent mental impairment may result. Later treatment cannot reverse these deficiencies (Dussault and Ruel, 1987). At the cellular level, thyroid hormone has many actions: in different systems it may promote or arrest cell proliferation, regulate cell death, or stimulate differentiation and migration.

In the 1960s and 1970s, the hormone was suggested to act at the level of transcription through hormonal binding sites in the cell nucleus (Tata, 1963; Oppenheimer et al., 1972). The levels of these binding sites could vary during development, suggesting that changes in the putative receptors provided a means of control (Schwartz and Oppenheimer, 1978; Perez-Castillo et al., 1985). The cloning of the thyroid hormone receptors (TRs) in 1986 (Sap et al., 1986; Weinberger et al., 1986) revealed the existence of a family of TR isoforms encoded by two genes, *Thra* (*Nr1a1*) and *Thrb* (*Nr1a2*) (Fig. 1).

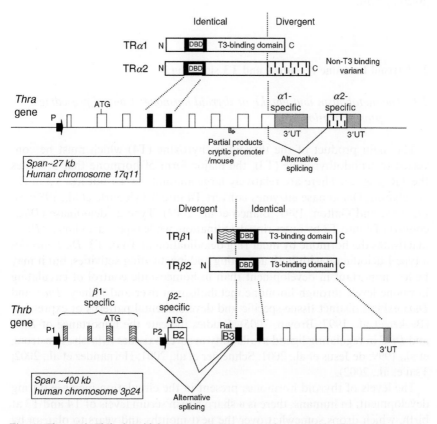

Fig. 1. *Thra* and *Thrb* thyroid hormone receptor genes in mammals. The major receptor products are indicated above each gene. The TRα2 product is not a T3 receptor. Exons are indicated by boxes and promoters by a black horizontal arrow (P, for *Thra*; P1, P2 for the dual promoters of *Thrb*). 3′UT, 3′ untranslated regions; DBD, DNA-binding domains.

This chapter reviews these genes and summarizes their functions based on targeted mutagenesis. A feature of TR signaling is its adaptability as will be illustrated in two developmental processes: (1) hearing exemplifies a complex function that is subject to multiple levels of control by thyroid hormone in different cell types in the auditory system and (2) the color visual system is unexpectedly dependent on TR signaling but unlike the auditory system, this largely centers around the differentiation of a single cell type, the cone photoreceptor.

Apart from its developmental actions, thyroid hormone also regulates homeostasis in adults. Although adult functions are not the topic of this chapter, they are touched upon, both for completeness and because it is difficult to distinguish between actions in late development and those at the onset of adult homeostasis. Many TR functions in development concern the maturational transition as function is switched on, or enhanced, within a given system.

2. Thyroid hormone receptors and T3 signaling

2.1. The metabolism and uptake of thyroid hormone: Control preceding receptor activation

The main product of the thyroid is thyroxine (T4) which must be converted to triiodothyronine (T3), the major form of hormone that activates the TR (Fig. 2). There are relatively large amounts of T4 but less T3 in the circulation. Deiodinase enzymes convert T4 into T3 (Köhrle et al., 1987; St. Germain and Galton, 1997; Bianco et al., 2002). Type 2 deiodinase (*Dio2*) converts T4 into T3 by outer ring deiodination, while type 3 deiodinase (*Dio3*) inactivates the hormone by inner ring deiodination of T4 or T3. *Dio1* encodes a type 1 deiodinase with both activating and inactivating activities, but it may be less important in development than in homeostatic control of circulating hormone levels through hormone metabolism in liver and kidney. *Dio2* and *Dio3* exhibit distinct tissue-specific and developmental patterns of expression (Becker et al., 1997; Brown, 2005). Studies indicate the importance of *Dio2* and *Dio3* in regulating ligand availability in several tissues (Marsh-Armstrong et al., 1999; de Jesus et al., 2001; Schneider et al., 2001; Hernandez et al., 2002; Tsai et al., 2002).

The levels of thyroid hormones present in the circulation increase during development. In humans, there is a sharp rise in serum levels of T4 and T3 at birth, which drops somewhat over the next months and years to plateau by adolescence (Fisher, 1996). In mice, which are less mature at birth than humans, there is a slower postnatal increase in both T4 and T3 with a peak at postnatal day 15 (P15), followed by a modest decline by weaning

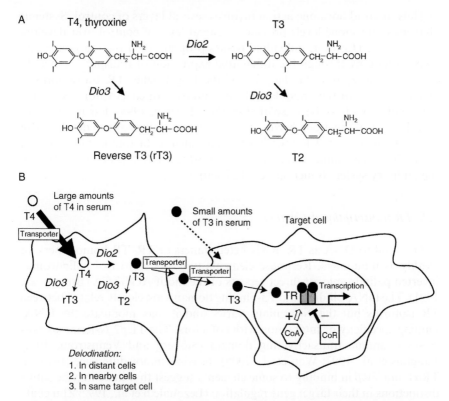

Fig. 2. Thyroid hormones and the control of TR activation. (A) Metabolic activation and inactivation of thyroid hormone. *Dio2* deiodinase converts T4 into the active ligand T3. *Dio3* mediates inner ring deiodination of T4 or T3 to form the largely inactive metabolites rT3 or T2. (B) Physiological control of TR activation in a target cell. In a target tissue, T3 levels might be amplified by *Dio2* in the TR-expressing target cell itself, or in adjacent or more distant cells, as has been suggested in the cochlea (Fig. 3). Transporters for uptake and release of hormone are likely to be important in the local hormonal communication between cells. In the target cell, TR activity is regulated by ligand and also by corepressors (CoR) or coactivators (CoA).

age (P21), then a gradual stabilization to adult levels (Campos-Barros et al., 2000).

Another means of control concerns transporters that regulate hormone uptake in specific tissues. This area has been relatively understudied but the importance of transporters was illustrated when the carboxylate transporter MCT8, a thyroid hormone transporter *in vitro*, was found to be mutated in human X-linked mental retardation and in Allan–Herndon–Dudley syndrome. The mutations are associated with abnormalities in serum thyroid hormones and neurological defects (Dumitrescu et al., 2004; Friesema et al., 2004; Maranduba et al., 2006).

Thus, thyroid hormone action involves several layers of control. Systemic changes in hormone levels provide a general level of control over developmental rates. However, individual tissues program their own response with some autonomy through the differential expression of transporters, deiodinases, and receptors. At the level of the target gene, TR transcriptional activity may be further modulated by corepressors or coactivators. Although many cofactors have been isolated *in vitro* (Koenig, 1998), less is known of their biological functions *in vivo*. There may be specific roles in development as has been suggested for the hairless and alien cofactors in brain (Potter et al., 2002; Tenbaum et al., 2003) or transducin β-like protein 1(TBL1) in the auditory system (Guenther et al., 2000).

2.2. TR transcriptional function

TRs bind to DNA as TR–RXR heterodimers or TR–TR homodimers, or weakly as monomers. Response elements typically consist of palindromic, inverted palindromic, or direct repeats of the motif AGGTCA (Brent et al., 1991). The DNA-binding and T3-binding domains are closely related between TR isoforms but the N-terminus varies and it may modulate the DNA-binding specificity or interaction with cofactors. The TRβ2 N-terminus possesses a unique transactivating domain (Sjöberg and Vennström, 1995; Langlois et al., 1997; Wan et al., 2005). *In vitro*, minor differences between TRα1 and TRβ in binding to some elements suggest that there may be subtle distinctions in their target gene regulation (Lezoualc'h et al., 1992; Zhu et al., 1997; Yang and Privalsky, 2001). However, TRα1, TRβ1, and TRβ2 isoforms transactivate similarly on most response elements.

TRs act as versatile regulators and mediate both positive and negative control of gene expression in response to T3. Perhaps contrary to preconceptions, several tissue screens suggest that there may be at least as many genes that are negatively regulated as there are genes positively regulated by T3 (Iglesias et al., 1996; Miller et al., 2001; Flores-Morales et al., 2002; Poguet et al., 2003; Yen et al., 2003).

2.3. Thra *and* Thrb *genes*

The *Thra* and *Thrb* genes are closely related in mammalian, avian, amphibian, and fish species (Forrest et al., 1990; Yaoita et al., 1990; Sakurai et al., 1992; Wood et al., 1994; Marchand et al., 2001), suggesting that both genes are fundamentally required for T3 signaling (Fig. 1). *Thrb* spans about 400 kb in mammals. A complex series of 5′ exons encode the N-termini of two T3 receptor isoforms, TRβ1 and TRβ2, which are generated by differential promoter usage and splicing (Hodin et al., 1989; Shi et al., 1992;

Sjöberg et al., 1992). Splicing of upstream, noncoding exons generates a variety of 5′ untranslated leader sequences that may influence translational efficiency (Yaoita and Brown, 1990; Frankton et al., 2004).

Thra spans less than 30 kb and encodes a single T3 receptor, TRα1. TRα1 is found in all mammalian and nonmammalian species examined to date. Mammals, but not birds, amphibia, or fish, coexpress a C-terminal splice variant that does not bind T3 or transactivate (Lazar et al., 1988). TRα2 binds DNA only weakly and has been reported to exert weak repression over normal TRs (Koenig et al., 1989; Tagami et al., 1998). The absence of TRα2 in nonmammalian species suggests that its role is not as fundamental as that of TRα1 or TRβ isoforms. In isolated species, an assortment of truncated TRα products have been found. These may arise by internal translation initiation in chick (Bigler and Eisenman, 1988) or through a weak, cryptic promoter in intron 7 in mouse *Thra* (Chassande et al., 1997). Some gene-targeting manipulations in mouse *Thra* promote expression of these partial products which are thought to worsen certain phenotypes (Plateroti et al., 2001). Most of the shorter products lack DNA-binding domains and it has been suggested that they somehow interfere with normal TR function. A TRβ3 N-terminal variant has been found in rat (Williams, 2000). TRβ3 transactivates in response to T3 but it has not been found in humans or other species (Forrest et al., 1990; Frankton et al., 2004). These atypical TRα and TRβ products are not universal in vertebrate species, nor even in mammals, and their functions remain subject to debate.

2.4. Differential expression of Thra and Thrb

Thra and *Thrb* are differentially expressed in development. Expression overlaps in many tissues but there are examples of cell types that express predominantly a single isoform. Although different TR isoforms have considerable functional overlap, the relative abundance of a given TR isoform in a tissue often correlates with its functional significance as discussed later.

2.4.1. Thra

TRα1 is the most widely expressed TR isoform and is near-ubiquitously present from the earliest stages of development (Forrest et al., 1990; Strait et al., 1990; Yaoita and Brown, 1990; Kawahara et al., 1991). It is also the major, if not sole, TR isoform in mouse embryonic stem cells (Liu et al., 2002). The mammalian variant TRα2 is widely coexpressed with TRα1 and is typically in 2- to 6-fold excess over TRα1 (Mitsuhashi and Nikodem, 1989).

The expression of TRα1 in the embryo is suggestive of early developmental functions. However, the phenotypes reported so far for TRα1-deficient mice are manifested postnatally arguing that early TRα1 functions may be substituted for by other means. Another possibility is that some postnatal functions, for example, in the brain, are programmed by TRα1 in advance

(Itoh et al., 2001; Guadaño-Ferraz et al., 2003). TRα1 in embryonic brain might serve a priming role without visible effect until the neuron later becomes functional. TRα1 can mediate a neuronal-like differentiation of embryonic stem cells in response to T3 (Liu et al., 2002).

2.4.2. Thrb

TRβ1 is expressed later in development in many of the same tissues where low levels of TRα1 already exist, including brain, pituitary, cochlea, liver, kidney, and heart. Levels can rise sharply at a particular time in development such as in the brain or lung at birth (Forrest et al., 1990; Thompson, 1996). TRβ2 is highly restricted to the neural retina, inner ear, pituitary gland, and the paraventricular nucleus of the hypothalamus, which releases thyrotropin-releasing hormone (TRH) (Hodin et al., 1990; Bradley et al., 1992; Sjöberg et al., 1992).

In summary, most if not all cells in the early embryo express low levels of TRα1. As development progresses, TRβ1 or TRβ2 become expressed in many of these same tissues. The net result is that each tissue carries a given total mass of receptors composed of varying amounts of TRα1, TRβ1, and/ or TRβ2. Since these isoforms behave largely similarly *in vitro*, the total receptor mass is probably a characteristic that determines the capacity of a cell to respond to T3. Subtle functional distinctions between receptor isoforms may provide an additional level of tuning of gene regulation. There are also rarer examples of tissues with unique receptor specificity such as TRβ2 in cone photoreceptors (Sjöberg et al., 1992) or TRα1 in red blood cells (Forrest et al., 1990; Schneider et al., 1993).

3. Phenotypes caused by mutations in *Thra* and *Thrb*

Deletions in *Thra* and *Thrb* produce distinct although limited phenotypes consistent with each gene mediating specific functions (Tables 1 and 2). A fuller range of phenotypes that resemble the more severe defects caused by developmental hypothyroidism are found only when *Thra* and *Thrb* mutations are combined (see later). This supports the view that TRα1 and TRβ have considerable overlap or interchangeability in function and coregulate a wide range of functions.

3.1. Thra *mutations*

The complex structure of the *Thra* 3′ region that carries the TRα1- and TRα2-specific exons (Fig. 1) presents complications in the design and in some cases interpretation of gene-targeting studies. The phenotypic severity can vary because of inadvertent consequences such as the generation of partial products or changes in the ratio of TRα1:TRα2 expressed (Fraichard

Table 1
Targeted mutations in *Thra*

Targeted allele[a]	Other name	Phenotypes[b]	Viability	Comments[c]	References
Deletions (knockout)					
Thra tm1Ven	TRα1$^{-/-}$	Bradycardia Low body temperature	Viable	Deletes TRα1	Wikström et al., 1998
Thra tm2Ven	TRα2$^{-/-}$	Tachycardia High body temperature Bone defects	Viable	Deletes TRα2 Concomitant TRα1 ↑	Saltó et al., 2001
Thra tm1Jas	TRα$^{-/-}$	Dwarfism Hypothyroid Intestinal defects	Lethal after weaning	Delete TRα1/TRα2 Express partial products	Fraichard et al., 1997
Thra tm2Jas	TRα0/0	Low body temperature Bone defects Intestinal defects	Viable	Delete TRα1/TRα2 No partial products	Gauthier et al., 2001
Thra tm3Jas	TRα7/7	Defects in intestinal markers	Viable	Reduces partial products	Plateroti et al., 2001
Change-of-function (knock-in)					
Thra tm1Syc	TRα1+/PV	Dwarfism Poor fertility Brain glucose uptake ↓ Bone defects	+/PV viable PV/PV lethal in embryo/neonate	Dominant negative Mutant TRα1 ↑ Allele loses TRα2	Itoh et al., 2001; Kaneshige et al., 2001; O'Shea et al., 2005
Thra tm3Ven	TRα1+/R384C	Dwarfism Bone defects Anxiety, learning defects	+/R384C viable R384C/R384C lethal by weaning	Dominant negative Mutant TRα1 ↑ Allele loses TRα2	Tinnikov et al., 2002; Venero et al., 2005
Thra tm1Brnt	TRα1+/P398H	Fat tissue ↑ Pituitary TSH ↑ Bradycardia Low body temperature	+/P398H viable P398H/P398H lethal in embryo/neonate	Dominant negative	Liu et al., 2003

[a]Nomenclature according to Mouse Genome Informatics (http://www.informatics.jax.org/).
[b]Main phenotypes as reported initially are listed.
[c]↑ indicates mutation causes elevated expression.

Table 2
Targeted mutations in *Thrb*

Targeted allele[a]	Other name	Phenotypes[b]	Viability	Comment	References
Deletions (knockout)					
Thrb tm1Df	TRβ⁻/⁻	Hyperactive pituitary–thyroid Deafness	Viable	Deletes all TRβ products	Forrest et al., 1996a,b
Thrb tm2Df	TRβ2⁻/⁻	Color blindness	Viable	Deletes only TRβ2	Ng et al., 2001a
Thrb tm1Jas	TRβ⁻/⁻	Hyperactive pituitary–thyroid	Viable	Deletes all TRβ products	Gauthier et al., 1999
Thrb tm1Few	TRβ2⁻/⁻	Hyperactive pituitary–thyroid	Viable	Deletes only TRβ2	Abel et al., 1999
Thrb tm4Few	TRβ⁻/⁻	Hyperactive pituitary–thyroid	Viable	Deletes all TRβ products	Shibusawa et al., 2003
Change-of-function (knock-in)					
Thrb tm1Syc	TRβ+/PV	Hyperactive pituitary–thyroid Bone defects	Viable (homozygote too)	Dominant negative	Kaneshige et al., 2000
Thrb tm2Few	TRβ+/Δ337T	Hyperactive pituitary–thyroid Behavior defects Cerebellar defects	Viable (homozygote too)	Dominant negative	Hashimoto et al., 2001
Thrb tm3Few	TRβ GS125/GS125	Hyperactive pituitary–thyroid Deafness Photoreceptor defects	Viable	TRβ products defective in DNA binding	Shibusawa et al., 2003
Thrb tm5Few	TRβ E457A/E457A	Hyperactive pituitary–thyroid	Viable	Defective AF2 activator domain in TRβ products	Ortiga-Carvalho et al., 2005

[a]Nomenclature according to Mouse Genome Informatics (http://www.informatics.jax.org/).
[b]Only main phenotypes as reported initially given here.

et al., 1997; Saltó et al., 2001) (Table 1). The deletion of either TRα1, the T3 receptor, or TRα2, the nonreceptor product of *Thra*, results in viable mice (Wikström et al., 1998). However, mutations that delete both TRα1 and TRα2 and which generate partial products are lethal by weaning age (Plateroti et al., 2001). The partial products may somehow exacerbate some phenotypes. Several TRα1 functions have been indicated despite these complexities. Mice lacking TRα1 have a reduced heart rate (Johansson et al., 1998, 1999, 2002), low body temperature (Wikström et al., 1998; Gauthier et al., 2001; Marrif et al., 2005), and intestinal defects with decreased numbers of crypt cells and reduced mucosal thickness (Fraichard et al., 1997; Plateroti et al., 1999, 2001). In the brain, *Thra* mutations influence cerebellar development (Heuer and Mason, 2003; Morte et al., 2004), mating behavior (Dellovade et al., 2000), anxiety, learning (Guadaño-Ferraz et al., 2003; Venero et al., 2005), and glucose utilization (Itoh et al., 2001). TRα1-deficient mice are fertile but *Thra* mutations cause some increases in Sertoli cell numbers, resembling a change that occurs with neonatal hypothyroidism (Holsberger et al., 2005). TRα1-deficient mice have changes in oligodendrocytes in the optic nerve suggesting that TRα1 mediates the timing role of T3 on differentiation of oligodendrocyte precursors (Billon et al., 2001; Baas et al., 2002; Billon et al., 2002). Consistent with the presence of TRα1 but not TRβ in hematopoetic tissues, *Thra* mutations produce some impairment in erythroid (Angelin-Duclos et al., 2005) and lymphoid (Arpin et al., 2000; Tinnikov et al., 2002) differentiation. Change-of-function (knock-in) mutations suggest that TRα1 has a major role in bone development (Tinnikov et al., 2002; O'Shea et al., 2003, 2005). Deletion of TRα2 produces a range of phenotypes including increases in body temperature and heart rate, and subtle abnormalities in thyroid hormone levels that in some ways show opposite trends to those caused by loss of TRα1 (Saltó et al., 2001). The manipulation that deletes TRα2 unavoidably increases TRα1 levels such that the phenotypes may be explained equally by an increase in TRα1 or loss of TRα2. Heterozygous mice show intermediate phenotypes. The possibility that TRα1 overexpression contributes to the phenotypes is supported by introducing this mutation onto the TRβ-deficient background. This substantially corrects the deafness and thyroid dysfunction of the TRβ-deficient mice, which is most simply explained by increased TRα1 levels being able to compensate for the absence of TRβ in several systems (Ng et al., 2001c). The function of TRα2 itself remains enigmatic.

3.2. Thrb *mutations*

Thrb mutations produce goiter and hyperactivity of the pituitary–thyroid axis, indicating that TRβ isoforms mediate the feedback regulation of thyroid hormone production (Forrest et al., 1996a; Abel et al., 1999; Gauthier et al., 1999) (Table 2), consistent with the expression of TRβ1 and TRβ2 in the anterior pituitary. These receptors also regulate the pituitary at the level of release of TRH by the hypothalamus (Abel et al., 2001; Dupre et al., 2004). The response of the liver to T3 is mainly mediated by TRβ1 (Weiss et al., 1998; Gullberg et al., 2000; Amma et al., 2001; Gullberg et al., 2002). However, the basal function of the unchallenged liver is relatively intact in TRβ-deficient mice (Flores-Morales et al., 2002; Yen et al., 2003), suggesting that the role of TRβ may be to enhance the ability of the liver to adapt to changing metabolic demands. *Thrb* mutations produce neurological phenotypes, including audiogenic seizure susceptibility (Ng et al., 2001b) and changes in mating behavior (Dellovade et al., 2000) and cerebellar development (Hashimoto et al., 2001). *Thrb* probably has other functions in brain, as is suggested by the marked increase in TRβ expression in many brain regions in the neonatal period in species as diverse as the chick and rat (Forrest et al., 1990; Thompson, 1996) and by the incidence of neurological defects in the human syndrome of resistance to thyroid hormone, which is associated with *THRB* mutations (Refetoff et al., 1993). The identification of these functions may require newer approaches to study behavior or neuronal physiology (Sui and Gilbert, 2003) since examination of TR-deficient mice to date has not revealed gross morphological abnormalities in brain. TRβ is also critical for the development of the senses of hearing and color vision (see later).

3.3. Life without nuclear T3 receptors

Mice devoid of T3 receptors (TRα1, TRβ1, and TRβ2) are viable and most have normal longevity. Tissues and organ systems form but in many cases do not acquire normal or adult function (Göthe et al., 1999; Gauthier et al., 2001) (Table 3). Phenotypes include the known single gene-specific phenotypes discussed earlier, as well as an exacerbated goiter, dwarfism, defects in bone development with reduced elongation and ossification, poor female fertility with irregular ovulatory cycling, and abnormalities of many anterior pituitary hormones (Göthe et al., 1999; Wang, Z. and Forrest, D., in preparation) and retarded gut maturation (Gauthier et al., 2001). Most of these phenotypes may be explained by loss of the cooperativity between TRα1 and TRβ or loss of the ability to substitute for each other's absence.

The general phenotype of TR-deficient mice suggests that TRs promote maturation as the fetus or newborn progresses to an adult life-form. Organs are present and acquire at least rudimentary function, while vital functions operate well enough for viability. It is interesting to consider that amphibian metamorphosis, the radical developmental progression from a larval to adult life-form, depends on T3 signaling mediated by similar α1 and β receptors as are found in mammals. Although mammals do not undergo metamorphosis,

Table 3
Some of the exacerbated phenotypes caused by combined deletions in *Thra* and *Thrb*

Phenotype	Receptor deficiency[a]		
	TRα1	TRβ	TRα1/TRβ
Deafness	−	++	+++
Color blindness	−	++	n.a.
Dwarfism	−	−	++
Immature bones	+/−	n.a.	++
Hyperactive pituitary–thyroid axis, goiter	−	+	+++
Infertility (female)	−	−	++
Intestinal defects	+	−	++
Sertoli cell defects	+	−	n.a.
Oligodendrocyte defects	+	−	+
Erythroid defects	+	−	n.a.
Lymphoid defects	+	−	n.a.
Liver, defective T3 response	−	++	+++
Low body temperature	+	−	+
Bradycardia	+	−	+
Anxiety, cerebellar defects	+	−	n.a.
Mating behavior defects	+	+	n.a.
Audiogenic seizures	−	++	n.a.

[a]Phenotypic severity: −, no phenotype; +, least; +++, most severe; n.a., data not available. Based primarily on knockout (not dominant-negative) mutations. For references, see description in the chapter.

there may be underlying similarities in the maturational transition that accompanies the progression to adult life-form in both cases.

3.3.1. Comparison of a developmental lack of T3 versus lack of T3 receptors

The finding that most mice lacking all T3 receptors (TRα1, TRβ1, and TRβ2) survive from conception into old age has prompted an ongoing debate about whether this phenotype is milder than expected. It has been suggested that hypothyroidism may be more debilitating and in some mouse models may be lethal after weaning (Marians et al., 2002). Several explanations for this ostensible discrepancy have been considered elsewhere and are not reviewed again here (Forrest and Vennström, 2000; Flamant and Samarut, 2003; Wondisford, 2003). *In vivo* evidence for some of these ideas is inconclusive (Flamant et al., 2002; Morte et al., 2002; Mittag et al., 2005). In some tissues, such as cerebellum, hypothyroidism produces a more severe phenotype than TR deficiency, which may relate to the TR acting as a chronic repressor in the absence of T3 ligand (Morte et al., 2002). For other phenotypes, such as deafness, there is no obvious discrepancy between the lack of hormone or lack of receptors (see later).

3.4. Change-of-function mutations

Knock-in mutations that change the function of TRα1 or TRβ products have been generated (Tables 1 and 2) (Kaneshige et al., 2000, 2001; Hashimoto et al., 2001; Tinnikov et al., 2002; Liu et al., 2003; Shibusawa et al., 2003; Ortiga-Carvalho et al., 2005). Several of these mutations mimic those found in human *THRB* in the syndrome of resistance to thyroid hormone, an autosomal dominant disease (Refetoff et al., 1993). The heterozygous mutations reside in the C-terminus and impair T3 binding or cofactor binding to create dominant negative proteins (Chatterjee and Beck-Peccoz, 2001). Most of the phenotypes in mice carrying these mutations resemble in kind those described in knockout mice. However, the interpretation is more complicated given that a dominant-negative activity can interfere with any wild-type receptors of either the TRα1 or TRβ class. These mutations provide valuable models for human disease but because of the wider dominant-negative activity are less informative about receptor-specific functions in development.

4. The auditory system

Hearing is one of the most sensitive functions under thyroid hormone control. Hypothyroidism in early development presents a serious risk to hearing and cochlear function. In humans, the most sensitive period is *in utero* with a continued risk in the postnatal period, although this is less well quantified (Bellman et al., 1996; Rovet et al., 1996; Wasniewska et al., 2002). In areas with endemic iodine deficiency, deafness can be common because of the early thyroid hormone deficiency *in utero* (Goslings et al., 1975). There are some auditory defects in the syndrome of resistance to thyroid hormone (Brucker-Davis et al., 1996). In mice (Deol, 1973; O'Malley et al., 1995; Christ et al., 2004) and rats (Uziel et al., 1980; Van Middlesworth and Norris, 1980; Hébert et al., 1985; Knipper et al., 2000) the corresponding window of sensitivity is in the first 2–3 postnatal weeks (Fig. 3). This period mainly consists of terminal differentiation with the onset of auditory function around P13.

4.1. Thrb *and cochlear maturation*

The major functions of thyroid hormone in hearing are mediated by *Thrb* rather than *Thra*. Pronounced deafness arises in TRβ-deficient but not TRα1-deficient mice (Forrest et al., 1996b; Gauthier et al., 2001; Griffith

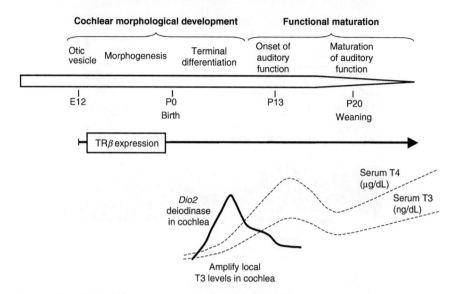

Fig. 3. TRs and cochlear development. Cochlear development may be viewed in two stages: morphogenesis, mainly *in utero*, and postnatal functional maturation. TR*β* is expressed from early stages but TR*β* deficiency causes relatively late defects in the postnatal period. The timing of development is probably conferred by the rise in *Dio2* activity by P8 that may generate a surge of T3 at the critical period, shortly before auditory function begins. *Dio2* allows the cochlea to stimulate its own response to T3 at a time when there are only low levels of T3 available in the circulation.

et al., 2002; Shibusawa et al., 2003). However, combined mutations exacerbate the phenotype, thereby unmasking an auxiliary role for TRα1 (Rüsch et al., 2001). In the cochlea, TR*β* and to a lesser extent TRα1 are expressed in the sensory epithelium that contains the sensory hair cells and supporting cells and in the adjacent zone of the greater epithelial ridge that secretes the tectorial membrane and which forms the inner sulcus (Bradley et al., 1994). Lower levels are detected in the stria vascularis, which regulates the endocochlear potential, the spiral ganglion, and other areas.

In TR-deficient mice, the cochlea has normal form at birth, but over the next 2–3 weeks the terminal differentiation of the inner sulcus, sensory epithelium, and opening of the tunnel of Corti are delayed, while the secreted tectorial membrane is malformed (Rüsch et al., 2001). The inner hair cells have delayed onset of a basolateral potassium current that is thought to allow the onset of high-frequency transduction at P13 (Rüsch et al., 1998). The outer hair cells have reduced nonlinear capacitance, an indicator of electromotility, and the endocochlear potential is reduced. This complex phenotype reflects the retardation of multiple cell types in the cochlea (Fig. 4).

Wild type

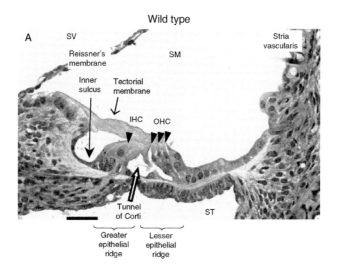

TR or *Dio2* deficient

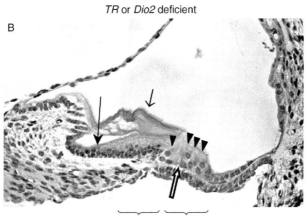

Fig. 4. Retarded cochlear development in TR- or *Dio2*-deficient mice. (A) In normal mice at P9, terminal differentiation of the greater and lesser epithelial ridges forms a thin inner sulcus, while the inner hair cells (IHCs) and outer hair cells (OHCs) are distinct in the sensory epithelium. The tectorial membrane is thin and extends to the hair cells. The tunnel of Corti is clearly open between the IHC and OHCs. Three fluid-filled compartments are marked: SV, scala vestibuli; SM, scala media; ST, scala tympani. Scale bar 20 μm. (B) In *Thrb*- or *Dio2*-deficient mice, differentiation is retarded. The inner sulcus is poorly formed and the dense underlying cell mass has not formed a thin epithelium. The tunnel of Corti is unopened and the tectorial membrane is enlarged. The IHC and OHCs are present but are poorly differentiated. In adults, many morphological features "catch up" with wild-type development, but the tectorial membrane remains malformed and there are permanent functional defects. Pictures generated in the authors' laboratory using wild type and *Dio*−/− samples.

The deafness caused by TRβ deficiency or lack of thyroid hormone is likely to involve impairment at other levels. Additional abnormalities have been observed in hypothyroid rodents in cochlear synaptogenesis (Uziel et al., 1983), myelination of the auditory nerve (Knipper et al., 1998), *Dio2* expression in the brainstem (Guadaño-Ferraz et al., 1999), and in morphology of spiral ganglion neurons (Rueda et al., 2003) and pyramidal cells of the auditory cortex (Ruiz-Marcos et al., 1983).

4.2. Target genes in the auditory system

Relatively little is known of the target genes underlying TR functions in the cochlea. Hypothyroidism in rats alters the expression profile of the tectorin glycoproteins, which may contribute to the malformation of the tectorial membrane (Knipper et al., 2001). The tectorial membrane has ultrastructural defects in TR-deficient mice (Rüsch et al., 2001) suggesting that there are defects in posttranslational processing. Outer hair cell electromotility is thought to be mediated in part by the motor protein prestin. Hypothyroid rats show retarded expression of prestin and a TR-binding site has been mapped in the 5′ region of the gene (Weber et al., 2002). A screen of cDNAs in wild-type and TRβ-deficient mice at P7 identified an extracellular matrix protein, emilin-2, that is expressed in the cochlear basilar membrane (Amma et al., 2003). This gene, however, is unlikely to be a direct target for TRβ and may be representative of other downstream genes that are indirectly regulated during cochlear maturation.

4.3. Thyroid hormone: A global timing signal in the cochlear microenvironment

The complex cochlear phenotype in TR-deficient mice raises a question: how does a single hormonal signal provided by T3 elicit the specialized differentiation of multiple different cell types? Perhaps the only common feature shared by these cell types is the time period of their differentiation prior to the onset of hearing. In the mouse cochlea at birth, morphogenic events, cell division, and the commitment to cell lineages are largely complete. Postnatally, therefore, TR signaling may trigger a common mechanism that pushes this diverse range of cell types into the final stages of differentiation in a timely manner. The nature of this mechanism is unknown. The necessity for timely maturation of the auditory system may be particularly important if hearing, like other senses, depends on sensory input for correct development. For example, the visual system depends on a critical period of sensory experience for correct maturation (Katz and Shatz, 1996).

Why might nature employ thyroid hormone as a key maturation factor for the onset of hearing? A clue may be that the various T3-sensitive cell types are spatially separated in the chambered structure of the cochlea. A diffusible hormone, such as T3, can provide a synchronized signal to cells in different locations at a coordinated time. In contrast, more local cell–cell-signaling mechanisms may be unable to communicate with more distant cells within the requisite time frame.

4.4. Deiodinases and receptors: Developmental integration

The timing of thyroid hormone action in the postnatal cochlea is evidently not regulated by the appearance of the receptor because TRβ is expressed as early as embryonic day 12 (E12) in the cochlea (Bradley et al., 1994). Instead, timing may be conferred by *Dio2* deiodinase, which is sharply induced in the first postnatal week in the cochlea, prior to the onset of hearing (Campos-Barros et al., 2000) (Fig. 3). *Dio2* is expressed in the chondrocytes in the surrounding structures of the cochlea that form the bony labyrinth and in the modiolus and septal divisions between the turns of the cochlea. This location is largely noncoincident with that of TRβ, which is mainly in the interior sensory tissues, suggesting that the separation of the cells that generate T3 and the internal cells that respond to T3 provides another level of control over TR activation. In the brain, *Dio2* is mainly expressed in glial cells rather than the adjacent T3-responsive neurons, suggesting that local cellular communication is an important feature of T3 action in target tissues *in vivo* (Guadaño-Ferraz et al., 1997). *Dio2*-deficient mice are deaf and have a similar cochlear phenotype as *Thrb*-deficient mice (Ng et al., 2004). Thus, *Dio2* may generate a surge of T3 for activation of TRβ at the critical postnatal period in the cochlea. *Dio2* may draw on the relatively abundant T4 in serum to amplify T3 levels in the cochlea. T3 levels in serum are low and are themselves inadequate for auditory development. The division of the cochlea into compartments suggests that transporters would be important in the uptake of T4 and release of T3 by the *Dio2*-expressing cells, a question that merits further study. A candidate intracellular carrier of thyroid hormones, mu-crystallin (CRYM), is abundant in the cochlea and is implicated in human nonsydromic deafness (Abe et al., 2003).

5. The color visual system

Until recently, the color visual system was not regarded as a target for T3 signaling. Despite numerous studies of mental retardation in human hypothyroidism, there had been little or no investigation of possible color blindness. Early investigations on the TR genes, however, indicated that *Thrb* was expressed in the chick embryonic eye (Sjöberg et al., 1992). TRβ2 was specifically expressed in the outer neuroblastic layer of the early retina, which contains newly generated photoreceptors, as early as E5 in the chick.

TRβ2 is also found in newly generated cones in the mouse eye, with a peak of expression around E18 (Ng et al., 2001a). This represents one of the most cell-specific expression patterns of any nuclear receptor.

5.1. Differentiation of cone photoreceptors

In most mammals, the color visual system is based on two opsin photo-pigments that confer sensitivity to short (S, blue) and medium–long (M, green) wavelengths of light. Color vision depends on M and S opsins being differentially expressed in cones to provide a system for discrimination between different light wavelengths (Nathans, 1999). Cones are generated *in utero* (Cepko et al., 1996) but do not express M and S opsins until later: in the mouse, S opsin appears in the neonate, while M opsin appears a week later at P9. In mice, M and S opsins are expressed in opposing gradients in cones across the superior–inferior axis of the retina (Szel et al., 1993) (Fig. 5). A fundamental question in color vision concerns the mechanisms that regulate the differential expression of M and S opsins. Multipotent progenitor cells give rise to different retinal cell types, including both rod and cone photoreceptors. Cone differentiation follows three stages: (1) The generation

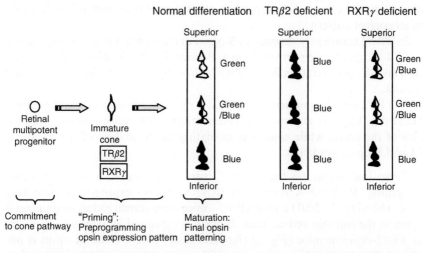

Fig. 5. Differentiation of cone photoreceptors in the mouse. Multipotent progenitors form committed cone precursors early in retinal development. Cone precursors express TRβ2, RXRγ, and other factors that are involved in priming the cells for correct patterning of green (M) and blue (S) opsins at postnatal stages. Green and blue cones show opposing gradients of distribution across the superior–inferior axis of the retina. Cones in midretinal regions express both green and blue opsins. In TRβ2-deficient mice, all cones are of the "blue" type. In RXR$\gamma^{-/-}$ mice, green opsin is normally expressed but blue opsin loses its gradient and is expressed in cones all around the retina.

of committed, postmitotic cones from the dividing multipotent progenitors. (2) The expression in cone precursors of transcription factors or other factors that determine the choice of M or S opsin expression. This represents a priming phase for opsin patterning but is poorly understood. It is in these cells that TRβ2 is expressed. (3) The terminal differentiation of cones, including the expression of opsins. The opsins are transported to the outer segments as the cones mature and become capable of phototransduction.

5.2. The role of TRβ2 in cones

TRβ2-deficient mice exhibit a form of color blindness with a selective loss of medium–long wave responses but retention of short wave responses (Ng et al., 2001a). The retina has a normal appearance with normal numbers of cones but fails to express M opsin (Fig. 5). S opsin is still present but with some changes in the gradient and timing of its expression: S opsin is expressed in all cones across the retina and it is also prematurely expressed. TRβ2 therefore has three distinct but still puzzling functions in opsin patterning:

1. TRβ2 induces M opsin. However, TRβ2 levels peak in the late embryo and declines in the neonate a week before M opsin is induced, making it unclear whether the activation is direct or indirect. It is not excluded that persistent low levels of TRβ2 directly induce M opsin. Alternatively, TRβ2 may prime the cone to be receptive to other signals that later induce M opsin in its correct pattern.

2. TRβ2 transiently suppresses S opsin in utero, minimizing its expression until the appropriate postnatal stage. Potentially, this involves direct repression and the postnatal decline of TRβ2 allows S opsin to be induced by other factors.

3. TRβ2 sets the gradients of M and S opsin expression across the retina. However, TRβ2 is expressed in cones around the entire superior–inferior axis of the retina while somehow mediating the formation of the opposing M and S opsin gradients.

These findings argue that other factors cooperate with TRβ2 in opsin patterning. Retinoid X receptor γ (RXRγ) is also expressed in immature cones (Mori et al., 2001) and RXR$\gamma^{-/-}$ mice were shown to fail to suppress S opsin in the superior retina. Thus, S opsin is expressed in all cones, as occurs in TRβ2-deficient mice (Fig. 5) (Roberts et al., 2005), while M opsin is unaltered. Potentially, this suppression of S opsin in the superior retina involves TRβ2–RXRγ heterodimers. However, in vitro studies so far do not indicate a simple model of TRβ2 or RXRγ binding to specific elements in the M or S opsin genes. Studies in our lab indicate that the orphan nuclear receptor RORβ induces S opsin in early postnatal cones (Srinivas et al., 2006). An interplay between TRβ2, which induces M opsin, and RORβ, which induces S opsin, may contribute to the control of the opposing gradients of M and S opsin.

5.3. The role of thyroid hormone ligand in cone differentiation

A systematic study of human patients with developmental thyroid disorders would be informative about the role of thyroid hormone in color vision. This question has been largely overlooked but infants exposed to prenatal or neonatal hypothyroidism have been reported to have defects in contrast sensitivity (Mirabella et al., 2005). Also, *in vitro* studies have suggested that T3 can induce markers of cone differentiation in cultures of human or rat fetal retina (Kelley et al., 1995a,b).

T3 can regulate opsins *in vivo* in neonatal mice (Roberts et al., 2006). T3 injections suppress S opsin and this suppression requires the presence of TRβ2. These studies support a model whereby TRβ2 has distinct functions at different stages: S opsin is sensitive to T3 in the neonate, while M opsin may be sensitive at later stages. Measurements of thyroid hormone content in different zones of the retina suggest that T3 may be present at slightly higher levels in the superior zone by P10. It is unknown how a hormonal gradient is formed but this could differentially regulate M and S opsins across the retina.

Cone opsins may be regulated by changes in T3 levels mediated by deiodinases. *Dio3* deiodinase is expressed in retina in *Xenopus* where it regulates asymmetric cell division in the dorsal–ventral axis during development (Marsh-Armstrong et al., 1999). It would be of interest to know if it also regulates cone opsins. Potentially, *Dio3,* which inactivates thyroid hormones, may protect the immature cones from excessive T3 stimulation. *Dio3* is also found in retina in the mouse (Ng, L., Forrest, D., and Srinivas, M., unpublished data) and chick (Harpavat and Cepko, 2003), suggestive of a conserved role. Further studies may reveal more precisely the role of T3 in photoreceptors (Sevilla-Romero et al., 2002), including its role *in utero*, where TRβ2 is most highly expressed in newly generated cones before S or M opsins are expressed.

5.4. TRβ2 and the color visual system of different species

TRβ2 is the most highly conserved TRβ isoform and it may regulate cone differentiation in a range of vertebrate species as well as mice. A rare recessive case of human resistance to thyroid hormone involving a homozygous deletion of *THRB* has been associated with monochromacy (Newell and Diddie, 1977). In the typical dominant form of this syndrome, there are only rare, anecdotal mentions of color visual defects (Lindstedt et al., 1982), and investigation is needed to determine the incidence and severity of possible color visual defects in this disease. TRβ2 is also found in retina in the chick and in the amphibian species *Rana catesbeiana* (Schneider, M., personal communication) and probably *Xenopus laevis* (Cossette and Drysdale, 2004).

A final reflection on the role of TRβ2 in color vision suggests that TRβ2 allows a primitive system adapt to more sophisticated demands. In TRβ2-deficient mice, the color visual system forms but expresses only S opsin. TRβ2 therefore provides a means to induce M opsin, and to control differential expression of both S and M opsins, the basis of color perception in mammals. Most amphibia, birds, and fish have a more complex range of cone opsins than mammals (Bowmaker, 1998). It is therefore of interest to know which opsins are under TRβ2 regulation in other species.

6. Concluding remarks

Extensive mutagenesis of *Thra* and *Thrb* in mouse models has identified a range of developmental and physiological defects. However, the cellular basis of many of these phenotypes is unclear. A challenge lies in elucidating the molecular and cellular mechanisms of TR-mediated control in these systems. The actions of TRα1 and TRβ are often additive, but may also be unique or antagonistic, probably reflecting differences in the expression of TRα1 and TRβ in distinct cell types within a complex organ such as in some brain regions. Finer level analyses and cell-specific deletions will be helpful in determining the roles of TRα1 and TRβ in individual cell types.

Although receptors occupy a key position in converting the hormonal signal into a cellular response, they do not function in isolation *in vivo*. A more complete picture of TR functions in development will require understanding how receptors, deiodinases, and transporters operate in harmony to provide integrated and timely regulation in a given tissue. Finally, given the unexpectedly critical role found for TRβ2 in the color visual system, further exploratory studies may identify other new roles for TRs in development.

Acknowledgments

This work was supported by the intramural research program at NIDDK/NIH.

References

Abe, S., Katagiri, T., Saito-Hisaminato, A., Usami, S., Inoue, Y., Tsunoda, T., Nakamura, Y. 2003. Identification of CRYM as a candidate responsible for nonsyndromic deafness, through cDNA microarray analysis of human cochlear and vestibular tissues. Am. J. Hum. Genet. 72, 73–82.

Abel, E.D., Boers, M.E., Pazos-Moura, C., Moura, E., Kaulbach, H., Zakaria, M., Lowell, B., Radovick, S., Liberman, M.C., Wondisford, F. 1999. Divergent roles for thyroid hormone receptor β isoforms in the endocrine axis and auditory system. J. Clin. Invest. 104, 291–300.

Abel, E.D., Ahima, R.S., Boers, M.E., Elmquist, J.K., Wondisford, F.E. 2001. Critical role for thyroid hormone receptor β2 in the regulation of paraventricular thyrotropin-releasing hormone neurons. J. Clin. Invest. 107, 1017–1023.

Amma, L.L., Campos-Barros, A., Wang, Z., Vennström, B., Forrest, D. 2001. Distinct tissue-specific roles for thyroid hormone receptors β and α1 in regulation of type 1 deiodinase expression. Mol. Endocrinol. 15, 467–475.

Amma, L.L., Goodyear, R., Faris, J.S., Jones, I., Ng, L., Richardson, G., Forrest, D. 2003. An emilin family extracellular matrix protein identified in the cochlear basilar membrane. Mol. Cell. Neurosci. 23, 460–472.

Angelin-Duclos, C., Domenget, C., Kolbus, A., Beug, H., Jurdic, P., Samarut, J. 2005. Thyroid hormone T3 acting through the thyroid hormone alpha receptor is necessary for implementation of erythropoiesis in the neonatal spleen environment in the mouse. Development 132, 925–934.

Arpin, C., Pihlgren, M., Fraichard, A., Aubert, D., Samarut, J., Chassande, O., Marvel, J. 2000. Effects of T3Rα1 and T3Rα2 gene deletion on T and B lymphocyte development. J. Immunol. 164, 152–160.

Baas, D., Legrand, C., Samarut, J., Flamant, F. 2002. Persistence of oligodendrocyte precursor cells and altered myelination in optic nerve associated to retina degeneration in mice devoid of all thyroid hormone receptors. Proc. Natl. Acad. Sci. USA 99, 2907–2911.

Becker, K., Stephens, K., Davey, J., Schneider, M., Galton, V. 1997. The type 2 and type 3 iodothyronine deiodinases play important roles in coordinating development in *Rana catesbeiana* tadpoles. Endocrinology 138, 2989–2997.

Bellman, S.C., Davies, A., Fuggle, P.W., Grant, D.B., Smith, I. 1996. Mild impairment of neuro-otological function in early treated congenital hypothyroidism. Arch. Dis. Child. 74, 215–218.

Bianco, A.C., Salvatore, D., Gereben, B., Berry, M.J., Larsen, P.R. 2002. Biochemistry, cellular and molecular biology, and physiological roles of the iodothyronine selenodeiodinases. Endocr. Rev. 23, 38–89.

Bigler, J., Eisenman, R.N. 1988. c-erbA encodes multiple proteins in chicken erythroid cells. Mol. Cell. Biol. 8, 4155–4161.

Billon, N., Tokumoto, Y., Forrest, D., Raff, M. 2001. Role of thyroid hormone receptors in timing oligodendrocyte differentiation. Dev. Biol. 235, 110–120.

Billon, N., Jolicoeur, C., Tokumoto, Y., Vennström, B., Raff, M. 2002. Normal timing of oligodendrocyte development depends on thyroid hormone receptor alpha 1 (TRalpha1). EMBO. J. 21, 6452–6460.

Bowmaker, J.K. 1998. Evolution of colour vision in vertebrates. Eye 12(Pt. 3b), 541–547.

Bradley, D.J., Towle, H.C., Young Iii, W.S. 1992. Spatial and temporal expression of α- and β-thyroid hormone receptor mRNAs, including the β2-subtype, in the developing mammalian nervous system. J. Neurosci. 12, 2288–2302.

Bradley, D.J., Towle, H.C., Young Iii, W.S. 1994. α and β thyroid hormone receptor (TR) gene expression during auditory neurogenesis: Evidence for TR isoform-specific transcriptional regulation *in vivo*. Proc. Natl. Acad. Sci. USA 91, 439–443.

Brent, G., Moore, D., Larsen, P. 1991. Thyroid hormone regulation of gene expression. Ann. Rev. Physiol. 53, 17–35.

Brown, D.D. 2005. The role of deiodinases in amphibian metamorphosis. Thyroid 15, 815–821.

Brucker-Davis, F., Skarulis, M.C., Pikus, A., Ishizawar, D., Mastroianni, M.-A., Koby, M., Weintraub, B.D. 1996. Prevalence and mechanisms of hearing loss in patients with resistance to thyroid hormone (RTH). J. Clin. Endocrinol. Metab. 81, 2768–2772.

Campos-Barros, A., Amma, L.L., Faris, J.S., Shailam, R., Kelley, M.W., Forrest, D. 2000. Type 2 iodothyronine deiodinase expression in the cochlea before the onset of hearing. Proc. Natl. Acad. Sci. USA 97, 1287–1292.

Cepko, C.L., Austin, C.P., Yang, X., Alexiades, M., Ezzeddine, D. 1996. Cell fate determination in the vertebrate retina. Proc. Natl. Acad. Sci. USA 93, 589–595.

Chassande, O., Fraichard, A., Gauthier, K., Flamant, F., Legrand, C., Savatier, P., Laudet, V., Samarut, J. 1997. Identification of transcripts initiated from an internal promoter in the cerbA alpha locus that encode inhibitors of retinoic acid receptor-alpha and triiodothyronine receptor activities. Mol. Endocrinol. 11, 1278–1290.

Chatterjee, V., Beck-Peccoz, P. 2001. Resistance to thyroid hormone. In: *Endocrinology* (L. Degroot, J. Jameson, Eds.), 4th edn., Philadelphia, PA: WB Saunders Company.

Christ, S., Biebel, U.W., Hoidis, S., Friedrichsen, S., Bauer, K., Smolders, J.W. 2004. Hearing loss in athyroid pax8 knockout mice and effects of thyroxine substitution. Audiol. Neurootol. 9, 88–106.

Cossette, S.M., Drysdale, T.A. 2004. Early expression of thyroid hormone receptor beta and retinoid X receptor gamma in the *Xenopus* embryo. Differentiation 72, 239–249.

de Jesus, L.A., Carvalho, S.D., Ribeiro, M.O., Schneider, M., Kim, S.W., Harney, J.W., Larsen, P.R., Bianco, A.C. 2001. The type 2 iodothyronine deiodinase is essential for adaptive thermogenesis in brown adipose tissue. J. Clin. Invest. 108, 1379–1385.

Dellovade, T.L., Chan, J., Vennström, B., Forrest, D., Pfaff, D.W. 2000. The two thyroid hormone receptor genes have opposite effects on estrogen-stimulated sex behaviors. Nat. Neurosci. 3, 472–475.

Deol, M.S. 1973. An experimental approach to the understanding and treatment of hereditary syndromes with congenital deafness and hypothyroidism. J. Med. Genetics 10, 235–242.

Dumitrescu, A.M., Liao, X.H., Best, T.B., Brockmann, K., Refetoff, S. 2004. A novel syndrome combining thyroid and neurological abnormalities is associated with mutations in a monocarboxylate transporter gene. Am. J. Hum. Genet. 74, 168–175.

Dupre, S.M., Guissouma, H., Flamant, F., Seugnet, I., Scanlan, T.S., Baxter, J.D., Samarut, J., Demeneix, B.A., Becker, N. 2004. Both thyroid hormone receptor (TR) beta 1 and TR beta 2 isoforms contribute to the regulation of hypothalamic thyrotropin-releasing hormone. Endocrinology 145, 2337–2345.

Dussault, J.H., Ruel, J. 1987. Thyroid hormones and brain development. Ann. Rev. Physiol. 49, 321–324.

Fisher, D. 1996. Thyroid physiology in the perinatal period and during childhood. In: *Werner and Ingbar's The Thyroid* (L. Braverman, R. Utiger, Eds.), 7th edn., Philadelphia: Lippincott-Raven.

Flamant, F., Samarut, J. 2003. Thyroid hormone receptors: Lessons from knockout and knock-in mutant mice. Trends Endocrinol. Metab. 14, 85–90.

Flamant, F., Poguet, A.L., Plateroti, M., Chassande, O., Gauthier, K., Streichenberger, N., Mansouri, A., Samarut, J. 2002. Congenital hypothyroid Pax8(–/–) mutant mice can be rescued by inactivating the TRalpha gene. Mol. Endocrinol. 16, 24–32.

Flores-Morales, A., Gullberg, H., Fernandez, L., Stahlberg, N., Lee, N.H., Vennström, B., Norstedt, G. 2002. Patterns of liver gene expression governed by TRβ. Mol. Endocrinol. 16, 1257–1268.

Forrest, D., Vennström, B. 2000. Functions of thyroid hormone receptors in mice. Thyroid 10, 41–52.

Forrest, D., Sjöberg, M., Vennström, B. 1990. Contrasting developmental and tissue-specific expression of α and β thyroid hormone receptor genes. EMBO J. 9, 1519–1528.

Forrest, D., Hanebuth, E., Smeyne, R.J., Everds, N., Stewart, C.L., Wehner, J.M., Curran, T. 1996a. Recessive resistance to thyroid hormone in mice lacking thyroid hormone receptor β: Evidence for tissue-specific modulation of receptor function. EMBO J. 15, 3006–3015.

Forrest, D., Erway, L.C., Ng, L., Altschuler, R., Curran, T. 1996b. Thyroid hormone receptor β is essential for development of auditory function. Nat. Genet. 13, 354–357.

Fraichard, A., Chassande, O., Plateroti, M., Roux, J., Trouillas, J., Dehay, C., Legrand, C., Gauthier, K., Kedinger, M., Malaval, L., Rousset, B., Samarut, J. 1997. The T3Rα gene encoding a thyroid hormone receptor is essential for post-natal development and thyroid hormone production. EMBO J. 16, 4412–4420.

Frankton, S., Harvey, C.B., Gleason, L.M., Fadel, A., Williams, G.R. 2004. Multiple messenger ribonucleic acid variants regulate cell-specific expression of human thyroid hormone receptor beta1. Mol. Endocrinol. 18, 1631–1642.

Friesema, E.C., Grueters, A., Biebermann, H., Krude, H., Von Moers, A., Reeser, M., Barrett, T.G., Mancilla, E.E., Svensson, J., Kester, M.H., Kuiper, G.G., Balkassmi, S., et al. 2004. Association between mutations in a thyroid hormone transporter and severe X-linked psychomotor retardation. Lancet 364, 1435–1437.

Gauthier, K., Chassande, O., Plateroti, M., Roux, J.-P., Legrand, C., Pain, B., Rousset, B., Weiss, R., Trouillas, J., Samarut, J. 1999. Different functions for the thyroid hormone receptors TRα and TRβ in the control of thyroid hormone production and post-natal survival. EMBO J. 18, 623–631.

Gauthier, K., Plateroti, M., Harvey, C.B., Williams, G.R., Weiss, R.E., Refetoff, S., Willott, J.F., Sundin, V., Roux, J.P., Malaval, L., Hara, M., Samarut, J., et al. 2001. Genetic analysis reveals different functions for the products of the thyroid hormone receptor α locus. Mol. Cell. Biol. 21, 4748–4760.

Goslings, B.M., Djokomoeljanto, R., Hoedijono, R., Soepardjo, H., Querido, A. 1975. Studies on hearing loss in a community with endemic cretinism in central Java, Indonesia. Acta Endocrinol. (Copenhagen) 78, 705–713.

Göthe, S., Wang, Z., Ng, L., Nilsson, J., Campos-Barros, A., Ohlsson, C., Vennström, B., Forrest, D. 1999. Mice devoid of all known thyroid hormone receptors are viable but exhibit disorders of the pituitary-thyroid axis, growth and bone maturation. Genes Dev. 13, 1329–1341.

Griffith, A.J., Szymko, Y.M., Kaneshige, M., Quinonez, R.E., Kaneshige, K., Heintz, K.A., Mastroianni, M.A., Kelley, M.W., Cheng, S.Y. 2002. Knock-in mouse model for resistance to thyroid hormone (RTH): An RTH mutation in the thyroid hormone receptor beta gene disrupts cochlear morphogenesis. J. Assoc. Res. Otolaryngol. 3, 279–288.

Guadaño-Ferraz, A., Obregón, M., St. Germain, D., Bernal, J. 1997. The type 2 iodothyronine deiodinase is expressed primarily in glial cells in the neonatal rat brain. Proc. Natl. Acad. Sci. USA 94, 10391–10396.

Guadaño-Ferraz, A., Escámez, M., Rausell, E., Bernal, J. 1999. Expression of type 2 iodothyronine deiodinase in hypothyroid rat brain indicates an important role of thyroid hormone in the development of specific primary sensory neurons. J. Neurosci. 19, 3430–3439.

Guadaño-Ferraz, A., Benavides-Piccione, R., Venero, C., Lancha, C., Vennström, B., Sandi, C., Defelipe, J., Bernal, J. 2003. Lack of thyroid hormone receptor alpha1 is associated with selective alterations in behavior and hippocampal circuits. Mol. Psychiatry 8, 30–38.

Gudernatsch, J. 1912. Feeding experiments on tadpoles. I. The influence of specific organs given as food on growth and differentiation. Roux Arch. Entwicklungsmechanik der Organismen 35, 457–483.

Guenther, M.G., Lane, W.S., Fischle, W., Verdin, E., Lazar, M.A., Shiekhattar, R. 2000. A core SMRT corepressor complex containing HDAC3 and TBL1, a WD40-repeat protein linked to deafness. Genes Dev. 14, 1048–1057.

Gullberg, H., Rudling, M., Forrest, D., Angelin, B., Vennström, B. 2000. Thyroid hormone receptor β-deficient mice show complete loss of normal cholesterol 7α-hydroxylase response to thyroid hormone but display enhanced resistance to dietary cholesterol. Mol. Endocrinol. 14, 1739–1749.

Gullberg, H., Rudling, M., Salto, C., Forrest, D., Angelin, B., Vennstrom, B. 2002. Requirement for thyroid hormone receptor beta in T3 regulation of cholesterol metabolism in mice. Mol. Endocrinol. 16, 1767–1777.

Harpavat, S., Cepko, C.L. 2003. Thyroid hormone and retinal development: An emerging field. Thyroid 13, 1013–1019.

Hashimoto, K., Curty, F.H., Borges, P.P., Lee, C.E., Abel, E.D., Elmquist, J.K., Cohen, R.N., Wondisford, F.E. 2001. An unliganded thyroid hormone receptor causes severe neurological dysfunction. Proc. Natl. Acad. Sci. USA 98, 3998–4003.

Hébert, R., Langlois, J.-M., Dussault, J.H. 1985. Permanent defects in rat peripheral auditory function following perinatal hypothyroidism: Determination of a critical period. Dev. Brain Res. 23, 161–170.

Hernandez, A., Fiering, S., Martinez, E., Galton, V.A., St. Germain, D. 2002. The gene locus encoding iodothyronine deiodinase type 3 (Dio3) is imprinted in the fetus and expresses antisense transcripts. Endocrinology 143, 4483–4486.

Heuer, H., Mason, C.A. 2003. Thyroid hormone induces cerebellar Purkinje cell dendritic development via the thyroid hormone receptor alpha1. J. Neurosci. 23, 10604–10612.

Hodin, R., Lazar, M., Chin, W. 1990. Differential and tissue-specific regulation of the multiple rat c-erbA messenger RNA species by thyroid hormone. J. Clin. Invest. 85, 101–105.

Hodin, R.A., Lazar, M.A., Wintman, B.I., Darling, D.S., Koenig, R.J., Larsen, P.R., Moore, D.D., Chin, W.W. 1989. Identification of a thyroid hormone receptor that is pituitary-specific. Science 244, 76–78.

Holsberger, D.R., Kiesewetter, S.E., Cooke, P.S. 2005. Regulation of neonatal Sertoli cell development by thyroid hormone receptor alpha1. Biol. Reprod. 73, 396–403.

Iglesias, T., Caubin, J., Stunnenberg, H.G., Zaballos, A., Bernal, J., Munoz, A. 1996. Thyroid hormone-dependent transcriptional repression of neural cell adhesion molecule during brain maturation. EMBO J. 15, 4307–4316.

Itoh, Y., Esaki, T., Kaneshige, M., Suzuki, H., Cook, M., Sokoloff, L., Cheng, S.Y., Nunez, J. 2001. Brain glucose utilization in mice with a targeted mutation in the thyroid hormone α or β receptor gene. Proc. Natl. Acad. Sci. USA 98, 9913–9918.

Johansson, C., Vennström, B., Thorén, P. 1998. Evidence that decreased heart rate in thyroid hormone receptor α1-deficient mice is an intrinsic defect. Am. J. Physiol. 275, R640–R646.

Johansson, C., Göthe, S., Forrest, D., Vennström, B., Thorén, P. 1999. Cardiovascular phenotype and temperature control in mice lacking thyroid hormone receptor β or both α1 and β. Am. J. Physiol. Heart Circ. Physiol. 276, H2006–H2012.

Johansson, C., Koopmann, R., Vennström, B., Benndorf, K. 2002. Accelerated inactivation of voltage-dependent K+ outward current in cardiomyocytes from thyroid hormone receptor alpha1-deficient mice. J. Cardiovasc. Electrophysiol. 13, 44–50.

Kaneshige, M., Kaneshige, K., Zhu, X., Dace, A., Garrett, L., Carter, T.A., Kazlauskaite, R., Pankratz, D.G., Wynshaw-Boris, A., Refetoff, S., Weintraub, B., Willingham, M.C., et al. 2000. Mice with a targeted mutation in the thyroid hormone β receptor gene exhibit impaired growth and resistance to thyroid hormone. Proc. Natl. Acad. Sci. USA 97, 13209–13214.

Kaneshige, M., Suzuki, H., Kaneshige, K., Cheng, J., Wimbrow, H., Barlow, C., Willingham, M.C., Cheng, S. 2001. A targeted dominant negative mutation of the thyroid hormone alpha 1 receptor causes increased mortality, infertility, and dwarfism in mice. Proc. Natl. Acad. Sci. USA 98, 15095–15100.

\Katz, L., Shatz, C. 1996. Synaptic activity and the construction of cortical circuits. Science 274, 1133–1138.

Kawahara, A., Baker, B.S., Tata, J.R. 1991. Developmental and regional expression of thyroid hormone receptor genes during *Xenopus* metamorphosis. Development 112, 933–943.

Kelley, M.W., Turner, J.K., Reh, T.A. 1995a. Ligands of steroid/thyroid receptors induce cone photoreceptors in vertebrate retina. Development 121, 3777–3785.

Kelley, M.W., Turner, J.K., Reh, T.A. 1995b. Regulation of proliferation and photoreceptor differentiation in fetal human retinal cell cultures. Invest. Ophthalmol. Vis. Sci. 36, 1280–1289.

Knipper, M., Bandtlow, C., Gestwa, L., Kopschall, I., Rohbock, K., Wiechers, B., Zenner, H.-P., Zimmermann, U. 1998. Thyroid hormone affects Schwann cell and oligodendrocyte gene expression at the glial transition zone of the VIIIth nerve prior to cochlea function. Development 125, 3709–3718.

Knipper, M., Zinn, C., Maier, H., Praetorius, M., Rohbock, K., Kopschall, I., Zimmermann, U. 2000. Thyroid hormone deficiency before the onset of hearing causes irreversible damage to peripheral and central auditory systems. J. Neurophysiol. 83, 3101–3112.

Knipper, M., Richardson, G., Mack, A., Muller, M., Goodyear, R., Limberger, A., Rohbock, K., Kopschall, I., Zenner, H.P., Zimmermann, U. 2001. Thyroid hormone-deficient period prior to the onset of hearing is associated with reduced levels of beta-tectorin protein in the tectorial membrane: Implication for hearing loss. J. Biol. Chem. 276, 39046–39052.

Koenig, R.J. 1998. Thyroid hormone receptor coactivators and corepressors. Thyroid 8, 703–713.

Koenig, R.J., Lazar, M.A., Hodin, R.A., Brent, G.A., Larsen, P.R., Chin, W.W., Moore, D.D. 1989. Inhibition of thyroid hormone action by a non-hormone binding c-erbA protein generated by alternative mRNA splicing. Nature 337, 659–661.

Köhrle, J., Brabant, G., Hesch, R.-D. 1987. Metabolism of thyroid hormones. Hormone Res. 26, 58–78.

Langlois, M.-F., Zanger, K., Mondem, T., Safer, J., Hollenberg, A., Wondisford, F. 1997. A unique role of the β-2 thyroid hormone receptor isoform in negative regulation by thyroid hormone. J. Biol. Chem. 272, 24927–24933.

Lazar, M., Hodin, R., Darling, D., Chin, W. 1988. Identification of a rat c-erbA α-related protein which binds deoxyribonucleic acid but does not bind thyroid hormone. Mol. Endocrinol. 2, 893–901.

Lezoualc'h, F., Hassan, A.H.S., Giraud, P., Loeffler, J.-P., Lee, S.L., Demeneix, B.A. 1992. Assignment of the β-thyroid hormone receptor to 3,5,3'-triiodothyronine-dependent inhibition of transcription from the thyrotropin-releasing hormone promoter in chick hypothalamic neurons. Mol. Endocrinol. 6, 1797–1804.

Lindstedt, G., Lundberg, P.A., Sjogren, B., Ernest, I., Sundquist, O. 1982. Thyroid hormone resistance in a 35-year old man with recurrent goitre. Scand. J. Clin. Lab. Invest. 42, 585–593.

Liu, Y.Y., Tachiki, K.H., Brent, G.A. 2002. A targeted thyroid hormone receptor alpha gene dominant-negative mutation (P398H) selectively impairs gene expression in differentiated embryonic stem cells. Endocrinology 143, 2664–2672.

Liu, Y.Y., Schultz, J.J., Brent, G.A. 2003. A thyroid hormone receptor alpha gene mutation (P398H) is associated with visceral adiposity and impaired catecholamine-stimulated lipolysis in mice. J. Biol. Chem. 278, 38913–38920.

Maranduba, C.M., Friesema, E.C., Kok, F., Kester, M.H., Jansen, J., Sertie, A.L., Passos-Bueno, M.R., Visser, T.J. 2006. Decreased cellular T3 uptake and metabolism in Allan–Herndon–Dudley syndrome (AHDS) due to a novel mutation in the MCT8 thyroid hormone transporter. J. Med. Genet. 43, 457–460.

Marchand, O., Safi, R., Escriva, H., Van Rompaey, E., Prunet, P., Laudet, V. 2001. Molecular cloning and characterization of thyroid hormone receptors in teleost fish. J. Mol. Endocrinol. 26, 51–65.

Marians, R.C., Ng, L., Blair, H.C., Unger, P., Graves, P.N., Davies, T.F. 2002. Defining thyrotropin-dependent and -independent steps of thyroid hormone synthesis by using thyrotropin receptor-null mice. Proc. Natl. Acad. Sci. USA 99, 15776–15781.

Marrif, H., Schifman, A., Stepanyan, Z., Gillis, M.A., Calderone, A., Weiss, R.E., Samarut, J., Silva, J.E. 2005. Temperature homeostasis in transgenic mice lacking thyroid hormone receptor-alpha gene products. Endocrinology 146, 2872–2884.

Marsh-Armstrong, N., Huang, H., Remo, B.F., Liu, T.T., Brown, D.D. 1999. Asymmetric growth and development of the *Xenopus laevis* retina during metamorphosis is controlled by type III deiodinase. Neuron 24, 871–878.

McCarrison, R. 1908. Observations of endemic cretinism in the chitral and gilgit valleys. Lancet 2, 1275–1280.

Miller, L.D., Park, K.S., Guo, Q.M., Alkharouf, N.W., Malek, R.L., Lee, N.H., Liu, E.T., Cheng, S.Y. 2001. Silencing of Wnt signaling and activation of multiple metabolic pathways in response to thyroid hormone-stimulated cell proliferation. Mol. Cell. Biol. 21, 6626–6639.

Mirabella, G., Westall, C.A., Asztalos, E., Perlman, K., Koren, G., Rovet, J. 2005. Development of contrast sensitivity in infants with prenatal and neonatal thyroid hormone insufficiencies. Pediatr. Res. 57, 902–907.

Mitsuhashi, T., Nikodem, V. 1989. Regulation of expression of the alternative mRNAs of the rat α-thyroid hormone receptor gene. J. Biol. Chem. 264, 8900–8904.

Mittag, J., Friedrichsen, S., Heuer, H., Polsfuss, S., Visser, T.J., Bauer, K. 2005. Athyroid Pax8–/– mice cannot be rescued by the inactivation of thyroid hormone receptor alpha1. Endocrinology 146, 3179–3184.

Mori, M., Ghyselinck, N.B., Chambon, P., Mark, M. 2001. Systematic immunolocalization of retinoid receptors in developing and adult mouse eyes. Invest. Ophthalmol. Vis. Sci. 42, 13128.

Morte, B., Manzano, J., Scanlan, T., Vennström, B., Bernal, J. 2002. Deletion of the thyroid hormone receptor alpha 1 prevents the structural alterations of the cerebellum induced by hypothyroidism. Proc. Natl. Acad. Sci. USA 99, 3985–3989.

Morte, B., Manzano, J., Scanlan, T.S., Vennstrom, B., Bernal, J. 2004. Aberrant maturation of astrocytes in thyroid hormone receptor alpha1 knock out mice reveals an interplay between thyroid hormone receptor isoforms. Endocrinology 145, 1386–1391.

Murray, G. 1891. Note on the treatment of myxoedema by hypodermic injections of an extract of the thyroid gland of a sheep. Br. Med. J. 2, 796–797.

Nathans, J. 1999. The evolution and physiology of human color vision: Insights from molecular genetic studies of visual pigments. Neuron 24, 299–312.

Newell, F.W., Diddie, K.R. 1977. Typical monochromacy, congenital deafness, and resistance to intracellular action of thyroid hormone. Klin. Monatsbl. Augenheilkd. 171, 731–734.

Ng, L., Hurley, J.B., Dierks, B., Srinivas, M., Saltó, C., Vennström, B., Reh, T.A., Forrest, D. 2001a. A thyroid hormone receptor that is required for the development of green cone photoreceptors. Nat. Genet. 27, 94–98.

Ng, L., Pedraza, P.E., Faris, J.S., Vennström, B., Curran, T., Morreale De Escobar, G., Forrest, D. 2001b. Audiogenic seizure susceptibility in thyroid hormone receptor β-deficient mice. Neuroreport 12, 2359–2362.

Ng, L., Rüsch, A., Amma, L., Nordström, K., Erway, L., Vennström, B., Forrest, D. 2001c. Suppression of the deafness and thyroid dysfunction in *Thrb*-null mice by an independent mutation in the *Thra* thyroid hormone receptor α gene. Hum. Mol. Genet. 10, 2701–2708.

Ng, L., Goodyear, R.J., Woods, C.A., Schneider, M.J., Diamond, E., Richardson, G.P., Kelley, M.W., Germain, D.L., Galton, V.A., Forrest, D. 2004. Hearing loss and retarded cochlear development in mice lacking type 2 iodothyronine deiodinase. Proc. Natl. Acad. Sci. USA 101, 3474–3479.

O'Malley, B.W., Li, D., Turner, D.S. 1995. Hearing loss and cochlear abnormalities in the congenital hypothyroid (hyt/hyt) mouse. Hearing Res. 88, 181–189.

O'Shea, P.J., Harvey, C.B., Suzuki, H., Kaneshige, M., Kaneshige, K., Cheng, S.Y., Williams, G.R. 2003. A thyrotoxic skeletal phenotype of advanced bone formation in mice with resistance to thyroid hormone. Mol. Endocrinol. 17, 1410–1424.

O'Shea, P.J., Bassett, J.H., Sriskantharajah, S., Ying, H., Cheng, S.Y., Williams, G.R. 2005. Contrasting skeletal phenotypes in mice with an identical mutation targeted to thyroid hormone receptor alpha1 or beta. Mol. Endocrinol. 19, 3045–3059.

Oppenheimer, J., Koerner, D., Schwartz, H., Surks, M. 1972. Specific nuclear triiodothyronine binding sites in rat liver and kidney. J. Clin. Endocrinol. Metab. 35, 330–333.

Ortiga-Carvalho, T.M., Shibusawa, N., Nikrodhanond, A., Oliveira, K.J., Machado, D.S., Liao, X.H., Cohen, R.N., Refetoff, S., Wondisford, F.E. 2005. Negative regulation by thyroid hormone receptor requires an intact coactivator-binding surface. J. Clin. Invest. 115, 2517–2523.

Perez-Castillo, A., Bernal, J., Ferreiro, B., Pans, T. 1985. The early ontogenesis of thyroid hormone receptor in the rat fetus. Endocrinology 117, 2457–2461.

Plateroti, M., Chassande, O., Fraichard, A., Gauthier, K., Freund, J.N., Samarut, J., Kedinger, M. 1999. Involvement of T3Ralpha-and beta-receptor subtypes in mediation of T3 functions during postnatal murine intestinal development. Gastroenterology 116, 1367–1378.

Plateroti, M., Gauthier, K., Domon-Dell, C., Freund, J.N., Samarut, J., Chassande, O. 2001. Functional interference between thyroid hormone receptor α and natural truncated TR delta α isoforms in the control of intestine development. Mol. Cell. Biol. 21, 4761–4772.

Poguet, A.L., Legrand, C., Feng, X., Yen, P.M., Meltzer, P., Samarut, J., Flamant, F. 2003. Microarray analysis of knockout mice identifies cyclin D2 as a possible mediator for the action of thyroid hormone during the postnatal development of the cerebellum. Dev. Biol. 254, 188–199.

Potter, G.B., Zarach, J.M., Sisk, J.M., Thompson, C.C. 2002. The thyroid hormone-regulated corepressor hairless associates with histone deacetylases in neonatal rat brain. Mol. Endocrinol. 16, 2547–2560.

Refetoff, S., Weiss, R.E., Usala, S.J. 1993. The syndromes of resistance to thyroid hormone. Endocrine. Rev. 14, 348–399.

Roberts, M.R., Hendrickson, A., Mcguire, C.R., Reh, T.A. 2005. Retinoid X receptor (gamma) is necessary to establish the S-opsin gradient in cone photoreceptors of the developing mouse retina. Invest. Ophthalmol. Vis. Sci. 46, 2897–2904.

Roberts, M.R., Srinivas, M., Forrest, D., Morreale De Escobar, G., Reh, T.A. 2006. Making the gradient: Thyroid hormone regulates cone opsin expression in the developing mouse retina. Proc. Natl. Acad. Sci. USA 103, 6218–6223.

Rovet, J., Walker, W., Bliss, B., Buchanan, L., Ehrlich, R. 1996. Long-term sequelae of hearing impairment in congenital hypothyroidism. J. Pediatr. 128, 776–783.

Rueda, J., Prieto, J.J., Cantos, R., Sala, M.L., Merchan, J.A. 2003. Hypothyroidism prevents developmental neuronal loss during auditory organ development. Neurosci. Res. 45, 401–408.

Ruiz-Marcos, A., Salas, J., Sanchez-Toscano, F., Escobar Del Rey, F., Morreale De Escobar, G. 1983. Effect of neonatal and adult-onset hypothyroidism on pyramidal cells of the rat auditory cortex. Dev. Brain Res. 9, 205–213.

Rüsch, A., Erway, L., Oliver, D., Vennström, B., Forrest, D. 1998. Thyroid hormone receptor β-dependent expression of a potassium conductance in inner hair cells at the onset of hearing. Proc. Natl. Acad. Sci. USA 95, 15758–15762.

Rüsch, A., Ng, L., Goodyear, R., Oliver, D., Lisoukov, I., Vennström, B., Richardson, G., Kelley, M.W., Forrest, D. 2001. Retardation of cochlear maturation and impaired hair cell function caused by deletion of all known thyroid hormone receptors. J. Neurosci. 21, 9792–9800.

Sakurai, A., Miyamoto, T., Degroot, L.J. 1992. Cloning and characterization of the human thyroid hormone receptor beta 1 gene promoter. Biochem. Biophys. Res. Commun. 185, 78–84.

Saltó, C., Kindblom, J.M., Johansson, C., Wang, Z., Gullberg, H., Nordström, K., Mansén, A., Ohlsson, C., Thorén, P., Forrest, D., Vennström, B. 2001. Ablation of TRα2 and a concomitant overexpression of α1 yields a mixed hypo- and hyperthyroid phenotype in mice. Mol. Endocrinol. 15, 2115–2128.

Sap, J., Muñoz, A., Damm, K., Goldberg, Y., Ghysdael, J., Leutz, A., Beug, H., Vennström, B. 1986. The c-*erb*A protein is a high affinity receptor for thyroid hormone. Nature 324, 635–640.

Schneider, M.J., Davey, J.C., Galton, V.A. 1993. *Rana catesbeiana* tadpole red blood cells express an alpha, but not a beta, c-*erb*A gene. Endocrinology 133, 2488–2495.

Schneider, M.J., Fiering, S.N., Pallud, S.E., Parlow, A.F., St. Germain, D.L., Galton, V.A. 2001. Targeted disruption of the type 2 selenodeiodinase gene (DIO2) results in a phenotype of pituitary resistance to T4. Mol. Endocrinol. 15, 2137–2148.

Schwartz, H.L., Oppenheimer, J.H. 1978. Ontogenesis of 3,5,3′-triiodothyronine receptors in neonatal rat brain: Dissociation between receptor concentration and stimulation of oxygen consumption by 3,5,3′-triiodothyronine. Endocrinology 103, 943–948.

Sevilla-Romero, E., Munoz, A., Pinazo-Duran, M.D. 2002. Low thyroid hormone levels impair the perinatal development of the rat retina. Ophthalmic. Res. 34, 181–191.

Shi, Y.B., Yaoita, Y., Brown, D.D. 1992. Genomic organization and alternative promoter usage of the two thyroid hormone receptor beta genes in *Xenopus laevis*. J. Biol. Chem. 267, 733–738.

Shibusawa, N., Hashimoto, K., Nikrodhanond, A.A., Liberman, M.C., Applebury, M.L., Liao, X. H., Robbins, J.T., Refetoff, S., Cohen, R.N., Wondisford, F.E. 2003. Thyroid hormone action in the absence of thyroid hormone receptor DNA-binding *in vivo*. J. Clin. Invest. 112, 588–597.

Sjöberg, M., Vennström, B. 1995. Ligand-dependent and -independent transactivation by thyroid hormone receptor β2 is determined by the structure of the hormone response element. Mol. Cell. Biol. 15, 4718–4726.

Sjöberg, M., Vennström, B., Forrest, D. 1992. Thyroid hormone receptors in chick retinal development: Differential expression of mRNAs for α and N-terminal variant β receptors. Development 114, 39–47.

Srinivas, M., Ng, L., Liu, H., Jia, L., Forrest, D. 2006. Activation of the blue opsin gene in cone photoreceptor development by retinoid-related orphan receptor β. Mol. Endocrinol. 20, 1728–1741.

St. Germain, D., Galton, V. 1997. The deiodinase family of selenoproteins. Thyroid 7, 655–668.

Strait, K.A., Schwartz, H.L., Perez-Castillo, A., Oppenheimer, J.H. 1990. Relationship of cerbA mRNA content to tissue triiodothyronine nuclear binding capacity and function in developing and adult rats. J. Biol. Chem. 265, 10514–10521.

Sui, L., Gilbert, M. 2003. Pre- and postnatal propylthiouracil-induced hypothyroidism impairs synaptic transmission and plasticity in area CA1 of the neonatal rat hippocampus. Endocrinology 144, 4195–4203.

Szel, A., Rohlich, P., Mieziewska, K., Aguirre, G., van Veen, T. 1993. Spatial and temporal differences between the expression of short- and middle-wave sensitive cone pigments in the mouse retina: A developmental study. J. Comp. Neurol. 331, 564–577.

Tagami, T., Kopp, P., Johnson, W., Arseven, O., Jameson, J. 1998. The thyroid hormone receptor variant α2 is a weak antagonist because it is deficient in interactions with nuclear receptor corepressors. Endocrinology 139, 2535–2544.

Tata, J.R. 1963. Inhibition of the biological action of thyroid hormones by actinomycin D and puromycin. Nature 197, 1167–1168.

Tenbaum, S.P., Juenemann, S., Schlitt, T., Bernal, J., Renkawitz, R., Munoz, A., Baniahmad, A. 2003. Alien/CSN2 gene expression is regulated by thyroid hormone in rat brain. Dev. Biol. 254, 149–160.

Thompson, C.C. 1996. Thyroid hormone-responsive genes in developing cerebellum include a novel synaptotagmin and a *hairless* homolog. J. Neurosci. 16, 7832–7840.

Tinnikov, A., Nordström, K., Thorén, P., Kindblom, J.M., Malin, S., Rozell, B., Adams, M., Rajanayagam, O., Pettersson, S., Ohlsson, C., Chatterjee, K., Vennström, B. 2002. Retardation of post-natal development caused by a negatively acting thyroid hormone receptor α1. EMBO J. 21, 5079–5087.

Tsai, C.E., Lin, S.P., Ito, M., Takagi, N., Takada, S., Ferguson-Smith, A.C. 2002. Genomic imprinting contributes to thyroid hormone metabolism in the mouse embryo. Curr. Biol. 12, 1221–1226.

Uziel, A., Rabie, A., Marot, M. 1980. The effect of hypothyroidism on the onset of cochlear potentials in developing rats. Brain Res. 182, 172–175.

Uziel, A., Pujol, R., Legrand, C., Legrand, J. 1983. Cochlear synaptogenesis in the hypothyroid rat. Dev. Brain Res. 7, 295–301.

Van Middlesworth, L., Norris, C. 1980. Audiogenic seizures and cochlear damage in rats after perinatal antithyroid treatment. Endocrinology 106, 1686–1690.

Venero, C., Guadano-Ferraz, A., Herrero, A.I., Nordstrom, K., Manzano, J., De Escobar, G.M., Bernal, J., Vennstrom, B. 2005. Anxiety, memory impairment, and locomotor dysfunction caused by a mutant thyroid hormone receptor alpha1 can be ameliorated by T3 treatment. Genes Dev. 19, 2152–2163.

Wan, W., Farboud, B., Privalsky, M.L. 2005. Pituitary resistance to thyroid hormone syndrome is associated with T3 receptor mutants that selectively impair beta2 isoform function. Mol. Endocrinol. 19, 1529–1542.

Wasniewska, M., De Luca, F., Siclari, S., Salzano, G., Messina, M.F., Lombardo, F., Valenzise, M., Ruggeri, C., Arrigo, T. 2002. Hearing loss in congenital hypothalamic hypothyroidism: A wide therapeutic window. Hear. Res. 172, 87–91.

Weber, T., Zimmermann, U., Winter, H., Mack, A., Kopschall, I., Rohbock, K., Zenner, H.P., Knipper, M. 2002. Thyroid hormone is a critical determinant for the regulation of the cochlear motor protein prestin. Proc. Natl. Acad. Sci. USA 99, 2901–2906.

Weinberger, C., Thompson, C.C., Ong, E.S., Lebo, R., Gruol, D.J., Evans, R.M. 1986. The *c-erb*-A gene encodes a thyroid hormone receptor. Nature 324, 641–646.

Weiss, R., Murata, Y., Cua, K., Hayashi, Y., Seo, H., Refetoff, S. 1998. Thyroid hormone action on liver, heart, energy expenditure in thyroid hormone receptor *β*-deficient mice. Endocrinology 139, 4945–4952.

Wikström, L., Johansson, C., Saltó, C., Barlow, C., Campos Barros, A., Baas, F., Forrest, D., Thorén, P., Vennström, B. 1998. Abnormal heart rate and body temperature in mice lacking thyroid hormone receptor α1. EMBO J. 17, 455–461.

Williams, G.R. 2000. Cloning and characterization of two novel thyroid hormone receptor *β* isoforms. Mol. Cell. Biol. 20, 8329–8342.

Wondisford, F.E. 2003. Thyroid hormone action: Insight from transgenic mouse models. J. Investig. Med. 51, 215–220.

Wood, W.M., Dowding, J.M., Haugen, B.R., Bright, T.M., Gordon, D.F., Ridgway, E.C. 1994. Structural and functional characterization of the genomic locus encoding the murine *β*2 thyroid hormone receptor. Mol. Endocrinol. 8, 1605–1617.

Yang, Z., Privalsky, M.L. 2001. Isoform-specific transcriptional regulation by thyroid hormone receptors: Hormone-independent activation operates through a steroid receptor mode of co-activator interaction. Mol. Endocrinol. 15, 1170–1185.

Yaoita, Y., Brown, D.D. 1990. A correlation of thyroid hormone receptor gene expression with amphibian metamorphosis. Genes Develop. 4, 1917–1924.

Yaoita, Y., Shi, Y.-B., Brown, D. 1990. *Xenopus laevis* a and b thyroid hormone receptors. Proc. Natl. Acad. Sci. USA 87, 7090–7094.

Yen, P.M., Feng, X., Flamant, F., Chen, Y., Walker, R.L., Weiss, R.E., Chassande, O., Samarut, J., Refetoff, S., Meltzer, P.S. 2003. Effects of ligand and thyroid hormone receptor isoforms on hepatic gene expression profiles of thyroid hormone receptor knockout mice. EMBO Rep. 4, 581–587.

Zhu, X.-G., Mcphie, P., Lin, K.-H., Cheng, S.-Y. 1997. The differential hormone-dependent transcriptional activation of thyroid hormone receptor isoforms is mediated by interplay of their domains. J. Biol. Chem. 272, 9048–9054.

PPARs in fetal and early postnatal development

Nicolas Rotman, Liliane Michalik,
Béatrice Desvergne and Walter Wahli

*Center for Integrative Genomics,
National Research Center "Frontiers in Genetics,"
University of Lausanne, Lausanne, Switzerland*

Contents

Advances in Developmental Biology
Volume 16 ISSN 1574-3349
DOI: 10.1016/S1574-3349(06)16002-0

Peroxisome proliferator-activated receptor (PPARα), PPARβ/δ, and PPARγ are ligand-activated transcription factors belonging to the hormone nuclear receptor family. They are implicated not only in many aspects of metabolism and physiology but also in embryonic developmental processes. Although the role of PPARs in embryonic development is less well studied, evidence has begun to accumulate that suggests a significant role in this process. PPARs play a role in placenta formation, adipogenesis, differentiation of the skin and intestinal epithelia, and may also act in neural development. These developmental processes involve PPARs in different cellular functions, such as the determination of cell identity, cell differentiation, cell death and survival, and cell migration. For instance, PPARs are implicated in the differentiation of trophoblast giant cells, adipocytes, keratinocytes, and intestinal epithelial cells, and PPARβ/δ protects keratinocytes from apoptosis during hair follicle formation. Finally, connections between PPAR signaling and other pathways are being intensely investigated, which should contribute to a better understanding of the coordinating roles of PPARs during development.

1. Introduction

The three peroxisome proliferator-activated receptors (PPARs) form a specific subgroup of the hormone nuclear receptor family. Their name refers to their implication in the mediation of a pleiotropic response to various chemicals, particularly peroxisome proliferation in rodents (Lee et al., 1995). In 15 years of intensive research since their discovery (Issemann and Green, 1990; Dreyer et al., 1992), PPARs have been implicated in many vital processes, such as glucose and lipid metabolism and inflammation, as well as in a variety of developmental programs that will be described herein (Desvergne and Wahli, 1999; Kersten et al., 2000). PPARα (or NR1C1), PPARβ/δ (or NR1C2, also known as FAAR or NUC1), and PPARγ (or NR1C3) are encoded by distinct genes at different loci. They were independently identified in the early 1990s in mice (Issemann and Green, 1990) and *Xenopus laevis* (Dreyer et al., 1992) and were subsequently also found in fish, chicken, hamster, rat, and human (Desvergne and Wahli, 1999). PPARγ1 and PPARγ2, the two isoforms of PPARγ that have been identified in mammals, are obtained by alternative promoter usage and differential splicing (Fajas et al., 1997). PPARs display the canonical structure of hormone nuclear receptors. The most N-terminal region or A/B region contains a weak transactivation domain called AF-1. It is followed by the C domain (two zinc finger motifs), which binds DNA, the D domain, which functions as a hinge region, and the E domain, which contains the ligand-binding pocket or domain (LBD) and the ligand-dependent transactivation function AF-2 (Fig. 1A).

The view on the mode of action of PPARs is schematized in Fig. 1B. PPARs are activated by various fatty acids and eicosanoids or by synthetic compounds that function as ligands entering and occupying the LBD pocket. This interaction stabilizes an activator complex consisting of a PPAR:retinoid X receptor (PPAR:RXR) heterodimer and coactivators that stimulate the transcription of target genes by binding to specific DNA elements, the peroxisome proliferator response elements (PPREs) located in the regulatory regions of these genes. The PPRE consensus sequence is 5'-AACT·AGGNCA·A·AGGTCA-3', and all the different elements described so far correspond in varying degrees to this sequence (Juge-Aubry et al., 1997). PPARs are most often located within the cell nucleus, therefore, it is thought that their ligands are carried from the cytoplasm to the nucleus and presented to them by fatty acid-binding proteins (FABPs) (Tan et al., 2002; Haunerland and Spener, 2004).

The first hint suggesting that PPARs have a role to play in developmental processes came from the analysis of their expression pattern in *X. laevis*. PPARα mRNA and even more PPARβ/δ mRNA were found at high levels during oogenesis and throughout early development (Dreyer et al., 1992). In rodents, all three PPAR proteins were detected as early as embryonic day 5 (E5) in mouse embryos (Keller et al., 2000; Table 1). However, the earliest time examined for transcript analysis is E8.5 in mouse and E9.5 in rat when PPARβ/δ mRNA is already abundant, whereas PPARα and PPARγ mRNAs are only detected later (E13.5) (Kliewer et al., 1994; Braissant and Wahli, 1998). In parallel to these expression data, functional studies have pointed to essential roles of PPARs in various aspects of development during both embryogenesis and early postnatal life. We will review PPAR functions in placenta formation, adipogenesis, skin development, and intestinal epithelium differentiation. We will also summarize the data available so far about the role of PPARs in neural development. Finally, we will discuss the specific or redundant involvement of the three PPARs during embryogenesis and early postnatal life.

2. PPARβ/δ and PPARγ are essential for placenta development

A possible role of PPARγ and PPARβ/δ in placenta formation was first suspected with the failure to obtain homozygous null mice for these genes due to an early embryonic lethality. Therefore, the focus will be first on mouse placenta development, followed by a discussion of functions conserved in the formation of the human placenta.

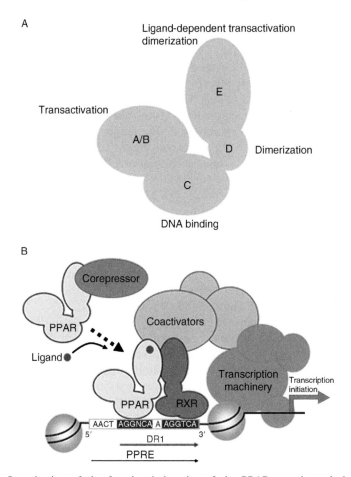

Fig. 1. Organization of the functional domains of the PPAR protein and the active transactivation PPAR:RXR complex. (A) The PPAR protein is organized in functional regions whose main functions are indicated. (B) On ligand binding, a complex formed by a PPAR:RXR dimer and coactivators is stabilized. The PPAR:RXR dimer recognizes specific DNA sequences on the promoter region of target genes and binds to them, which favors transcription initiation on recruitment of the transcription machinery and chromatin remodeling. The PPREs consist of the direct repeat of six nucleotides spaced by one nucleotide (direct repeat 1 or DR1 core motif) flanked by a specific 5′ sequence. The PPRE consensus sequence is indicated.

2.1. Role of PPARγ and PPARβ/δ in mouse placenta development

In mice, two distinct cell lineages form within the embryo by E3.5: the outer trophectoderm epithelium and the inner cell mass. Most of the placenta cell types derive from the trophectoderm, which starts to differentiate around the time of embryo implantation (E4.5). By E9.5, the placenta is composed of three main layers: (1) the trophoblast giant cells that lie across

Table 1

Timing of expression of the three PPAR isotypes during fetal and early postnatal development in the rodent (See Color Insert.)

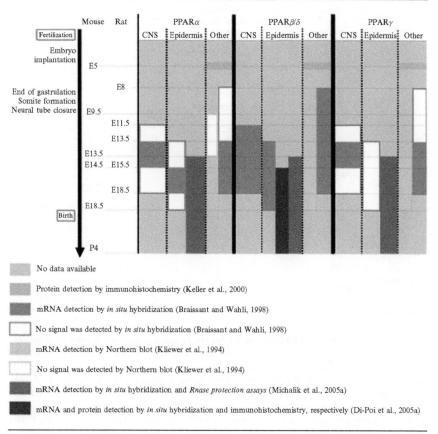

No data available

Protein detection by immunohistochemistry (Keller et al., 2000)

mRNA detection by *in situ* hybridization (Braissant and Wahli, 1998)

No signal was detected by *in situ* hybridization (Braissant and Wahli, 1998)

mRNA detection by Northern blot (Kliewer et al., 1994)

No signal was detected by Northern blot (Kliewer et al., 1994)

mRNA detection by *in situ* hybridization and *Rnase protection assays* (Michalik et al., 2005a)

mRNA and protein detection by *in situ* hybridization and immunohistochemistry, respectively (Di-Poi et al., 2005a)

Expression data within the placenta are not summarized in the table. CNS, central nervous system.

the maternal decidua, (2) the spongiotrophoblast that supports (3) the labyrinth (Fig. 2). The latter contains numerous villi filled with maternal blood and abundant fetal vasculature connected to the developing fetus by the umbilical cord. This system ensures an exchange of material between the mother and the embryo. PPAR*β*/*δ* and PPAR*γ* are expressed throughout the above mentioned three layers of the mouse placenta at E9.5 (Barak et al., 1999; Nadra et al., 2006), but genetic studies have shown no redundant essential functions for these two genes, as detailed later.

PPAR*γ* is important for the differentiation of the labyrinth layer in the mouse placenta. Homozygous PPAR*γ*-null embryos die during gestation (between E9.5 and E12.5), whereas heterozygous embryos have no obvious

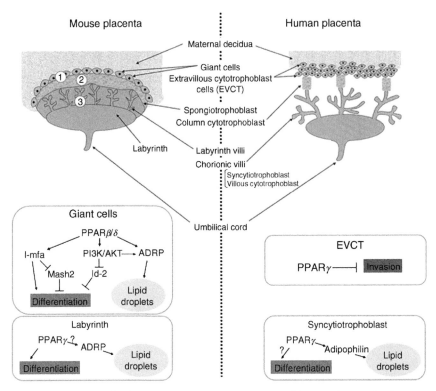

Fig. 2. Comparative role of PPARβ/δ and PPARγ in mouse and human placenta development. Schematic representations of the mouse and human differentiated placenta are shown. Structures appearing in light grey in the mouse placenta derive from the trophectoderm, while the labyrinth layer derives both from the trophectoderm and from the inner cell mass. The functions of PPARs in the differentiation of placenta structures are summarized in the boxes. Arrows indicate promotion while T-shape connectors indicate inhibition. Question marks identify uncertain mechanisms. ADRP, adipocyte differentiation-related protein; EVCT, extravillous cytotrophoblast cells; PI3K, phosphatidylinositol-3 kinase. 1: trophoblast giant cells; 2: spongiotrophoblast; 3: labyrinth.

phenotype (Barak et al., 1999). Before E9.5, a Mendelian distribution of the different genotypes (WT, $+/+$ and $-/-$) was observed within the offspring of crosses between heterozygous parents, and until E9.5, homozygous PPARγ-null embryos are externally indistinguishable from their wild-type (WT) littermates. Thus, PPARγ becomes essential at E9.5, when placenta defects are generally revealed. This is consistent with its high expression levels within this organ starting from E8.5 onward, while its expression remains below detection levels using Northern blot and *in situ* hybridization in the embryo proper at E9.5 (Kliewer et al., 1994; Braissant and Wahli, 1998; Barak et al., 1999). Immunohistochemistry analyses confirmed these expression data, revealing very low amounts of PPARγ protein in E5 and E11 (no data at

E9.5) embryos (Keller et al., 2000; Table 1). Thus, expression of PPARγ is much higher in the placenta than in the embryo. In addition, histological analyses of PPARγ-null placentas performed at E9.5 pointed to various anomalies mainly affecting the labyrinth. Inappropriate formation of vascular lakes together with delayed labyrinth differentiation were severe enough to impair proper material exchanges between the mother and the fetus and may be the cause of embryo lethality. This idea is reinforced by the fact that viable PPARγ-null embryos were recovered at E9.5 and at later stages up to birth, when the PPARγ-null fetuses were experimentally associated with a wild-type placenta. This was achieved by making chimeric aggregates between wild-type tetraploid morulas and diploid embryos obtained from crosses with heterozygous parents (Barak et al., 1999). In such developing aggregates, the trophoblast-derived compartments of the placenta are entirely of tetraploid origin, whereas the embryo and the mesodermal-derived compartments of the placenta are only diploid (Rossant and Cross, 2001). The fact that PPARγ-null embryos are viable when fed through a wild-type tetraploid placenta demonstrates unambiguously the essential role of the wild-type PPARγ allele in mouse placenta development and function. More precisely, these data show that PPARγ is essential for the components of the placenta that have a trophoblastic but not an embryonic (allantois) origin. However, little is known about the molecular mechanisms involved. Mucin 1 (Muc1), which is localized exclusively around maternal sinuses in the labyrinth, is encoded in a direct PPARγ target gene, but does not seem to be instrumental in the defect observed in PPARγ-null placentas (Shalom-Barak et al., 2004). In-depth analyses that describe the regulatory role of PPARγ in mouse placenta development are still lacking.

PPARβ/δ is also essential for placenta development in mice, as shown in two studies (Barak et al. 2002; Nadra et al., 2006). Similarly to PPARγ-null mice, most of the PPARβ/δ-null fetuses die by E10.5, whereas heterozygous animals are viable and generally healthy under standard conditions. This was observed with two mutant constructs independently generated (by us and others) that lack the DNA-binding domain of the PPARβ/δ gene (Barak et al., 2002; Nadra et al., 2006). A third mutant line lacking the end of the ligand-binding domain displayed no embryonic lethality (Peters et al., 2000; Table 2). One can speculate that this latter PPARβ/δ mutant allele retains the ability to play some role in placenta formation, or alternatively that some compensatory genes enabling the placenta to overcome the absence of PPARβ/δ are present in the genetic background of this third mutant line, but are absent in the two former lines.

While we have been able to generate a PPARβ/δ-null line (see later) by intercrossing the few PPARβ/δ-null mice obtained from heterozygous crosses (Nadra et al., 2006), we performed an in-depth analysis of the causes of the high PPARβ/δ-null embryonic lethality. PPARβ/δ-null placentas obtained from crosses with heterozygous parents exhibit severe abnormalities (Barak

Table 2
Summary of the developmental phenotypes observed in homozygous mutant mice lacking one given functional PPAR gene

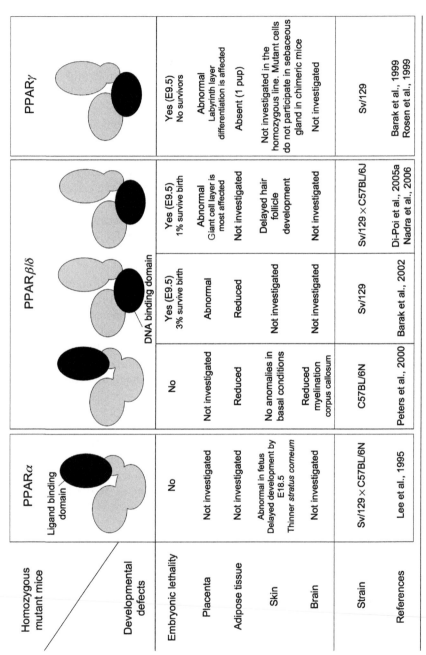

Homozygous mutant mice / Developmental defects	PPARα	PPARβ/δ			PPARγ
	Ligand binding domain		DNA binding domain		
Embryonic lethality	No	No	Yes (E9.5) 3% survive birth	Yes (E9.5) 1% survive birth	Yes (E9.5) No survivors
Placenta	Not investigated	Not investigated	Abnormal	Abnormal Giant cell layer is most affected	Abnormal Labyrinth layer differentiation is affected
Adipose tissue	Not investigated	Reduced	Reduced	Not investigated	Absent (1 pup)
Skin	Abnormal in fetus Delayed development by E18.5 Thinner *stratus corneum*	No anomalies in basal conditions	Not investigated	Delayed hair follicle development	Not investigated in the homozygous line. Mutant cells do not participate in sebaceous gland in chimeric mice
Brain	Not investigated	Reduced myelination corpus callosum	Not investigated	Not investigated	Not investigated
Strain	Sv/129 × C57BL/6N	C57BL/6N	Sv/129	Sv/129 × C57BL/6J	Sv/129
References	Lee et al., 1995	Peters et al., 2000	Barak et al., 2002	Di-Poi et al., 2005a Nadra et al., 2006	Barak et al., 1999 Rosen et al., 1999

In the receptor scheme, the black parts represent the receptor domains whose genomic region has been deleted in the mutant animals.

et al., 2002; Nadra et al., 2006). At E9.5 and E10.5, all three layers of the placenta of PPARβ/δ-null concepti are reduced in size. However, molecular analyses by *in situ* hybridization with marker genes for each of these layers showed that the giant cell layer is the most affected. This led to the hypothesis that PPARβ/δ regulates the differentiation of trophoblast stem cells into giant cells. Rcho-1 trophoblast cells in culture differentiate massively into giant cells when treated with L165041, a selective PPARβ/δ agonist. This effect is abolished when PPARβ/δ expression is knocked down by an anti-PPARβ/δ siRNA. Notably, while these cells express PPARγ, no differentiation occurs when they are treated with rosiglitazone, a specific agonist of PPARγ, confirming the involvement of PPARβ/δ in giant cell differentiation in this cell culture model. It is implicated in the control of at least two pathways. First, activated PPARβ/δ induces Akt phosphorylation in Rcho-1 cells, via the upregulation of 3-phosphoinositide-dependent kinase-1 (PDK1) and integrin-linked kinase (ILK), which are encoded in direct PPARβ/δ target genes (Section 3.4). This induced Akt activity promotes giant cell differentiation (Kamei et al., 2002; Nadra et al., 2006). Second, activation of PPARβ/δ increases the expression of I-mfa, an inhibitor of basic helix-loop-helix (bHLH) transcription factors, such as Mash-2, whose inhibition is necessary for giant cell differentiation (Nakayama et al., 1997; Kraut et al., 1998). These observations in Rcho-1 cells are corroborated *in vivo*, at E9.5, when I-mfa mRNA and phospho-Akt levels are decreased in PPARβ/δ-null compared with wild-type placenta. In addition, immunohistodetection of phospho-Akt in the wild-type placenta at E9.5 reveals that the phosphoprotein localizes essentially in the nuclei of the giant cells. These data demonstrate that PPARβ/δ acts through activation of phosphatidylinositol-3 kinase/Akt (PI3K/Akt) and induction of I-mfa to promote giant cell differentiation *in vivo*. They also point to an unexplored role of nuclear Akt in this process.

A feature of giant cell differentiation that depends on PPARβ/δ, both *in vitro* and *in vivo*, is the accumulation of lipid droplets in these cells. In the wild-type placenta, lipid droplet accumulation parallels the expression of adipocyte differentiation-related protein (ADRP), a lipid droplet-associated protein coded by a gene that contains a PPRE in its promoter region. At E9.5, ADRP is mainly expressed in the giant cell layer under the control of PPARβ/δ and, as expected, its expression is abolished in PPARβ/δ-null placentas but is unchanged in PPARγ-null placentas. At E16.5, ADRP was found mainly in the labyrinth layer of wild-type placentas, which is also the most affected region in PPARγ-null placentas at E9.5. Thus, one can speculate that early accumulation of lipid droplets takes place in the giant cell layer by E9.5 under the control of PPARβ/δ, while later accumulation of lipids in the labyrinth occurs under the control of PPARγ (Fig. 2).

Besides being important for placenta development through the secretion of hormones and the accumulation of lipids, giant cells are also important for embryo implantation. It is noteworthy that while PPARβ/δ is essential for proper giant cell differentiation and placenta development, it does not seem to play a major role in embryo implantation. Both the total number of embryos and the proportion of PPARβ/δ-null embryos recovered a few days after implantation, but before E9.5, are normal in crosses between heterozygous parents (Barak et al., 2002; Nadra et al., 2006). Giant cells are generated from two different cell populations in two timely separated waves of differentiation. The first precedes implantation and is required for its success, while the second follows this event (Rossant and Cross, 2001). Thus, the data described earlier indicate that PPARβ/δ only intervenes during the second wave of giant cell differentiation, in a mechanism distinct from the first wave. Alternatively, there may be some redundancy between PPARs, or other pathways, at the time of the first wave of giant cell differentiation.

In summary, both PPARγ and PPARβ/δ are essential for placenta development. PPARβ/δ controls giant cell differentiation through activation of PI3K/Akt and induction of I-mfa. It also mediates early accumulation of lipid droplets in these cells through the control of ADRP expression. PPAR is essential for proper labyrinth development and might intervene in late lipid accumulation through the control of ADRP expression (Fig. 2).

2.2. Is there a role for PPARγ in human trophoblast cell differentiation and invasion?

Given the essential functions of PPARγ and PPARβ/δ in murine placenta development, one can speculate that they have a similar importance in human placenta formation since mouse and human placentas have similarities, despite the fact that their structures are not obviously homologous. As shown in Fig. 2, the invasion of the maternal tissue by murine trophoblast giant cells is somewhat similar to the invasion of the uterus by the human extravillous cytotrophoblast (EVCT) cells. Another parallel concerns the labyrinth layer of the mouse, which corresponds to the chorionic villi of the human placenta covered by syncytiotrophoblast cells. Little data is available to compare the role of PPARβ/δ in murine and human placenta development, while a few comparisons are possible concerning PPARγ. First, PPARγ is expressed in villous syncytiotrophoblast cells in the human placenta, as well as in primary cultures of human cytotrophoblast cells (Schaiff et al., 2000). Second, treatment of primary human cytotrophoblast cells with troglitazone, a specific PPARγ agonist, enhances their differentiation into syncytiotrophoblast cells, as evidenced by the increased secretion of human chorionic gonadotropin and the acceleration of syncytia formation (Schaiff et al., 2000). Surprisingly, treatment with the natural PPARγ agonist 15-deoxy-$\Delta^{12,14}$-prostaglandin J_2 (15dPGJ2) shows the opposite effect (Schaiff et al., 2000). This apparent contradiction may be explained by the fact that 15dPGJ2 was used at concentrations (10 μM) that also activate PPARγ-independent pathways (Berry et al., 2005). The expression of adipophilin, which is the human orthologue of ADRP and is expressed in human chorionic villi, is enhanced on treatment with troglitazone in primary human cytotrophoblast cells (Bildirici et al., 2003). Finally, treatment with the PPARγ agonist GW1929 enhances the uptake of fatty acids in primary human cytotrophoblast cells (Schaiff et al., 2005). In summary, PPARγ may promote differentiation of human syncytiotrophoblast cells as it does it for the mouse labyrinth, at least in part via enhancing the expression of adipophilin and ADRP, respectively, and facilitating fatty acid uptake (Fig. 2).

PPARγ expression is also detected in EVCT cells, both *in vivo* and in cell culture (Waite et al., 2000; Fournier et al., 2002). EVCT cells can invade a matrigel matrix *in vitro,* a property that is markedly reduced on treatment with the PPARγ ligand troglitazone, in a dose-dependent manner. Conversely, treatment with a PPARγ antagonist has a stimulating effect (Fournier et al., 2002). Whether this *in vitro* assay is informative about the invasion of the maternal tissues *in vivo* remains to be established. It would also be interesting to use this assay to analyze the effect of PPARβ/δ selective agonists, since PPARβ/δ is essential for murine trophoblast giant cell differentiation.

3. Roles of PPARγ and PPARβ/δ in adipogenesis

Following a significant transient expression in the developing central nervous system (CNS) of the rodent embryos, the first robust and sustained expression of PPARγ occurs in the developing adipose tissue (Braissant et al., 1996; Barak et al., 1999). This pattern of expression is in itself suggestive of a key role of PPARγ in adipose tissue, without minimizing its thus far unexplored function in other developing tissues, where it is expressed at lower levels. In this section, we describe PPARγ as a major regulator of adipogenesis and the molecular pathways controlling its activity and mediating its effects. We finally comment on the specific implication of PPARβ/δ in that same adipogenic process.

The cell lineage leading to adipocytes is not yet clearly established *in vivo*. It is assumed that adipocytes arise from mesenchymal precursors of mesodermal origin that have the ability to develop into myocytes, chondrocytes, osteoblasts, or adipocytes. This hypothesis is supported by the observation that various multipotent cell lines can differentiate into these four cell types (Pittenger et al., 1999). In fact, most of our knowledge on adipocyte development comes from cell culture studies based either on immortal cell lines or on primary cultures (Gregoire et al., 1998). *In vitro*, the differentiation process leading to adipocytes can be grossly divided into two phases. During the first phase, the mesenchymal precursor cell becomes committed to its adipocyte fate, which means that the cell loses its capacity to develop into a cell other than adipocyte. During the second phase, the committed cell differentiates

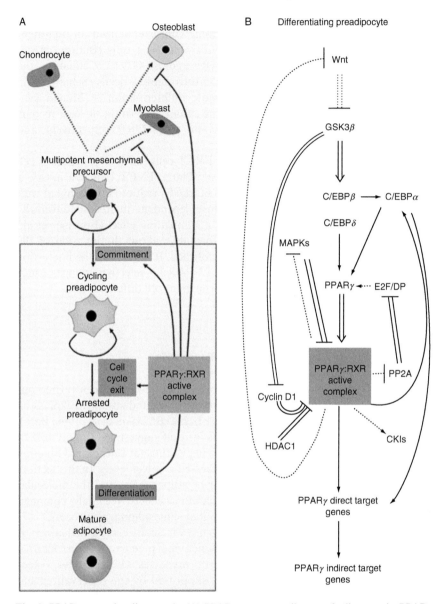

Fig. 3. PPARγ controls adipogenesis. (A) PPARγ promotes all steps of adipogenesis. PPARγ controls the commitment of precursor cells to preadipocytes, mediates the exit of the cell cycle that is required for differentiation into adipocytes, and finally for terminal differentiation into mature adipocytes. (B) PPARγ is at the center of the molecular interactions controlling differentiation of preadipocytes into adipocytes. C/EBPs and PPARγ mutually regulate their expression levels and promote adipocyte differentiation. They are regulated by complex phosphorylation cascades some of which are controlled by the Wnt pathway. The later is active in early adipogenic stages, when it prevents the differentiation of preadipocytes. It is subsequently shut off, which retrieves

into a mature adipocyte by acquiring its corresponding morphological and biochemical features (Gregoire et al., 1998). The transition between these two phases is marked by a withdrawal from the cell cycle (Shao and Lazar, 1997; Morrison and Farmer, 1999).

3.1. PPARγ controls all steps of adipogenesis

PPARγ has been implicated in all aspects of adipogenesis mentioned above (Fig. 3A). PPARγ controls the commitment of a precursor cell to become a preadipocyte. When transfected into multipotent mouse fibroblasts, PPARγ2 stimulates adipogenesis in a dose-dependent manner (Tontonoz et al., 1994). Multipotent mouse fibroblasts are also committed to adipocyte differentiation when treated with the PPARγ selective agonist rosiglitazone, but not when treated with the PPARγ antagonist GW0072 (Oberfield et al., 1999). Similarly, wild-type but not PPARγ-null mouse embryonic stem (ES) cells can differentiate into adipocytes (Rosen et al., 1999). The myogenic differentiation process of G8 myoblasts into myotubes is strongly inhibited by transfection of PPARγ or CCAAT/enhancer-binding protein α (C/EBPα, see later). Moreover, cotransfection of both PPARβ and C/EBPα results in myoblast "transdifferentiation" into adipocytes (Hu et al., 1995). Along the same lines, murine primary mesenchymal stem cells (MSCs) cannot spontaneously differentiate into osteoblasts but show enhanced ability to become adipocytes when TAZ (transcriptional coactivator with PDZ-binding motif) is silenced by siRNA. As TAZ can directly bind to PPARγ and function as a corepressor that inhibits the transcription of PPARγ target genes, these adipogenic effects are likely due to the release of PPARγ inhibition (Hong et al., 2005). *In vivo*, young mice treated with the PPARγ agonist rosiglitazone show decreased osteoblast differentiation and increased number of adipocytes in the bone marrow (Ali et al., 2005). In summary, all these observations demonstrate that the presence of active PPARγ in multipotent mesenchymal precursor cells impairs their ability to develop into myoblasts or osteoblasts while adipogenesis is promoted. However, the role of PPARγ in the commitment into the adipocyte lineage is still debated (Vernochet et al., 2002). More precisely, it has been proposed that the culture media used in the cell culture experiments induced the commitment to preadipocytes and that

GSK3β inhibition and facilitates PPARγ activation via phosphorylation of C/EBPβ and C/EBPα and degradation of cyclin D1. In a feedback control loop, PPARγ inhibits Wnt and MAPK signaling. Arrows indicate activation, while T-shape connectors indicate inhibition. Simple lines are for regulations at the transcriptional level, while double lines show posttranscriptional regulations. Plain lines (simple as well as double) illustrate direct effects while broken lines are for indirect effects. C/EBPs, CCAAT/enhancer-binding proteins; CKI, cyclin kinase inhibitors; E2F/DP, E2 transcription factor/dipeptidyl peptidase; HDAC1, histone deacetylase 1; MAPKs, mitogen-activated protein kinases; PP2A, protein phosphatase 2A.

PPARγ then induced the differentiation program of already committed cells. In support of this hypothesis, PPARγ-null ES cells express high mobility group protein I-C (HMGIC) and lipoprotein lipase (LPL), which are considered as preadipocyte markers. In other words, ES cells could become preadipocytes without PPARγ. However, the expression of HMGIC and LPL in preadipocytes is not sufficient to conclude that any cell expressing these two genes is a preadipocyte. For instance, HMGIC is expressed in mesenchymal derivatives that can differentiate in chondrocytes or in myoblasts (Hirning-Folz et al., 1998), and LPL is found in myocytes. Therefore, based on all the experiments described above, which used different multipotent cell lines and different culture media, it is reasonable to argue that PPARγ is involved in the first phase of adipocyte commitment, although its expression at that stage is low (Gregoire et al., 1998).

The transition from preadipocyte to mature adipocyte is then marked by an exit from the cell cycle. PPARγ also participates in this step by inhibiting the DNA-binding activity of E2 transcription factor/dipeptidyl peptidase (E2F/DP) via posttranscriptional inhibition of protein phosphatase 2A (PP2A) (Altiok et al., 1997) and upregulation of the cyclin kinase inhibitors (CKIs) p18 and p21 in a ligand-dependent manner (Morrison and Farmer, 1999).

Finally, PPARγ is required for terminal differentiation of preadipocytes into mature adipocytes. Thiazolidinediones (TZDs) that were shown to induce terminal differentiation of murine or human preadipocytes (Sandouk et al. 1993; Adams et al., 1997) were independently characterized as PPARγ ligands (Forman et al., 1995; Lehmann et al., 1995). Conversely, ectopic expression of a dominant negative mutant of PPARγ in these same human preadipocytes markedly reduced their TZD-mediated differentiation (Gurnell et al., 2000). This clearly establishes the need for PPARγ activity also in the final steps of the differentiation process from preadipocyte into adipocyte.

Even though the fact that homozygous PPARγ-null mice die during embryogenesis (Table 2) complicated the *in vivo* analysis of the role of PPARγ in the later steps of adipocyte differentiation, different lines of evidence indicate that PPARγ is required for *in vivo* adipogenesis. The placenta rescue experiment mentioned earlier provided one PPARγ-null pup that survived up to birth, but lacked adipose tissue, which suggested that PPARγ is required for adipose tissue formation in the developing mice (Barak et al., 1999). Another study, using chimeric mice comprising PPARγ-null and wild-type cells, revealed that PPARγ-null cells were absent from adipose tissue, while they were present in other tissues of mesodermal origin (Rosen et al., 1999). This shows that cells lacking a functional PPARγ, while present in mesodermal precursor cell populations, do not participate in adipose tissue formation. Skeletal muscles were enriched in PPARγ-null cells, which supports the idea that PPARγ attenuates myogenesis and that adipocyte and myocyte development are mutually exclusive.

Both PPARγ1 and PPARγ2 isoforms contribute to the development of the adipose tissue *in vivo*. A mouse mutant line deprived of the PPARγ2 isoform was generated. These mice reach birth but have reduced adipose mass (Zhang et al., 2004), which reveals a contribution of PPARγ2 to full adipose tissue development. The fact that the expression of the PPARγ1 isoform is almost unaffected by this mutation, together with the fact that some adipose differentiation remains, suggests that PPARγ1 has also an important and specific role in adipogenesis, *in vivo*. Noteworthy, genetic ablation of PPARγ, selectively induced in the mature adipose tissue of adult mice, triggers the death of the adipocytes within days (Imai et al., 2004). Therefore, it appears that PPARγ is essential not only for adipogenesis but also for adipocyte survival and maintenance of the differentiated state.

3.2. Molecular pathways controlling PPARγ expression and activity during adipogenesis

Taken together, the *in vivo* and cell culture data discussed earlier demonstrate that PPARγ is a major regulator of all aspects of adipogenesis. We will now focus our attention on the molecular mechanisms that regulate PPARγ expression and/or activity within this context (Fig. 3B). The CCAAT/enhancer-binding protein (C/EBP) transcription factors play a major role in adipogenesis and have been shown to activate PPARγ transcription in cultured cells. The view is that C/EBPβ and C/EBPδ are expressed transiently in preadipocytes following adipogenic stimulation. These two factors stimulate the expression of low levels of PPARγ and C/EBPα proteins, which then mutually reinforce their synthesis. Therefore, it seems that during early adipocyte differentiation, C/EBPα does not play a major role by itself but participates indirectly by increasing PPARγ expression (Rosen et al., 2002). However, C/EBPα might then be essential to activate a subset of genes (e.g., adiponectin) during terminal differentiation (Park et al., 2004).

In addition to the C/EBP family of transcription factors, other regulators control PPARγ expression and activity. E2Fs induce PPARγ expression in preadipocytes and inhibit this expression after terminal differentiation (Fajas et al., 2002). Phosphorylation cascades also directly or indirectly regulate PPARγ expression and activity in preadipocytes. Mitogen-activated ERK-activating kinase/extracellular signal-regulated kinase (MEK/ERK) signaling and glycogen synthase kinase 3β (GSK3β) promote the phosphorylation of C/EBPα and -β, which contributes to the induction of PPARγ expression (Ross et al., 1999; Prusty et al., 2002; Park et al., 2004; Tang et al., 2005). However, high mitogen-activated protein kinase (MAPK) activity leads to direct phosphorylation of PPARγ, which inhibits its transactivation properties (Adams et al., 1997; Kortum et al., 2005; Tang et al., 2006).

GSK3β not only indirectly promotes PPARγ expression but also increases its activity by inhibiting cyclin D1. Cyclin D1 has been shown to mediate the direct recruitment of histone deacetylase 1 (HDAC1) on the PPRE of PPARγ target genes (Fu et al., 2005). This stabilizes the compact conformation of the chromatin in that region and inhibits transcription initiation. Thereby, GSK3β, which phosphorylates cyclin D1 and promotes its degradation (Ryves and Harwood, 2003), relieves the cyclin D1-mediated inhibition of PPARγ activity. The importance of GSK3β in the regulation of PPARγ expression and activity links PPARγ to the Wnt signaling because GSK3β and Wnt signaling mutually inhibit each other. Wnt signaling is active in early adipogenic stages, during which it prevents preadipocyte differentiation (Ross et al., 2000). Its subsequent shut off (Bennett et al., 2002) relieves GSK3β inhibition and facilitates PPARγ activation via the phosphorylation of C/EBPβ and C/EBPα and degradation of cyclin D1. In turn, PPARγ inhibits Wnt and MAPK signaling via a feedback control loop (Moldes et al., 2003; Tang et al., 2006). These complex reciprocal interactions are likely to ensure the fine tuning of the major players of adipogenesis in the different phases of adipocyte development, as illustrated in Fig. 3.

3.3. Direct interaction between PPARγ and PGC-1 promotes brown fat development

Brown adipocytes differ from white adipocytes in that they are enriched in mitochondria and express uncoupling protein 1 (UCP1). UCP1 allows uncoupling of respiration from ATP synthesis, which results in the dissipation of energy as heat. Thus, the function of the brown adipose tissue (BAT) on cold exposure is to transform energy from food into heat through the catabolism of stored lipids. BAT development has not been as intensively studied as that of WAT, but it appears that the two programs share common features (reviewed in Rosen and Spiegelman, 2000). In particular, both PPARγ and Wnt signaling regulate BAT development. The direct interaction of PPARγ with its coactivator PPAR gamma coactivator 1α (PGC-1α) is essential to promote BAT development and UCP1 production. Conversely, Wnt signaling inhibits BAT formation (Kang et al., 2005). The requirement for thermogenesis through BAT is maximal at birth and decreases during aging. This decrease correlates with a change of the subcellular localization of several transcription factors including PPARγ, which is already excluded from the nuclei of this tissue in 3-month-old mice. This is therefore another way to regulate PPARγ activity *in vivo*, by preventing its binding to the promoter region of target genes (Rim et al., 2004).

3.4. PPARβ/δ is important for the early steps of adipogenesis

The role of the PPARβ/δ isotype in adipogenesis is considerably less understood than that of PPARγ. However, PPARβ/δ has some importance in this process. First, homozygous PPARβ/δ-null mice show markedly reduced adipose stores in the early postnatal period (Peters et al., 2000; Barak et al., 2002; Table 2), suggesting that PPARβ/δ is important for adipose tissue development *in vivo*. However, because PPARβ/δ is ubiquitously expressed, it might act on adipogenesis via either cell autonomous or systemic mechanisms. In an attempt to solve this question, knockout mice that display a selective ablation of PPARβ/δ in the adipose tissue have been produced (Barak et al., 2002). These mice do not show a reduced adipose mass compared to their control littermates, suggesting that PPARβ/δ expression may not be important in adipose tissue itself but may have an indirect role via another organ. However, the aP2 promoter, used to drive the production of the Cre recombinase for adipose tissue-specific PPARβ/δ ablation, is active relatively late during adipocyte differentiation (Hunt et al., 1986). Therefore, the selective deletion of PPARβ/δ only occurred relatively late, and thus does not provide clues on a possible direct role of PPARβ/δ in early adipogenesis. Cell culture data are in favor of such a direct early role. Multipotent fibroblasts that overexpress PPARβ/δ differentiate into adipocytes on treatment with the selective PPARβ/δ ligand, L165041 (Hansen et al., 2001). In addition, preadipocyte terminal differentiation into adipocyte is reduced in stably transfected cells expressing a dominant negative PPARβ/δ construct (Bastie et al., 2000). Finally, differentiation of preadipocytes obtained from PPARβ/δ-null mice is markedly reduced on treatment with the selective PPARβ/δ agonist L165041 compared with wild-type preadipocytes, whereas they fully differentiate on treatment with a PPARγ ligand (Matsusue et al., 2004). Overall, these cell culture data show that PPARβ/δ can promote adipogenesis in culture, but that it is not essential for terminal differentiation of adipocytes, when PPARγ is highly active. Therefore, PPARβ/δ may play a role only when the PPARγ signaling is not yet at its maximum, as occurs during the early steps of the adipocyte differentiation process (Amri et al., 1995). In order to further clarify the role of PPARβ/δ in the determination of the adipocyte lineage, it would be interesting to assess whether multipotent fibroblasts, or ES cells, derived from PPARβ/δ-null mice can be committed to the adipogenic pathway.

In conclusion, PPARγ is a major component of adipocyte development, from commitment to terminal differentiation and maintenance of the differentiated state, while PPARβ/δ is likely to be important especially for the very early differentiation steps.

4. PPARs and skin development

The skin is a highly specialized organ that functions as a barrier between the body and its external environment. With lipids being essential constituents of this barrier, the implication of PPARs in the regulation of skin development and homeostasis does not come as a surprise. Mammalian skin is composed of two main layers, the outermost epidermis and the dermis, below which the subcutaneous layer or hypodermis is found. The mature epidermis is a multistratified epithelium mainly composed of keratinocytes and the dermis is a collagen-rich mesenchyme that contains blood vessels, nerve endings, and epidermal appendages derived from the epidermis. These include hair follicles and sebaceous glands. The hypodermis is mainly composed of adipocytes. In this part, we will concentrate on the role of PPARs in the development of the epidermis and its appendages.

4.1. Expression profile of PPARs in mouse skin development

The interfollicular epidermis, that is, the epidermis in between the hair follicles, is composed of several keratinocyte layers. The basal layer consists of proliferating undifferentiated cells and the outermost layer, the *stratum corneum*, comprises the terminal differentiated keratinocytes. The vectorial differentiation program of keratinocytes unfolds between these two layers (Fig. 4). The epidermis forms at the end of fetal development from a single basal cell layer topped by a superficial layer, the periderm. Skin appendages arise from the same epithelial progenitor cells that give rise to the interfollicular epidermis. In particular, hair follicles originate from invaginations into the dermis of a proliferative epithelium of epidermal origin. Later on, sebocytes differentiate and develop into sebaceous glands connected to the mature hair follicles (Fig. 4).

In rodents, the three PPARs are expressed in the interfollicular epidermis and in the developing hair follicles from the time these structures arise until birth (Braissant and Wahli, 1998; Michalik et al., 2001; Di-Poi et al., 2005b). PPARs are also found in sebocytes in culture, as well as in sebaceous glands *in vivo* (Rosen et al., 1999; Rosenfield et al., 1999; Kim et al., 2001; Di-Poi et al., 2005b). Analyses of mutant mice for PPARs gave insights into the different contributions of each isotype in the development of the mouse skin as presented in a later section.

4.2. PPARα involvement in keratinocyte differentiation

PPARα-null mice were first reported to have no obvious skin phenotype (Lee et al., 1995), based on observations made after birth. However, *in utero* analyses showed that these mutant concepti display a delayed skin development and an abnormal *stratum corneum* at E18.5. In addition, various PPARα ligands were shown to promote the maturation of the epidermal barrier in fetal rat skin *in vitro* and *in vivo* (Hanley et al., 1997; Komuves et al., 1998; Hanley et al., 1999). Transgenic mice that express a constitutively active form of PPARα have been generated. In these mice, a VP16:PPARα fusion protein

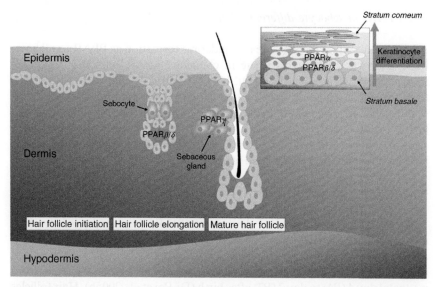

Fig. 4. PPARs and mouse skin development. The structure of mouse skin is schematized and the role of PPARs in its development is summarized. Mouse skin is composed of two main layers, the outermost epidermis and the dermis, below which the subcutaneous layer or hypodermis is found. The mature epidermis is a multistratified epithelium mainly composed of keratinocytes that undergo a vectorial differentiation program to form the *stratum corneum*, the outermost layer of the epidermis. Hair follicles arise from the invagination of the epithelium of epidermal origin. PPARα and PPARβ/δ promote keratinocyte vectorial differentiation. PPARβ/δ is important for hair follicle elongation and PPARγ is essential for sebaceous gland formation.

is produced under the control of the keratin K5 promoter that is active in the epidermis and other ectodermally derived epithelia (Yang et al., 2006). The expression of this transgenic PPARα greatly enhances the expression of the endogenous PPARα gene, confirming an autoregulatory mechanism of this gene. The overexpression of both the native and exogenous ligand-independent PPARα protein had dramatic consequences for skin development, thinning the interfollicular epidermis, and decreasing the number of hair follicles. In addition, primary keratinocytes derived from these pups displayed reduced proliferation, as evidenced by a reduced incorporation of BrdU, and enhanced differentiation as shown by the analysis of molecular markers. It is thus tempting to conclude that PPARα plays a direct role in cell cycle exit and differentiation, and hair follicle development. However, overexpression of the constitutively active form of PPARα may have perturbed other pathways possibly through competition for the heterodimerization partner RXR (Juge-Aubry et al., 1995; Wang et al., 2005). In summary, it seems that PPARα is important for late fetal differentiation of the keratinocytes, but its disruption has no obvious consequences in adulthood under normal conditions.

4.3. PPARs in sebocyte differentiation

PPARγ activity is dispensable for epidermis formation as PPARγ-null keratinocytes participate in this tissue in chimeric "PPARγ-null/WT" mice described earlier (Rosen et al., 1999). However, PPARγ-null cells were absent from sebaceous glands, suggesting that PPARγ is required for sebocyte development. The involvement of PPARα and PPARβ/δ in sebocyte differentiation has not been assessed so far. The only available information comes from sebocyte cultures in which treatment with PPARα or PPARβ/δ activators induces differentiation (Rosenfield et al., 1999).

4.4. PPARβ/δ promotes keratinocyte differentiation and survival

We have extensively studied the role of PPARβ/δ in epidermal development using the PPARβ/δ-null mouse line we have generated (Section 1.1.2). Although the interfollicular epidermis does not seem to be affected in these null mice, we observed a delay in hair follicle morphogenesis that is apparent from postnatal day 4 (P4) to day 7 (P7) after birth (Di-Poi et al., 2005b). Hair follicles are less developed in the skin of PPARβ/δ-null mice compared to wild-type skin. This is most likely the result of an increased apoptosis in the follicular keratinocyte population as PPARβ/δ enhances follicular keratinocytes survival at the time of proliferation, which promotes the elongation of the invaginating hair follicles. PPARβ/δ mediates its antiapoptotic effect by increasing expression of ILK and PDK1, thus resulting in the activation of the Akt1-dependent survival pathway (Di-Poi et al., 2002, 2005b). In addition to its protective effects against apoptosis, PPARβ/δ might also act on proliferation *in vivo* since a decreased expression of cyclin D1 was observed in the skin of null animals at P4 (Di-Poi et al., 2005b). It is noteworthy that PPARβ/δ is important relatively late in hair follicle morphogenesis in spite of high expression at earlier stages. This might be due to the fact that the PPARβ/δ endogenous ligands are produced quite late. We demonstrated that hepatocyte growth factor (HGF) produced by dermal cells acts as a paracrine signal that triggers cyclooxygenase-type-2 (COX-2) production in keratinocytes. COX-2 activity in turn leads to the production of so far unidentified PPARβ/δ ligands at the time of hair follicle elongation (Di-Poi et al., 2005b). Activated PPARβ/δ then favors hair follicle elongation.

Treatment of cultured keratinocytes with PPARβ/δ agonists also suggests that this isotype might be important for keratinocyte terminal differentiation (Tan et al., 2001; Kim et al., 2006). This is supported by *in vivo* mouse studies using a topical skin treatment with PPARβ/δ agonists, which accelerates the reestablishment of the skin barrier function after physical alteration (Schmuth et al., 2004).

In rodents, PPARβ/δ expression is not detected by *in situ* hybridization in adult interfollicular epidermis, while it is highly expressed in this tissue in fetal stages. The repression observed in adulthood is most likely mediated by C/EBPs. In adult keratinocytes of the interfollicular epidermis, C/EBPα and C/EBPβ repress PPARβ/δ expression through the recruitment of HDAC1 at a specific C/EBP-binding site in PPARβ/δ promoter (Di-Poi et al., 2005a).

In conclusion, both PPARα and PPARβ/δ appear to be important for keratinocyte differentiation. In addition, PPARβ/δ plays a role in hair follicle development by protecting keratinocytes from cell death through the Akt pathway. Finally, PPARγ is essential for sebocyte differentiation (Fig. 4).

5. PPARs promote cell differentiation in the gut epithelium

PPARs are involved in the differentiation of the intestinal epithelium (Lefebvre et al., 1999; Koeffler, 2003). The large amount of data obtained in the context of cancer has been described elsewhere (Michalik et al., 2004). Herein, we concentrate on the normal differentiation process of gut epithelial cells.

5.1. PPARγ is important for the differentiation of the gut epithelium

The expression of marker genes for gut epithelium differentiation has been analyzed in PPARγ heterozygous mice (Drori et al., 2005). In the small intestine of these animals, the expression of Kruppel-like transcription factor 4 (KLF4), keratin 20, and keratin 19 is reduced by 50%, which suggests a physiological role for PPARγ in the differentiation program of the small intestine epithelium. In the same study, a new coactivator of PPARγ named Hic-5 has been identified, which is involved in the control of the differentiation program together with PPARγ. Within the embryonic mouse intestine, Hic-5 and PPARγ colocalize specifically in the epithelial layer and their expression is high during epithelium development. When Hic-5 is overexpressed experimentally in 3T3-L1 preadipocytes, the adipogenic program is partially repressed and several marker genes of the intestinal epithelium are induced. Thus, it is likely that the outcome of the PPARγ-mediated differentiation program is modulated by coactivators, Hic-5 being involved in the commitment of precursor cells to the intestinal epithelium program (Drori et al., 2005).

5.2. PPARβ/δ promotes Paneth cell differentiation

The small intestine epithelium harbors, among other epithelial cell types, the Paneth cells. They arise from multipotent stem cells localized within the intestinal crypt and are specialized in the antipathogen defense of the intestine (Karam, 1999). PPARβ/δ is expressed in every part of the small intestine but at particularly high levels within the intestinal crypts (Varnat et al., 2006).

Paneth cells only appear during postnatal development, and their number dramatically increases at P21 concomitantly with PPARβ/δ expression. This increase is strongly attenuated in PPARβ/δ-null mice. In adult PPARβ/δ-null mice, the number of Paneth cells is markedly reduced compared to the wild-type animals. In addition, the morphology of the remaining Paneth cells is affected as observed by electron microscopy. The endoplasmic reticulum of the mutated cells is less compact and the number of large secretory granules is decreased, suggesting a role for PPARβ/δ in the secretory pathway. The direct implication of PPARβ/δ in this Paneth cell differentiation was demonstrated in wild-type mice where a treatment with a PPARβ/δ-specific agonist (L165041) increased the number of Paneth cells, while the same treatment had no effect in PPARβ/δ-null mice. The molecular pathway of Paneth cell differentiation involves the inhibition by PPARβ/δ of the Indian Hedgehog (Ihh) signaling pathway. Indeed Ihh expression is increased in PPARβ/δ-null mice compared with wild-type animals, and is decreased in wild-type mice on L165041 treatment. Importantly, inhibition of Hedgehog signaling by treatment of wild-type mice with cyclopamine also resulted in an increase of Paneth cell number. Thus Ihh, which represents the main mediator of the Hedgehog pathway in the gut (Wang et al., 2002; Van Den Brink et al., 2004), is negatively controlled by PPARβ/δ and is a negative regulator of Paneth cell differentiation. Analyses of the localization of Ihh and its direct targets Patched-1 (Ptch-1) and Hedgehog interacting protein (Hip) detected by immunohistochemistry revealed that Ihh is produced by mature Paneth cells, while Ptch-1 and Hip are located in discrete cells that may correspond to Paneth cell precursors since they coexpress lysozyme.

On the basis of these findings, we proposed the model illustrated in Fig. 5. Mature Paneth cells limit their own formation by producing and secreting Ihh that interacts with Ptch-1 on the surface of Paneth cell precursors. The subsequent activation of the Hedgehog pathway within Paneth cell precursors inhibits their differentiation. This effect is balanced by PPARβ/δ, which inhibits Ihh production in the mature Paneth cells. Thus, PPARβ/δ indirectly promotes Paneth cell differentiation and its absence leads to an overproduction of Ihh and to a strong inhibition of the differentiation of Paneth cell precursors into mature Paneth cells (Varnat et al., 2006).

In conclusion, both PPARγ and PPARβ/δ promote the differentiation of different intestinal epithelial cells in mice, that is, enterocytes and Paneth cells, respectively.

6. Do PPARs play a role in neural development?

During rat embryonic development, all three PPARs are detected at relatively high levels within the CNS at E13.5. At E15.5, both PPARα and PPARγ expression is decreased while that of PPARβ/δ is still high. Finally at E18.5, only PPARβ/δ is detected, at lower levels than at earlier stages

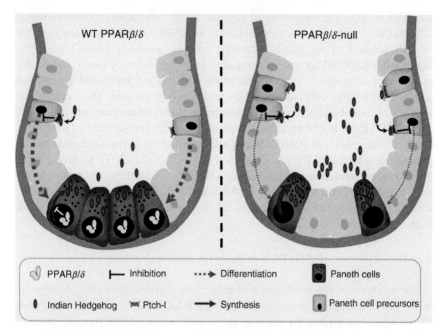

Fig. 5. PPARβ/δ promotes Paneth cell differentiation. An intestinal crypt scheme is shown in the wild type (left panel) or in the PPARβ/δ-null mouse genotypes (right panel). Paneth cell precursors differentiate into mature Paneth cells that lie in the bottom of the crypt. Mature cells limit their own formation by producing and secreting Indian Hedgehog (Ihh) that interacts with Patched-1 (Ptch-1) on the surface of Paneth cell precursors to inhibit their differentiation. This effect is balanced by PPARβ/δ that inhibits Ihh production. Thus, PPARβ/δ indirectly promotes Paneth cell differentiation; its absence leads to an overproduction of Ihh and to a strong inhibition of Paneth cell precursors differentiation into mature Paneth cells that become underrepresented. (See Color Insert.)

(Braissant and Wahli, 1998; Table 1). This remarkable transient expression of PPARα and PPARγ and the high levels of PPARβ/δ in the CNS during development can be regarded as hints for a role of PPARs in embryonic neural development.

6.1. Potential role of PPARs in myelination and neuronal differentiation

Neural development has not yet been studied in detail in PPAR-null mice. While no defect has been reported in PPARα-null mice, PPARβ/δ-null mice show an alteration in myelination of the corpus callosum, but with no accompanying change in the expression of molecular markers for myelination (Peters et al., 2000, Table 2). Myelination within the CNS is ensured by oligodendrocytes. In mice, PPARβ/δ is highly expressed in these cells in culture and *in vivo* (Granneman et al., 1998; Saluja et al., 2001; Woods et al., 2003). Treatment of primary glial cell cultures with the PPARβ/δ agonist

L165041 induces the formation of gigantic membrane sheets characteristic of differentiated oligodendrocytes (Saluja et al., 2001). In addition, the number of cells labeled by an oligodendrocyte specific antibody (A007) was 30% higher after L165041 treatment compared with the control. Finally, molecular markers of oligodendrocyte differentiation displayed increased expression levels in the treated cells, as judged by *in situ* hybridizations. Thus, consistent with the *in vivo* observations discussed above, these cell culture data indicate a role of PPARβ/δ in the myelination process, via the promotion of oligodendrocyte differentiation. PPARs might also be involved in the differentiation of neurons in which PPARβ/δ is expressed (Woods et al., 2003). In rats, PPAR proteins are detected in cortical neurons in culture and their expression levels vary as differentiation proceeds. PPARα and PPARγ protein levels increase, while PPARβ/δ levels are decreased at the time of terminal differentiation (Cimini et al., 2005).

6.2. Potential role of PPARβ/δ in early neural development

The analysis of the teratogenic effects of the antiepileptic drug, valproic acid (VPA), suggested a potential role of PPARβ/δ in neural development. VPA, when administered during early embryonic development, induces defects in neural tube closure both in mice and in humans, possibly via disruption of preexisting signal(s) required for proper neural tube closure. In addition, VPA effects have also been studied *in vitro* in the teratocarcinoma F9 cell line, in which the drug induces a neuron-like differentiation program, together with some characteristics of neural crest cells (NCCs) (Werling et al., 2001; Lampen et al., 2005). PPARβ/δ is expressed in F9 cells and is necessary for VPA-induced differentiation. Differentiation is abolished in F9 cells expressing an anti-PPARβ/δ siRNA, while it is enhanced if PPARβ/δ is overexpressed (Werling et al., 2001; Lampen et al., 2005). Consistent with these cell culture observations, the expression of PPARβ/δ in mouse embryos at the time of NCC migration presents a pattern that recapitulates the path followed by NCC from the neural tube toward the branchial arches (Werling et al., 2001). This suggests a function of PPARβ/δ in early neural development, possibly in NCC migration, which deserves further in-depth analysis.

In brief, the patterns of PPAR expression suggest an implication of the three PPARs in CNS formation, but the only functional data available indicate a specific role of PPARβ/δ in myelination and in early neural development.

7. Concluding remarks

Even if our understanding of the roles of PPARs in development is still at its infancy, it nonetheless appears that PPARs can be implicated in virtually all cellular events that define development: (1) PPARs control the determination of cell identity, as illustrated by the role of PPARγ in the commitment of multipotent mesenchymal stem cells into the adipocyte lineage, both *in vitro* and *in vivo*; (2) PPARs promote cell differentiation of various cell types, such as keratinocytes by PPARα; (3) PPARs influence cell death and survival, as evidenced in the context of hair follicle elongation involving PPARβ/δ; (4) PPARs interact with regulators of the cell cycle, exemplified by E2F or CKI and PPARγ; and finally (5) PPARs may promote cell migration, since PPARβ/δ is important for wound healing and plays a key role in hair follicle elongation that both require cell movements.

Our knowledge in mice points to interesting connections between PPARs and pathways well established in developmental processes. For instance, PPARβ/δ interplays with the PI3K/Akt pathway in placenta and skin development, and PPARγ interplays with the MEK/ERK pathway as well as with Wnt signaling. Expression of both PPARγ and PPARβ/δ is dependent on the C/EBP family of transcription factors in adipogenesis and in interfollicular epidermis, respectively. In addition, PPARβ/δ is connected with the Hedgehog pathway in Paneth cell differentiation.

Importantly, PPARs have nonredundant functions during development since the PPAR-null mice for each of the isotypes show different developmental defects (Table 2). This is well illustrated in placenta development, where PPARβ/δ and PPARγ expression overlap but both are independently essential (Section 1). This is also true in epidermis development, during which all three PPARs are present in embryogenenesis and early postnatal development, while PPARα, PPARβ/δ, and PPARγ have specific roles in keratinocyte differentiation, hair follicle maturation, and sebaceous gland formation, respectively (Section 3).

What are the mechanisms that ensure the isotype specificity of PPARs? At this point, it should be underscored that all three PPARs share the same heterodimeric partner: RXR. They also recognize the same response element (PPRE) in the regulatory regions of target genes. However, if the three different PPAR subtypes can bind to a same subset of target genes, preferential binding occurs to other target promoters by means of both the variation in the DNA sequence just upstream of the DR1 motif and degree of similarity with the consensus sequence (Juge-Aubry et al., 1997; Fig. 1). In addition, much of the PPAR isotype specificity might essentially be determined at the level of the ligands and of the cofactors participating in promoter stimulation or repression. Our knowledge concerning the endogenous PPAR ligands is still very modest (Lathion et al., 2006), and those regulating the developmental processes reviewed herein have not been identified yet. However, it is already clear that their production is highly regulated (Section 3.4). Concerning the coactivators, association of PPARγ with PGC-1 triggers brown adipocyte formation, while its association with Hic-5 promotes gut epithelial differentiation, in support of the theory that regulations by coactivators participate in the determination of cell fate. One can imagine that coactivators selective of a given

PPAR isotype may exist, even if this is not likely. Additional levels of regulation, such as phosphorylation or segregation of the PPAR proteins in different subcellular regions, might also be important (Section 2.2).

The fact that PPARs display specific roles does not exclude the possibility that they are redundant for some other functions during development. It would be interesting to generate inducible double or triple loss of function mutant mouse lines. Functional studies on the roles of PPARs in development should also be conducted in other species than rodents, for instance birds, fish, and frogs. Such studies might contribute to revealing new functions of a given PPAR isotype, either because the putative redundancy rules would not be the same in different species or because the interplay with other developmental pathways would be different. Given the broad influence of the signaling pathways connected to PPARs and the wide expression pattern of PPARs (particularly PPARβ/δ) it is likely that new important roles of PPARs in development will be unveiled in the near future in mice or in other species.

References

Adams, M., Reginato, M.J., Shao, D., Lazar, M.A., Chatterjee, V.K. 1997. Transcriptional activation by peroxisome proliferator-activated receptor gamma is inhibited by phosphorylation at a consensus mitogen-activated protein kinase site. J. Biol. Chem. 272, 5128–5132.

Ali, A.A., Weinstein, R.S., Stewart, S.A., Parfitt, A.M., Manolagas, S.C., Jilka, R.L. 2005. Rosiglitazone causes bone loss in mice by suppressing osteoblast differentiation and bone formation. Endocrinology 146, 1226–1235.

Altiok, S., Xu, M., Spiegelman, B.M. 1997. PPARgamma induces cell cycle withdrawal: Inhibition of E2F/DP DNA-binding activity via down-regulation of PP2A. Genes Dev. 11, 1987–1998.

Amri, E.Z., Bonino, F., Ailhaud, G., Abumrad, N.A., Grimaldi, P.A. 1995. Cloning of a protein that mediates transcriptional effects of fatty acids in preadipocytes. Homology to peroxisome proliferator-activated receptors. J. Biol. Chem. 270, 2367–2371.

Barak, Y., Nelson, M.C., Ong, E.S., Jones, Y.Z., Ruiz-Lozano, P., Chien, K.R., Koder, A., Evans, R.M. 1999. PPARgamma is required for placental, cardiac, and adipose tissue development. Mol. Cell 4, 585–595.

Barak, Y., Liao, D., He, W., Ong, E.S., Nelson, M.C., Olefsky, J.M., Boland, R., Evans, R.M. 2002. Effects of peroxisome proliferator-activated receptor delta on placentation, adiposity, and colorectal cancer. Proc. Natl. Acad. Sci. USA 99, 303–308.

Bastie, C., Luquet, S., Holst, D., Jehl-Pietri, C., Grimaldi, P.A. 2000. Alterations of peroxisome proliferator-activated receptor delta activity affect fatty acid-controlled adipose differentiation. J. Biol. Chem. 275, 38768–38773.

Bennett, C.N., Ross, S.E., Longo, K.A., Bajnok, L., Hemati, N., Johnson, K.W., Harrison, S.D., Macdougald, O.A. 2002. Regulation of Wnt signaling during adipogenesis. J. Biol. Chem. 277, 30998–31004.

Berry, E.B., Keelan, J.A., Helliwell, R.J., Gilmour, R.S., Mitchell, M.D. 2005. Nanomolar and micromolar effects of 15-deoxy-delta 12,14-prostaglandin J2 on amnion-derived WISH epithelial cells: Differential roles of peroxisome proliferator-activated receptors gamma and delta and nuclear factor kappa. B. Mol. Pharmacol. 68, 169–178.

Bildirici, I., Roh, C.R., Schaiff, W.T., Lewkowski, B.M., Nelson, D.M., Sadovsky, Y. 2003. The lipid droplet-associated protein adipophilin is expressed in human trophoblasts and is regulated by peroxisomal proliferator-activated receptor-gamma/retinoid X receptor. J. Clin. Endocrinol. Metab. 88, 6056–6062.

Braissant, O., Wahli, W. 1998. Differential expression of peroxisome proliferator-activated receptor-alpha, -beta, and -gamma during rat embryonic development. Endocrinology 139, 2748–2754.

Braissant, O., Foufelle, F., Scotto, C., Dauca, M., Wahli, W. 1996. Differential expression of peroxisome proliferator-activated receptors (PPARs): Tissue distribution of PPAR-alpha, -beta, and -gamma in the adult rat. Endocrinology 137, 354–366.

Cimini, A., Benedetti, E., Cristiano, L., Sebastiani, P., D'amico, M.A., D'angelo, B., Di Loreto, S. 2005. Expression of peroxisome proliferator-activated receptors (PPARs) and retinoic acid receptors (RXRs) in rat cortical neurons. Neuroscience 130, 325–337.

Desvergne, B., Wahli, W. 1999. Peroxisome proliferator-activated receptors: Nuclear control of metabolism. Endocr. Rev. 20, 649–688.

Di-Poi, N., Tan, N.S., Michalik, L., Wahli, W., Desvergne, B. 2002. Antiapoptotic role of PPARbeta in keratinocytes via transcriptional control of the Akt1 signaling pathway. Mol. Cell 10, 721–733.

Di-Poi, N., Desvergne, B., Michalik, L., Wahli, W. 2005a. Transcriptional repression of peroxisome proliferator-activated receptor beta/delta in murine keratinocytes by CCAAT/enhancer-binding proteins. J. Biol. Chem. 280, 38700–38710.

Di-Poi, N., Ng, C.Y., Tan, N.S., Yang, Z., Hemmings, B.A., Desvergne, B., Michalik, L., Wahli, W. 2005b. Epithelium-mesenchyme interactions control the activity of peroxisome proliferator-activated receptor beta/delta during hair follicle development. Mol. Cell. Biol. 25, 1696–1712.

Dreyer, C., Krey, G., Keller, H., Givel, F., Helftenbein, G., Wahli, W. 1992. Control of the peroxisomal beta-oxidation pathway by a novel family of nuclear hormone receptors. Cell 68, 879–887.

Drori, S., Girnun, G.D., Tou, L., Szwaya, J.D., Mueller, E., Xia, K., Shivdasani, R.A., Spiegelman, B.M. 2005. Hic-5 regulates an epithelial program mediated by PPARgamma. Genes Dev. 19, 362–375.

Fajas, L., Auboeuf, D., Raspe, E., Schoonjans, K., Lefebvre, A.M., Saladin, R., Najib, J., Laville, M., Fruchart, J.C., Deeb, S., Vidal-Puig, A., Flier, J., et al. 1997. The organization, promoter analysis, and expression of the human PPARgamma gene. J. Biol. Chem. 272, 18779–18789.

Fajas, L., Landsberg, R.L., Huss-Garcia, Y., Sardet, C., Lees, J.A., Auwerx, J. 2002. E2Fs regulate adipocyte differentiation. Dev. Cell 3, 39–49.

Forman, B.M., Tontonoz, P., Chen, J., Brun, R.P., Spiegelman, B.M., Evans, R.M. 1995. 15-Deoxy-delta 12, 14-prostaglandin J2 is a ligand for the adipocyte determination factor PPAR gamma. Cell 83, 803–812.

Fournier, T., Pavan, L., Tarrade, A., Schoonjans, K., Auwerx, J., Rochette-Egly, C., Evain-Brion, D. 2002. The role of PPAR-gamma/RXR-alpha heterodimers in the regulation of human trophoblast invasion. Ann. NY Acad. Sci. 973, 26–30.

Fu, M., Rao, M., Bouras, T., Wang, C., Wu, K., Zhang, X., Li, Z., Yao, T.P., Pestell, R.G. 2005. Cyclin D1 inhibits peroxisome proliferator-activated receptor gamma-mediated adipogenesis through histone deacetylase recruitment. J. Biol. Chem. 280, 16934–16941.

Granneman, J., Skoff, R., Yang, X. 1998. Member of the peroxisome proliferator-activated receptor family of transcription factors is differentially expressed by oligodendrocytes. J. Neurosci. Res. 51, 563–573.

Gregoire, F.M., Smas, C.M., Sul, H.S. 1998. Understanding adipocyte differentiation. Physiol. Rev. 78, 783–809.

Gurnell, M., Wentworth, J.M., Agostini, M., Adams, M., Collingwood, T.N., Provenzano, C., Browne, P.O., Rajanayagam, O., Burris, T.P., Schwabe, J.W., Lazar, M.A., Chatterjee, V.K.

2000. A dominant-negative peroxisome proliferator-activated receptor gamma (PPARgamma) mutant is a constitutive repressor and inhibits PPARgamma-mediated adipogenesis. J. Biol. Chem. 275, 5754–5759.

Hanley, K., Jiang, Y., Crumrine, D., Bass, N.M., Appel, R., Elias, P.M., Williams, M.L., Feingold, K.R. 1997. Activators of the nuclear hormone receptors PPARalpha and FXR accelerate the development of the fetal epidermal permeability barrier. J. Clin. Invest. 100, 705–712.

Hanley, K., Komuves, L.G., Bass, N.M., He, S.S., Jiang, Y., Crumrine, D., Appel, R., Friedman, M., Bettencourt, J., Min, K., Elias, P.M., Williams, M.L., et al. 1999. Fetal epidermal differentiation and barrier development *in vivo* is accelerated by nuclear hormone receptor activators. J. Invest. Dermatol. 113, 788–795.

Hansen, J.B., Zhang, H., Rasmussen, T.H., Petersen, R.K., Flindt, E.N., Kristiansen, K. 2001. Peroxisome proliferator-activated receptor delta (PPARdelta)-mediated regulation of pre-adipocyte proliferation and gene expression is dependent on cAMP signaling. J. Biol. Chem. 276, 3175–3182.

Haunerland, N.H., Spener, F. 2004. Fatty acid-binding proteins-insights from genetic manipulations. Prog. Lipid Res. 43, 328–349.

Hirning-Folz, U., Wilda, M., Rippe, V., Bullerdiek, J., Hameister, H. 1998. The expression pattern of the Hmgic gene during development. Genes Chromosomes Cancer 23, 350–357.

Hong, J.H., Hwang, E.S., Mcmanus, M.T., Amsterdam, A., Tian, Y., Kalmukova, R., Mueller, E., Benjamin, T., Spiegelman, B.M., Sharp, P.A., Hopkins, N., Yaffe, M.B. 2005. TAZ, a transcriptional modulator of mesenchymal stem cell differentiation. Science 309, 1074–1078.

Hu, E., Tontonoz, P., Spiegelman, B.M. 1995. Transdifferentiation of myoblasts by the adipogenic transcription factors PPAR gamma and C/EBP alpha. Proc. Natl. Acad. Sci. USA 92, 9856–9860.

Hunt, C.R., Ro, J.H., Dobson, D.E., Min, H.Y., Spiegelman, B.M. 1986. Adipocyte P2 gene: Developmental expression and homology of 5'-flanking sequences among fat cell-specific genes. Proc. Natl. Acad. Sci. USA 83, 3786–3790.

Imai, T., Takakuwa, R., Marchand, S., Dentz, E., Bornert, J.M., Messaddeq, N., Wendling, O., Mark, M., Desvergne, B., Wahli, W., Chambon, P., Metzger, D. 2004. Peroxisome proliferator-activated receptor gamma is required in mature white and brown adipocytes for their survival in the mouse. Proc. Natl. Acad. Sci. USA 101, 4543–4547.

Issemann, I., Green, S. 1990. Activation of a member of the steroid hormone receptor superfamily by peroxisome proliferators. Nature 347, 645–650.

Juge-Aubry, C.E., Gorla-Bajszczak, A., Pernin, A., Lemberger, T., Wahli, W., Burger, A.G., Meier, C.A. 1995. Peroxisome proliferator-activated receptor mediates cross-talk with thyroid hormone receptor by competition for retinoid X receptor. Possible role of a leucine zipper-like heptad repeat. J. Biol. Chem. 270, 18117–18122.

Juge-Aubry, C., Pernin, A., Favez, T., Burger, A.G., Wahli, W., Meier, C.A., Desvergne, B. 1997. DNA binding properties of peroxisome proliferator-activated receptor subtypes on various natural peroxisome proliferator response elements. Importance of the 5'-flanking region. J. Biol. Chem. 272, 25252–25259.

Kamei, T., Jones, S.R., Chapman, B.M., Kl, M.C., Dai, G., Soares, M.J. 2002. The phosphatidylinositol 3-kinase/Akt signaling pathway modulates the endocrine differentiation of trophoblast cells. Mol. Endocrinol. 16, 1469–1481.

Kang, S., Bajnok, L., Longo, K.A., Petersen, R.K., Hansen, J.B., Kristiansen, K., Macdougald, O.A. 2005. Effects of Wnt signaling on brown adipocyte differentiation and metabolism mediated by PGC-1alpha. Mol. Cell. Biol. 25, 1272–1282.

Karam, S.M. 1999. Lineage commitment and maturation of epithelial cells in the gut. Front. Biosci. 4, D286–D298.

Keller, J.M., Collet, P., Bianchi, A., Huin, C., Bouillaud-Kremarik, P., Becuwe, P., Schohn, H., Domenjoud, L., Dauca, M. 2000. Implications of peroxisome proliferator-activated receptors (PPARS) in development, cell life status and disease. Int. J. Dev. Biol. 44, 429–442.

Kersten, S., Desvergne, B., Wahli, W. 2000. Roles of PPARs in health and disease. Nature 405, 421–424.

Kim, D.J., Bility, M.T., Billin, A.N., Willson, T.M., Gonzalez, F.J., Peters, J.M. 2006. PPAR-beta/delta selectively induces differentiation and inhibits cell proliferation. Cell Death Differ. 13, 53–60.

Kim, M.J., Deplewski, D., Ciletti, N., Michel, S., Reichert, U., Rosenfield, R.L. 2001. Limited cooperation between peroxisome proliferator-activated receptors and retinoid X receptor agonists in sebocyte growth and development. Mol. Genet. Metab. 74, 362–369.

Kliewer, S.A., Forman, B.M., Blumberg, B., Ong, E.S., Borgmeyer, U., Mangelsdorf, D.J., Umesono, K., Evans, R.M. 1994. Differential expression and activation of a family of murine peroxisome proliferator-activated receptors. Proc. Natl. Acad. Sci. USA 91, 7355–7359.

Koeffler, H.P. 2003. Peroxisome proliferator-activated receptor gamma and cancers. Clin. Cancer Res. 9, 1–9.

Komuves, L.G., Hanley, K., Jiang, Y., Elias, P.M., Williams, M.L., Feingold, K.R. 1998. Ligands and activators of nuclear hormone receptors regulate epidermal differentiation during fetal rat skin development. J. Invest. Dermatol. 111, 429–433.

Kortum, R.L., Costanzo, D.L., Haferbier, J., Schreiner, S.J., Razidlo, G.L, Wu, M.H., Volle, D.J., Mori, T., Sakaue, H., Chaika, N.V., Chaika, O.V., Lewis, R.E. 2005. The molecular scaffold kinase suppressor of Ras 1 (KSR1) regulates adipogenesis. Mol. Cell. Biol. 25, 7592–7604.

Kraut, N., Snider, L., Chen, C.M., Tapscott, S.J., Groudine, M. 1998. Requirement of the mouse I-mfa gene for placental development and skeletal patterning. EMBO J. 17, 6276–6288.

Lampen, A., Grimaldi, P.A., Nau, H. 2005. Modulation of peroxisome proliferator-activated receptor delta activity affects neural cell adhesion molecule and polysialyltransferase ST8SiaIV induction by teratogenic valproic acid analogs in F9 cell differentiation. Mol. Pharmacol. 68, 193–203.

Lathion, C., Michalik, L., Wahli, W. 2006. Physiological ligands of PPARs in inflammation and lipid homeostasis. Future Lipidol. 1, 191–201.

Lee, S.S., Pineau, T., Drago, J., Lee, E.J., Owens, J.W., Kroetz, D.L., Fernandez-Salguero, P.M., Westphal, H., Gonzalez, F.J. 1995. Targeted disruption of the alpha isoform of the peroxisome proliferator-activated receptor gene in mice results in abolishment of the pleiotropic effects of peroxisome proliferators. Mol. Cell. Biol. 15, 3012–3022.

Lefebvre, M., Paulweber, B., Fajas, L., Woods, J., Mccrary, C., Colombel, J.F., Najib, J., Fruchart, J.C., Datz, C., Vidal, H., Desreumaux, P., Auwerx, J. 1999. Peroxisome proliferator-activated receptor gamma is induced during differentiation of colon epithelium cells. J. Endocrinol. 162, 331–340.

Lehmann, J.M., Moore, L.B., Smith-Oliver, T.A., Wilkison, W.O., Willson, T.M., Kliewer, S.A. 1995. An antidiabetic thiazolidinedione is a high affinity ligand for peroxisome proliferator-activated receptor gamma (PPAR gamma). J. Biol. Chem. 270, 12953–12956.

Matsusue, K., Peters, J.M., Gonzalez, F.J. 2004. PPARbeta/delta potentiates PPARgamma-stimulated adipocyte differentiation. FASEB J. 18, 1477–1479.

Michalik, L., Desvergne, B., Tan, N.S., Basu-Modak, S., Escher, P., Rieusset, J., Peters, J.M., Kaya, G., Gonzalez, F.J., Zakany, J., Metzger, D., Chambon, P., et al. 2001. Impaired skin wound healing in peroxisome proliferator-activated receptor (PPAR)alpha and PPARbeta mutant mice. J. Cell Biol. 154, 799–814.

Michalik, L., Desvergne, B., Wahli, W. 2004. Peroxisome-proliferator-activated receptors and cancers: Complex stories. Nat. Rev. Cancer 4, 61–70.

Moldes, M., Zuo, Y., Morrison, R.F., Silva, D., Park, B.H., Liu, J., Farmer, S.R. 2003. Peroxisome-proliferator-activated receptor gamma suppresses Wnt/beta-catenin signalling during adipogenesis. Biochem. J. 376, 607–613.

Morrison, R.F., Farmer, S.R. 1999. Role of PPARgamma in regulating a cascade expression of cyclin-dependent kinase inhibitors, p18(INK4c) and p21(Waf1/Cip1), during adipogenesis. J. Biol. Chem. 274, 17088–17097.

Nadra, K., Anghel, S.I., Joye, E., Tan, N.S., Basu-Modak, S., Trono, D., Wahli, W., Desvergne, B. 2006. Differentiation of trophoblast giant cells and their metabolic functions are dependent on PPARβ/δ. Mol. Cell. Biol. 26, 3266–3281.

Nakayama, H., Liu, Y., Stifani, S., Cross, J.C. 1997. Developmental restriction of Mash-2 expression in trophoblast correlates with potential activation of the notch-2 pathway. Dev. Genet. 21, 21–30.

Oberfield, J.L., Collins, J.L., Holmes, C.P., Goreham, D.M., Cooper, J.P., Cobb, J.E., Lenhard, J.M., Hull-Ryde, E.A., Mohr, C.P., Blanchard, S.G., Parks, D.J., Moore, L. B., et al. 1999. A peroxisome proliferator-activated receptor gamma ligand inhibits adipocyte differentiation. Proc. Natl. Acad. Sci. USA 96, 6102–6106.

Park, B.H., Qiang, L., Farmer, S.R. 2004. Phosphorylation of C/EBPbeta at a consensus extracellular signal-regulated kinase/glycogen synthase kinase 3 site is required for the induction of adiponectin gene expression during the differentiation of mouse fibroblasts into adipocytes. Mol. Cell. Biol. 24, 8671–8680.

Peters, J.M., Lee, S.S., Li, W., Ward, J.M., Gavrilova, O., Everett, C., Reitman, M.L., Hudson, L.D., Gonzalez, F.J. 2000. Growth, adipose, brain, and skin alterations resulting from targeted disruption of the mouse peroxisome proliferator-activated receptor beta (delta). Mol. Cell. Biol. 20, 5119–5128.

Pittenger, M.F., Mackay, A.M., Beck, S.C., Jaiswal, R.K., Douglas, R., Mosca, J.D., Moorman, M.A., Simonetti, D.W., Craig, S., Marshak, D.R. 1999. Multilineage potential of adult human mesenchymal stem cells. Science 284, 143–147.

Prusty, D., Park, B.H., Davis, K.E., Farmer, S.R. 2002. Activation of MEK/ERK signaling promotes adipogenesis by enhancing peroxisome proliferator-activated receptor gamma (PPARgamma) and C/EBPalpha gene expression during the differentiation of 3T3-L1 pre-adipocytes. J. Biol. Chem. 277, 46226–46232.

Rim, J.S., Xue, B., Gawronska-Kozak, B., Kozak, L.P. 2004. Sequestration of thermogenic transcription factors in the cytoplasm during development of brown adipose tissue. J. Biol. Chem. 279, 25916–25926.

Rosen, E.D., Spiegelman, B.M. 2000. Molecular regulation of adipogenesis. Annu. Rev. Cell Dev. Biol. 16, 145–171.

Rosen, E.D., Sarraf, P., Troy, A.E., Bradwin, G., Moore, K., Milstone, D.S., Spiegelman, B.M., Mortensen, R.M. 1999. PPARgamma is required for the differentiation of adipose tissue *in vivo* and *in vitro*. Mol. Cell 4, 611–617.

Rosen, E.D., Hsu, C.H., Wang, X., Sakai, S., Freeman, M.W., Gonzalez, F.J., Spiegelman, B.M. 2002. C/EBPalpha induces adipogenesis through PPARgamma: A unified pathway. Genes Dev. 16, 22–26.

Rosenfield, R.L., Kentsis, A., Deplewski, D., Ciletti, N. 1999. Rat preputial sebocyte differentiation involves peroxisome proliferator-activated receptors. J. Invest. Dermatol. 112, 226–232.

Ross, S.E., Erickson, R.L., Hemati, N., Macdougald, O.A. 1999. Glycogen synthase kinase 3 is an insulin-regulated C/EBPalpha kinase. Mol. Cell. Biol. 19, 8433–8441.

Ross, S.E., Hemati, N., Longo, K.A., Bennett, C.N., Lucas, P.C., Erickson, R.L., Macdougald, O.A. 2000. Inhibition of adipogenesis by Wnt signaling. Science 289, 950–953.

Rossant, J., Cross, J.C. 2001. Placental development: Lessons from mouse mutants. Nat. Rev. Genet. 2, 538–548.

Ryves, W.J., Harwood, A.J. 2003. The interaction of glycogen synthase kinase-3 (GSK-3) with the cell cycle. Prog. Cell Cycle Res. 5, 489–495.

Saluja, I., Granneman, J.G., Skoff, R.P. 2001. PPAR delta agonists stimulate oligodendrocyte differentiation in tissue culture. Glia 33, 191–204.

Sandouk, T., Reda, D., Hofmann, C. 1993. Antidiabetic agent pioglitazone enhances adipocyte differentiation of 3T3-F442A cells. Am. J. Physiol. 264, C1600–C1608.

Schaiff, W.T., Carlson, M.G., Smith, S.D., Levy, R., Nelson, D.M., Sadovsky, Y. 2000. Peroxisome proliferator-activated receptor-gamma modulates differentiation of human trophoblast in a ligand-specific manner. J. Clin. Endocrinol. Metab. 85, 3874–3881.

Schaiff, W.T., Bildirici, I., Cheong, M., Chern, P.L., Nelson, D.M., Sadovsky, Y. 2005. Peroxisome proliferator-activated receptor-gamma and retinoid X receptor signaling regulate fatty acid uptake by primary human placental trophoblasts. J. Clin. Endocrinol. Metab. 90, 4267–4275.

Schmuth, M., Haqq, C.M., Cairns, W.J., Holder, J.C., Dorsam, S., Chang, S., Lau, P., Fowler, A.J., Chuang, G., Moser, A.H., Brown, B.E., Mao-Qiang, M., et al. 2004. Peroxisome proliferator-activated receptor (PPAR)-beta/delta stimulates differentiation and lipid accumulation in keratinocytes. J. Invest. Dermatol. 122, 971–983.

Shalom-Barak, T., Nicholas, J.M., Wang, Y., Zhang, X., Ong, E.S., Young, T.H., Gendler, S.J., Evans, R.M., Barak, Y. 2004. Peroxisome proliferator-activated receptor gamma controls Muc1 transcription in trophoblasts. Mol. Cell. Biol. 24, 10661–10669.

Shao, D., Lazar, M.A. 1997. Peroxisome proliferator activated receptor gamma, CCAAT/ enhancer-binding protein alpha, and cell cycle status regulate the commitment to adipocyte differentiation. J. Biol. Chem. 272, 21473–21478.

Tan, N.S., Michalik, L., Noy, N., Yasmin, R., Pacot, C., Heim, M., Fluhmann, B., Desvergne, B., Wahli, W. 2001. Critical roles of PPAR beta/delta in keratinocyte response to inflammation. Genes Dev. 15, 3263–3277.

Tan, N.S., Shaw, N.S., Vinckenbosch, N., Liu, P., Yasmin, R., Desvergne, B., Wahli, W., Noy, N. 2002. Selective cooperation between fatty acid binding proteins and peroxisome proliferator-activated receptors in regulating transcription. Mol. Cell. Biol. 22, 5114–5127.

Tang, Q.Q., Gronborg, M., Huang, H., Kim, J.W., Otto, T.C., Pandey, A., Lane, M.D. 2005. Sequential phosphorylation of CCAAT enhancer-binding protein beta by MAPK and glycogen synthase kinase 3beta is required for adipogenesis. Proc. Natl. Acad. Sci. USA 102, 9766–9771.

Tang, X., Guilherme, A., Chakladar, A., Powelka, A.M., Konda, S., Virbasius, J.V., Nicoloro, S.M., Straubhaar, J., Czech, M.P. 2006. An RNA interference-based screen identifies MAP4K4/NIK as a negative regulator of PPAR{gamma}, adipogenesis, and insulin-responsive hexose transport. Proc. Natl. Acad. Sci. USA 103, 2087–2092.

Tontonoz, P., Hu, E., Spiegelman, B.M. 1994. Stimulation of adipogenesis in fibroblasts by PPAR gamma 2, a lipid-activated transcription factor. Cell 79, 1147–1156.

Van Den Brink, G.R., Bleuming, S.A., Hardwick, J.C., Schepman, B.L., Offerhaus, G.J., Keller, J.J., Nielsen, C., Gaffield, W., Van Deventer, S.J., Roberts, D.J., Peppelenbosch, M.P. 2004. Indian Hedgehog is an antagonist of Wnt signaling in colonic epithelial cell differentiation. Nat. Genet. 36, 277–282.

Varnat, F., Heggeler, B.B., Grisel, P., Boucart, N., Corthesy-Theulaz, I., Wahli, W., Desvergne, B. 2006. PPARβ/δ regulates Paneth cell differentiation via controlling the hedgehog signaling pathway. Gastroenterology 131, 538–553.

Vernochet, C., Milstone, D.S., Iehle, C., Belmonte, N., Phillips, B., Wdziekonski, B., Villageois, P., Amri, E.Z., O'donnell, P.E., Mortensen, R.M., Ailhaud, G., Dani, C. 2002. PPARgamma-dependent and PPARgamma-independent effects on the development of adipose cells from embryonic stem cells. FEBS Lett. 510, 94–98.

Waite, L.L., Person, E.C., Zhou, Y., Lim, K.H., Scanlan, T.S., Taylor, R.N. 2000. Placental peroxisome proliferator-activated receptor-gamma is up-regulated by pregnancy serum. J. Clin. Endocrinol. Metab. 85, 3808–3814.

Wang, L., Shao, Y.Y., Ballock, R.T. 2005. Peroxisome proliferator activated receptor-gamma (PPARgamma) represses thyroid hormone signaling in growth plate chondrocytes. Bone 37, 305–312.

Wang, L.C., Nassir, F., Liu, Z.Y., Ling, L., Kuo, F., Crowell, T., Olson, D., Davidson, N.O., Burkly, L.C. 2002. Disruption of hedgehog signaling reveals a novel role in intestinal morphogenesis and intestinal-specific lipid metabolism in mice. Gastroenterology 122, 469–482.

Werling, U., Siehler, S., Litfin, M., Nau, H., Gottlicher, M. 2001. Induction of differentiation in F9 cells and activation of peroxisome proliferator-activated receptor delta by valproic acid and its teratogenic derivatives. Mol. Pharmacol. 59, 1269–1276.

Woods, J.W., Tanen, M., Figueroa, D.J., Biswas, C., Zycband, E., Moller, D.E., Austin, C.P., Berger, J.P. 2003. Localization of PPARdelta in murine central nervous system: Expression in oligodendrocytes and neurons. Brain Res. 975, 10–21.

Yang, Q., Yamada, A., Kimura, S., Peters, J.M., Gonzalez, F.J. 2006. Alterations in skin and stratified epithelia by constitutively activated PPARalpha. J. Invest. Dermatol. 126, 374–385.

Zhang, J., Fu, M., Cui, T., Xiong, C., Xu, K., Zhong, W., Xiao, Y., Floyd, D., Liang, J., Li, E., Song, Q., Chen, Y.E. 2004. Selective disruption of PPARgamma 2 impairs the development of adipose tissue and insulin sensitivity. Proc. Natl. Acad. Sci. USA 101, 10703–10708.

Regulation of murine embryonic patterning and morphogenesis by retinoic acid signaling

Tracie Pennimpede,[1,]* Don Cameron[1,]* and Martin Petkovich[1,2]

[1]*Department of Biochemistry, Division of Cancer Biology and Genetics, Cancer Research Institute, Queen's University, Kingston, Ontario, Canada K7L 3N6*
[2]*Department of Pathology and Molecular Medicine, Division of Cancer Biology and Genetics, Cancer Research Institute, Queen's University, Kingston, Ontario, Canada K7L 3N6*

Contents

---

*These authors contributed equally to this work.

Advances in Developmental Biology
Volume 16 ISSN 1574-3349
DOI: 10.1016/S1574-3349(06)16003-2

1. Introduction

Embryonic organogenesis and developmental patterning are highly complex processes, which require the precise coordinated expression of various genes and their products within very specific spatiotemporal constraints. Both experimental and clinical evidence has reinforced that the actions of the fat-soluble vitamin A and its active derivatives, known as retinoids, are extremely important throughout vertebrate development. Retinoids exert biological effects on cell differentiation, proliferation, and morphogenesis by binding to their cognate receptors, the retinoic acid receptors (RARs) and retinoid X receptors (RXRs), resulting in the subsequent transactivation of retinoid-responsive genes. This chapter will focus largely on the requirement of the retinoid receptors during murine development by discussing the many RAR- and RXR-null genetic models that have been created to date. In addition, the overarching importance of controlling the bioavailability of RA within specific tissues will be examined. There are many levels of control within retinoid signaling such that the availability of RA is as essential as the presence of the receptors themselves. In this respect, work that has been performed by our laboratory and others toward understanding the role of the enzymes that control the distribution of RA throughout embryonic development will be reviewed.

1.1. Vitamin A intake and metabolism

Vitamin A, or retinol, and its active derivatives are required at the cellular level for growth, apoptosis, differentiation, and proliferation. Within the whole organism, regulation of these processes is essential for the maintenance of reproduction, vision, embryogenesis, and tissue homeostasis (Blomhoff et al., 1990; Collins and Mao, 1999; Morriss-Kay and Ward,

1999; Mark et al., 2006). Despite being physiologically essential, retinol cannot be *de novo* synthesized in mammals and is therefore obtained from the diet in the form of carotenoids or retinyl esters. The intake, transport, and metabolism of vitamin A have been extensively studied (Allen and Bloxham, 1989; Collins and Mao, 1999; Penniston and Tanumihardjo, 2006). Briefly, ingested retinyl esters are hydrolyzed to retinol within the intestinal lumen, and carotenoids are metabolized to retinal. Retinol and retinal are then absorbed by intestinal epithelial cells wherein they bind to cellular retinol-binding protein (CRBP) II to prevent their oxidation (Collins and Mao, 1999). This complex becomes a substrate for the enzyme lecithin: retinol acyltransferase, which reesterifies retinol/retinal to long-chain fatty acids and packages them into chylomicrons (Collins and Mao, 1999; Penniston and Tanumihardjo, 2006). These chylomicrons circulate first in the intestinal lymph and then move throughout the general circulation to be taken up by hepatic cells. Up to 80% of a vertebrate's total vitamin A is stored in lipid droplets within hepatic stellate cells (Collins and Mao, 1999). When required by tissues, the retinyl esters are hydrolyzed to retinol, which must be bound to retinol-binding protein (RBP) for transport in the circulation due to its hydrophobicity. Once within target tissues, retinol undergoes the various enzymatic reactions which convert it to many different products, including the principal transcriptionally active metabolites all-*trans* and 9-*cis* retinoic acid (atRA and 9-*cis*-RA). Consequently, these metabolites are produced in tissues when and where they are needed, and these same tissues have the capability of inactivating vitamin A derivatives through spatiotemporally regulated cytochrome P450-mediated oxidation. Finally, the resultant oxidized polar metabolites are excreted from the body (Collins and Mao, 1999).

1.2. Nutritional vitamin A homeostasis

Controlling the intake and availability of dietary vitamin A has been shown to be important throughout life. Although there are homeostatic mechanisms in place to manage the storage and distribution of vitamin A, there have been many documented cases of vitamin A deficiency (VAD), especially in developing nations (Penniston and Tanumihardjo, 2006). More recently, the dietary effects of toxic levels of vitamin A are also becoming of greater interest due to an increase in the amount of preformed vitamin A available from supplemented foods (Penniston and Tanumihardjo, 2006). VAD has been shown to lead to degeneration of many tissues including the retina and testes (Wolbach and Howe, 1978), whereas excess intake has been shown to result in hypervitaminosis A, with characteristic effects on the liver, skin, bone, and central nervous system (Teelmann, 1989).

Maintaining proper levels of vitamin A is especially important during pregnancy in order to avoid abnormal embryonic morphogenesis. Studies on VAD rats have shown a resultant high rate of fetal resorption, which could be reversed by supplementing the dams with retinyl acetate, retinol, or large doses of atRA (although in the latter case the pups died shortly after birth due to atRA toxicity) (White et al., 1998). As early as the 1950s, Wilson and Warkany showed that pups from rats fed a VAD diet exhibited malformations of the heart, lung, eye, and urogenital tract (Wilson and Warkany, 1950; Wilson et al., 1953). On the other hand, studies on vitamin A teratogenicity also illustrated a variety of congenital defects, including exencephaly and instances of spina bifida, cleft palate, and various eye defects (Cohlan, 1954; Lammer et al., 1985). These early observations led to the awareness that vitamin A plays an important role in fetal development; determining how it was able to elicit its profound pleiotropic cellular and tissue-specific effects subsequently became a substantial focus of interest.

1.3. Discovery of the retinoid receptors

A major breakthrough in the field of retinoids came in 1987 with the cloning of the first member of the retinoid nuclear receptor family, RARα (designated hRAR) (Giguere et al., 1987; Petkovich et al., 1987) which was shown to belong to the nuclear hormone family of receptors and to bind RA with high specificity. This finding established that a nuclear hormone superfamily existed and that these structurally similar receptors had very distinct ligands. Subsequently, the characterization of the first RXR family member in 1990 (Mangelsdorf et al., 1990), the heterodimeric binding partner of the RARs (Leid et al., 1992b), greatly furthered our understanding of the structural and functional mechanisms tying together vitamin A and hormone signaling.

One of the main sources of the diversity and complexity of signaling through the retinoid receptors is their multiplicity. Within each of the two receptor subfamilies (RAR and RXR) there are three isotypes (α, β, and γ), each encoded by separate genes and with multiple isoforms, resulting from alternative splicing and the differential use of two promoters (reviewed in Leid et al., 1992a; Chambon, 1996). The full family of retinoid receptors represents in total 14 different isoforms: RARα1–2, RARβ1–4, RARγ1–2, RXRα1–2, RXRβ1–2, and RXRγ1–2 (Leid et al., 1992a; Chambon, 1996). In general, these receptors are able to act as ligand-activated transcription factors that modulate the activity of RA target gene promoters by binding to specific DNA motifs called RA response elements (RAREs) (Chambon, 1996; Gronemeyer et al., 2004). The presumptive natural ligand for the RXRs is 9-*cis*-RA, whereas RARs are able to bind both atRA and 9-*cis*-RA (Allenby et al., 1993; Chambon, 1996). A wide variety of synthetic ligands have been developed for the RARs (Beckett-Jones, 1996; Klaholz et al., 1998; Resche-Rigon and Gronemeyer, 1998; Gehin et al., 1999; Germain et al., 2004) with a myriad of functions, including isotype- and isoform-specific agonism and antagonism. In contrast, no isotype-selective agonists for RXRs currently exist—the difficulty in designing these is compounded by a lack of divergent residues within the ligand-binding pockets of the isotypes (Wurtz et al., 1996; Gronemeyer et al., 2004), whereas the RAR isotypes exhibit ligand specificity as the result of three divergent residues within their respective ligand-binding pockets (Renaud et al., 1995).

1.4. Mechanisms for controlling RA distribution within the developing embryo

As the actions of retinoids, particularly RA, were becoming better understood, it became apparent that restricting RA levels within developing tissues was an essential part of proper embryonic patterning. Various studies have suggested that RA can act as a morphogen, providing positional information to cells within developing tissues (Maden, 2002). RA has historically been shown to have a posteriorizing and/or duplicating effect on developing structures, and prime examples of this include its effects during amphibian limb regeneration, following RA application to the chick limb bud, and within the developing hindbrain region of vertebrate embryos (Maden, 1998 and references therein).

In light of these morphogenetic actions of RA, and knowing from earlier studies that radiolabeled retinoic acid could only be recovered in the urine as water-soluble metabolic products, even following a large intrajugular dose (Emerick et al., 1967), it was postulated that RA action was limited by its availability to tissues (i.e., its synthesis and catabolism) (Frolik et al., 1979). The enzymes performing these functions were later discovered to be the retinaldehyde dehydrogenases (Raldh1–Raldh4) and cytochrome P450 enzymes (Cyp26A1, Cyp26B1, and Cyp26C1), respectively. Raldhs control the production of RA by catalyzing the irreversible oxidation of retinal to retinoic acid (Duester, 1996; Zhao et al., 1996; Fischer et al., 1999; Niederreither et al., 2002a; Lin et al., 2003) and the Cyp26s limit the availability of RA by oxidizing it to polar metabolites, such as 4-*oxo*-RA, 4-OH-RA, 18-OH-RA, and 5,6-*epoxy*-RA, which are then excreted (Fujii et al., 1997; White et al., 1997, 2000; Taimi et al., 2004). The discovery of these enzymes, and their complementary expression patterns, revealed a means for controlling RA distribution in developing embryos within the strict spatiotemporal constraints that are present during embryogenesis.

Although an examination of all of the complexities of RA signaling is beyond the scope of this chapter, the importance of maintaining appropriate levels of active RA within tissues should be noted, especially with respect to the action of retinoid receptors within differentiating cells, tissues, organs,

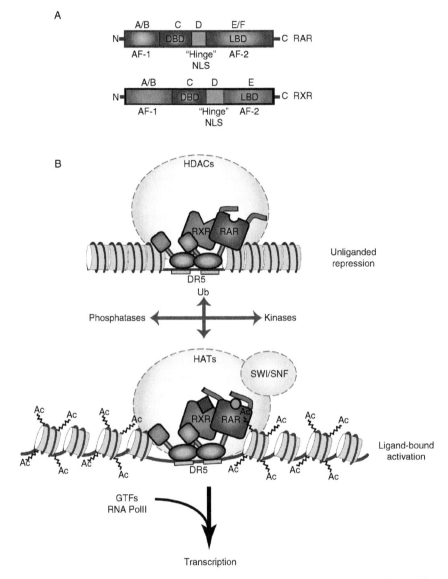

Fig. 1. Retinoid receptor structure and regulation of RA-responsive genes. (A) The modular A–F domain structure of RARs and RXRs. The A/B region contains the AF-1 domain, the C region contains the DNA-binding domain (DBD), D is the "hinge" region containing a nuclear localization signal (NLS), the E region contains the ligand-binding domain (LBD) and AF-2. The F region is not found in RXRs. (B) A schematic representation of the current model of retinoid-regulated gene activation showing RAR/RXR heterodimer bound to a DR5 response element. Gene repression via chromatin compaction results from aporeceptor association with a large protein complex containing various corepressors with HDAC activity. Ligand binding then induces transcriptional activation by the recruitment of various coactivator proteins with

and systems. Just as limiting the distribution of the RARs and RXRs exerts a level of control for the pleiotropic effects of RA, so too does the distribution of RA itself provide an additional method of limiting its actions on target cells.

2. RAR and RXR structure and function

RAR and RXR proteins exhibit characteristics common to members of the nuclear hormone receptor superfamily. The RAR/RXR protein domains and their specific functions, as well as the basic mode of target gene activation, and methods of receptor activation and repression will be examined.

2.1. Retinoid receptor domains and their functions

Homologues for RARs and RXRs exist across all vertebrate species studied to date. The amino acid sequences of RARs and RXRs can be broken down to display the classical A–F regions of homology with each other and other members of the nuclear receptor superfamily (Fig. 1A) (Leid et al., 1992a; Chambon, 1994; Chambon, 1996; Gronemeyer and Miturski, 2001; Gronemeyer et al., 2004). The N-terminal portion of the sequence is often grouped as the A/B region. The A region is isoform specific, whereas the B region is moderately well conserved among all isoforms (Leid et al., 1992a; Chambon, 1996). This A/B region contains a ligand-independent transcriptional activation function (AF-1) domain, which has been shown to interact with coregulators and other transcription factors (Gronemeyer et al., 2004). The central C domain contains the DNA-binding domain (DBD) core, consisting of a 66 amino acid-long region including two zinc finger-like motifs, as well as a surface involved in RAR/RXR interactions (Chambon, 1996). This region is highly conserved among the RAR and RXR isotypes and less conserved with the other nuclear receptors, which is logical considering that it confers recognition of the RAREs within RA-responsive genes. Region D contains a span of basic amino acids which function as a nuclear localization signal (Leid et al., 1992a; Gronemeyer et al., 2004), and represents the hinge region located between the DBD of the C region and the ligand-binding domain (LBD) of the 220 amino acid-long E

HAT activity and ATP-dependent chromatin remodelers like SWI/SNF. This leads to chromatin decompaction and the recruitment of the basal transcription machinery. Various phosphorylation events and the ubiquitin-proteasome pathway have been suggested to play a role in transcriptional regulation, although the *in vivo* sequence of events is still unclear. Ub, ubiquitin; Ac, acetyl groups; GTFs, general transcription factors, HATs, histone acetyltransferases; HDACs, histone deacetylases. (See Color Insert.)

region (Chambon, 1996). Apart from the LBD, the E domain also holds surfaces for dimerization and binding of coregulators as well as a ligand-dependent activation function region (AF-2) (Mark et al., 2006). *In vitro*, the AF-1 and AF-2 within a specific receptor isoform have been shown to be able to cooperate in a cell- and promoter-specific manner (Nagpal et al., 1992; Nagpal et al., 1993). The function, if any, of the completely noncon-served C-terminal F domain, which is absent from the RXRs, is presently unknown.

The LBD-containing E region has been extremely well characterized. Crystallographic models of apo (unliganded) and holo (ligand-bound) LBDs revealed a canonical three-dimensional fold made up of 12 α-helices, known as an antiparallel α-helical sandwich. This fold is made up of three layers with helix (H) 4, H5, H8, H9, and H11 sandwiched between H1, H2, and H3 on one side, and H6, H7, and H10 on the other side (Renaud and Moras, 2000). This structure forms a compact binding pocket with a conserved β-turn between H5 and H6 (Wei, 2003), and the amphipathic H12, located within the AF-2 core domain at the C-terminal end of the LBD, acts as a moveable lid for this pocket (Gronemeyer et al., 2004). Upon ligand binding, the LBD undergoes a conformational change, making it much more compact which stabilizes the ligand–receptor interaction while providing a surface for coactivator binding. It is this "molecular switch" that moves the receptor into a transcriptionally active state, and it is often referred to as the "mouse-trap" model because the conformational change causes H12 to snap back onto the entrance of the ligand-binding pocket (Bastien and Rochette-Egly, 2004; Gronemeyer et al., 2004).

It should be noted that the amino acid sequences for each isoform are better conserved between species than the three isotypes are within a given species (Leid et al., 1992a). This strict interspecies sequence conservation, along with the unique spatiotemporal expression patterns for each tran-script, suggested that there must be some degree of inimitability of function among the receptors. In fact, although the functions of some of the receptor isoforms have been shown to be indispensable for the formation of certain tissues, for the most part their activities have turned out to be incredibly redundant, adding a further degree of perplexity to this complicated signal-ing system, which will be elaborated upon in the subsequent sections.

2.2. Mechanisms of RAR/RXR transcriptional regulation and target gene activation

The functional unit of retinoid signaling is the RAR/RXR heterodimer, of which there are 48 different possible combinations. *In vitro* experiments have shown that liganded RAR alone is sufficient to activate transcription but, upon addition of liganded RXR, a synergistic transcriptional activation can be observed (Minucci et al., 1997), suggesting that both receptors are important in the transcriptional process. This was supported by an increase in teratogenic effects on late blastula stage zebrafish and *Xenopus* embryos following coadministration of agonists for both receptor types (Minucci et al., 1997). Heterodimerization represents a way to create combinatorial diversity, while maintaining a high degree of evolutionary economics, from a limited number of receptors both at response elements and at the level of transactivation (Leid et al., 1992a).

The classic model of steroid hormone action occurs in several stages. First, lipophilic hormones, such as RA, pass through the lipid bilayer of the cell membrane. Some evidence suggests that RA is then shuttled to the nucleus by CRABPII where it is channeled and presented to RAR (Budhu and Noy, 2002), however, knockout models of CRABPI and -II have shown that this is not physiologically essential (de Bruijn et al., 1994; Gorry et al., 1994; Lampron et al., 1995). Within the nucleus, the RAR/RXR receptor complex binds to RAREs in a head-to-tail arrangement and with high affinity (Bastien and Rochette-Egly, 2004). The RA bipartite response element consists of two direct hexameric PuG(G/T)TCA core repeats (Leid et al., 1992a; Mangelsdorf and Evans, 1995) separated by a variable spacer. The most common spacer for RAR/RXR binding is 5 bp (DR5) but DR2 and DR1 elements have also been found (Bastien and Rochette-Egly, 2004). On DR5 and DR2 elements, the receptors bind as RXR/RAR heterodimers with RXR occupying the upstream hexamer (the 5′ motif), and on DR1 the receptors bind with reverse polarity, or with RAR at the 5′ position (the 3′ motif) (Kurokawa et al., 1994; Zechel et al., 1994; Bastien and Rochette-Egly, 2004). A novel RARE configuration designated by two half-sites with an 87 bp spacer was discovered within the Fgf8 promoter (Brondani et al., 2002). This response element was shown to be able to bind unliganded, phosphorylated RARα, suggesting that DNA folding is also important within the RARE environment.

The next step in RA-activated transcription is hormone binding, which induces an allosteric change in the LBD, including the repositioning of H12, and favors RAR and RXR interactions with each other and with DNA (Depoix et al., 2001). Transcriptional procession of RA-responsive genes is a result of ensuing conformational changes, and associated interactions between the receptors and large complexes of regulator proteins (Fig. 1B). In fact, a specific RAR/RXR heterodimer may facilitate bending of DNA such that the proteins it can interact with are limited in a way that is dependent on the context of the RARE.

2.3. The role of coregulators in retinoid receptor activation and repression

The function of nuclear receptors involves many other proteins including other transcription factors and coregulators, which are able to bind the receptors and create (or block) alternative binding surfaces through allosteric interactions. Thus, in a given cell, the response of these receptors to various ligands, or to changes in the cellular environment, is dependent on the particular complement of proteins with which it interacts.

Our current understanding of retinoid-regulated gene activation denotes that unliganded receptors repress transcription when bound to DNA (Dilworth and Chambon, 2001) by interacting with corepressors such as NCoR (Horlein et al., 1995), SMRT (Chen and Evans, 1995), and LCoR (Fernandes et al., 2003). This causes proteins with histone deacetylase (HDAC) activity to be recruited, keeping the chromatin in a compacted form. The binding of activating ligands then leads to a conformational change that causes the dissociation of corepressors and the concomitant recruitment of coactivators such as SRC-1/p160 or p300/CBP (Gronemeyer et al., 2004). These coactivators associate with the receptor AF-2 domain through the nuclear receptor boxes, which have an LXXLL signature motif (Heery et al., 1997; Gronemeyer et al., 2004). Some corepressors are also known to contain LXXLL (Fernandes et al., 2003) and/or L/IXXI/VI sites (Wei, 2003). Several coactivators exhibit intrinsic histone acetylase (HAT) or histone methyltransferase activity, and others engage protein complexes which elicit the requisite chromatin decompaction such as ATP-dependent chromatin remodelers like SWI/SNF (Kwon et al., 1994; Bastien and Rochette-Egly, 2004; Gronemeyer et al., 2004). Opening of the chromatin environment allows for the preinitiation complex to become poised and facilitates the binding of basal transcription machinery to the promoter. From recent evidence, it appears that a single coactivator molecule can interact with both heterodimeric partners and that coactivators and corepressors may compete for overlapping binding sites (Germain et al., 2002). Although both liganded heterodimeric partners are able to recruit coactivators, RXR is subordinated to RAR in that it cannot dissociate corepressors and cannot be activated by RXR-selective agonists in the absence of liganded RAR (Chen et al., 1996; Germain et al., 2002).

2.4. Regulation of the retinoid receptors by phosphorylation and proteasomal degradation

Numerous kinases have been shown to regulate RAR and RXR proteins by phosphorylation. RARα and RARγ are phosphorylated at AF-1 by the cyclin-dependent kinase activity of TFIIH (Rochette-Egly et al., 1997; Bastien et al., 2000) and it has been hypothesized that this phosphorylation is somehow related to the ability of the receptors to complex with the basal transcription machinery or with coregulators. RARα has been the most widely studied receptor in terms of phosphorylative regulation. PKA phosphorylation at serine 369 transcriptionally activates RARα, and PKC has been shown to phosphorylate serine 157 within the DBD, favoring dimerization and DNA binding (Delmotte et al., 1999). RARα (and to a lesser extent

RARβ) can also be phosphorylated at the conserved serine 96 by Akt, resulting in inhibition of receptor transactivation (Srinivas et al., 2006). In the case of RXRα, phosphorylation by members of the JNK family at several residues within the A/B region is involved in its ability to associate with its heterodimeric partner and activate transcription (Gianni et al., 2003), whereas phosphorylation at serine 265 within the Ω-loop between H1 and H3 inhibits activation of RA-target genes (Bruck et al., 2005). For RARγ, both transactivation and ubiquitination/degradation require phosphorylation by TFIIH and p38MAPK, respectively (Bastien et al., 2000; Gianni et al., 2002), although the mechanism of these events is mostly undetermined. The search for proteins that interact with both the phosphorylated and unphosphorylated AF-1 domain will unquestionably lead to a better understanding of the regulation of RAR/RXR function.

Signal transduction pathways regulating retinoid receptor activation can also exhibit cross talk via the phosphorylation of coregulators like SMRT, SRC-1, and TIF-2 (Hong and Privalsky, 2000; Rowan et al., 2000; Gerlitz et al., 2002). *In vitro* evidence illustrates that, in response to RA, the p160 coactivator SRC-3 is phosphorylated by p38MAPK which first allows it to associate with RARα and activate gene transcription and then later results in degradation of SRC-3 and subsequent transcriptional inhibition (Gianni et al., 2006). This response is specific to transcription directed by RARα (Gianni et al., 2006) which suggests that either the protein complex, the cell type, or the promoter context can influence transcriptional response at the level of phosphorylation. There is an emerging concept that, in order for transcription to proceed, there needs to be a continuous, combinatorial, and orchestrated exchange of receptors, coregulators, and transcription factors within a large protein complex—the so-called "transcriptional clock" (McKenna and O'Malley, 2002; Metivier et al., 2003; Dennis and O'Malley, 2005; Metivier et al., 2006). Phosphorylation would certainly provide a rapid mechanism to alter the binding surfaces to allow for this protein cycling; the availability of such kinases and the various members that comprise the protein complex would undoubtedly also play an important role in regulating transcription.

The variety and number of proteins involved in the regulation of gene transcription are bound to make the promoter a very crowded place. Evidence has shown that the ubiquitin-proteasome pathway is involved in transcriptional regulation, and it is tempting to speculate that its role is to target proteins for degradation, so as to limit the transcriptional response (Dennis and O'Malley, 2005). Although the 26S proteasome does target some coactivators and corepressors for degradation (Dennis and O'Malley, 2005; Gianni et al., 2006 and references therein), it seems that the relationship between ubiquitination and transcription is much more complex than initially thought.

Evidence suggests that the transcriptional activity of the retinoid receptors can also be controlled by the ubiquitin-proteasome pathway. In response to retinoids, RAR/RXR heterodimers bound to RAREs are degraded by the proteasome following ubiquitination of RAR (Zhu et al., 1999) and proteasome recruitment to AF-2 through the 26S proteasome ATPase subunit SUG-1 (Gianni et al., 2002). Proteasomal degradation of the receptors themselves may limit the duration of the transcriptional response following activation, as seems to be the case for RARγ but not RARα (Gianni et al., 2002). The ubiquitin-proteasome system also has the ability to directly target transcriptional machinery, histones (Dhananjayan et al., 2005), and various coregulators (Yan et al., 2003), thus affecting components of the retinoid-signaling machinery both directly and indirectly.

The rapidly changing and complex regulation of genes during embryonic development requires very precise mechanisms of control. Accuracy of this transcription is most likely facilitated by regulating phosphorylation and ubiquitination, resulting in the modification of binding surfaces, and thus the interacting proteins, leading to an appropriate transcriptional response.

3. Murine receptor expression patterns and knockout models

3.1. Embryonic expression of RARs and RXRs

RAR transcripts are widely distributed throughout the embryo during early morphogenesis (summarized in Table 1; see references Ruberte et al., 1991; Mollard et al., 2000). RARα shows nearly ubiquitous expression at

Table 1
Expression of RAR transcripts during early development

	E8.5[a]	E9.5	E10.5
RARα	Ubiquitous	Anterior limit at r3/r4[b] boundary	As at E9.5
RARβ	Foregut endoderm, neural epithelium (caudal neuropore to caudal hindbrain)	As at E8.5 plus limb proximal mesenchyme, frontonasal mesenchyme, anterior limit at r6/r7[a] boundary	As at E9.5 plus periocular and proximal limb mesenchyme, lung bud, septum transversum and peritoneal-pericardial canal
RARγ	Neural epithelium (open neural tube), presomitic mesenchyme	As at E8.5 plus branchial arch mesenchyme	As at E9.5 plus prevertebral and forelimb precartilaginous condensations

[a]E, embryonic day.
[b]r, rhombomere.

these stages, while the RARβ and RARγ expression domains are significantly more restricted. In the case of RXR expression, RXRα and RXRβ are also widely expressed during murine development, while RXRγ is restricted to developing muscle and regions of the central nervous system (Dolle et al., 1994).

3.2. Disruption of RARs in the developing embryo

Defects of the fetal VAD syndrome have been shown to include abnormal development of eye, heart, lung, kidney, and genitourinary tract, as well as neonatal growth retardation and emaciation, and male sterility (Wilson et al., 1953). Studies on knockouts of particular RAR isotypes, or one isoform of a particular isotype, have been performed and it was expected that embryos would recapitulate the malformations associated with VAD. For the most part, however, the ablation of a single receptor isotype did not mirror the defects seen in VAD syndrome (Lohnes et al., 1993; Lufkin et al., 1993; Kastner et al., 1996; Krezel et al., 1996; Ghyselinck et al., 1997). Animals that were compound null for several of the receptor isoforms or isotypes, however, collectively exhibited a variety of congenital malformations that typified those associated with VAD syndrome, and some presented defects not previously observed in VAD animals, such as craniofacial defects and abnormal patterning of spinal column and limb skeletal elements (Lohnes et al., 1994; Mendelsohn et al., 1994a; Grondona et al., 1996; Luo et al., 1996; Ghyselinck et al., 1997, and reviewed in Lohnes et al., 1995; see Mark et al., 2006 for more extensive discussion).

Mice containing disruptions in specific RAR isoforms, or all isoforms of a given subtype, were viable and displayed only some defects characteristic of fetal VAD syndrome (Li et al., 1993; Lohnes et al., 1993; Lufkin et al., 1993; Mendelsohn et al., 1994b; Ghyselinck et al., 1997). Embryos lacking RARα1, the most ubiquitously expressed isoform, developed normally and were indistinguishable from their wild-type littermates (Li et al., 1993; Lufkin et al., 1993). However, mice lacking both isoforms of RARα ($\alpha 1^{-/-}/\alpha 2^{-/-}$) showed a high rate of postnatal lethality and testis degeneration, both characteristic of VAD syndrome, indicating that the presence of at least one isoform of RARα is required for some aspects of development (Lufkin et al., 1993).

The RARβ gene yields four different isoforms. Mutant mice lacking two of four RARβ isoforms were viable and showed only subtle congenital defects related to the fetal VAD syndrome. The eyes of RARβ2/4-null mice displayed retrolenticular membrane (although this was not completely penetrant) resulting from persistence and hyperplasia of the primary vitreous body (PHPV) (Grondona et al., 1996), whereas mice lacking RARβ1/3 isoforms appeared unaffected (Ghyselinck et al., 1998). Disruption of all

RARβ isoforms resulted in mice that were viable, although growth deficient, and also displayed retrolenticular membrane, but with higher penetrance than was seen in the RARβ2/4 mutants. Additionally, loss of all RARβ isoforms resulted in homeotic transformations and malformations of cervical vertebrae (Ghyselinck et al., 1997).

Mice lacking RARγ2 developed normally and appeared indistinguishable from their wild-type littermates. However, loss of both isoforms of the RARγ gene resulted in some defects reminiscent of VAD as well as others not previously associated with VAD (Lohnes et al., 1993). Males lacking RARγ were sterile, likely due to keratinizing squamous metaplasia, characteristic of VAD, in prostate and seminal vesicle epithelia. Several homeotic transformations of cervical vertebrae were also found in RARγ-null mutants, for example, the transformation of C2 to a C1 identity (axis to atlas), an anterior transformation reminiscent of mice lacking the Hoxb4 gene (Ramirez-Solis et al., 1993). RAR$\gamma1^{-/-}$ embryos displayed some of the defects seen in RARγ-null animals, such as growth deficiency and malformed cervical vertebrae, but these defects were generally less severe and less penetrant (Subbarayan et al., 1997). Table 2 summarizes the developmental malformations observed in the various compound mutants. The most severely affected were those containing null mutations of RARα and RARγ, many of which died at various times during gestation and displayed a spectrum of developmental defects (Lohnes et al., 1994).

3.3. Disruption of RXRs in the developing embryo

The genes encoding the three RXRs have also been disrupted in the mouse, providing information about the developmental roles of these receptors. RXRα was shown to be the only isoform absolutely required for development, as RXR$\alpha^{-/-}$ embryos died during gestation and displayed defects in development of the eye and heart (Kastner et al., 1994; Sucov et al., 1994). Furthermore, mice harboring a single allele of RXRα (i.e., RXR$\alpha^{+/-}/\beta^{-/-}/\gamma^{-/-}$) were growth deficient but viable, suggesting that virtually all developmental and postnatal roles of the RXRs can be accomplished with a single copy of RXRα (Krezel et al., 1996). This implies that RXRα is the main heterodimerization partner of the RARs involved in RA signaling throughout development. When the RXRα-null mutation was introduced into an RARα-, RARβ-, or RARγ-null background, nearly all of the defects seen in the compound RAR-null mutants were reproduced (Kastner et al., 1997), including the full spectrum of VAD defects. However, this effect was not apparent when RXRβ- or RXRγ-null mutations were crossed into an RAR mutant background. RXRγ-null mutants appeared unaffected, and compound RXR$\alpha\gamma$ mutants displayed the same defects of the heart and eye as the RXR$\alpha^{-/-}$ mutants (Krezel et al., 1996) further suggesting that RXRγ

Table 2
Summary of congenital defects observed in compound RAR-null mutant mice

VAD associated	Mutant genotype
Eye	
Retinal/optic nerve coloboma	$\alpha\gamma$, $\beta\gamma$
Retrolenticular membrane	$\alpha\beta$, $\beta\gamma$, $\alpha\gamma$
Unfused eyelids	$\alpha\gamma$
Retinal dysplasia	$\beta\gamma$
Lack of anterior chamber	$\alpha\gamma$
Heart	
PTA[a]	$\alpha\beta$
Thin myocardium	$\alpha\gamma$
Aortic arch defect	$\alpha\beta$, $\alpha\gamma$, $\beta\gamma$
Ventricular septum defect	$\alpha\beta$, $\alpha\gamma$
Respiratory	
Lung hypoplasia	$\alpha\beta$
Hypoplasia of kidney	$\alpha\beta$, $\alpha\gamma$
Ureter defect	$\alpha\beta$, $\alpha\gamma$, $\beta\gamma$
Genital tract	
Male	$\alpha\gamma$
Female	$\alpha\beta$, $\alpha\gamma$

Non-VAD associated	Mutant genotype
Limb	
Forelimb	$\alpha\gamma$
Hindlimb	$\alpha\gamma$
Digital webbing	$\beta\gamma$, $\beta^{+/-}\gamma$
Axial skeleton	
Vertebral homeotic	$\alpha\beta$, $\alpha\gamma$
transformations	
Craniofacial skeletal malformation	$\alpha\gamma$
Eye	
Corneal lenticular stalk	$\alpha\gamma$
Lens agenesis	$\alpha\gamma$
Glandular defects	
Sub-maxillary, sublingual	$\alpha\gamma$, $\beta\gamma$
Harderian	$\alpha\gamma$, $\beta\gamma$
Thymus, thyroid, parathyroid	$\alpha\beta$, $\alpha\gamma$

[a]PTA, persistent truncus arteriosus.

is largely dispensable for development. Similarly, mice retaining only RXRγ as the functional isotype (i.e., RXRαβ compound mutants) died early in gestation (embryonic day (E)9.5–E10.5) and exhibited a variety of severe malformations such as axial truncation, abnormal turning, dilated heart cavity, open neural tube, lack of branchial arches (BAs) 2 and 3, and hypoplastic frontonasal process (Wendling et al., 1999). Approximately half of RXRβ-null mutants died before or at birth, and the surviving males were sterile resulting from defective spermatogenesis and progressive degeneration of the germinal epithelium (Kastner et al., 1996).

4. The function of RA and retinoid receptors within developing systems

The gene ablation studies described above represent investigations into the function of the retinoid receptors during development. It was the hope that the developmental effects observed would provide insight into the pleiotropic effects of retinoids during critical stages of embryogenesis and organ formation. Extensive research has been performed on RAR, RXR, and compound mutants as far as characterizing the associated malformations, and these have been discussed at length elsewhere (Kastner et al., 1994; Lohnes et al., 1994; Mendelsohn et al., 1994a; Kastner et al., 1995; Lohnes et al., 1995; Mark et al., 2006 and references therein). The emphasis here will be on the major defects associated with development of the heart, eye, limb, and hindbrain region. A number of studies focused on these tissues have revealed an important connection between metabolic RA availability and RA signaling for controlling the spatiotemporal expression of genes implicated in organogenesis. In addition, evidence that supports a role for active repression by unliganded receptors, and studies on conditional retinoid receptor knockout mice will be discussed.

4.1. Heart

The embryonic heart is the first organ to form (Buckingham et al., 2005) which permits study even in mutants displaying early embryonic lethal phenotypes. Also, malformations of the heart and aortic arch derivatives are often seen in fetuses from VAD dams (Wilson et al., 1953), in RXRα$^{-/-}$ embryos, and in many of the RAR compound mutants (Kastner et al., 1994; Mendelsohn et al., 1994a and discussed later).

Cardiogenesis in the mouse begins with the formation of the presumptive heart region around E6.5 when myocardial progenitor cells occupy the primitive streak (see reviews by Kelly and Buckingham, 2002; Zaffran and Frasch, 2002; Buckingham et al., 2005). These cells then migrate to the anterior portion of the embryo under the head folds, forming the cardiac crescent. At E7.5, differentiated myocardial cells can be observed, and at E8.0 fusion of the cardiac crescent at the midline forms the cardiac tube, which is made up of external myocardial and internal endocardial layers (Zaffran and Frasch, 2002). The heart tube then undergoes rightward looping around E8.5. Well defined chambers can be observed by E10.5 and septation of the embryonic heart is completed by E14.5, separating the chambers with connections to the aorta and pulmonary trunk (Buckingham et al., 2005).

Expression of RXRs and RARs has been examined within the context of the developing heart. RXRα and RXRβ are found ubiquitously, unlike RXRγ which is absent even though it is expressed in other muscle primordia (Dolle et al., 1994). RARβ1/3 transcripts were detected at E11.5 in the heart conotruncal mesenchyme, whereas RARα1 and RARβ2/4 were moderately expressed in the myocardium at E13.5 (Mollard et al., 2000) and RARγ was present in the endocardial cushion tissue (Dolle et al., 1990).

The embryonic lethality resulting from ablation of RXRα was suspected to be the result of heart abnormalities, although some embryos survived to E16.5 (Kastner et al., 1994). These mice exhibited a variable "thin myocardium" phenotype which was characterized by abnormal thinness of the compact layer, the ventricular septum, and atrium walls (Kastner et al., 1994). Similar phenotypes were observed in RXR$\alpha^{+/-}$/RAR$\gamma^{-/-}$, RXR$\alpha^{-/-}$/RAR$\gamma^{-/-}$, and RXR$\alpha^{-/-}$/RAR$\alpha^{-/-}$ compound mutants, but with increased frequency of the defects in the ventricular septum and additional defects of the aorticopulmonary septum and aortic arches (Kastner et al., 1994). In the RAR compound mutants, persistent truncus arteriosus (PTA), which is the failure of aortic sac division by the aorticopulmonary septum, was frequently observed in E18.5 RAR$\alpha\gamma$, RAR$\alpha\beta$2 (Mendelsohn et al., 1994a), RAR$\alpha\beta$ (Ghyselinck et al., 1997), and RARα1β (Luo et al., 1996) null animals, and sometimes seen in RAR$\alpha^{-/-}$/β2$^{+/-}$ and α1β2 mutant fetuses (Mendelsohn et al., 1994a), as well as in RXRα/RARβ- and RXRα/RARγ-null embryos (Kastner et al., 1997). All of these compound mutants also displayed some form(s) of aortic arch abnormality at E18.5 (see the original publications for a detailed description). Abnormalities in heart formation were not seen in α1γ or β2γ mutants (Mendelsohn et al., 1994a).

The parameters of retinoid formation and activity have also been studied in the developing mouse heart. Using an antibody against Raldh2, as well as a reporter mouse line that expresses β-galactosidase under the control of the hsp68 promoter containing three RAREs (RARE-lacZ), it was shown that the production of RA and the activation of receptors occurred in the same regions of the heart and, moreover, that this corresponded to regions that did not express Cyp26A1 (Moss et al., 1998). In fact, later analysis of Cyp26A1/Cyp26B1 expression showed that Cyp26A1 was present in the early embryo (E7.5–E8.0) within the cardiac crescent, heart outflow tract, atria, and sinus venosus (MacLean et al., 2001), specific areas wherein activation of the RARE-lacZ transgene was not observed (Moss et al., 1998). Cyp26A1 and Cyp26B1 were also expressed during organogenesis (E14.5) within the atrioventricular valve and outflow tract endocardial cells, respectively (Abu-Abed et al., 2002).

The importance of RA distribution within the heart was verified by the disruption of Raldh2, which resulted in the heart forming as a single, dilated cavity with no particular left–right organization (Niederreither et al., 1999). These mutants also showed abnormal looping (Niederreither et al., 2001), but heart abnormalities have not been reported in mice with disruptions in Cyp26A1 or Cyp26B1 genes (Abu-Abed et al., 2001; Sakai et al., 2001; Yashiro et al., 2004). In addition, an early role for RA signaling in defining populations of cells that give rise to the heart has been established in zebrafish development (Keegan et al., 2005).

4.2. Eye

RA has been shown to be an indispensable molecule with respect to eye development, as conceptuses from VAD dams exhibit various eye defects. The localization of the RAR and RXR isoforms has been characterized in both adult and developing mouse eyes (for an in-depth discussion of this, refer to Mori et al., 2001). Each of the receptor isotypes has been shown to be present in the prenatal eye and throughout its development. RARβ, RARγ, and RXRα, in particular, are abundantly expressed within the periocular mesenchyme (PM) (Mori et al., 2001), suggesting that these receptors are extremely important for transducing retinoid signals within this tissue and play a major role in ocular development in general.

The formation of the eye in mice begins at late gastrulation with the specification of bilateral fields within the neuroepithelium of the anteriorly forming neural tube (Fini et al., 1997). The optic pit is then created by the invagination of the neural epithelium and, following the closure of the neural tube between E8.5 and E9.5, the optic pits are pushed outward to form the optic vesicle. The lens is then derived as a thickened disk from surface ectoderm-derived cells. The optic vesicle collapses inward to form the optic cup around E12.5–E13.5 (Cvekl and Tamm, 2004), where the inner layer represents neural epithelia, which will further differentiate to become various neural and glial cell types, and the outer layer becomes retinal pigmented epithelia (RPE), which will become the iris and ciliary body at approximately E15.5 (Fini et al., 1997; Cvekl and Tamm, 2004). Once lens vesicle formation is complete, the remaining surface ectoderm cells become corneal epithelia. Then a wave of migrating neural crest cells (NCCs) forms the corneal stroma

(E14.5–E15.5) (Cvekl and Tamm, 2004), as well as the sclera, vitreous body, and anterior chamber (the space between the cornea and the lens) (Fini et al., 1997). PM cells are derived from migrating neural mesenchyme and are in close contact with the RPE cells throughout eye development for the formation of the choroid and sclera (Zhao and Overbeek, 2001). The eyelids begin to develop at E13.5 and fuse by E15.5–E16.5 (Kawada et al., 2000), and the optic nerve is formed from the outgrowth of axons derived from the ganglia of the innermost layer of the neural retina, which eventually meet at the back of the eye (Fini et al., 1997).

Ocular defects seen in $RXR\alpha^{-/-}$ mice included abnormal thickening of the corneal stroma and coloboma of the optic nerve (Kastner et al., 1994). Similar malformations were seen in $RXR\alpha^{+/-}/RAR\gamma^{-/-}$ fetuses, in addition to shortening of the ventral retina. These defects were often less severe than in the $RXR\alpha^{-/-}$ single mutant, although neither $RXR\alpha^{+/-}$ nor $RAR\gamma^{-/-}$ alone exhibited any ocular defects (Kastner et al., 1994). The severity of these defects was increased in $RXR\alpha^{-/-}/RAR\gamma^{+/-}$ and especially in $RXR\alpha^{-/-}/RAR\gamma^{-/-}$ mutants, with a thick layer of mesenchyme burying the eye, bilateral eversion of the retina, and defects of the iris (coloboma and absence of the ventral iris). $RXR\alpha^{-/-}/RAR\alpha^{-/-}$ mutants exhibited retinal eversion, retinal hypoplasia, and coloboma of the iris (Kastner et al., 1994). A conditional ablation of $RXR\alpha$ in mouse RPE has been performed using the Cre/loxP system under the control of the tyrosinase-related protein 1 promoter. The resulting mutants displayed a relatively mild phenotype with retinal dystrophy and abnormal photoreceptor histology and function (Mori et al., 2004). The phenotype does not recapitulate the ocular defects seen in $RXR\alpha$-null mice, suggesting that the cells of the RPE are not the main mediators of retinoid signaling in the murine eye.

RAR compound knockout mice presented with a number of ocular defects at E18.5. The $RAR\alpha\gamma$ mutants displayed the widest range and most conspicuous defects including coloboma of the retina and optic nerve, unfused eyelids, and lack of anterior chamber (Lohnes et al., 1994), all of which were completely penetrant. The only defects observed in $RAR\alpha1^{-/-}/\alpha2^{+/-}/\gamma^{-/-}$ mutants were unfused eyelids and a corneal-lenticular stalk (characterized by continuity of corneal and lens epithelia) in two of five fetuses. Mutants for $RAR\beta2\gamma$, on the other hand, exhibited all of the aforementioned defects with the exception of unfused eyelids and corneal-lenticular stalk (Lohnes et al., 1994). At E18.5, $RAR\beta\gamma$ mutants showed abnormal retrolenticular membrane along with various mesenchymal defects of the stroma, sclera, and anterior chamber (Ghyselinck et al., 1997). Eyes of $RAR\alpha1\gamma$ and $RAR\alpha\beta2^{+/-}$ mutants were unaffected and, with the exception of a fibrous retrolenticular membrane, the eyes of $RAR\alpha1\beta$ (Luo et al., 1996), $RAR\alpha\beta2$, and $RAR\alpha1\beta2$ fetuses were also normal (Lohnes et al., 1994). Retinal dysplasia was observed in adult $RAR\beta2/\gamma2$ mutants (Grondona et al., 1996).

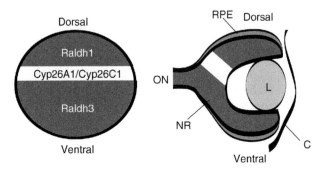

Fig. 2. RA distribution within the retina around E15.5. Diagrammatic representation of a frontal view of the retina (left) showing dorsal Raldh1 and ventral Raldh3 expression. Cyp26A1 and Cyp26C1 are expressed in the area between Raldh1 and Raldh3. On the right is a side profile view of the eye further illustrating the domains of expression. It should be noted that Raldh2 is also expressed at this stage within the periocular mesenchyme which is closely associated with the retinal pigmented epithelia (RPE). L, lens; C, cornea; ON, optic nerve; NR, neural retina.

Distribution of RA within the developing eye (especially the retina) represents a striking example of the coordinated expression of Raldh and Cyp26 enzymes. Raldh1 and Raldh3 are the main sources of RA in the eye, with Raldh1 being expressed in the dorsal retina and Raldh3 ventrally (Fig. 2) (Li et al., 2000), although Raldh2 has been shown to be expressed in the PM between E14.5 and E15.5 (Niederreither et al., 1997) and transiently at E8.5 in the optic vessel (Wagner et al., 2000). In contrast, Cyp26A1 is expressed within the equatorial region of the retina, where none of the Raldh transcripts are observed, and is followed by the expression of Cyp26C1 (Sakai et al., 2004). Single knockouts of Raldh1 (Fan et al., 2003) or Raldh3 (Dupe et al., 2003) have failed to show significant ocular defects, but Raldh1/Raldh3 compound mutants display severe ocular malformations (Matt et al., 2005). More importantly, these defects were similar to those found following specific ablation of RARβ/RARγ in the NCC-derived PM (Matt et al., 2005). These data, and the fact that most of the ocular defects observed in retinoid receptor null fetuses appear to correspond to defects in neural crest-derived structures, suggest that NCCs are responsible for mediating retinoid signals within the developing eye.

4.3. Limb

A role for RA in limb morphogenesis had been implied from early studies on RA teratogenesis. In 1973, Kochhar showed that administration of exogenous RA to pregnant dams could induce limb deformities in developing embryos (Kochhar, 1973). Early patterning of the limb is accomplished by two signaling centers (reviewed in Niswander, 2003). The first is the zone of polarizing activity (ZPA) located at the posterior base of the limb bud. The ZPA controls anterior–posterior (A–P) patterning through secretion of Sonic Hedgehog (Shh), which acts as a diffusible morphogen (Harfe et al., 2004). The second is the apical ectodermal ridge (AER), a strip of cells along the distal edge of the limb bud that controls limb outgrowth through fibroblast growth factor (Fgf) signaling.

Early experiments using the chick limb bud model showed that ectopic administration of RA to the anterior portion of the early limb bud could induce mirror image pattern duplications of the A–P axis, identical to that seen upon grafting cells from the ZPA onto the anterior limb bud (Tickle et al., 1982). These early studies suggested that RA must play an important role in proper limb development.

Although the physiological role of RA signaling in limb patterning and morphogenesis is not entirely clear, changes in the distribution of RA or its ability to signal in the developing limb by genetic or pharmacological means have had dramatic consequences. In early stages of limb development (E10.5), both RARα and RARγ transcripts are distributed throughout the limb bud, whereas RARβ transcripts are present only in the flanking trunk mesoderm (Dolle et al., 1989). At E11.5, RARβ is also expressed in the AER. As the limb continues to grow, RARγ becomes expressed in central precartilaginous condensations, while maintaining expression in the undifferentiated mesenchyme, and RARβ is also detectable in the interdigital mesenchyme (Dolle et al., 1989). Both RARα- and RARγ-null mice have normal limbs, indicating that these receptors can functionally compensate for one another. Compound RAR$\alpha\gamma$ mutants, however, show several limb malformations (Lohnes et al., 1994). In these mutants, the forelimbs displayed a variety of defects such as malformation of the scapula, agenesis of the radius, D1 and central carpal bones, and syndactyly (Lohnes et al., 1994). The hindlimbs of these mutants were slightly less affected, showing only malformations of the tibia (Lohnes et al., 1994). Despite the specific expression of RARβ in the interdigital mesenchyme, mice lacking this receptor do not exhibit any limb defects. RAR$\beta^{+/-}$/RAR$\gamma^{-/-}$ and RAR$\beta^{-/-}$/RAR$\gamma^{-/-}$ compound mutants, however, display a completely penetrant interdigital webbing phenotype, which is occasionally seen in other mutant backgrounds (i.e., RAR$\alpha^{-/-}$, RAR$\gamma^{-/-}$, RXR$\alpha^{+/-}$, and RXR$\alpha^{+/-}$/RAR$\alpha^{+/-}$, RXR$\alpha^{+/-}$/RAR$\gamma^{+/-}$ compound heterozygotes) (Ghyselinck et al., 1997; Kastner et al., 1997). This webbing defect results from reduced apoptosis and increased proliferation in mesenchymal cells of the interdigital necrotic zones (Dupe et al., 1999).

Given the dramatic effects of excess RA and/or loss of specific RARs in the developing limb, it is apparent that the distribution of RA must be finely controlled for proper morphogenesis to occur. Null mutations of RA-synthesizing and RA-metabolizing enzymes also result in malformations

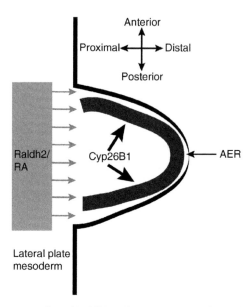

Fig. 3. Schematic representation of Raldh2 and Cyp26B1 expression in the early limb bud. At E10.5, RA produced in the flanking trunk region by Raldh2 diffuses into the proximal limb bud. Cyp26B1 is expressed in distal undifferentiated mesenchyme and excluded from the AER.

within the limb. Although there is no Raldh expression in the developing limb itself, the high expression of Raldh2 in the trunk somites during limb bud initiation and outgrowth acts as the presumptive source of RA (Fig. 3). Consistent with this is the observation that Raldh2$^{-/-}$ embryos completely lack limb buds (Niederreither et al., 1999; Niederreither et al., 2002b). Although these Raldh2-null mutants die between E9.5 and E10.5, corresponding to the initial period of limb bud development, administration of RA through the maternal food supply can rescue development in a stage- and dose-dependent manner such that effects of loss of Raldh2 later in limb development can be examined (Niederreither et al., 2002b; Mic and Duester, 2003; Mic et al., 2004). Rescued embryos display markedly reduced forelimb growth while the hindlimbs appear unaffected, possibly due to expression of Raldh3 in the mesonephric region, located near the base of the hindlimb bud (Niederreither et al., 2002b). Expression analysis of various genes involved in limb morphogenesis indicated that normal limb patterning was impaired in these mutants. Fgf8, which is normally evenly distributed along the AER, was upregulated in the anterior portion of the AER in Raldh2$^{-/-}$ embryos and, in extreme cases, two distal outgrowths were evident, whereas Shh, normally expressed in the ZPA, was also notably reduced and sometimes abnormally expressed in the anterior portion of the limb bud (Niederreither et al., 2002b). Further analysis of RA-rescued Raldh2 mutant embryos established that RA synthesis was required for both the initiation of the forelimb bud and the establishment of a functional AER (Mic et al., 2004). While RA synthesis by Raldh2 is absolutely required for normal limb morphogenesis, local metabolism of RA in the growing limb bud is also necessary for proper proximodistal (P–D) patterning and outgrowth. Cyp26B1 is specifically expressed in distal limb mesenchyme, but is absent from the AER and the hand plate at later stages (Fig. 3) (MacLean et al., 2001). Mice lacking functional Cyp26B1 display severely truncated limbs and only two or three digits per limb in both fore and hindlimbs. These defects resemble those resulting from excess RA (Yashiro et al., 2004). Although the mutants maintained an intact ZPA and AER, as seen by apparently unaffected expression of Shh and Fgf8, respectively, the expression of genes involved in P–D limb patterning was altered. Distal Hox genes (Hoxd10, Hoxd12, and Hoxd13) were downregulated, while proximal Meis2 expression extended abnormally into the distal limb bud. This suggests that, in the absence of RA metabolism by Cyp26B1, the distal limb bud mesenchyme adopts a more proximal identity leading to a shortening of the P–D axis (Yashiro et al., 2004).

4.4. Hindbrain

The hindbrain is transiently divided into eight distinct segments, known as rhombomeres (r), between E8 and E10.5 (reviewed in Gavalas, 2002; Santagati and Rijli, 2003). These rhombomeres each express a unique combination of genes according to their position along the A–P axis. Patterning of the hindbrain is important for the proper organization of cranial nerves, as well as for the development of craniofacial structures. As the cranial neural tube closes in the mouse, cranial NCCs undergo an epithelial to mesenchymal transition and migrate toward the BAs. These migratory NCCs will form the majority of the skeletal, muscle, and connective tissues of the face (Santagati and Rijli, 2003). Patterning and specification of the rhombomeres are thought to provide NCCs with some degree of positional information that they retain during migration and population of the BAs. However, once these cells reach their final destination, signals from the local environment are critical for facial patterning (reviewed in Trainor and Krumlauf, 2001). Numerous studies have indicated a critical role for RA signaling in the hindbrain, which is necessary for proper rhombomere patterning as well as subsequent development of BAs.

During hindbrain development, RARα is expressed in the spinal cord and caudal hindbrain up to the level of the r3/r4 boundary, while RARβ is expressed to the r6/r7 boundary (Mollard et al., 2000) RAR$\alpha\gamma$ mutants show an expansion of r3- and r4-specific genes Krox-20 and Hoxb1, respectively, and do not express more posterior rhombomere-specific genes such as

Kreisler (r5/r6-specific gene) (Wendling et al., 2001). RARαβ mutants, on the other hand, exhibit an enlarged region corresponding to r5 and r6 (Dupe et al., 1999). These effects on hindbrain patterning can be mimicked by administration of a pan-RAR antagonist, BMS493, to wild-type embryos. BMS493 treatment at E7.0 results in a phenotype resembling that of RARαγ$^{-/-}$ mice, while treatment at E8.0 results in an RARαβ mutant-like hindbrain phenotype (Wendling et al., 2001), indicating that different receptors are required at different stages of hindbrain development and that RA signaling is required to specify posterior rhombomeric identity. In addition, loss of Raldh2 dramatically affects posterior hindbrain development. Embryos lacking Raldh2 showed normally patterned and specified anterior rhombomeres, but the region from r3 to r8 was severely affected (Niederreither et al., 2000). Krox-20 was present in a broad domain in the caudal hindbrain, as were its downstream targets Hoxa2, Hoxb2, and EphA4, while Kreisler was not expressed at all in these mutants (Niederreither et al., 2000). These results suggest that RA is required for the specification of caudal rhombomere identity and that, in its absence, the region caudal to r2 is present as a broad domain of r3/r4 specified cells (Niederreither et al., 2000). Moreover, experiments in chick embryos treated with a pan-RAR antagonist at different stages and doses during hindbrain development showed that more posterior rhombomeres require increasing levels of RA for their specification (Dupe and Lumsden, 2001). Furthermore, evidence has shown that the posterior rhombomeres are exposed to RA for longer periods of time than those located more anteriorly (Sirbu et al., 2005).

It has been shown that RA can affect pattern formation in the hindbrain through the transcriptional regulation of Hox genes, a family of evolutionarily conserved transcription factors involved in specifying positional identity during development (reviewed in Krumlauf, 1994; Deschamps and van Nes, 2005). These genes are unique in that they are arranged in four separate clusters (Hoxa, Hoxb, Hoxc, and Hoxd), and their spatiotemporal expression pattern reflects their chromosomal position within the cluster such that 3′ genes are expressed early and with more anterior limits, while 5′ genes are expressed later and in more posterior regions (Deschamps and van Nes, 2005). Several studies have shown that administration of exogenous RA can alter rhombomere-specific Hox gene expression, such that anterior rhombomeres will express Hox genes normally found in more caudal rhombomeres (Conlon and Rossant, 1992; Marshall et al., 1992; Oosterveen et al., 2003). Moreover, enhancer elements containing RAREs have been discovered within the regulatory regions of Hox clusters that are necessary for establishment of proper expression domains of several genes, including Hoxa1, Hoxb1, Hoxb4, and Hoxd4 (Marshall et al., 1994; Frasch et al., 1995; Morrison et al., 1996; Dupe et al., 1997; Gould et al., 1998; Zhang et al., 2000).

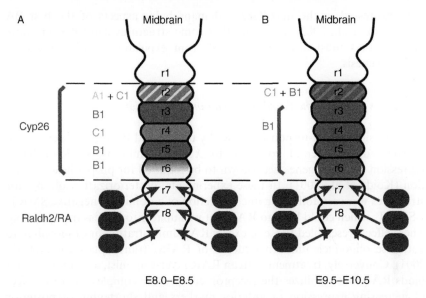

Fig. 4. Schematic diagram of Raldh2 and Cyp26 expression during hindbrain patterning. (A) At E8.5, Cyp26B1 is expressed in r3 and r5 and extends partially into r6, while Cyp26C1 is expressed in r2 and r4. Cyp26A1 is also transiently expressed in r2. RA is produced by Raldh2 in the flanking cervical somites. (B) At E9.5, Cyp26B1 is expressed from r2 to r6, and Cyp26C1 is expressed in r2. This Cyp26 expression presumably protects anterior rhombomeres from the posteriorizing effects of RA signaling.

Hindbrain development is acutely sensitive to disturbances in RA distribution. Cyp26A1 is transiently expressed in r2 at E8.5 (Fig. 4) and Cyp26A1$^{-/-}$ embryos display a partial posteriorization of this rhombomere, as evidenced by ectopic expression of r4-specific Hoxb1 in some r2 cells (Abu-Abed et al., 2001). Cyp26B1 and Cyp26C1 are also expressed in the hindbrain during segmentation. At E8.0, Cyp26B1 is expressed in r3 and r5, while Cyp26C1 is expressed in r2 and r4 (Fig. 4) (MacLean et al., 2001; Tahayato et al., 2003). This likely indicates that these anterior rhombomeres must be protected from RA signaling for proper pattern formation, although the effect of a loss of these enzymes on hindbrain development has yet to be reported.

Hindbrain rhombomeres are direct targets of RA, and this signaling must be finely controlled for proper patterning. Although migratory NCCs are not direct RA targets, the caudal BAs 3–6 and their associated branchial pouches (BPs) require RA for their formation and development of derived structures (reviewed in Mark et al., 2004). Compound mutant mice lacking RARα and RARβ, or RARα and RARγ, both show hypoplasia of BAs 3–6 and their corresponding BPs, which is also seen in mice lacking RXRα and RARα, RARβ, or RARγ (Kastner et al., 1997). Cyp26A1 and Cyp26C1 are both expressed in the maxillary and mandibular aspects of the first BA (Tahayato et al., 2003) indicating that some structures derived from more anterior BAs must also be protected from exposure to RA for proper morphogenesis.

4.5. Roles for RAR-mediated transcriptional repression

Repression of RA-responsive genes by unliganded RARs has been shown in some developmental contexts. In *Xenopus* embryos, RAR-mediated repression of target genes was shown to be required for proper head formation (Koide et al., 2001). In these experiments, posteriorization of anterior structures was seen following microinjection of a dominant negative SMRT, c-SMRT, which could bind to RAR but was unable to repress transcription. These effects resembled those of excess RA administration or of morpholino antisense oligonucleotide knockdown of RARα transcripts (Koide et al., 2001). Conversely, treatment with an RAR inverse agonist, a compound that binds RAR but stabilizes the receptor–corepressor complex, had the effect of increasing expression of anterior markers and shortening of posterior structures.

A similar requirement for RAR-mediated gene repression has been observed in mouse skeletal differentiation (Weston et al., 2002). Primary limb bud mesenchyme undergoes differentiation into chondroblasts upon removal of RA in culture, which can be seen by the appearance of cartilaginous nodules. Treatment of these cells with RAR antagonist, or introduction of a dominant negative RARα, enhanced chondroblast differentiation. This effect was lost when cells were also treated with the HDAC inhibitor trichostatin A (Weston et al., 2002). In addition, expression of the transcription factor Sox9, which is involved in chondroblast differentiation and is upregulated in the absence of RA, could be attenuated on introduction of a dominant negative NCoR mutant unable to recruit HDACs (Weston et al., 2002). These results suggest that RAR-mediated transcriptional repression of RA-responsive genes in the absence of RA may be required for some developmental programs.

4.6. Conditional RXR and RAR knockouts

Various RARs and/or RXRs have been conditionally disrupted in basal keratinocytes, NCCs, and retinal pigment epithelia (as discussed above in Section 4.2). An examination of RA-induced proliferation of keratinocytes was performed in mice with selective disruptions of RXRα, RARγ, RXRα/RARα, or RXRα/RARγ in epidermal keratinocytes (where RARβ is not expressed) (Chapellier et al., 2002). This study provided genetic and pharmacological evidence that RXRα/RARγ heterodimers within suprabasal keratinocytes are responsible for the epidermal hyperplasia that is seen following RA treatment (Fisher and Voorhees, 1996; Chapellier et al., 2002), but that RAR signaling is not directly required for the self-renewal of epidermal keratinocytes (Chapellier et al., 2002).

Disruption of NCCs within selected mutant backgrounds showed some of the major defects associated with the knockout phenotypes themselves. For example, conditional ablation of NCCs within RAR$\beta\gamma$ mutants was shown to result in various eye defects (Matt et al., 2005, and discussed earlier), and disruption of cardiac NCCs within RAR$\alpha1\beta$ mutant embryos resulted in failure of aorticopulmonary septum formation (PTA), but otherwise normal NCC fate (e.g., in the hindbrain) (Jiang et al., 2002). These experiments show that, although retinoid receptor signaling is extremely complex, selective and tissue-specific ablation of receptor subtypes and isoforms is a powerful approach for unraveling the intricacies of the stage-specific morphogenic actions of RA.

5. RA synthesis and catabolism play a dominant role in controlling RA-mediated morphogenesis

In the course of studying the function of the retinoid receptors over the past 25 years, it has become apparent that retinoid signaling requires not only the presence of the receptors themselves but also the coordinated production of RA. The discovery of nuclear receptors that transduce the retinoid signal and the systematic analyses of mice lacking various receptor isotypes and isoforms have been instrumental in defining tissues that require RA for development and that are sensitive to either RA deficiency or excess. Furthering our understanding of the role of RA in development has been the discovery of the enzymes involved in the synthesis (Raldhs) and catabolism (Cyp26s) of RA *in vivo*. During early development Raldh2, the enzyme primarily responsible for RA synthesis during this stage, and Cyp26A1 are expressed in complementary nonoverlapping domains along the extending body axis, which exquisitely parallels the domains of RA observed in mice by RARE-lacZ reporter activity (Fig. 5) (Rossant et al., 1991). Raldh2 transcripts are present in the trunk somites, while Cyp26A1 is expressed in undifferentiated tail bud tissues (Fujii et al., 1997; Niederreither et al., 1997; MacLean et al., 2001). Loss of embryonic RA synthesis by disruption of Raldh2 results in early embryonic lethality and severe malformations including hypoplasia of frontonasal process and BAs, dilated heart cavity, and severe truncation (Niederreither et al., 1999). These embryos also showed virtually no reporter transgene activity, indicating that no RA was present, except in the developing eye, where Raldh1 and Raldh3 are expressed (Li et al., 2000).

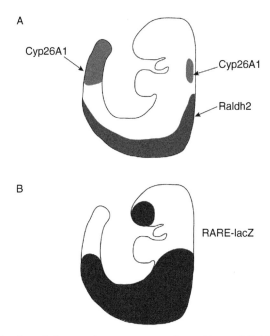

Fig. 5. Schematic view of the complementary expression patterns of Raldh2 and Cyp26A1 genes during embryonic axis extension. (A) Between E8.5 and E10.5, Cyp26A1 is specifically expressed in tail bud and cervical mesenchyme, while Raldh2 is present in trunk somites. (B) A RARE-lacZ reporter transgene shows regions in the embryo where RA is present. RA is produced in the trunk region where Raldh2 is expressed and can diffuse into surrounding tissue. It is excluded from the developing tail bud due to the presence of Cyp26A1. RA is also detected in the developing eye region due to Raldh1 and Raldh3 expression.

Loss of Cyp26A1 results in defects of the posterior embryo such as caudal truncation, fusion of hindlimbs, and malformations of the kidneys, urogenital tract, and hindgut. These defects are accompanied by altered expression of tail bud-specific signaling molecules (Fgf8, Wnt3a) and transcription factors (Brachyury, Hoxd11, Cdx4), which are also observed upon treatment of embryos with excess RA (Iulianella et al., 1999).

A principal example of the necessity for maintaining appropriate RA levels, and having the appropriate receptor mediators, is the rescue of Cyp26A1-null embryos following an ablation of RARγ (Abu-Abed et al., 2003). The original observation that RARγ-null (and to a lesser degree RARγ$^{+/-}$) embryos were resistant to caudal truncations normally caused by exogenous RA administration to pregnant dams at E9.0 (Kessel, 1992; Lohnes et al., 1993) suggested that RARγ might mediate the RA signal within the posterior embryo. Our laboratory had previously shown that the ablation of murine Cyp26A1 resulted in various caudal truncation phenotypes (Abu-Abed et al., 2001; Sakai et al., 2001), leading to the postulation that the role of Cyp26A1 might be to protect the caudal portion of the embryo from ectopic signaling by RA. The fact that RARγ ablation in the Cyp26A1 null background could rescue not only the lethality but also the posterior defects associated with Cyp26A1 disruption illustrates the importance of both the receptors and RA distribution for proper patterning and development.

Another example of the importance of maintaining appropriate RA levels for receptor-mediated signaling comes from studies performed in Raldh2-null mice. Targeted disruption of Raldh2 results in embryonic lethality at midgestation accompanied by arrested development at around E8.75 (Niederreither et al., 1999). In these studies, where the retinoid receptors are expressed normally, administration of exogenous RA during critical stages of development (E6.75–E10.25 or E6.75–E8.25) could conditionally rescue the developmental defects associated with the ablation of Raldh2 (Niederreither et al., 1999; Mic et al., 2004) and, more importantly, could trigger expression of RARE-lacZ, illustrating signaling via the retinoid receptors (Mic et al., 2002). It is clear from these examples that although the mechanisms behind retinoid receptor signaling and the intricacies of pattern formation within the developing embryo are highly complex, there are some very basic requirements that must be met for the induction of receptor-mediated transcription, and these rely on a balance between the presence of the receptors themselves and/or their signaling molecules.

6. Conclusions and perspectives

RA participates in the continuum of morphogenetic events that give rise to pattern at all stages of development. The systematic analyses of single and compound retinoid receptor mutants have clearly demonstrated that RA action through these receptors is essential for proper embryogenesis. The unexpected functional redundancy that has been observed among the RARs, as compared to the spectrum of defects observed in the compound mutants, suggests that the overlapping spatial organization of RARs (and RXRs) within the developing organism establishes a complex and intricate control platform to mediate RA signaling. The role of RA is to synchronously activate genetic algorithms within nascent cell fields, triggering the initiation of morphogenetic events that give rise to pattern (Fig. 6). It is becoming clear that this requires precisely timed and localized production of RA which is restricted in its ability to diffuse across the target cell field by the presence of catabolic boundaries. More recently, we are beginning to understand how such localized pools of RA are generated and dispersed, in a tightly controlled manner, during early stages of tissue formation. As development proceeds, the coupling of retinoic acid synthesis (source) and catabolism (sink) provides the requisite amount of RA at precise times within various

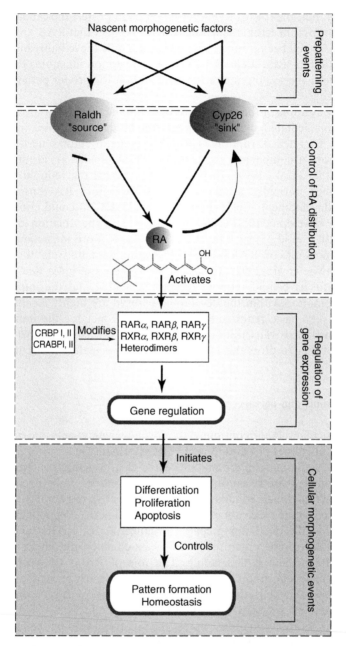

Fig. 6. Retinoid control of embryonic patterning. The RA-synthesizing (Raldh) and catabolizing (Cyp26) enzymes define the embryonic distribution of this morphogen, leading to pattern formation and morphogenesis. Although RA itself can induce Cyp26A1 expression and downregulate Raldh2 expression in some tissues, the factors that initiate and maintain the stage- and tissue-specific expression of these enzymes remain largely unknown.

tissue primordia. This triggers morphogenetic events in systems as diverse as hindbrain, limb, eyes, teeth, and tailbud, among others.

In order to understand the role of RA in developmental processes, research efforts are currently moving in several directions:

1. *Establishing morphogenetic fields:* The factors that initiate and define the expression of the Raldh and Cyp26 enzymes is unknown, however, identification of these factors will uncover the earliest determinants of tissue fate.

2. *Selective perturbation of domains of RA distribution:* Conditional deletion of genes encoding the Raldh and Cyp26 enzymes will establish the consequences of disrupting RA dispersal on patterning. In addition, using specific inhibitors to chemically ablate their function will permit precise manipulation of RA fields.

3. *Genetic and pharmacological ablation of selected receptor isoforms:* The complete genetic ablation of different receptor subtypes precludes analysis of their later subordinate roles within developing tissues and organs. Therefore, research focus is now shifting toward understanding the stage- and tissue-specific roles of these receptors by generating conditional gene disruptions and pharmacological knockout models. For example, a recent study used isotype-specific antagonist treatments to show that there were stage-specific requirements for RARα and RARβ in lung morphogenesis (Desai et al., 2006). This result illustrates that it may be necessary to express different receptor subtypes (or potentially isoforms) at different stages of development within particular organs.

4. *Establishing genetic pathways downstream of RA signaling:* Numerous genes have been identified as either direct or indirect targets of RA. How these targets fit into genetic pathways that regulate apoptosis, differentiation, and/or proliferation, as well as how these pathways are coordinately regulated during organogenesis, are important research aims requiring further investigation.

Although it is clear that the RA signal transduction system is complex, we have described a simplified view of the role of RA signaling and metabolism in morphogenesis. Inputs from many different signaling pathways must be integrated to properly orchestrate the emergence of structure and function within tissues. It is clear that RA, through its receptors, plays a central role in both initiating and integrating the signaling events that culminate in the formation of an embryo.

Acknowledgments

We apologize to those whose work has not been cited due to space constraints or oversight. The authors would like to thank Glenn MacLean for critical reading of the chapter as well as many useful discussions. Thanks also to Cory Dean for assistance with figures.

References

Abu-Abed, S., Dolle, P., Metzger, D., Beckett, B., Chambon, P., Petkovich, M. 2001. The retinoic acid-metabolizing enzyme, CYP26A1, is essential for normal hindbrain patterning, vertebral identity, and development of posterior structures. Genes Dev. 15, 226–240.

Abu-Abed, S., MacLean, G., Fraulob, V., Chambon, P., Petkovich, M., Dolle, P. 2002. Differential expression of the retinoic acid-metabolizing enzymes CYP26A1 and CYP26B1 during murine organogenesis. Mech. Dev. 110, 173–177.

Abu-Abed, S., Dolle, P., Metzger, D., Wood, C., MacLean, G., Chambon, P., Petkovich, M. 2003. Developing with lethal RA levels: Genetic ablation of Rarg can restore the viability of mice lacking Cyp26a1. Development 130, 1449–1459.

Allen, J.G., Bloxham, D.P. 1989. The pharmacology and pharmacokinetics of the retinoids. Pharmacol. Ther. 40, 1–27.

Allenby, G., Bocquel, M.T., Saunders, M., Kazmer, S., Speck, J., Rosenberger, M., Lovey, A., Kastner, P., Grippo, J.F., Chambon, P., Levin, A.A. 1993. Retinoic acid receptors and retinoid X receptors: Interactions with endogenous retinoic acids. Proc. Natl. Acad. Sci. USA 90, 30–34.

Bastien, J., Rochette-Egly, C. 2004. Nuclear retinoid receptors and the transcription of retinoid-target genes. Gene 328, 1–16.

Bastien, J., Adam-Stitah, S., Riedl, T., Egly, J.M., Chambon, P., Rochette-Egly, C. 2000. TFIIH interacts with the retinoic acid receptor gamma and phosphorylates its AF-1-activating domain through cdk7. J. Biol. Chem. 275, 21896–21904.

Beckett-Jones, B., Petkovich, M. 1996. Targeting transcription through nuclear receptors. Curr. Pharm. Des. 2, 155–168.

Blomhoff, R., Green, M.H., Berg, T., Norum, K.R. 1990. Transport and storage of vitamin A. Science 250, 399–404.

Brondani, V., Klimkait, T., Egly, J.M., Hamy, F. 2002. Promoter of FGF8 reveals a unique regulation by unliganded RARalpha. J. Mol. Biol. 319, 715–728.

Bruck, N., Bastien, J., Bour, G., Tarrade, A., Plassat, J.L., Bauer, A., Adam-Stitah, S., Rochette-Egly, C. 2005. Phosphorylation of the retinoid x receptor at the omega loop, modulates the expression of retinoic-acid-target genes with a promoter context specificity. Cell. Signal. 17, 1229–1239.

Buckingham, M., Meilhac, S., Zaffran, S. 2005. Building the mammalian heart from two sources of myocardial cells. Nat. Rev. Genet. 6, 826–835.

Budhu, A.S., Noy, N. 2002. Direct channeling of retinoic acid between cellular retinoic acid-binding protein II and retinoic acid receptor sensitizes mammary carcinoma cells to retinoic acid-induced growth arrest. Mol. Cell. Biol. 22, 2632–2641.

Chambon, P. 1994. The retinoid signaling pathway: Molecular and genetic analyses. Semin. Cell Biol. 5, 115–125.

Chambon, P. 1996. A decade of molecular biology of retinoic acid receptors. FASEB J. 10, 940–954.

Chapellier, B., Mark, M., Messaddeq, N., Calleja, C., Warot, X., Brocard, J., Gerard, C., Li, M., Metzger, D., Ghyselinck, N.B., Chambon, P. 2002. Physiological and retinoid-induced proliferations of epidermis basal keratinocytes are differently controlled. EMBO J. 21, 3402–3413.

Chen, J.D., Evans, R.M. 1995. A transcriptional co-repressor that interacts with nuclear hormone receptors. Nature 377, 454–457.

Chen, J.Y., Clifford, J., Zusi, C., Starrett, J., Tortolani, D., Ostrowski, J., Reczek, P.R., Chambon, P., Gronemeyer, H. 1996. Two distinct actions of retinoid-receptor ligands. Nature 382, 819–822.

Cohlan, S.Q. 1954. Congenital anomalies in the rat produced by excessive intake of vitamin A during pregnancy. Pediatrics 13, 556–567.

Collins, M.D., Mao, G.E. 1999. Teratology of retinoids. Annu. Rev. Pharmacol. Toxicol. 39, 399–430.

Conlon, R.A., Rossant, J. 1992. Exogenous retinoic acid rapidly induces anterior ectopic expression of murine Hox-2 genes *in vivo*. Development 116, 357–368.

Cvekl, A., Tamm, E.R. 2004. Anterior eye development and ocular mesenchyme: New insights from mouse models and human diseases. Bioessays 26, 374–386.

de Bruijn, D.R., Oerlemans, F., Hendriks, W., Baats, E., Ploemacher, R., Wieringa, B., Geurts van Kessel, A. 1994. Normal development, growth and reproduction in cellular retinoic acid binding protein-I (CRABPI) null mutant mice. Differentiation 58, 141–148.

Delmotte, M.H., Tahayato, A., Formstecher, P., Lefebvre, P. 1999. Serine 157, a retinoic acid receptor alpha residue phosphorylated by protein kinase C *in vitro*, is involved in RXR. RARalpha heterodimerization and transcriptional activity. J. Biol. Chem. 274, 38225–38231.

Dennis, A.P., O'Malley, B.W. 2005. Rush hour at the promoter: How the ubiquitin-proteasome pathway polices the traffic flow of nuclear receptor-dependent transcription. J. Steroid. Biochem. Mol. Biol. 93, 139–151.

Depoix, C., Delmotte, M.H., Formstecher, P., Lefebvre, P. 2001. Control of retinoic acid receptor heterodimerization by ligand-induced structural transitions. A novel mechanism of action for retinoid antagonists. J. Biol. Chem. 276, 9452–9459.

Desai, T.J., Chen, F., Lu, J., Qian, J., Niederreither, K., Dolle, P., Chambon, P., Cardoso, W.V. 2006. Distinct roles for retinoic acid receptors alpha and beta in early lung morphogenesis. Dev. Biol. 291, 12–24.

Deschamps, J., van Nes, J. 2005. Developmental regulation of the Hox genes during axial morphogenesis in the mouse. Development 132, 2931–2942.

Dhananjayan, S.C., Ismail, A., Nawaz, Z. 2005. Ubiquitin and control of transcription. Essays Biochem. 41, 69–80.

Dilworth, F.J., Chambon, P. 2001. Nuclear receptors coordinate the activities of chromatin remodeling complexes and coactivators to facilitate initiation of transcription. Oncogene 203047–203054.

Dolle, P., Ruberte, E., Kastner, P., Petkovich, M., Stoner, C.M., Gudas, L.J., Chambon, P. 1989. Differential expression of genes encoding alpha, beta and gamma retinoic acid receptors and CRABP in the developing limbs of the mouse. Nature 342, 702–705.

Dolle, P., Ruberte, E., Leroy, P., Morriss-Kay, G., Chambon, P. 1990. Retinoic acid receptors and cellular retinoid binding proteins. I. A systematic study of their differential pattern of transcription during mouse organogenesis. Development 110, 1133–1151.

Dolle, P., Fraulob, V., Kastner, P., Chambon, P. 1994. Developmental expression of murine retinoid X receptor (RXR) genes. Mech. Dev. 45, 91–104.

Duester, G. 1996. Involvement of alcohol dehydrogenase, short-chain dehydrogenase/reductase, aldehyde dehydrogenase, and cytochrome P450 in the control of retinoid signaling by activation of retinoic acid synthesis. Biochemistry 35, 12221–12227.

Dupe, V., Lumsden, A. 2001. Hindbrain patterning involves graded responses to retinoic acid signalling. Development 128, 2199–2208.

Dupe, V., Davenne, M., Brocard, J., Dolle, P., Mark, M., Dierich, A., Chambon, P., Rijli, F.M. 1997. *In vivo* functional analysis of the Hoxa-1 3′ retinoic acid response element (3′RARE). Development 124, 399–410.

Dupe, V., Ghyselinck, N.B., Wendling, O., Chambon, P., Mark, M. 1999. Key roles of retinoic acid receptors alpha and beta in the patterning of the caudal hindbrain, pharyngeal arches and otocyst in the mouse. Development 126, 5051–5059.

Dupe, V., Matt, N., Garnier, J.M., Chambon, P., Mark, M., Ghyselinck, N.B. 2003. A newborn lethal defect due to inactivation of retinaldehyde dehydrogenase type 3 is prevented by maternal retinoic acid treatment. Proc. Natl. Acad. Sci. USA 100, 14036–14041.

Emerick, R.J., Zile, M., DeLuca, H.F. 1967. Formation of retinoic acid from retinol in the rat. Biochem. J. 102, 606–611.

Fan, X., Molotkov, A., Manabe, S., Donmoyer, C.M., Deltour, L., Foglio, M.H., Cuenca, A.E., Blaner, W.S., Lipton, S.A., Duester, G. 2003. Targeted disruption of Aldh1a1 (Raldh1) provides evidence for a complex mechanism of retinoic acid synthesis in the developing retina. Mol. Cell. Biol. 23, 4637–4648.

Fernandes, I., Bastien, Y., Wai, T., Nygard, K., Lin, R., Cormier, O., Lee, H.S., Eng, F., Bertos, N.R., Pelletier, N., Mader, S., Han, V.K., et al. 2003. Ligand-dependent nuclear receptor corepressor LCoR functions by histone deacetylase-dependent and -independent mechanisms. Mol. Cell 11, 139–150.

Fini, M.E., Strissel, K.J., West-Mays, J.A. 1997. Perspectives on eye development. Dev. Genet. 20, 175–185.

Fischer, A.J., Wallman, J., Mertz, J.R., Stell, W.K. 1999. Localization of retinoid binding proteins, retinoid receptors, and retinaldehyde dehydrogenase in the chick eye. J. Neuro-cytol. 28, 597–609.

Fisher, G.J., Voorhees, J.J. 1996. Molecular mechanisms of retinoid actions in skin. FASEB J. 10, 1002–1013.

Frasch, M., Chen, X., Lufkin, T. 1995. Evolutionary-conserved enhancers direct region-specific expression of the murine Hoxa-1 and Hoxa-2 loci in both mice and *Drosophila*. Development 121, 957–974.

Frolik, C.A., Roberts, A.B., Tavela, T.E., Roller, P.P., Newton, D.L., Sporn, M.B. 1979. Isolation and identification of 4-hydroxy- and 4-oxoretinoic acid. *In vitro* metabolites of all-trans-retinoic acid in hamster trachea and liver. Biochemistry 18, 2092–2097.

Fujii, H., Sato, T., Kaneko, S., Gotoh, O., Fujii-Kuriyama, Y., Osawa, K., Kato, S., Hamada, H. 1997. Metabolic inactivation of retinoic acid by a novel P450 differentially expressed in developing mouse embryos. EMBO J. 16, 4163–4173.

Gavalas, A. 2002. Arranging the hindbrain. Trends Neurosci. 25, 61–64.

Gehin, M., Vivat, V., Wurtz, J.M., Losson, R., Chambon, P., Moras, D., Gronemeyer, H. 1999. Structural basis for engineering of retinoic acid receptor isotype-selective agonists and antagonists. Chem. Biol. 6, 519–529.

Gerlitz, G., Jagus, R., Elroy-Stein, O. 2002. Phosphorylation of initiation factor-2 alpha is required for activation of internal translation initiation during cell differentiation. Eur. J. Biochem. 269, 2810–2819.

Germain, P., Iyer, J., Zechel, C., Gronemeyer, H. 2002. Co-regulator recruitment and the mechanism of retinoic acid receptor synergy. Nature 415, 187–192.

Germain, P., Kammerer, S., Perez, E., Peluso-Iltis, C., Tortolani, D., Zusi, F.C., Starrett, J., Lapointe, P., Daris, J.P., Marinier, A., de Lera, A.R., Rochel, N., et al. 2004. Rational design of RAR-selective ligands revealed by RARbeta crystal stucture. EMBO Rep. 5, 877–882.

Ghyselinck, N.B., Dupe, V., Dierich, A., Messaddeq, N., Garnier, J.M., Rochette-Egly, C., Chambon, P., Mark, M. 1997. Role of the retinoic acid receptor beta (RARbeta) during mouse development. Int. J. Dev. Biol. 41, 425–447.

Ghyselinck, N.B., Wendling, O., Messaddeq, N., Dierich, A., Lampron, C., Decimo, D., Viville, S., Chambon, P., Mark, M. 1998. Contribution of retinoic acid receptor beta isoforms to the formation of the conotruncal septum of the embryonic heart. Dev. Biol. 198, 303–318.

Gianni, M., Bauer, A., Garattini, E., Chambon, P., Rochette-Egly, C. 2002. Phosphorylation by p38MAPK and recruitment of SUG-1 are required for RA-induced RAR gamma degrada-tion and transactivation. EMBO J. 21, 3760–3769.

Gianni, M., Tarrade, A., Nigro, E.A., Garattini, E., Rochette-Egly, C. 2003. The AF-1 and AF-2 domains of RAR gamma 2 and RXR alpha cooperate for triggering the transactiva-tion and the degradation of RAR gamma 2/RXR alpha heterodimers. J. Biol. Chem. 278, 34458–34466.

Gianni, M., Parrella, E., Raska, I., Gaillard, E., Nigro, E.A., Gaudon, C., Garattini, E., Rochette-Egly, C. 2006. P38MAPK-dependent phosphorylation and degradation of SRC-3/AIB1 and RARalpha-mediated transcription. EMBO J. 25, 739–751.

Giguere, V., Ong, E.S., Segui, P., Evans, R.M. 1987. Identification of a receptor for the morphogen retinoic acid. Nature 330, 624–629.

Gorry, P., Lufkin, T., Dierich, A., Rochette-Egly, C., Decimo, D., Dolle, P., Mark, M., Durand, B., Chambon, P. 1994. The cellular retinoic acid binding protein I is dispensable. Proc. Natl. Acad. Sci. USA 91, 9032–9036.

Gould, A., Itasaki, N., Krumlauf, R. 1998. Initiation of rhombomeric Hoxb4 expression requires induction by somites and a retinoid pathway. Neuron 21, 39–51.

Grondona, J.M., Kastner, P., Gansmuller, A., Decimo, D., Chambon, P., Mark, M. 1996. Retinal dysplasia and degeneration in RARbeta2/RARgamma2 compound mutant mice. Development 122, 2173–2188.

Gronemeyer, H., Miturski, R. 2001. Molecular mechanisms of retinoid action. Cell. Mol. Biol. Lett. 6, 3–52.

Gronemeyer, H., Gustafsson, J.A., Laudet, V. 2004. Principles for modulation of the nuclear receptor superfamily. Nat. Rev. Drug. Discov. 3, 950–964.

Harfe, B.D., Scherz, P.J., Nissim, S., Tian, H., McMahon, A.P., Tabin, C.J. 2004. Evidence for an expansion-based temporal Shh gradient in specifying vertebrate digit identities. Cell 118, 517–528.

Heery, D.M., Kalkhoven, E., Hoare, S., Parker, M.G. 1997. A signature motif in transcriptional co-activators mediates binding to nuclear receptors. Nature 387, 733–736.

Hong, S.H., Privalsky, M.L. 2000. The SMRT corepressor is regulated by a MEK-1 kinase pathway: Inhibition of corepressor function is associated with SMRT phosphorylation and nuclear export. Mol. Cell. Biol. 20, 6612–6625.

Horlein, A.J., Naar, A.M., Heinzel, T., Torchia, J., Gloss, B., Kurokawa, R., Ryan, A., Kamei, Y., Soderstrom, M., Glass, C.K., Rosenfeld, M.G. 1995. Ligand-independent repression by the thyroid hormone receptor mediated by a nuclear receptor co-repressor. Nature 377, 397–404.

Iulianella, A., Beckett, B., Petkovich, M., Lohnes, D. 1999. A molecular basis for retinoic acid-induced axial truncation. Dev. Biol. 205, 33–48.

Jiang, X., Choudhary, B., Merki, E., Chien, K.R., Maxson, R.E., Sucov, H.M. 2002. Normal fate and altered function of the cardiac neural crest cell lineage in retinoic acid receptor mutant embryos. Mech. Dev. 117, 115–122.

Kastner, P., Grondona, J.M., Mark, M., Gansmuller, A., LeMeur, M., Decimo, D., Vonesch, J.L., Dolle, P., Chambon, P. 1994. Genetic analysis of RXR alpha developmental function: Convergence of RXR and RAR signaling pathways in heart and eye morphogenesis. Cell 78, 987–1003.

Kastner, P., Mark, M., Chambon, P. 1995. Nonsteroid nuclear receptors: What are genetic studies telling us about their role in real life? Cell 83, 859–869.

Kastner, P., Mark, M., Leid, M., Gansmuller, A., Chin, W., Grondona, J.M., Decimo, D., Krezel, W., Dierich, A., Chambon, P. 1996. Abnormal spermatogenesis in RXR beta mutant mice. Genes Dev. 10, 80–92.

Kastner, P., Mark, M., Ghyselinck, N., Krezel, W., Dupe, V., Grondona, J.M., Chambon, P. 1997. Genetic evidence that the retinoid signal is transduced by heterodimeric RXR/RAR functional units during mouse development. Development 124, 313–326.

Kawada, T., Kamei, Y., Fujita, A., Hida, Y., Takahashi, N., Sugimoto, E., Fushiki, T. 2000. Carotenoids and retinoids as suppressors on adipocyte differentiation via nuclear receptors. Biofactors 13, 103–109.

Keegan, B.R., Feldman, J.L., Begemann, G., Ingham, P.W., Yelon, D. 2005. Retinoic acid signaling restricts the cardiac progenitor pool. Science 307, 247–249.

Kelly, R.G., Buckingham, M.E. 2002. The anterior heart-forming field: Voyage to the arterial pole of the heart. Trends Genet. 18, 210–216.

Kessel, M. 1992. Respecification of vertebral identities by retinoic acid. Development 115, 487–501.

Klaholz, B.P., Renaud, J.P., Mitschler, A., Zusi, C., Chambon, P., Gronemeyer, H., Moras, D. 1998. Conformational adaptation of agonists to the human nuclear receptor RAR gamma. Nat. Struct. Biol. 5, 199–202.

Kochhar, D.M. 1973. Limb development in mouse embryos. I. Analysis of teratogenic effects of retinoic acid. Teratology 7, 289–298.

Koide, T., Downes, M., Chandraratna, R.A., Blumberg, B., Umesono, K. 2001. Active repression of RAR signaling is required for head formation. Genes Dev. 15, 2111–2121.

Krezel, W., Dupe, V., Mark, M., Dierich, A., Kastner, P., Chambon, P. 1996. RXR gamma null mice are apparently normal and compound RXR alpha +/−/RXR beta −/−/RXR gamma −/− mutant mice are viable. Proc. Natl. Acad. Sci. USA 93, 9010–9014.

Krumlauf, R. 1994. Hox genes in vertebrate development. Cell 78, 191–201.

Kurokawa, R., DiRenzo, J., Boehm, M., Sugarman, J., Gloss, B., Rosenfeld, M.G., Heyman, R.A., Glass, C.K. 1994. Regulation of retinoid signalling by receptor polarity and allosteric control of ligand binding. Nature 371, 528–531.

Kwon, H., Imbalzano, A.N., Khavari, P.A., Kingston, R.E., Green, M.R. 1994. Nucleosome disruption and enhancement of activator binding by a human SWI/SNF complex. Nature 370, 477–481.

Lammer, E.J., Chen, D.T., Hoar, R.M., Agnish, N.D., Benke, P.J., Braun, J.T., Curry, C.J., Fernhoff, P.M., Grix, A.W., Jr., Lott, I.T., Richard, J.M., Sun, S.C. 1985. Retinoic acid embryopathy. N. Engl. J. Med. 313, 837–841.

Lampron, C., Rochette-Egly, C., Gorry, P., Dolle, P., Mark, M., Lufkin, T., LeMeur, M., Chambon, P. 1995. Mice deficient in cellular retinoic acid binding protein II (CRABPII) or in both CRABPI and CRABPII are essentially normal. Development 121, 539–548.

Leid, M., Kastner, P., Chambon, P. 1992a. Multiplicity generates diversity in the retinoic acid signalling pathways. Trends Biochem. Sci. 17, 427–433.

Leid, M., Kastner, P., Lyons, R., Nakshatri, H., Saunders, M., Zacharewski, T., Chen, J.Y., Staub, A., Garnier, J.M., Mader, S., Chambon, P. 1992b. Purification, cloning, and RXR identity of the HeLa cell factor with which RAR or TR heterodimerizes to bind target sequences efficiently. Cell 68, 377–395.

Li, E., Sucov, H.M., Lee, K.F., Evans, R.M., Jaenisch, R. 1993. Normal development and growth of mice carrying a targeted disruption of the alpha 1 retinoic acid receptor gene. Proc. Natl. Acad. Sci. USA 90, 1590–1594.

Li, H., Wagner, E., McCaffery, P., Smith, D., Andreadis, A., Drager, U.C. 2000. A retinoic acid synthesizing enzyme in ventral retina and telencephalon of the embryonic mouse. Mech. Dev. 95, 283–289.

Lin, M., Zhang, M., Abraham, M., Smith, S.M., Napoli, J.L. 2003. Mouse retinal dehydrogenase 4 (RALDH4), molecular cloning, cellular expression, and activity in 9-cis-retinoic acid biosynthesis in intact cells. J. Biol. Chem. 278, 9856–9861.

Lohnes, D., Kastner, P., Dierich, A., Mark, M., LeMeur, M., Chambon, P. 1993. Function of retinoic acid receptor gamma in the mouse. Cell 73, 643–658.

Lohnes, D., Mark, M., Mendelsohn, C., Dolle, P., Dierich, A., Gorry, P., Gansmuller, A., Chambon, P. 1994. Function of the retinoic acid receptors (RARs) during development (I). Craniofacial and skeletal abnormalities in RAR double mutants. Development 120, 2723–2748.

Lohnes, D., Mark, M., Mendelsohn, C., Dolle, P., Decimo, D., LeMeur, M., Dierich, A., Gorry, P., Chambon, P. 1995. Developmental roles of the retinoic acid receptors. J. Steroid Biochem. Mol. Biol. 53, 475–486.

Lufkin, T., Lohnes, D., Mark, M., Dierich, A., Gorry, P., Gaub, M.P., LeMeur, M., Chambon, P. 1993. High postnatal lethality and testis degeneration in retinoic acid receptor alpha mutant mice. Proc. Natl. Acad. Sci. USA 90, 7225–7229.

Luo, J., Sucov, H.M., Bader, J.A., Evans, R.M., Giguere, V. 1996. Compound mutants for retinoic acid receptor (RAR) beta and RAR alpha 1 reveal developmental functions for multiple RAR beta isoforms. Mech. Dev. 55, 33–44.

MacLean, G., Abu-Abed, S., Dolle, P., Tahayato, A., Chambon, P., Petkovich, M. 2001. Cloning of a novel retinoic-acid metabolizing cytochrome P450, Cyp26B1, and comparative expression analysis with Cyp26A1 during early murine development. Mech. Dev. 107, 195–201.

Maden, M. 1998. The role of retinoids in developmental mechanisms in embryos. Subcell. Biochem. 30, 81–111.

Maden, M. 2002. Retinoid signalling in the development of the central nervous system. Nat. Rev. Neurosci. 3, 843–853.

Mangelsdorf, D.J., Evans, R.M. 1995. The RXR heterodimers and orphan receptors. Cell 83, 841–850.

Mangelsdorf, D.J., Ong, E.S., Dyck, J.A., Evans, R.M. 1990. Nuclear receptor that identifies a novel retinoic acid response pathway. Nature 345, 224–229.

Mark, M., Ghyselinck, N.B., Chambon, P. 2004. Retinoic acid signalling in the development of branchial arches. Curr. Opin. Genet. Dev. 14, 591–598.

Mark, M., Ghyselinck, N.B., Chambon, P. 2006. Function of retinoid nuclear receptors: Lessons from genetic and pharmacological dissections of the retinoic acid signaling pathway during mouse embryogenesis. Annu. Rev. Pharmacol. Toxicol. 46, 451–480.

Marshall, H., Nonchev, S., Sham, M.H., Muchamore, I., Lumsden, A., Krumlauf, R. 1992. Retinoic acid alters hindbrain Hox code and induces transformation of rhombomeres 2/3 into a 4/5 identity [see comments]. Nature 360, 737–741.

Marshall, H., Studer, M., Popperl, H., Aparicio, S., Kuroiwa, A., Brenner, S., Krumlauf, R. 1994. A conserved retinoic acid response element required for early expression of the homeobox gene Hoxb-1. Nature 370, 567–571.

Matt, N., Dupe, V., Garnier, J.M., Dennefeld, C., Chambon, P., Mark, M., Ghyselinck, N.B. 2005. Retinoic acid-dependent eye morphogenesis is orchestrated by neural crest cells. Development 132, 4789–4800.

McKenna, N.J., O'Malley, B.W. 2002. Combinatorial control of gene expression by nuclear receptors and coregulators. Cell 108, 465–474.

Mendelsohn, C., Lohnes, D., Decimo, D., Lufkin, T., LeMeur, M., Chambon, P., Mark, M. 1994a. Function of the retinoic acid receptors (RARs) during development (II). Multiple abnormalities at various stages of organogenesis in RAR double mutants. Development 120, 2749–2771.

Mendelsohn, C., Mark, M., Dolle, P., Dierich, A., Gaub, M.P., Krust, A., Lampron, C., Chambon, P. 1994b. Retinoic acid receptor beta 2 (RAR beta 2) null mutant mice appear normal. Dev. Biol. 166, 246–258.

Metivier, R., Penot, G., Hubner, M.R., Reid, G., Brand, H., Kos, M., Gannon, F. 2003. Estrogen receptor-alpha directs ordered, cyclical, and combinatorial recruitment of cofactors on a natural target promoter. Cell 115, 751–763.

Metivier, R., Reid, G., Gannon, F. 2006. Transcription in four dimensions: Nuclear receptor-directed initiation of gene expression. EMBO Rep. 7, 161–167.

Mic, F.A., Duester, G. 2003. Patterning of forelimb bud myogenic precursor cells requires retinoic acid signaling initiated by Raldh2. Dev. Biol. 264, 191–201.

Mic, F.A., Haselbeck, R.J., Cuenca, A.E., Duester, G. 2002. Novel retinoic acid generating activities in the neural tube and heart identified by conditional rescue of Raldh2 null mutant mice. Development 129, 2271–2282.

Mic, F.A., Sirbu, I.O., Duester, G. 2004. Retinoic acid synthesis controlled by Raldh2 is required early for limb bud initiation and then later as a proximodistal signal during apical ectodermal ridge formation. J. Biol. Chem. 279, 26698–26706.

Minucci, S., Leid, M., Toyama, R., Saint-Jeannet, J.P., Peterson, V.J., Horn, V., Ishmael, J.E., Bhattacharyya, N., Dey, A., Dawid, I.B., Ozato, K. 1997. Retinoid X receptor (RXR) within the RXR-retinoic acid receptor heterodimer binds its ligand and enhances retinoid-dependent gene expression. Mol. Cell. Biol. 17, 644–655.

Mollard, R., Viville, S., Ward, S.J., Decimo, D., Chambon, P., Dolle, P. 2000. Tissue-specific expression of retinoic acid receptor isoform transcripts in the mouse embryo. Mech. Dev. 94, 223–232.

Mori, M., Ghyselinck, N.B., Chambon, P., Mark, M. 2001. Systematic immunolocalization of retinoid receptors in developing and adult mouse eyes. Invest. Ophthalmol. Vis. Sci. 42, 1312–1318.

Mori, M., Metzger, D., Picaud, S., Hindelang, C., Simonutti, M., Sahel, J., Chambon, P., Mark, M. 2004. Retinal dystrophy resulting from ablation of RXR alpha in the mouse retinal pigment epithelium. Am. J. Pathol. 164, 701–710.

Morrison, A., Moroni, M.C., Ariza-McNaughton, L., Krumlauf, R., Mavilio, F. 1996. *In vitro* and transgenic analysis of a human HOXD4 retinoid-responsive enhancer. Development 122, 1895–1907.

Morriss-Kay, G.M., Ward, S.J. 1999. Retinoids and mammalian development. Int. Rev. Cytol. 188, 73–131.

Moss, J.B., Xavier-Neto, J., Shapiro, M.D., Nayeem, S.M., McCaffery, P., Drager, U.C., Rosenthal, N. 1998. Dynamic patterns of retinoic acid synthesis and response in the developing mammalian heart. Dev. Biol. 199, 55–71.

Nagpal, S., Saunders, M., Kastner, P., Durand, B., Nakshatri, H., Chambon, P. 1992. Promoter context- and response element-dependent specificity of the transcriptional activation and modulating functions of retinoic acid receptors. Cell 70, 1007–1019.

Nagpal, S., Friant, S., Nakshatri, H., Chambon, P. 1993. RARs and RXRs: Evidence for two autonomous transactivation functions (AF-1 and AF-2) and heterodimerization *in vivo*. EMBO J. 12, 2349–2360.

Niederreither, K., McCaffery, P., Drager, U.C., Chambon, P., Dolle, P. 1997. Restricted expression and retinoic acid-induced downregulation of the retinaldehyde dehydrogenase type 2 (RALDH-2) gene during mouse development. Mech. Dev. 62, 67–78.

Niederreither, K., Subbarayan, V., Dolle, P., Chambon, P. 1999. Embryonic retinoic acid synthesis is essential for early mouse post-implantation development [see comments]. Nat. Genet. 21, 444–448.

Niederreither, K., Vermot, J., Schuhbaur, B., Chambon, P., Dolle, P. 2000. Retinoic acid synthesis and hindbrain patterning in the mouse embryo. Development 127, 75–85.

Niederreither, K., Vermot, J., Messaddeq, N., Schuhbaur, B., Chambon, P., Dolle, P. 2001. Embryonic retinoic acid synthesis is essential for heart morphogenesis in the mouse. Development 128, 1019–1031.

Niederreither, K., Fraulob, V., Garnier, J.M., Chambon, P., Dolle, P. 2002a. Differential expression of retinoic acid-synthesizing (RALDH) enzymes during fetal development and organ differentiation in the mouse. Mech. Dev. 110, 165–171.

Niederreither, K., Vermot, J., Schuhbaur, B., Chambon, P., Dolle, P. 2002b. Embryonic retinoic acid synthesis is required for forelimb growth and anteroposterior patterning in the mouse. Development 129, 3563–3574.

Niswander, L. 2003. Pattern formation: Old models out on a limb. Nat. Rev. Genet. 4, 133–143.

Oosterveen, T., van Vliet, P., Deschamps, J., Meijlink, F. 2003. The direct context of a hox retinoic acid response element is crucial for its activity. J. Biol. Chem. 278, 24103–24107.

Penniston, K.L., Tanumihardjo, S.A. 2006. The acute and chronic toxic effects of vitamin A. Am. J. Clin. Nutr. 83, 191–201.

Petkovich, M., Brand, N.J., Krust, A., Chambon, P. 1987. A human retinoic acid receptor which belongs to the family of nuclear receptors. Nature 330, 444–450.

Ramirez-Solis, R., Zheng, H., Whiting, J., Krumlauf, R., Bradley, A. 1993. Hoxb-4 (Hox-2.6) mutant mice show homeotic transformation of a cervical vertebra and defects in the closure of the sternal rudiments. Cell 73, 279–294.

Renaud, J.P., Moras, D. 2000. Structural studies on nuclear receptors. Cell. Mol. Life Sci. 57, 1748–1769.

Renaud, J.P., Rochel, N., Ruff, M., Vivat, V., Chambon, P., Gronemeyer, H., Moras, D. 1995. Crystal structure of the RAR-gamma ligand-binding domain bound to all-trans retinoic acid. Nature 378, 681–689.

Resche-Rigon, M., Gronemeyer, H. 1998. Therapeutic potential of selective modulators of nuclear receptor action. Curr. Opin. Chem. Biol. 2, 501–507.

Rochette-Egly, C., Adam, S., Rossignol, M., Egly, J.M., Chambon, P. 1997. Stimulation of RAR alpha activation function AF-1 through binding to the general transcription factor TFIIH and phosphorylation by CDK7. Cell 90, 97–107.

Rossant, J., Zirngibl, R., Cado, D., Shago, M., Giguere, V. 1991. Expression of a retinoic acid response element-hsplacZ transgene defines specific domains of transcriptional activity during mouse embryogenesis. Genes Dev. 5, 1333–1344.

Rowan, B.G., Weigel, N.L., O'Malley, B.W. 2000. Phosphorylation of steroid receptor coactivator-1. Identification of the phosphorylation sites and phosphorylation through the mitogen-activated protein kinase pathway. J. Biol. Chem. 275, 4475–4483.

Ruberte, E., Dolle, P., Chambon, P., Morriss-Kay, G. 1991. Retinoic acid receptors and cellular retinoid binding proteins. II. Their differential pattern of transcription during early morphogenesis in mouse embryos. Development 111, 45–60.

Sakai, Y., Meno, C., Fujii, H., Nishino, J., Shiratori, H., Saijoh, Y., Rossant, J., Hamada, H. 2001. The retinoic acid-inactivating enzyme CYP26 is essential for establishing an uneven distribution of retinoic acid along the anterioposterior axis within the mouse embryo. Genes Dev. 15, 213–225.

Sakai, Y., Luo, T., McCaffery, P., Hamada, H., Drager, U.C. 2004. CYP26A1 and CYP26C1 cooperate in degrading retinoic acid within the equatorial retina during later eye development. Dev. Biol. 276, 143–157.

Santagati, F., Rijli, F.M. 2003. Cranial neural crest and the building of the vertebrate head. Nat. Rev. Neurosci. 4, 806–818.

Sirbu, I.O., Gresh, L., Barra, J., Duester, G. 2005. Shifting boundaries of retinoic acid activity control hindbrain segmental gene expression. Development 132, 2611–2622.

Srinivas, H., Xia, D., Moore, N.L., Uray, I.P., Kim, H., Ma, L., Weigel, N.L., Brown, P.H., Kurie, J.M. 2006. Akt phosphorylates and suppresses the transactivation of retinoic acid receptor alpha. Biochem. J. 395, 653–662.

Subbarayan, V., Kastner, P., Mark, M., Dierich, A., Gorry, P., Chambon, P. 1997. Limited specificity and large overlap of the functions of the mouse RAR gamma 1 and RAR gamma 2 isoforms. Mech. Dev. 66, 131–142.

Sucov, H.M., Dyson, E., Gumeringer, C.L., Price, J., Chien, K.R., Evans, R.M. 1994. RXR alpha mutant mice establish a genetic basis for vitamin A signaling in heart morphogenesis. Genes Dev. 8, 1007–1018.

Tahayato, A., Dolle, P., Petkovich, M. 2003. Cyp26C1 encodes a novel retinoic acid-metabolizing enzyme expressed in the hindbrain, inner ear, first branchial arch and tooth buds during murine development. Gene Expr. Patterns 3, 449–454.

Taimi, M., Helvig, C., Wisniewski, J., Ramshaw, H., White, J., Amad, M., Korczak, B., Petkovich, M. 2004. A novel human cytochrome P450, CYP26C1, involved in metabolism of 9-cis and all-trans isomers of retinoic acid. J. Biol. Chem. 279, 77–85.

Teelmann, K. 1989. Retinoids: Toxicology and teratogenicity to date. Pharmacol. Ther. 40, 29–43.

Tickle, C., Alberts, B., Wolpert, L., Lee, J. 1982. Local application of retinoic acid to the limb bond mimics the action of the polarizing region. Nature 296, 564–566.

Trainor, P.A., Krumlauf, R. 2001. Hox genes, neural crest cells and branchial arch patterning. Curr. Opin. Cell Biol. 13, 698–705.

Wagner, E., McCaffery, P., Drager, U.C. 2000. Retinoic acid in the formation of the dorsoventral retina and its central projections. Dev. Biol. 222, 460–470.

Wei, L.N. 2003. Retinoid receptors and their coregulators. Annu. Rev. Pharmacol. Toxicol. 43, 47–72.

Wendling, O., Chambon, P., Mark, M. 1999. Retinoid X receptors are essential for early mouse development and placentogenesis. Proc. Natl. Acad. Sci. USA 96, 547–551.

Wendling, O., Ghyselinck, N.B., Chambon, P., Mark, M. 2001. Roles of retinoic acid receptors in early embryonic morphogenesis and hindbrain patterning. Development 128, 2031–2038.

Weston, A.D., Chandraratna, R.A., Torchia, J., Underhill, T.M. 2002. Requirement for RAR-mediated gene repression in skeletal progenitor differentiation. J. Cell Biol. 158, 39–51.

White, J.A., Beckett-Jones, B., Guo, Y.D., Dilworth, F.J., Bonasoro, J., Jones, G., Petkovich, M. 1997. cDNA cloning of human retinoic acid-metabolizing enzyme (hP450RAI) identifies a novel family of cytochromes P450. J. Biol. Chem. 272, 18538–18541.

White, J.A., Ramshaw, H., Taimi, M., Stangle, W., Zhang, A., Everingham, S., Creighton, S., Tam, S.P., Jones, G., Petkovich, M. 2000. Identification of the human cytochrome P450, P450RAI-2, which is predominantly expressed in the adult cerebellum and is responsible for all-trans-retinoic acid metabolism. Proc. Natl. Acad. Sci. USA 97, 6403–6408.

White, J.C., Shankar, V.N., Highland, M., Epstein, M.L., DeLuca, H.F., Clagett-Dame, M. 1998. Defects in embryonic hindbrain development and fetal resorption resulting from vitamin A deficiency in the rat are prevented by feeding pharmacological levels of all-trans-retinoic acid. Proc. Natl. Acad. Sci. USA 95, 13459–13464.

Wilson, J.G., Warkany, J. 1950. Congenital anomalies of heart and great vessels in offspring of vitamin A-deficient rats. Am. J. Dis. Child. 79, 963.

Wilson, J.G., Roth, C.B., Warkany, J. 1953. An analysis of the syndrome of malformations induced by maternal vitamin A deficiency. Effects of restoration of vitamin A at various times during gestation. Am. J. Anat. 92, 189–217.

Wolbach, S.B., Howe, P.R. 1978. Nutrition classics. The Journal of Experimental Medicine 42, 753–777, 1925. Tissue changes following deprivation of fat-soluble A vitamin. S. Burt Wolbach and Percy R. Howe. Nutr. Rev. 36, 16–19.

Wurtz, J.M., Bourguet, W., Renaud, J.P., Vivat, V., Chambon, P., Moras, D., Gronemeyer, H. 1996. A canonical structure for the ligand-binding domain of nuclear receptors. Nat. Struct. Biol. 3, 206.

Yan, F., Gao, X., Lonard, D.M., Nawaz, Z. 2003. Specific ubiquitin-conjugating enzymes promote degradation of specific nuclear receptor coactivators. Mol. Endocrinol. 17, 1315–1331.

Yashiro, K., Zhao, X., Uehara, M., Yamashita, K., Nishijima, M., Nishino, J., Saijoh, Y., Sakai, Y., Hamada, H. 2004. Regulation of retinoic acid distribution is required for proximodistal patterning and outgrowth of the developing mouse limb. Dev. Cell. 6, 411–422.

Zaffran, S., Frasch, M. 2002. Early signals in cardiac development. Circ. Res. 91, 457–469.

Zechel, C., Shen, X.Q., Chen, J.Y., Chen, Z.P., Chambon, P., Gronemeyer, H. 1994. The dimerization interfaces formed between the DNA binding domains of RXR, RAR and TR determine the binding specificity and polarity of the full-length receptors to direct repeats. EMBO J. 13, 1425–1433.

Zhang, F., Nagy Kovacs, E., Featherstone, M.S. 2000. Murine hoxd4 expression in the CNS requires multiple elements including a retinoic acid response element. Mech. Dev. 96, 79–89.

Zhao, D., McCaffery, P., Ivins, K.J., Neve, R.L., Hogan, P., Chin, W.W., Drager, U.C. 1996. Molecular identification of a major retinoic-acid-synthesizing enzyme, a retinaldehyde-specific dehydrogenase. Eur. J. Biochem. 240, 15–22.

Zhao, S., Overbeek, P.A. 2001. Regulation of choroid development by the retinal pigment epithelium. Mol. Vis. 7, 277–282.

Zhu, J., Gianni, M., Kopf, E., Honore, N., Chelbi-Alix, M., Koken, M., Quignon, F., Rochette-Egly, C., de The, H. 1999. Retinoic acid induces proteasome-dependent degradation of retinoic acid receptor alpha (RARalpha) and oncogenic RARalpha fusion proteins. Proc. Natl. Acad. Sci. USA 96, 14807–14812.

Molecular mediators of retinoic acid signaling during development

Karen Niederreither[1,2] and Pascal Dollé[3]

[1]*Department of Medicine, Center for Cardiovascular Development,*
Baylor College of Medicine, Houston, Texas
[2]*Department of Molecular and Cellular Biology,*
Center for Cardiovascular Development,
Baylor College of Medicine, Houston, Texas
[3]*Institut de Génétique et de Biologie Moléculaire et Cellulaire,*
UMR 7104 du CNRS, U. 596 de l'INSERM,
Université Louis Pasteur, BP 10142, 67404 Illkirch Cedex,
CU de Strasbourg, France

Contents

Advances in Developmental Biology
Volume 16 ISSN 1574-3349
DOI: 10.1016/S1574-3349(06)16004-4

Like adults, the developing vertebrate embryo and fetus requires sufficient amounts of vitamin A for healthy survival. Vitamin A (retinol) and its derivatives (collectively known as retinoids) are small lipophilic molecules containing a hexacycle, an unsaturated side chain, and several methyl groups (Fig. 1A). Vitamin A actions in the adult include prevention of night blindness, maintenance of fertility, and prevention of premalignant cell growth. Since the 1930s, it has also been recognized that this nutrient is required throughout pregnancy and that its deficiency causes a wide range of fetal defects (Wilson et al., 1953, and references therein). Its primary dietary sources are beta-carotene obtained from plants and retinyl esters from meats. After beta-carotene cleavage or retinyl ester hydrolysis, retinol is reesterified in enterocytes and incorporated into chylomicrons to be transported to the liver, its main site of storage (Napoli, 1996 for review). Whereas storage occurs in the form of retinyl esters in stellate cells, hepatocytes synthesize a retinol-binding protein (RBP), which acts as a carrier protein for retinol

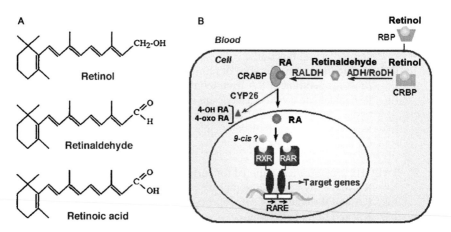

Fig. 1. Chemical structure of all-*trans* retinol, retinaldehyde, and retinoic acid (A) and scheme of the intracellular pathways leading RA-mediated gene regulation (B). Nonlabeled bars in (A) represent methyl groups. ADH/RoDH, alcohol (retinol) dehydrogenase; CRBP, cellular retinol-binding protein; CRABP, cellular retinoic acid-binding protein; RALDH, retinaldehyde dehydrogenase; RARE, retinoic acid response element; RBP, retinol-binding protein.

released in the circulation. Homeostatic controls exist to maintain steady plasma levels of retinol. However, this molecule is not directly active and the actions of vitamin A are regulated at tissue- and cell-specific levels via enzymatic production of bioactive derivatives. With the important exception of photoreceptor cells in the retina, in which retinaldehyde is required as a prosthetic group for opsins in the phototransduction process (Lamb and Pugh, 2004 for review), most if not all other vitamin A functions are mediated by its acidic derivative retinoic acid (RA). In this chapter, we address the general question of how RA signaling is activated or restricted during embryonic and fetal development, and what has been learned from work performed on a variety of RA deficiencies *in vivo* or *ex vivo* animal models. Emphasis will be placed on mouse models generated by gene targeting techniques, as these allow to study the function of specific molecular players in the context of the whole animal. Knowledge on the developmental functions of proteins implicated in the maternal–embryonic transfer of vitamin A, the intracellular metabolism of RA, and its signaling by nuclear receptors is thus reviewed (Section 1). We then consider a number of developmental events in which the involvement of RA has been established. All of these involve region-specific patterns of embryonic RA synthesis by a single enzyme, the retinaldehyde dehydrogenase 2 (RALDH2). Data obtained through the analysis of *Raldh2*-null (Section 2) or "conditional" (Section 3) mouse mutants are reviewed in the context of additional knowledge obtained by experimentally interfering with RA levels and/or signaling in other vertebrate species.

1. Molecular mediators of vitamin A functions

1.1. Animal models of vitamin A deficiency

One method to uncover the importance of retinoid signaling throughout development is to produce animal models in which the conceptus is deprived from maternal retinol. This can be achieved in mammals by an extended vitamin A-deficient (VAD) regime. Maternal dietary VAD in mice or rats leads to a well-defined spectrum of fetal abnormalities affecting numerous organs including the heart, lungs, kidneys, and reproductive tract (Wilson et al., 1953). However, this nutritional method cannot give the most severe cases of embryonic retinoid deficiency because a full vitamin A deficiency is incompatible with maternal fertility. In order for the embryos to implant and the pregnancy to proceed in fully VAD mothers, the trick of supplying the active ligand RA at early stages of pregnancy has been used. Then RA is either withdrawn or given at low levels, resulting in dosage-dependent alterations of early embryonic processes, for example, in heart, hindbrain, and limb development (Dickman et al., 1997; White et al., 2000). These defects are quite comparable to those observed in another vertebrate model, the VAD quail embryos (Heine et al., 1985;

Dersch and Zile, 1993; Zile, 2001), pointing to the conserved nature of vitamin A signaling pathways throughout species. The latter system is much more amenable for obtaining severe embryonic vitamin A deficiency, as *in vivo* development can occur with a completely retinol-deficient yolk. Furthermore, retinoids provided to the vitamin A-deficient embryo up to the five-somite stage of development, but not later, rescued heart morphogenesis and embryonic development, thus defining a critical retinoid-sensitive time point (Kostetskii et al., 1998). Numerous studies have been performed on quail embryos since the initial reports describing VAD-induced defects (Gale et al., 1999; Zile et al., 2000; Halilagic et al., 2003). This system can be compared in its severity to mouse models with genetically engineered mutations disrupting enzymatic RA synthesis, and collectively this work has enhanced our understanding of the evolutionarily conserved mechanisms involving RA during morphogenesis.

1.2. Molecular mechanisms of RA action

Most of the biological functions of vitamin A are mediated by its acidic derivative RA (Fig. 1A). RA acts as a ligand for a subfamily of nuclear receptors, the RA receptors (RARs) α, β, and γ. The RARs heterodimerize with members of another subfamily, the "retinoid X" receptors (RXRs) α, β, and γ. The RXRs are heterodimerization partners for several other nuclear receptors, including the thyroid hormone (TR), vitamin D (VDR), and peroxisome proliferator-activated receptors (PPARs). In the absence of ligand, RAR/RXR heterodimers bind to RA-response elements (RAREs) and RARs recruit corepressors. Corepressors mediate their negative transcriptional effects by recruiting histone deacetylase complexes, which by removing acetyl groups stabilize the nucleosome structure so that DNA becomes inaccessible for transcription. Binding of RA leads to a conformational change of the RAR ligand-binding domain, releasing corepressors and recruiting coactivators. While some coactivators interact with the basal transcriptional machinery, others induce chromatin remodeling which activates transcription (Jepsen and Rosenfeld, 2002 for review). An additional level of regulation of RARs and RXRs exists through the existence of N-terminal isoform variants generated by alternative splicing and/or promoter usage (Chambon, 1996 for review). While these isoforms differ in their transcriptional activation properties, they share the same DNA-binding domain. Specific actions of RAR and RXR splice isoforms have yet to be defined in animal models. Potentially important roles in the adult have been suggested, as reduced RARβ2 expression appears to correlate with increased oncogenic transformation (Sun and Lotan, 2002 for review). Another level by which retinoid signaling might be fine-tuned could be through the action of the 9-*cis*-RA stereoisomer which, unlike all-*trans*-RA, has been shown to bind and activate RXRs in cell culture (Heyman et al., 1992; Levin et al., 1992).

However, while all-*trans*-RA and some other derivatives can be detected by high performance liquid chromatography (HPLC) methods in embryonic extracts (Maden et al., 1998; Ulven et al., 2000), the 9-*cis* isomer has never been detected endogenously in developing or adult tissues under normal physiological conditions. As neither 9-*cis*-RA nor RXR-selective synthetic ligands can rescue the lethal embryonic defects due to defective RA synthesis, whereas RAR-selective ligands are capable of (Mic et al., 2003), the significance of 9-*cis*-RA in development (and other biological) processes is questionable.

The pleiotropic functions of RA are enacted by its ability to control gene expression at the level of chromatin. One clue as to which genes can be regulated by RA is whether they contain RAR/RXR-binding sites (RAREs). To date, over 500 genes have been characterized as being RA-responsive (Balmer and Blomhoff, 2002), but only a minority of these have been shown to contain consensus RAR/RXR-binding site(s). While for many of these genes RA-responsiveness has been observed in cell culture systems, for example, during the process of differentiation of F9 or P19 embryonic carcinoma cells (Jonk et al., 1994; Bouillet et al., 1995), few genes have been proved to be RA-dependent *in vivo* at the level of the whole organism. This requires demonstrating altered expression levels within one of the RA-deficiency animal models. Among the confirmed direct targets of RA are several *Hox* genes, a family of clustered genes encoding homeodomain proteins involved in patterning along the embryonic anterior–posterior axis. These genes were shown to require RA because selective mutation of their RAREs does produce phenotypic defects and alters some aspects of their developmental expression (Dupé et al., 1997; Studer et al., 1998), as first shown by studies in which similar mutations of the RAREs had been introduced in reporter transgenic constructs (Marshall et al., 1994). Clearly, one of the main challenges when studying the involvement of RA in a given developmental phenomenon is to distinguish early (i.e., possibly direct) regulatory events from secondary changes in gene expression. Along with use of RA-deficiency models and RARE mutations in whole animals, a better understanding of the extent of "true" RA-responsive genes *in vivo* will involve genome-wide screenings to determine which genes contain RAREs in evolutionarily conserved regions.

1.3. RA receptors have pleiotropic, although essentially redundant, developmental functions

Gene targeting techniques in the mouse became available soon after the cloning of RARs and RXRs almost 20 years ago. RARs exhibit differential, tissue-specific patterns of expression during development (Mollard et al., 2000, and references therein). It was found, however, that null mutations of each of the RARs lead to viable mice with phenotypic defects limited to specific organ systems. Hence, *Rarα*-null male mutants undergo postnatal testis degeneration (Lufkin et al., 1993), *Rarβ* mice exhibit transformation of cervical vertebrae and the presence of a retrolenticular membrane (Ghyselinck et al., 1997), whereas *Rarγ* mutants have some glandular defects and premalignant lesions (Lohnes et al., 1993). *Rxrγ*-null mutants are viable and show no detectable defect, whereas *Rxrβ* mutants have a selective abnormality of the testis Sertoli cells (Kastner et al., 1996). RXRα is the only retinoid receptor whose null mutation is lethal before birth, due to both placental and heart defects (Kastner et al., 1994; Sapin et al., 1997). Mutant fetuses exhibit reduced numbers of cardiomyocytes in the compact zone of the myocardium (the layer adjacent to the epicardium undergoing rapid growth). Ultrastuctural studies showed premature differentiation of subepicardial cardiomyocytes, possibly linked to a reduced cell proliferation in this layer (Kastner et al., 1997a). As RXRs heterodimerize with various nuclear receptors (including PPARγ, whose mutation causes placental defects also leading to heart failure), it has been questioned whether part of the $Rxrα^{-/-}$ phenotype may not reflect a defect in retinoid signaling. Conditional mutagenesis of *Rxrα* in epicardial cells was found to produce similar myocardial defects as observed in the $Rxrα^{-/-}$-null mutants (Merki et al., 2005). Other work has shown that selective interference with RA signaling in epicardial cells by overexpression of a dominant negative RAR also leads to a $Rxrα^{-/-}$-like myocardial phenotype (Chen et al., 2002). Furthermore, it was shown that primary cultures of $Rxrα^{-/-}$ epicardial cells fail to secrete a trophic activity that stimulates proliferation of myocardial cells (Chen et al., 2002). These data indicate that RA would act within the epicardium to induce some secreted factor(s), rather than by diffusion to elicit gene responses in myocardial cells. Reductions in FGF and Wnt signaling observed in the epicardial-specific *Rxrα* conditional mutant provide candidates for such RA-regulated signals (Merki et al., 2005).

To unravel possible redundant RAR developmental functions, an important approach was to generate various combinations of double mutants through intercrosses (Mark et al., 1999, and references therein). Many of these compound mutant mice died just prior to birth. When analyzed histologically (Lohnes et al., 1994; Mendelsohn et al., 1994), the *Rar* double mutants exhibited a wide range of defects including all abnormalities previously described in the rodent fetal VAD syndrome, along with several novel alterations. The resulting abnormalities clearly showed that RA signaling was affecting selective developmental processes. Severe craniofacial skeletal defects found in *Rarα;Rarγ* double mutants indicated that retinoids may regulate cranial neural crest migration or survival (Lohnes et al., 1994). Defective remodeling of the aortic arch arteries and persistent truncus arteriosus, which are most frequent in *Rarα; Rarβ* compound mutants, suggested an effect on cardiac neural crest migration (Section 3.1). A neural crest-specific combined inactivation of *Rarβ* and *Rarγ* was shown to phenocopy the severe ocular abnormalities found in the corresponding double null mutants, incriminating the neural crest-derived periocular mesenchyme as the primary target of RA action (Matt et al., 2005a). Exception for a limited number of studies, the phenotypic heterogeneity and the difficulty to obtain large numbers of compound mutants made molecular analysis of the various developmental defects burdensome. Interesting data have nonetheless been obtained concerning sequential roles of *Rarγ* and *Rarβ* during hindbrain development, which had to be analyzed in compound mutants with *Rarα* (Dupé et al., 1999; Wendling et al., 2001). *Rarα; Rarβ* mutants have also been particularly useful for investigating the actions of RA during growth and branching morphogenesis of the kidney, which led to the identification of the receptor tyrosine kinase Ret as a target of RA regulation (Batourina et al., 2001).

Intercrosses have also been produced between *Rar* and *Rxr* mutants to assess the developmental significance of specific RAR/RXR heterodimers (Kastner et al., 1997b; Mark et al., 1999, and references therein). These studies have implicated RXRα as the main heterodimeric partner for RARs during development: the phenotypic defects found in *Rar* double mutants were recapitulated in various *Rar;Rxrα* mutant combinations (Kastner et al., 1997b). The overall conclusion of these genetic studies is that RARs mediate the pleiotropic effects of RA on developmental processes although, due to large overlaps in their expression, functional compensation can occur when one receptor is invalidated. Furthermore, there appears to be RAR-independent developmental functions of RXRs that would involve other heterodimerization partner(s), for instance PPARσ during placentogenesis (Barak et al., 1999; Wendling et al., 1999).

1.4. Developmental significance of retinol-binding proteins

Mammalian embryos receive maternal retinol via transplacental transfer. The retinol-binding protein (RBP), a protein involved in postnatal transport of retinol in blood plasma, also potentially plays a role in the transfer of retinol to the embryo. This protein is specifically produced by the embryonic visceral yolk sac endoderm, the layer which mediates maternal–embryonic exchanges before establishment of a functional placenta (Ward et al., 1997). The functional significance of yolk sac RBP is not clear, as $Rbp^{-/-}$ knockout mice are viable (Vogel et al., 2002). It was found, however, that $Rbp^{-/-}$ embryos are prone to major defects characteristic of defective retinoid signaling when generated from $Rbp^{-/-}$ mothers maintained on a VAD diet (Quadro et al., 2005). In the adult, RBP participates to retinol storage functions, and $Rbp^{-/-}$ mutants become vitamin A deficient much more rapidly than wild-type (WT) mice because they are unable to mobilize their liver retinoid stores (Quadro et al., 1999). The conclusion from these studies is that both maternal and embryonic RBP may be dispensable when the animals are maintained under a vitamin A-sufficient diet—most likely because the embryo receives enough retinol from circulating maternal retinyl esters, but that $Rbp^{-/-}$ embryos are highly sensitive to limiting conditions of maternal retinoids, and furthermore are unable to build retinoid stores during fetal development (Quadro et al., 2005).

Additional proteins act as cytosolic-binding proteins for retinol (the cellular retinol-binding proteins, CRBPI–III) and RA (the cellular RA-binding proteins, CRABPI and -II). While CRBPII is essentially an intestinal protein, many cells express CRBPI, which helps the uptake of retinol into cells and the conversion to retinyl esters for storage (Ghyselinck et al., 1999, and references therein). Although CRBPI is expressed in many tissues during development, its inactivation by gene targeting does not lead to detectable changes in the levels of RA-regulated genes in the embryo (Matt et al., 2005b). $CrbpI^{-/-}$ mutants are viable, yet CRBPI deficiency results in an approximately 50% reduction of retinyl ester accumulation in hepatic stellate cells. $CrbpI^{-/-}$ mice fed a vitamin A-deficient diet fully exhaust their RE stores within 5 months, and eventually develop severe manifestations of the vitamin A deficiency syndrome (Ghyselinck et al., 1999). CRBPIII is dispensable for development and normal life, but its inactivation markedly impairs retinoid incorporation into milk (Piantedosi et al., 2005). CRABPI and -II exhibit differential expression patterns during development (Ruberte et al., 1992), and in vitro studies presented these proteins as critical regulators of intracellular RA signaling. Combined targeted mutation of CRABPI and -II does not result in any defined fetal or postnatal alterations, yet neonatal deaths are increased from 2% to 9% in these mutants (Lampron et al., 1995).

Genetic studies have therefore led to the conclusion that none of the retinol-binding proteins are strictly necessary for mouse development to proceed normally. Although their presence may be dispensable in laboratory animals fed with a vitamin A-sufficient diet, they are involved in aspects of retinoid storage, metabolism, or transport (Napoli, 1999 for review) that might become critical within the developing embryo under limiting conditions, as exemplified by the case of $Rbp^{-/-}$ mutants produced from VAD mothers (Quadro et al., 2005).

1.5. In vivo generation of RA from retinol

RA is generated intracellularly from retinol by two oxidative reactions involving the transformation of the alcohol moiety of retinol into an aldehyde (retinaldehyde) and a carboxylic acid (Fig. 1B). Several distinct families of dehydrogenases have been implicated in the generation of RA in embryonic or adult tissues (Duester, 2000; Duester et al., 2003 for reviews). Some members of the alcohol dehydrogenase (ADH) family, especially ADH1, ADH3, and ADH4, can efficiently metabolize retinol into retinaldehyde in vitro (Molotkov and Duester, 2002; Duester et al., 2003, and references therein). These enzymes can also metabolize ethanol into acetaldehyde. As the retinol/retinaldehyde conversion is rate-limiting *in vivo*, it has been suggested that ethanol ingested during pregnancy could compete with endogenous retinol for ADH activity, thereby decreasing the rate of RA production in embryonic tissues (Duester, 1991; Deltour et al., 1996; Henderson et al., 1999; Molotkov and Duester, 2002). This might explain some similarities between the vitamin A deficiency and fetal alcohol syndromes (Cartwright and Smith, 1995; Dickman et al., 1997, and references therein).

Whereas ADH3 is a ubiquitously expressed enzyme, ADH1 and ADH4 exhibit differential temporal and spatial expression profiles in the mouse embryo (Ang et al., 1996, and references therein). $Adh3^{-/-}$ mice have reduced viability and growth when fed on a standard diet, which can be improved by dietary retinol supplementation (Molotkov et al., 2002a). $Adh1^{-/-}$ and $Adh4^{-/-}$ mice are phenotypically normal, suggesting redundant functions between these dehydrogenases (Deltour et al., 1999). $Adh1/Adh4$ double null mutants, while surviving postnatally, are impaired in their ability to produce retinaldehyde following exogenous administration of retinol, demonstrating that the corresponding enzymes act *in vivo* as retinol dehydrogenases (Molotkov et al., 2002b). Furthermore, $Adh4^{-/-}$ embryos exhibit an increased rate of lethality when generated from vitamin A-deficient mothers, indicating that this enzyme participates in retinol metabolism during development and that its activity becomes critical under limiting conditions (Deltour et al., 1999). However, the evidence clearly suggests that ADHs are not responsible for the tissue-specificity of RA synthesis. Also, there is insufficient evidence to support a developmental function of short-chain dehydrogenase/reductases (SDR), another class of enzymes which can act in retinol oxidation (Napoli, 2001 for review).

1.6. Retinaldehyde dehydrogenases are responsible for the tissue-specific patterns of embryonic RA production

Three cytosolic aldehyde dehydrogenases (ALDHs) have been shown to have a high catalytic activity on retinaldehyde. These are known as retinaldehyde dehydrogenases (RALDHs) 1–3, although the corresponding genes are named *Aldh1a1–4* (Duester, 2000). Two of them (RALDH1 and -3) were originally characterized as human ALDH isoenzymes, and were subsequently found to correspond to the RA-synthesizing activities present in the dorsal and ventral embryonic mouse retina (Li et al., 2000; Mic et al., 2000). RALDH2 was identified through PCR cloning as a novel enzyme accounting for RA production in the early embryonic trunk and eye region (Zhao et al., 1996). Crystallographic analysis revealed a "disordered loop" structure in the substrate access channel of RALDH2, capable of discriminating between retinaldehyde and short-chain aldehydes (Lamb and Newcomer, 1999). This might explain its

high catalytic activity on retinaldehyde; in this enzyme substrate binding promotes an "induced fit" prior to catalysis, wherein a disordered to ordered
transition occurs. RALDH1 has an ordered loop, and as a consequence a larger
substrate access channel allowing other aldehydes to fit (Moore et al., 1998).
It is also at least 10-fold less efficient than RALDH2 in metabolizing retinaldehyde (Grun et al., 2000; Gagnon et al., 2002). Structural studies have not
yet been performed on RALDH3, but the kinetics of this enzyme for retinaldehyde are similar to those of RALDH2 (Grun et al., 2000). A fourth retinaldehyde dehydrogenase (RALDH4–*Aldh8a1*) has been characterized (Lin et al.,
2003). This enzyme differs from the previously known RALDHs in that it has a
high catalytic activity on 9-*cis* retinaldehyde, while acting weakly on all-*trans*
retinaldehyde. Its expression appears to be restricted in adult mice to the liver
and kidney (Lin et al., 2003), and has not been detected during embryogenesis
(Dollé, P., unpublished data). Its *in vivo* function remains to be established.

The *Raldh1*, *Raldh2*, and *Raldh3* genes exhibit distinct patterns of expression
during mouse development, which collectively correlate with the dynamics of
RA production and the patterns of activity of RA-responsive reporter transgenes (Rossant et al., 1991). *Raldh1* expression starts at about midgestation
(E9.5). This enzyme is strongly expressed in the dorsal—and to a lesser extent
the ventral—retina, as well as other eye structures including the lens epithelium.
It is also expressed in various other developing organs such as the lung, liver,
intestine, and kidney (Haselbeck et al., 1999; Niederreither et al., 2002a).
Raldh1$^{-/-}$ mouse mutants are viable and do not display any detectable defect,
including in the retina and visual system (Fan et al., 2003). This could be
because of functional compensation by other RALDHs, especially RALDH3
which is also expressed in the developing eye. Study of *Raldh1/Raldh3* double
mutants has demonstrated redundant functions in eye morphogenesis, which
could only be assessed at prenatal stages because of the lethality of *Raldh3*$^{-/-}$
mutants (Matt et al., 2005a). More generally, the apparently normal phenotype
of *Raldh1*$^{-/-}$ mice could reflect the fact that this enzyme is only marginally
involved in endogenous RA synthesis, but may be required for the metabolism
of other aldehyde(s), for example, in the liver where it is highly expressed during
adult life.

Raldh3 expression, on the other hand, could account for most of the RA
synthesis in embryonic craniofacial structures from ~E8.5. This enzyme is first
induced in the frontonasal ectoderm, and eventually its expression extends to
the developing lens and nasal placodes (Li et al., 2000; Mic et al., 2000). While
expression remains strong in the nasal epithelium, it eventually disappears
from the lens epithelium and appears in the developing ventral retina (Wagner
et al., 2000). Few other embryonic structures express *Raldh3*; these include the
mesonephros (embryonic kidney) and later, the renal excretory ducts (Li et al.,
2000; Mic et al., 2000; Niederreither et al., 2002a). *Raldh3*$^{-/-}$ knockout mutants
die at birth due to abnormal development of the nasal cavities, with an atresia of
the choanae preventing them to breathe normally (Dupé et al., 2003). They also

display mild ocular abnormalities, with a shortening of the ventral retina. No abnormality of the kidney or excretory ducts has been detected in these mutants. While the study of compound *Raldh3/Raldh1* mutants has revealed some redundant functions in eye development (Matt et al., 2005a; Molotkov et al., 2006), redundant roles with RALDH2 are also being investigated, especially during development of the forebrain and kidney (Niederreither, K. and P.D., unpublished data).

Raldh2 is the first RALDH to be expressed during embryogenesis. Its induction during gastrulation (~E7.0 in the mouse) in the posterior embryonic mesoderm (Fig. 2A) correlates with the stage at which RA can first be detected by HPLC analysis or enzymatic bioassays (McCaffery and Dräger, 1997; Ulven et al., 2000). From E8.5, *Raldh2* is expressed in a dynamic and spatially restricted manner in many mesodermal derivatives including the somites, the cervical and posterior branchial arch mesenchyme, the posteriormost region of the heart tube, and the lateral plate mesenchyme of the limb-forming region (Fig. 2B) (Niederreither et al., 1997). During development of the first 10–12 somite pairs, *Raldh2* expression progresses in the rostral presomitic mesoderm in concert with the onset of somite formation (Vermot et al., 2005a; Section 2.2). These various sites of expression correlate with most of the embryonic regions in which RA signaling activity had been mapped through the analysis of reporter transgene activity (Rossant et al., 1991). *Raldh2* is also transiently expressed in the rostral forebrain neuroepithelium, including the eye field, by ~E8.0–E8.5 (Wagner et al., 2000; Ribes et al., 2006). During late embryonic and fetal development, *Raldh2* is expressed in various tissues, such as the meninges, pleura, pericardium and epicardium, stomach and intestine mesenchyme, kidney cortex, and interdigital mesenchyme of the developing limbs (Niederreither et al., 1997; Fig. 2B). It is also expressed in discrete spinal cord cell populations at brachial and lumbar levels (Zhao et al., 1996; Section 3.3), that have been identified as lateral motor colums giving birth to motor neuron pools innervating the developing limbs (Sockanathan and Jessell, 1998). As described in the next sections, studies performed on *Raldh2* mouse mutants have characterized a number of important functions during morphogenetic processes.

1.7. Region-specific metabolism of embryonic RA by CYP26 cytochromes

Another level of regulation of embryonic RA activity comes from its tissue-specific metabolism by a subfamily of cytochromes P450s, the CYP26 enzymes. The three members of this subfamily (CYP26A1, -B1, and -C1) are known to act on RA and metabolize it into more polar derivatives (4-OH, 4-oxo RA), and genetic studies have provided evidence that their activity is required to prevent RA to act in inappropriate embryonic regions. Hence,

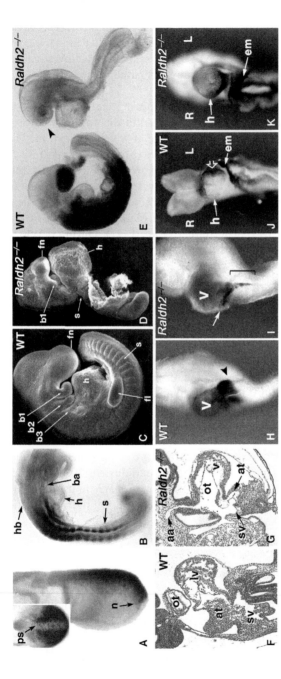

Fig. 2. Developmental expression of the retinoic acid-synthesizing enzyme RALDH2, and early embryonic abnormalities in corresponding null mutants. (A) In wild-type embryos, *Raldh2* gene expression is first seen during gastrulation in the posterior embryonic mesoderm up to the node (n) region (main panel: profile view of an E7.5 embryo; inset: posterior view showing expression on each side of the primitive streak, ps). (B) Eventually at E8.5, *Raldh2* is expressed at highest levels in the cervical region, with sharp boundaries of expression at the level of the posterior hindbrain (hb), branchial arches (ba), and heart tube (h) mesoderm. Expression is also seen in the epithelialized somites (s). (C and D) *Raldh2⁻ᐟ⁻* null mutants exhibit severe morphogenetic defects, characterized at E9.5 by a shortening of the cervical and trunk region with truncated somites (s; Fig. 3), a markedly dilated heart (h) cavity, a truncated forebrain and frontonasal (fn) region and a lack of visible branchial arches (b2, b3), except for the first arch (b1). (E) *Raldh2⁻ᐟ⁻* embryos at E8.5 show no detectable activity of the RA-sensitive RARE-*lacZ* transgene, except in the facial region which expresses *Raldh3* (arrowhead). (F and G) Histological analysis of the heart at E9.5 shows highly hypoplastic inflow cavities (atrium and sinus venosus, at and sv, respectively), defective trabeculation of the ventricular (v) myocardium and lack of morphogenesis of ventricular and outflow tract (ot) chambers, resulting in an abnormal juxtaposition of inflow and

CYP26A1 is required to restrict RA activity within the hindbrain and the posterior region of the embryo (Abu-Abed et al., 2001; Sakai et al., 2001; Sirbu et al., 2005), whereas CYP26B1 is necessary to regulate RA action in the developing limbs (Yashiro et al., 2004). Additional functions of CYP26A1 have also been documented during development of the retina (Sakai et al., 2004) and in the prospective head region of the gastrulating embryo (P.D., unpublished observations). The function of CYP26C1 has not yet been investigated genetically.

As a general trend, the CYP26 enzymes are often expressed in embryonic regions or tissue-layers adjacent to those producing RA. This led to the suggestion that local gradients or RA may be generated in areas located between the RA-producing regions (sources) and those metabolizing it (sinks) (Reijntjes et al., 2004, and references therein). However, evidence would suggest that, rather than setting RA gradients, activity of CYP26 enzymes is required in a dynamic manner to establish sharp boundaries to RA activity, for example, within the developing hindbrain (Sirbu et al., 2005; Section 2.3).

2. RALDH2 generates RA required for early embryonic development

2.1. RA and early heart development

Raldh2 targeted null mutation is lethal by midgestation at E9.5–E10.5 (Niederreither et al., 1999; Mic et al., 2002). The mutant embryos exhibit complex morphological defects. Their most striking features are an abnormal, bulging heart tube, a shortening of the embryonic anteroposterior axis with small and compact somites, and a mild frontonasal truncation with apparent lack of posterior (second to sixth) branchial arches (Fig. 2C and D). The *Raldh2*-null allele was crossed with a RA-responsive transgenic line (RARE-*lacZ*; Rossant et al., 1991), and it was found that transgenic null mutant embryos display almost no activation of the reporter transgene, except in the

outflow (aortic arches, aa) circulations in the *Raldh2*$^{-/-}$ mutants. (H and I) Abnormal distribution of *Tbx5* transcripts in the heart of E9.5 *Raldh2*$^{-/-}$ embryos. While this gene is normally strongly expressed in the posterior heart chambers of wild-type embryos (H), which a sharp boundary at the level of the sinus venosus (arrowhead), its expression in mutants is limited to a narrow band of cells (white arrow), with an ill-defined posterior boundary (bracket). (J and K) An abnormal posterior location of *Hand1* transcripts, which normally become lateralized along the prospective left wall of the left ventricle during left–right (L–R) heart looping (arrow in J), indicates an impaired looping morphogenesis in *Raldh2*$^{-/-}$ embryos (K). *Hand1* expression in extraembryonic membranes (em) is unaltered. Panels A and B from Niederreither et al. (1997), Mech. Dev. 62, 67; C–E and J and K from Niederreither et al. (1999), Nat. Genet. 21, 444; F–I from Niederreither et al. (2001), Development 128, 1019, with permission. (See Color Insert.)

craniofacial region that expresses *Raldh3* (Fig. 2E). In addition, the ability of maternally administered RA to partially rescue the *Raldh2*$^{-/-}$ phenotype (Section 3.1) established that the underlying abnormalities result from a deficiency in endogenous RA production.

Molecular studies performed on *Raldh2*$^{-/-}$ mutants allowed an understanding of the basis of the abnormal heart phenotype in the context of a variety of events that may be RA-regulated (Niederreither et al., 2001). Previous work performed in other species has shown that conditions of RA deficiency or excess can alter cardiac progenitor cell migration, chamber patterning, looping morphogenesis and differentiation. One of the earliest effects of RA deficiency was reported in zebrafish mutants for *Raldh2*, and following administration of RAR antagonists to wild-type embryos, which led to an enlargement of the cardiac field at the gastrula stage (viewed as an expansion of the *Nkx2.5*-expressing domain). The earliest role of RA might be to restrict cardiac progenitor populations, possibly by imposing a limit on the density of progenitor cells within a competent domain (Keegan et al., 2005). Dose-dependent effects were reported when interfering with RA signaling in *Xenopus*, by the use of pharmacological RAR antagonists. At high doses, these blocked fusion of the lateral fields of myocardium. Intermediate doses allowed heart field fusion but not tube formation, whereas low doses allowed tube formation but disrupted heart looping. Only when RARs were inhibited prior to gastrulation was an expansion of the early heart field detected (Collop et al., 2006).

Another postulated role of RA once the heart tube is formed is to regulate its anterior–posterior growth and/or patterning, which manifests in region-specific heart abnormalities resulting from conditions of RA excess or deficiency in zebrafish, chick, and mouse (Stainier and Fishman, 1992; Yutzey et al., 1994; Iulianella and Lohnes, 2002; Hochgreb et al., 2003). *Raldh2* expression and RA-reporter transgene activity are restricted to the posterior (sinoatrial) region of the mouse heart (Moss et al., 1998; Xavier-Neto et al., 1999, 2001), leading to the hypothesis that RA may be required for sinoatrial growth. Supporting this hypothesis, excess RA treatments during gastrulation in the chick (Drysdale et al., 1997), zebrafish (Stainier and Fishman, 1992), and mouse (Chazaud et al., 1999; K.N., unpublished results) lead to an "atrialization" (an expansion of posterior heart structures at the expense of the prospective ventricular region). The phenotypes observed in *Raldh2*$^{-/-}$ mouse mutants (Niederreither et al., 2001) and vitamin A-deficient quail embryos (Zile et al., 2000) also support such a role. There is a poor histological development of posterior heart chambers in *Raldh2*$^{-/-}$ mutants (Fig. 2F and G). Notably, expression of *Tbx5* (a major determinant of posterior heart chambers), *Tbx20* (another T-box factor required to restrict the formation of the cardiac cushions), and *Gata4* are all spatially reduced in *Raldh2*$^{-/-}$ embryos (Fig. 2H and I, and data not shown). The posterior heart phenotype of VAD quail embryos is even more severe, as their inflow tract is fully closed and expression of *Gata4* is reduced to a greater extent (Kostetskii et al., 1999).

Additionally to the sinoatrial defects, *Raldh2*$^{-/-}$ hearts have abnormal ventricular trabeculae and their ventricular cardiomyocytes appear to undergo premature differentiation in the subepicardial compact zone (Niederreither et al., 2001). This suggests a role of RA in regulating myocardial cell proliferation. A similar phenotype has been observed in *Rar/Rxr* mouse mutants at later stages (Kastner et al., 1997a), and from the study of *Rxrα*$^{-/-}$ mutants it has been proposed that RA produced by the epicardium would stimulate proliferation in the myocardial compact zone by inducing growth factor-dependent pathways that remain to be fully characterized (Chen et al., 2002; Stuckmann et al., 2003; Merki et al., 2005; Section 1.3). At early stages, RA could also act by regulating growth factor expression in the pharyngeal region (Section 3.1), which would signal in the adjacent heart field.

Another role of RA may be to regulate left–right (L–R) axis determination, which is responsible for the directional looping morphogenesis of the heart tube. When wild-type mouse embryos were administered excess RA in culture, L–R axis determination was perturbed, leading to a bilateral expression of left-side specific determinants such as *Nodal* or *Pitx2* and random L–R heart looping (Chazaud et al., 1999). Converse experiments in which endogenous RA signaling was blocked using synthetic RAR inhibitors led to a downregulation of *Nodal* and *Pitx2* expression, which again randomized or blocked heart looping (Chazaud et al., 1999). Analysis of genes involved in early heart regionalization, such as *Hand1*, indicated that the abnormal heart morphology in *Raldh2*$^{-/-}$ mutants reflects a lack of L–R looping morphogenesis (Fig. 2J and K). However, expression of left-side specific determinants (*Nodal* and *Pitx2*) was not altered in these mutants, indicating that the looping defect was not primarily linked to abnormal embryonic L–R axis (*situs*) determination. In the VAD quail embryos (where yolk retinoids are fully depleted) left-sided initiation of *nodal* and *Pitx2* also occurs and there is some evidence that the heart tube initiates looping, which then cannot proceed due to severe cardiac malformations (Zile et al., 2000).

The exact implication of RA in the events leading to L–R axis determination (Raya and Izpisùa-Belmonte, 2006 for review) remains unclear. Both RA and Sonic Hedgehog (Shh) were found to be present in vesicular structures generated by cells along the right side of Hensen's node and transported by a directional flow of extraembryonic fluid (the "nodal flow"; Nonaka et al., 2005, and references therein) toward the left side (Tanaka et al., 2005). Such directional transport of signaling molecules may be one of the earliest events breaking the L–R symmetry of the early embryo, allowing to trigger side-specific gene expression cascades (Raya and Izpisùa-Belmonte, 2006, and references therein). Although the embryonic node region contains high amounts of RA (Hogan et al., 1992), various studies involving analysis of a RA-responsive transgene (Fig. 3A) or of putative RA-target genes

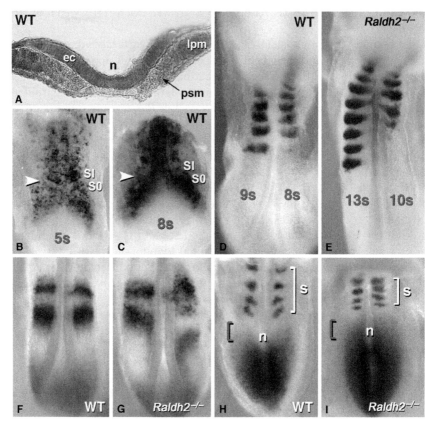

Fig. 3. Retinoic acid deficiency in *Raldh2$^{-/-}$* mutants affects the bilateral symmetry of somite development. (A–C) In E8 (two somite stage) wild-type mouse embryos, RA signaling as detected by the activity of the RARE-*lacZ* reporter transgene occurs symmetrically in the left and right presomitic mesoderm, and is seen in the ectodermal layer of the node. During the first cycles of somitogenesis, the RA signal progresses caudally along the rostral presomitic mesoderm, as RA-responding cells are consistently seen at a comparable distance from the last formed intersomitic boundary (arrowheads) and epithelialized somite (SI, S0 designating the somite undergoing epithelialization; somite stages are indicated in red). (D and E) Asymmetric L-R somitic development in *Raldh2$^{-/-}$* embryos. Two examples are shown, with the presence of one (D) or three (E) additional somites formed on the left side (somite numbers are indicated in red). Somites are marked by the expression of *Uncx4.1*. (F and G) Desynchronized waves of expression of an oscillating gene (*Lunatic fringe*) along the left and right presomitic mesoderm of *Raldh2$^{-/-}$* embryos. (H and I) Combined detection of *Uncx4.1* and *Fgf8* transcripts in six somite stage embryos reveals an abnormal compaction of somites (s), as well as anteriorization of the *Fgf8* mRNA gradient along both sides of the node (n; see red brackets), in *Raldh2$^{-/-}$* embryos. Panel A from Sirbu and Duester (2006), Nat. Cell Biol. 8, 271; B–I from Vermot and Pourquié (2005), Science 308, 563, with permission. (See Color Insert.)

have not yielded any evidence for a L–R asymmetry in RA signaling around this region—or more generally within the embryo (Chazaud et al., 1999; Sirbu and Duester, 2006; Niederreither, K. and P.D., unpublished observations). Hence, an involvement of RA as being instrumental in setting L–R asymmetric gene responses remains questionable, and it may be more likely that RA, although possibly involved as a coregulator of some asymmetrically expressed genes, needs to be present in symmetric cell populations to regulate various morphogenetic processes.

2.2. RA and the control of somitogenesis

Another role of RA linked to L–R asymmetry was discovered in the context of somitogenesis. Somites are transient epithelial structures that are formed sequentially along the left and right paraxial mesoderm during extension of the body axis and that will give rise to bilaterally symmetric structures including the axial skeleton (vertebrae and ribs) and limb and body wall muscles. Yet these structures are formed while the early embryo is exposed to the signals determining L–R axis asymmetry. *Raldh2* is dynamically expressed during development of the first somite pairs (Niederreither et al., 1997). Activation of retinoid signaling (as observed by the activity of the RARE-*lacZ* reporter transgene) is observed in the ectodermal layer of the node (Fig. 3A), as well as in the rostral presomitic mesoderm where it progresses along with the formation of the 10–12 first somite pairs (Fig. 3B and C; Vermot et al., 2005a; Sirbu and Duester, 2006). Eventually this activity recedes, and at later stages of somitogenesis both *Raldh2* and RARE-*lacZ* expression are induced in maturing somites, rather than in presomitic mesoderm (Fig. 2B).

As $Raldh2^{-/-}$ mutants were initially reported to have abnormally small somites (Niederreither et al., 1999), they were analyzed in detail with various markers of early somite development. This analysis revealed a striking abnormality in the dynamics of somite formation (Vermot et al., 2005a). At the ~8–12 somite stages, most mutants exhibited asymmetrical somite numbers, in which one to three additional somites were formed on the left side of the embryo (Fig. 3D and E). This abnormal phenotype prompted us to analyze the behavior of the "molecular oscillator" in the RA-deficient embryos. Several genes mostly belonging to the Notch and Wnt pathways exhibit oscillating patterns of mRNA expression along the presomitic mesoderm, leading to apparent waves of gene expression progressing in a caudal to rostral direction, after which mRNA expression is usually stabilized in a given compartment of the forming somite (Saga and Takeda, 2001; Dubrulle and Pourquié, 2004a for reviews). These cyclic patterns are repeated during each cycle of somite formation, and their tight progression is necessary for the correct alignment and size of somites. In $Raldh2^{-/-}$ embryos, the oscillatory waves of expression eventually became desynchronized along the left and right presomitic mesoderm after formation of a few somite pairs (Fig. 3F and G). Furthermore, genes transiently expressed in the forming somites (such as *Mesp2*) exhibited asymmetric patterns, sometimes with an additional stripe on the left or right side. These patterns strongly suggested that the speed of the oscillatory waves, and thus of the cycles of somite formation, eventually became uncoordinated along the left and right presomitic mesoderm of the RA-deficient embryos (Vermot et al., 2005a). Further understanding of these dynamic abnormalities will await the availability of *in vivo* fluorescent reporter transgenes allowing to follow by time-lapse analysis the progression of the oscillatory waves in mutants.

Another important component regulating somite development is a posterior (high) to anterior (low) gradient of FGF signaling, resulting from a *Fgf8* mRNA and protein gradient that regresses posteriorly—that is, maintained at the same relative levels—during extension of the body axis along both the presomitic mesoderm and the newly formed neurectoderm. During each cycle of somite formation, a critical value of FGF signaling (the "determination front") is believed to set the position of the future intersomitic boundary and stabilize the oscillations in order to allow stable gene expression in specific somite compartments (Dubrulle et al., 2001; Dubrulle and Pourquié, 2004b). In addition, it was found that the posterior region of high *Fgf8* expression marked a neural stem cell zone, in which neural progenitor cells were maintained in an undetermined state (Diez del Corral et al., 2003; Diez del Corral and Storey, 2004). Experimental work suggested that RA produced in somites and rostral presomitic mesoderm may normally act to restrict the *Fgf8* gradient (Diez del Corral et al., 2003). This is supported by the observation of an anteriorization of the *Fgf8* mRNA gradient, both in the $Raldh2^{-/-}$ mouse embryos (Fig. 3H and I; Vermot et al., 2005a; Sirbu and Duester, 2006) and in the VAD quail model (Diez del Corral et al., 2003). A detailed analysis of *Fgf8* mRNA showed that the abnormal gradient distribution mainly occurred in the ectodermal layer, rather than in presomitic mesoderm, of $Raldh^{-/-}$ mutants (Sirbu and Duester, 2006). Further work by these authors provided evidence that RA activity is required only prior to the formation of ~6–8 somite pairs in order to play its role in stabilizing the molecular oscillations.

Thus, embryonic RA is transiently required at early somite stages, most likely as a symmetrical signal that protects the synchrony of the first molecular oscillations from destabilizing effects of L–R asymmetric signals, thereby ensuring a tight bilateral synchrony in somite development. The molecular nature of the events regulated by RA remains unknown. However, these are clearly linked to the function of L–R determinants, since by crossing $Raldh2^{-/-}$ mutants with a mutant (*Inverted situs*, *Inv*) exhibiting a randomization of the L–R asymmetry, Vermot and Pourquié (2005) showed that the sidedness of the asymmetric somite phenotypes correlated with the embryonic *situs* or L–R sidedness. These authors, as well as Kawakami et al. (2005), also showed that the involvement of RA in controlling the bilateral synchrony of somito-genesis is evolutionarily conserved in chick and zebrafish. It will clearly be of interest to further dissect the regulatory interactions between RA and FGF8 during embryonic axis elongation, as the same signals regulate events as different as mesoderm segmentation and cell fate determination in the neural plate.

2.3. RA and brain patterning

$Raldh2^{-/-}$ mouse mutants exhibit distinct neuroepithelial patterning and growth defects. A model for regionalization of the central nervous system (CNS) was proposed by Nieuwkoop et al. (1952) and refined by Foley et al. (2000). This model states that neural tissue initially has an anterior character and that posteriorizing signals are required to induce posterior neural cell fates. Exposure to exogenous RA was first shown in *Xenopus* embryos to result in truncation of anterior brain structures with an enlargement of posterior (hindbrain and spinal cord) regions (Durston et al., 1989). This led to the idea that embryonic RA may be one of the "posteriorizing" signals involved in anterior–posterior CNS patterning. The hindbrain has been extensively studied to examine such a possible role. This region of the brain undergoes a transient segmentation into (according to the species) seven or eight rhombomeres, that is required for proper patterning of the brain stem, organization of the cranial nerves and their nuclei, as well as segmental pathways of neural crest cell migration toward the branchial arches. Trans-genic and knockout studies have shown RA directly regulates genes (e.g., *Hoxa1* and *Hoxb1)* involved in rhombencephalic growth and patterning (Dupé et al., 1997; Studer et al., 1998). Explants and transplantation studies indicate that the RA activity inducing *Hox* gene expression originates from the somitic mesoderm adjacent to posterior rhombomeres (Gould et al., 1998), consistent with the pattern of expression of *Raldh2* (Fig. 2B).

$Raldh2^{-/-}$ mouse mutants exhibit growth and segmentation defects of the hindbrain (Niederreither et al., 2000). These appear to result from early changes in the distribution of molecular determinants of anterior rhombomere identity (such as *Krox20* and *Hoxb1*), which extend abnormally throughout the poste-rior hindbrain at the expense of more caudal (r5–r8) rhombomere determi-nants. In other words, the posterior hindbrain of $Raldh2^{-/-}$ mutants is molecularly "transformed" toward a more anterior identity. As a result, neural crest cells produced from the affected hindbrain region are severely deficient and fail to colonize the posterior (second to sixth) branchial arch region. Dietary models of severe RA deficiency in rat (Dickman et al., 1997; White et al., 2000) and quail (Gale et al., 1999) show hindbrain defects consistent with those observed in $Raldh2^{-/-}$ mutants. RA can therefore be considered as a diffusible "posteriorizing" signal produced mesodermally, and which regulates the expression of posterior rhombomeric determinants in the hindbrain neuroepithelium (Gould et al., 1998; Dupé et al., 1999; Wendling et al., 2001; Serpente et al., 2005). Gavalas (2002) has reviewed the RA-deficiency pheno types obtained in various animal models, which altogether show increasing degrees of anteriorization correlating with the severity of the deficiency, being partly consistent with the posterior neural transformation model of Nieuwkoop. These data have often been interpreted in the context of a RA diffusion gradient along the hindbrain (Maden, 2002). It was found that CYP26 enzymes have complex and rapidly evolving expression patterns during hindbrain segmentation (MacLean et al., 2001; Tahayato et al., 2003; Reijntjes et al., 2004). Their metabolizing activity is therefore expected to create zones of RA enrichment or absence in specific (pre-)rhombomeres, which may be important to establish segmental patterns of gene expression. For example, the control of *Hoxb1* rhombomeric expression may require both the presence of CYP26A1 in pre-rhombomere 2 and the induction of CYP26C1 in rhombomere 4, which altogether would generate "pulses" of RA action of different lengths (Sirbu et al., 2005).

The influence of RA on neuroepithelial growth has also been demonstrated in the embryonic forebrain. Here RA acts with other secreted molecules, such as Shh, FGFs, and Wnts, to set up patterning zones and influence neuronal differentiation (reviewed in Echevarria et al., 2003; Wilson and Houart, 2004; Wilson and Maden, 2005). Studies on VAD quail embryos (Halilagic et al., 2003) and chick embryos treated with RAR/RXR antagonists (Schneider et al., 2001) have implicated RA as an early signal required for the survival of anterior head mesenchyme and the development of telencephalic structures. *Raldh2* expression is seen transiently in the rostral and ventral forebrain neuroepithelium at the 8–15 somite stages, suggesting a localized role in growth regulation. $Raldh2^{-/-}$ embryos are deficient in RARE-*lacZ* reporter transgene activation in the rostral head region, especially in the forebrain neuroepithelium where no activation of the transgene occur until their lethality at E10, despite a normal onset of *Raldh3* expression in the adjacent facial ectoderm (Ribes et al., 2006). Thus, although RALDH3 activity has been proposed as the critical player to supply the developing brain with RA, this enzyme cannot significantly activate RA signaling in the forebrain neuroepithelium of *Raldh2* mutants. Morphologically, $Raldh2^{-/-}$ mutants have marked forebrain defects, with a lack of evagination of telencephalic vesicles, hypoplastic optic vesicles, and diencephalic region (Fig. 4A and B; Mic et al., 2004a; Ribes et al., 2006).

The signaling networks regulating forebrain growth are thought to be comparable to those regulating early limb development, as both systems are controlled by a positive feedback loop between *Fgfs* and *Shh*, which directs organ outgrowth (Marcucio et al., 2005; Tickle, 2006 for reviews). Studies in chick have shown that locally applied RAR/RXR antagonists or RALDH inhibitors were capable of reducing expression of both *Fgf8* and *Shh*, producing craniofacial and beak truncations (Schneider et al., 2001).

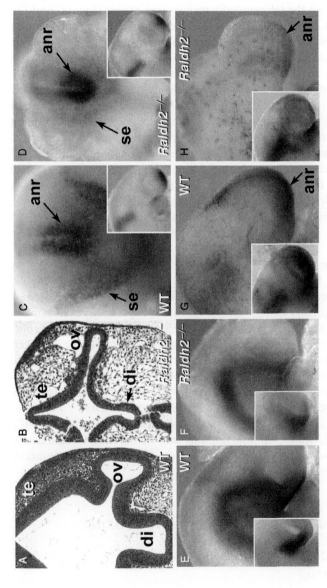

Fig. 4. Abnormal development and patterning of the embryonic forebrain in *Raldh2*⁻/⁻ mutants. (A and B) Histological analysis of E9.5 embryos shows an abnormally thin forebrain neuroepithelium, with a lack of formation of the telencephalic (te) vesicles and of optic vesicle (ov) morphogenesis. di, diencephalon. (C and D) *Fgf8* expression is not affected in the anterior neural ridge (or in other embryonic areas, for example, at the midhindbrain boundary: compare insets), but is defective in the facial surface ectodem, of *Raldh2*⁻/⁻ embryos (main panels: E9.5; insets: E8.5). (E and F) Decreased mRNA expression of *Gli1* along the ventral forebrain neuroepithelium (main panels) and of *Nkx2.1* in the ventral diencephalon and telencephalon (insets) indicates defective Shh signaling in *Raldh2*⁻/⁻ embryos. (G and H) Likewise, analysis of phosphorylated ERK1/2 proteins (main panels) and *Mkp3* mRNA (insets) indicate a decrease in response to FGF signaling in the rostral forebrain of the RA-deficient embryos. From Ribes et al. (2006), Development 133, 351, with permission. (See Color Insert.)

Surprisingly, no alteration in *Shh* expression was observed in the prechordal plate or along the ventral neuroepithelium of *Raldh2*$^{-/-}$ mutants until after the 16 somite stage (Ribes et al., 2006). Likewise, *Fgf8* expression in the anterior neural ridge was maintained in the *Raldh2*$^{-/-}$ mutants. Only *Fgf8* in ectoderm overlying the telencephalic vesicles was absent (Fig. 4C and D). Further analysis of genes downstream of these signaling pathways was performed, which showed reduced expression of *Shh* targets including *Patched1* and *Gli1* (Figure and data not shown). Functional evidence that *Shh* action was impaired came from the observation of diminished domains of *Nkx2.1* (Fig. 4E and F, insets) and of the oligodendrocyte-inducing transcription factor *Olig2* (data not shown). An indicator of effective levels of FGF signaling in the forebrain is the presence of phosphorylated ERK1 and -2 (Corson et al., 2003). Reduction in FGF signaling was apparent in *Raldh2*$^{-/-}$ mutants, as both pERK1/2 immunostaining and *Mkp3* expression (the latter functioning as an FGF-dependent feedback modulator) were reduced at the 14 somite stage (Fig. 4G and H). Region-specific defects in cell proliferation, as well as increased cell death, were also seen as a consequence of the RA deficiency. This was accompanied by an abnormal accumulation of what appeared to be "misdirected" neural crest cells in the preoptic frontonasal mesenchyme (Ribes et al., 2006).

Altogether, these results lead to a model where a localized burst of RA produced by RALDH2 normally stimulates growth of the telencephalic vesicles, optic vesicles, and diencephalon (Ribes et al., 2006). *Raldh2* deficiency first regionally offsets ventral forebrain expression, then impairs Shh response, and at later stages reduces FGF signaling. As a result, proliferation and expansion of neuronal progenitors is affected and excess cell death occurs. Eventually, from ~E9, the action of RA expands to more broadly affecting the overall growth of the forebrain. RALDH3 activity is likely to relay that of RALDH2 in generating RA for this purpose. However, genetic studies have not confirmed such a role, as the *Raldh3*$^{-/-}$ knockout mutants do not have obvious forebrain deficiencies (Dupé et al., 2003). The study of *Raldh2/Raldh3* double mutants may be required to understand the overall contribution of RA to early forebrain development (Molotkov et al., 2006; N.K. and P.D., unpublished data).

3. Additional approaches to investigate later roles of RALDH2

3.1. Rescue of Raldh2$^{-/-}$ embryos reveals functions in developing limbs and branchial arches

Several strategies have been used to bypass the early lethality of *Raldh2*$^{-/-}$ embryos, in order to investigate possible later roles of this RA-generating enzyme. One of these came from the observation that the morphological phenotype of *Raldh2$^{-/-}$* embryos could be improved by supplying RA to the pregnant mothers (Niederreither et al., 1999). RA is a diffusible molecule that is easily absorbed and transferred to the embryo through the yolk sac or placenta. In fact, RA is a potent teratogen, and a human RA embryopathy syndrome has unfortunately been reported (Lammer et al., 1985). The effects of exogenous RA have been extensively studied in rodent models with respect to various developmental processes, including craniofacial, hindbrain, and axial skeletal patterning (Kessel and Gruss, 1991; Simeone et al., 1995). We used subteratogenic doses of RA to attempt to rescue the *Raldh2$^{-/-}$* embryos. When provided during a critical developmental period (E7.5–E8.5), this supplementation extended the viability of null mutants until fetal stages, due to restoration of cardiac looping. Prolonged administration until E9.5 or E10.5 further increased the percentage of *Raldh2$^{-/-}$* fetuses recovered, most likely by improving heart chamber formation and myocardial function. None of the rescued mutants were ever viable, however, due to the presence of outflow tract abnormalities (Niederreither et al., 2001). In our first experiments, RA was administered by oral gavage (Niederreither et al., 1999; Mic et al., 2002). We then established conditions in which RA was supplied as a suspension mixed with powdered food. This more steady mode of administration allowed a more reliable phenotypic rescue, while preventing any potential teratogenic effect due to bolus administration. The doses that rescued *Raldh2$^{-/-}$* embryo survival were comparable to those used in studies of VAD rat embryos (Dickman et al., 1997; White et al., 2000; Section 1.1), and caused no detectable defects in control littermate embryos (Niederreither et al., 2002b,c).

The rescue of *Raldh2$^{-/-}$* embryos allowed to identify novel phenotypic defects, which in some cases could be reverted by maternally supplied RA in a stage- and dose-dependent manner. After a minimal rescue from E7.5 to E8.5, all *Raldh2$^{-/-}$* mutants had highly hypoplastic forelimbs (Niederreither et al., 2002a; Mic et al., 2004b). By extending the supplementation and/or increasing the amounts of RA provided, forelimb growth could be efficiently rescued in the mutants (Fig. 5A). Yet even the best rescued forelimbs showed alterations of the anterior–posterior patterning of skeletal elements (viz., mirror-image polydactyly). Molecular analysis showed that the mutant forelimb buds failed to develop a functional polarizing region. This signaling center, normally located in the posterodistal limb bud mesenchyme, produces Shh (Fig. 5B) that acts to coordinate both the growth and patterning of the limb bud by controlling expression of specific FGFs in the surface ectoderm and of various transcriptional regulators (including *Hox* genes) in the mesoderm (Tickle, 2006 for review). While the most severely growth-deficient *Raldh2$^{-/-}$* forelimb buds failed to activate *Shh*, in other instances a weak, ectopic *Shh* expression was seen along the whole distal limb margin, including in anterior regions (Fig. 5C and D). This correlated with abnormal expression of various genes (including *Hoxd* genes) in the anterior limb mesenchyme (Niederreither et al., 2002b).

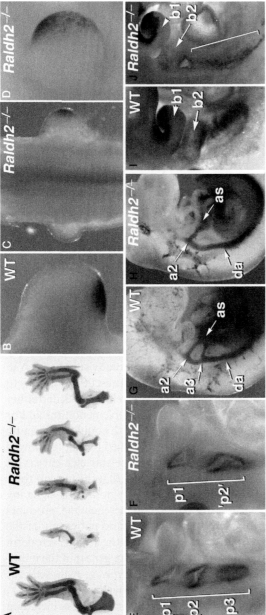

Fig. 5. Rescue of the lethality of *Raldh2⁻/⁻* embryos unveils novel developmental defects. (A) *Raldh2⁻/⁻* fetuses at E14.5 after a minimal RA-rescue from E7.5 to E8.5 have highly hypoplastic forelimbs (left-most sample, compare to the adjacent wild-type forelimb). By extending the duration and the doses of RA supplied, the growth and patterning of the *Raldh2⁻/⁻* forelimbs could be improved, although the best rescued limbs displayed abnormal anterior–posterior patterning (notice a six-digit polydactyly in the right-most sample). (B–D) At E10.5, the rescued *Raldh2⁻/⁻* embryos fail to activate *Sonic Hedgehog* (*Shh*), or express it in a inappropriate distal/anterior location within the forelimb buds (compare with the posteriorly restricted *Shh* domain in the wild-type limb bud). (E and F) *Pax9* specific expression in pharyngeal endoderm reveals abnormal development of branchial pouches (p1–p3) in rescued E9.5 *Raldh2⁻/⁻* mutants. The mutants exhibit a normal first pouch (p1) between the first and second branchial arches, and posteriorly have a single pouch-like structure (p2). (G and H) Impaired development of posterior branchial arches is also seen by the lack of third aortic arches (a3) following intracardiac ink injections (as, aortic sac; da, dorsal aorta). (I and J) *Fgf8* is expressed at low levels in a patchy domain along the pharyngeal endoderm of *Raldh2⁻/⁻* embryos (bracket), whereas it is normally expressed at high levels in the branchial pouch endoderm of wild-type embryos. In contrast, *Fgf8* expression levels are normal in the first arch (b1) ectoderm in mutants. A–D from Niederreither et al. (2002), Development 129, 3563; E–J from Niederreither et al., Development 130, 2525, with permission. (See Color Insert.)

These results show that embryonic RA, which is normally produced by RALDH2 throughout the lateral plate mesoderm before and during limb outgrowth, is required to achieve a strong, specific activation of *Shh* in posterior limb bud cells. This regulation could be direct, as the *Shh* promoter region contains a RARE (Chang et al., 1997). The basic helix loop helix factor Hand2 has also been implicated in the region-specific induction of *Shh* (Charité et al., 2000; Fernandez-Teran et al., 2000). *Hand2* expression is defective in the lateral plate mesoderm of unrescued *Raldh2$^{-/-}$* embryos (Mic et al., 2004b), although on RA-rescue its expression is only marginally altered within the abnormal limb buds (Niederreither et al., 2002b). These data suggest that the combinatory influences of a posteriorly restricted factor such as Hand2 and RA as an additional stimulatory influence are necessary for a full activation of *Shh* in the appropriate posterior limb domain, thus generating a functional polarizing region.

By analyzing activity of the RARE-*lacZ* reporter transgene in the rescued *Raldh2$^{-/-}$* mutants, we and others showed that maternally administered RA never allowed to restore normal levels of signaling as observed in wild-type littermates (Mic et al., 2002; Niederreither et al., 2002c, 2003). Thus, developmental events that may require a spatially restricted (or a highly regionally enriched) source of RA are unlikely to be rescued in the mutants, regardless of the amount of RA provided maternally. One consistent abnormality in the rescued mutants was an impaired development of the posterior (third to sixth) branchial arches (Fig. 5E and F) and of the corresponding aortic arteries (Fig. 5G and H; Niederreither et al., 2003). The second branchial arches developed normally in the rescued *Raldh2$^{-/-}$* embryos (Fig. 5E–J), whereas they did not form in the unrescued mutants (Fig. 2D). These results indicate that the pharyngeal mesendoderm requires high levels of RA to ensure normal development of the posterior branchial region, although the second branchial arches have a lower requirement which can be achieved on maternal RA supplementation. Several molecular targets were downregulated in the rescued *Raldh2$^{-/-}$* mutants, and some of them were also affected in wild-type embryos cultured in the presence of a RAR antagonist (Wendling et al., 2000). Reductions in *Fgf3*, *Fgf8*, *Hoxa1*, and *Hoxb1* were thus observed in various tissues of the posterior branchial region (ectoderm, mesenchyme, and foregut endoderm) of the *Raldh2$^{-/-}$* embryos (Fig. 3I and J, and data not shown), indicating that RA normally produced by RALDH2 in the mesoderm acts as a diffusible signal regulating gene expression in adjacent tissue layers.

Growth factor signaling in the pharyngeal region is also required to regulate cell populations that will contribute to the anterior pole of the heart, a region termed the secondary heart field (Kelly and Buckingham, 2002, and references therein). The pharyngeal defects in the rescued *Raldh2$^{-/-}$* embryos were found to alter the migratory pathways of postotic hindbrain neural crest cells (Niederreither et al., 2003). A subset of neural crest cells (the "cardiac" neural crest) normally migrate along the posterior aortic arches to reach the heart outflow tract, where their presence is required for the process of septation of the outflow tract into aorta and pulmonary artery (Waldo et al., 2005, and references therein). The rescued $Raldh2^{-/-}$ mutants consistently exhibit a lack of septation of the outflow tract, leading to a persistent truncus arteriosus, and their neural crest cells were found to colonize the outflow tract region less efficiently than in wild-type embryos (Niederreither et al., 2001). It remains to be seen whether the outflow tract defects in mutants are mostly due to this cell deficiency, and/or to molecular alterations in other populations derived from the anterior heart field. Another consequence of the pharyngeal defect in the rescued $Raldh2^{-/-}$ mutants is a severely impaired development of posterior cranial nerves. In particular, lack of vagal nerve outgrowth leads to aganglionosis in the developing stomach and gut wall, a condition that mimics human Hirschprung's disease (Niederreither et al., 2003).

3.2. Decreased RALDH2-mediated RA synthesis leads a phenocopy of DiGeorge syndrome abnormalities

There are several examples of engineered mouse lines in which a *neo* selectable marker has been inserted in an intronic location, and which behave as hypomorphic mutants. Reduced expression may be due to cryptic splice sites within the *neo* sequence leading to aberrant splicing and/or destabilization of the transcript product (Meyers et al., 1998). A *Raldh2* construct containing an intronic *neo* insertion was generated as a tool for conditional somatic mutagenesis (Section 3.3), and mice harboring the novel $Raldh2^{neo}$ allele were produced (either as homozygotes or heterozygotes with one null allele). At embryonic stages, the $Raldh2^{neo}$ mutants had overall decreases in *Raldh2* transcript and protein levels, as expected for a hypomorphic allele (Vermot et al., 2003). Most of these mutants died at perinatal stages due to a lack of septation of the heart outflow tract (persistent truncus arteriosus), a malformation which is lethal at birth because it does not allow separation of the pulmonary and systemic blood circulations. Additional abnormalities were observed, including an agenesis or hypoplasia of the thymus and parathyroid glands and the presence of abnormal laryngeal cartilages. All these structures are derivatives from the posterior branchial arches, yet no other organ system appeared to be affected in the hypomorphic mutants (Vermot et al., 2003). Analyses performed at early embryonic stages revealed a hypoplasia of the third to sixth branchial arches and a range of local molecular alterations that were reminiscent, although somewhat less severe, than those found in the RA-rescued $Raldh2^{-/-}$ embryos (Section 3.1).

The restricted set of developmental defects found in the hypomorphic *Raldh2* mutants have revealed a particular sensitivity of the developing posterior branchial arch region to decreased endogenous RA levels. Other organ systems seem to be more resistant to fluctuations of RA levels. Defects seen in the

Raldh2^{neo} mutants phenocopy the abnormalities characteristic of the human DiGeorge syndrome (DGS). This dominant syndrome is caused in most cases by a microdeletion of chomosome 22, leading to haploinsufficiency of a set of genes including *TBX1* and *CRKL* (Baldini, 2005, and references therein). Paradoxically, it was shown in *Crkl* and *Tbx1* mouse mutants that each individual mutation leads to locally increased RA signaling in the pharyngeal region (as seen by the activity of a RA-responsive transgene) and that *Raldh2* haploinsufficiency can rescue some of their phenotypic defects (Guris et al., 2006). These results highlight complex genetic interactions between retinoid signaling and the DGS-causing genes. In any events, the phenotype observed in *Raldh2* hypomorphic mutants suggest that conditions affecting embryonic RA levels, or its signaling ability, might play a role in the generation of human DGS-like abnormalities. Alcohol consumption during pregnancy has been implicated as one of the causes of DGS-like abnormalities. This could be because ethanol, by acting as a competitive substrate for the ADHs involved in the first reaction of RA synthesis (Section 1.5), could lead to a state of RA deficiency if present in the embryo during stages crucial for posterior pharyngeal arch development.

3.3. Conditional mutagenesis of Raldh2: A model to study RA function in brachial motor neuron specification

To study additional RALDH2 functions, we have generated a novel "floxed" (*loxP*-flanked) allele allowing tissue-specific somatic inactivation following Cre-mediated recombination. The excised allele was shown to behave as a null allele through a cross with transgenic mice expressing Cre in the germ line. Progeny from this cross exhibited early embryonic abnormalities similar to those of the *Raldh2$^{-/-}$* mutants (Vermot et al., 2006). We took advantage of this novel genetic tool to investigate the role of RALDH2 in the ontogenesis of limb-innervating neurons. RALDH2 has been shown to be expressed in specific pools of developing motor neurons at brachial and lumber levels of the spinal cord, which correspond to lateral motor columns (LMCs) that will innervate the limb muscles (Sockanathan and Jessell, 1998). By crossing the conditional mutants with a Cre line expressing the recombinase in the early neural tube (the *RARβ-Cre* line; Moon et al. 2000), a complete depletion of RALDH2 protein was achieved in the spinal cord LMC at brachial levels (Vermot et al., 2005b; Fig. 6A–D). Scattered RALDH2-expressing cells were detected in the adjacent meninges and mesenchymal cell populations at the level of the forelimb, indicating that the *RARβ-Cre* transgene-mediated incomplete excision outside of the neural tube (Fig. 6A and B). The resulting mice were viable, but displayed an abnormal flexure of the forelimb digits (Fig. 6E and F) possibly reflecting a neuromuscular defect. Whole-mount antineurofilament staining of E13.5

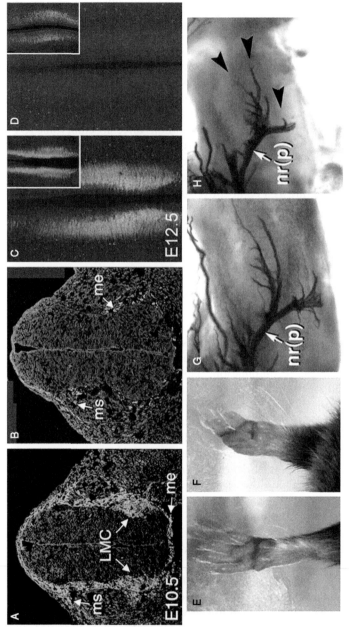

Fig. 6. Conditional mutagenesis of *Raldh2* leads to viable mutants with a loss of function in forelimb-innervating developing motor neurons. (A–D) Immunodetection of RALDH2 protein on transverse sections at brachial levels of the spinal cord of E10.5 wild-type and *RARβ-Cre;Raldh2* embryos (A and B) showing lack of protein in the lateral motor columns of the mutant, but persistent expression in adjacent meningeal (me) and mesenchymal (ms) cells. (C and D) are flat-mounts of E12.5 whole spinal cords analyzed by confocal microscopy (main panels: brachial levels; insets: lumbar levels). (E and F) Abnormal forelimb phenotype of adult *RARβ-Cre;Raldh2* mice. (G and H) Deficient outgrowth of the extensor branch (ramus profundus, p) of the nerve radialis (nr) in E13.5 *RARβ-Cre;Raldh2* embryos (arrowheads), as seen by whole-mount antineurofilament staining. From Vermot et al. (2005b), *Development* 132, 1611, with permission. (See Color Insert.)

RARβ-Cre;Raldh2 mutants showed atrophy and/or disorganization of one of the branches (ramus profundus) of the developing radialis nerve, whose axons normally innervate most of the extensor muscles of the forepaw (Fig. 6G and H). Unlike at brachial levels, excision of the *Raldh2* allele occurred in a mosaic pattern in the lumbar LMC, as seen by the presence of cells expressing the protein at E12.5 (compare Fig. 6C and D, insets), which could explain the absence of an abnormal hindlimb phenotype in the *RARβ-Cre;Raldh2* mutants.

The cellular basis of the forelimb defect was investigated, using markers of developing brachial motor neuron subpopulations. A slight ($\sim$10%) reduction of the overall number of motor neurons (assessed by counting Islet1/2$^+$ cells on serial spinal cord sections) was found in E11.5 mutants. The reduction mostly concerned a subset of the LMC neurons (the lateral LMC), which were identified by counting the numbers of Lim1$^+$/Islet2$^+$ neurons. Further analyses revealed an abnormal distribution of specific motor pools along the LMC (Vermot et al., 2005b). As RA has been implicated in regulating *Hox* gene expression, including within the developing spinal cord (Dasen et al., 2003), we analyzed whether *Hox* genes normally expressed at brachial levels may be altered in their expression in the *Raldh2* conditional mutants. It was thus found that both *HoxC6* and *HoxC8*, whose expression boundaries lie, respectively, at the rostral edge or within the wild-type *Raldh2* expression domain, were locally downregulated in the brachial LMC of *RARβ-Cre;Raldh2* mutants. Altogether, these mutants provide a viable animal model allowing to study the consequences of a lack of RA production with respect to brachial motor neuron development. This RA function had so far been only investigated in the chick embryo, using a combination of spinal cord explant cultures and blockade of RA signaling by *in vivo* electroporation of dominant negative RARs (Novitch et al., 2003; Sockanathan et al., 2003). The conditional mouse mutants also confirmed the importance of RA for the local control of *Hox* gene expression in the developing spinal cord.

4. Conclusions

Genetically engineered mice are now available with targeted loss of functions of most of the proteins acting in the RA signaling pathway. While these have sometimes led to unexpected results with respect to a lack of strict developmental requirement for suspected critical players (e.g., RBP or CRBPI), they have also revealed a high degree of functional redundancy between family members, for example, for the RA receptors or the ADHs that catalyze the first step in RA biosynthesis. In contrast, the study of *Raldh2* knockout mutants has revealed that a single enzyme is responsible for the tissue-specific production of RA required for a variety of processes in the early embryo. Additional functions of the same enzyme during late development and organogenesis have probably not been entirely characterized. Although systems such as phenotypic rescue by maternal RA have been useful, the generation of novel conditional mutants using Cre-mediated inactivation will be required to further characterize tissue-specific functions. Successful conditional mutagenesis could be complicated, however, by the fact that *Raldh2* has relatively widespread expression patterns and that RA diffusing from regions that were not targeted for loss of function may "artifactually" rescue some developmental events.

As most of the work achieved so far has involved analysis of specific molecular markers or candidate target genes, there is still a poor overall knowledge of the regulatory events that are RA dependent, and to what extent these are shared by—or specific to—certain developing systems. Furthermore, the mechanisms implicated in the RA-deficiency phenotypes have not always been characterized at the cellular level. Future work may benefit from the use of *in vivo* reporter mouse lines with fluorescent-tagged proteins, which can be crossed with RA-deficient mutants, thereby allowing the sorting of specific cell populations for molecular analyses, or the visualization of dynamic processes by time-lapse methods (Batourina et al., 2002, 2005). Additionally, genome-wide microarray approaches will be required for analysis of the transcriptome in various RA-target tissues. A better understanding of RA actions during embryonic and fetal development may potentially lead to applications in preventing congenital defects in humans. Furthermore, understanding how RA interacts with other growth factor or signaling pathways, such as FGFs and Shh, should extend the ability to use retinoids to regulate stem cell fate—including for therapeutic purposes.

Acknowledgments

We would like to thank Professors P. Chambon for his constant interest and support throughout these studies, G. Duester for a critical reading of the chapter, and all past and present members of our laboratories for their outstanding contributions. Work in the authors' laboratories is funded by the American Heart Association, the National Institute of Health (K.N.), the CNRS, INSERM, Hôpitaux Universitaires de Strasbourg, and Institut Universitaire de France and European Union (EVI-GENORET: LSHG-CT-2005–512036) (P.D.).

References

Abu-Abed, S., Dollé, P., Metzger, D., Beckett, B., Chambon, P., Petkovich, M. 2001. The retinoic acid-metabolizing enzyme, CYP26A1, is essential for normal hindbrain patterning, vertebral identity, and development of posterior structures. Genes Dev. 15, 226–240.

Ang, H.L., Deltour, L., Hayamizu, T.F., Zgombic-Knight, M., Duester, G. 1996. Retinoic acid synthesis in mouse embryos during gastrulation and craniofacial development linked to class IV alcohol dehydrogenase gene expression. J. Biol. Chem. 271, 9526–9534.

Baldini, A. 2005. Dissecting contiguous gene defects: TBX1. Curr. Opin. Genet. Dev. 15, 279–284.

Balmer, J.E., Blomhoff, R. 2002. Gene expression regulation by retinoic acid. J. Lipid Res. 43, 1773–1808.

Barak, Y., Nelson, M.C., Ong, E.S., Jones, Y.Z., Ruiz-Lozano, P., Chien, K.R., Koder, A., Evans, R.M. 1999. PPAR gamma is required for placental, cardiac, and adipose tissue development. Mol. Cell 4, 585–595.

Batourina, E., Gim, S., Bello, N., Shy, M., Clagett-Dame, M., Srinivas, S., Costantini, F., Mendelsohn, C. 2001. Vitamin A controls epithelial/mesenchymal interactions through Ret expression. Nat. Genet. 27, 74–78.

Batourina, E., Choi, C., Paragas, N., Bello, N., Hensle, T., Costantini, F.D., Schuchardt, A., Bacallao, R.L., Mendelsohn, C.L. 2002. Distal ureter morphogenesis depends on epithelial cell remodeling mediated by vitamin A and Ret. Nat. Genet. 32, 109–115.

Batourina, E., Tsai, S., Lambert, S., Sprenkle, P., Viana, R., Dutta, S., Hensle, T., Wang, F., Niederreither, K., McMahon, A.P., Carroll, T.J., Mendelsohn, C.L. 2005. Apoptosis induced by vitamin A signaling is crucial for connecting the ureters to the bladder. Nat. Genet. 37, 1082–1089.

Bouillet, P., Oulad-Abdelghani, M., Vicaire, S., Garnier, J.M., Schuhbaur, B., Dollé, P., Chambon, P. 1995. Efficient cloning of cDNAs of retinoic acid-responsive genes in P19 embryonal carcinoma cells and characterization of a novel mouse gene, Stra1 (mouse LERK-2/Eplg2). Dev. Biol. 170, 420–433.

Cartwright, M.M., Smith, S.M. 1995. Stage-dependent effects of ethanol on cranial neural crest cell development: Partial basis for the phenotypic variations observed in fetal alcohol syndrome. Alcohol Clin. Exp. Res. 19, 1454–1462.

Chambon, P. 1996. A decade of molecular biology of retinoic acid receptors. FASEB J. 10, 940–954.

Chang, B.E., Blader, P., Fischer, N., Ingham, P.W., Strahle, U. 1997. Axial (HNF3beta) and retinoic acid receptors are regulators of the zebrafish sonic hedgehog promoter. EMBO J. 16, 3955–3964.

Charité, J., McFadden, D.G., Olson, E.N. 2000. The bHLH transcription factor dHAND controls Sonic hedgehog expression and establishment of the zone of polarizing activity during limb development. Development 127, 2461–2470.

Chazaud, C., Chambon, P., Dollé, P. 1999. Retinoic acid is required in the mouse embryo for left-right asymmetry determination and heart morphogenesis. Development 126, 2589–2596.

Chen, T.H., Chang, T.C., Kang, J.O., Choudhary, B., Makita, T., Tran, C.M., Burch, J.B., Eid, H., Sucov, H.M. 2002. Epicardial induction of fetal cardiomyocyte proliferation via a retinoic acid-inducible trophic factor. Dev. Biol. 250, 198–207.

Collop, A.H., Broomfield, J.A., Chandraratna, R.A., Yong, Z., Deimling, S.J., Kolker, S.J., Weeks, D.L., Drysdale, T.A. 2006. Retinoic acid signaling is essential for formation of the heart tube in *Xenopus*. Dev. Biol. 291, 96–109.

Corson, L.B., Yamanaka, Y., Lai, K.M., Rossant, J. 2003. Spatial and temporal patterns of ERK signaling during mouse embryogenesis. Development 130, 4527–4537.

Dasen, J.S., Liu, J.P., Jessell, T.M. 2003. Motor neuron columnar fate imposed by sequential phases of Hoc-c activity. Nature 425, 926–933.

Deltour, L., Ang, H.L., Duester, G. 1996. Ethanol inhibition of retinoic acid synthesis as a potential mechanism for fetal alcohol syndrome. FASEB J. 10, 1050–1057.

Deltour, L., Foglio, M.H., Duester, G. 1999. Metabolic deficiencies in alcohol dehydrogenase Adh1, Adh3, and Adh4 null mutant mice. Overlapping roles of Adh1 and Adh4 in ethanol clearance and metabolism of retinol to retinoic acid. J. Biol. Chem. 274, 16796–16801.

Dersch, H., Zile, M.H. 1993. Induction of normal cardiovascular development in the vitamin A-deprived quail embryo by natural retinoids. Dev. Biol. 160, 424–433.

Dickman, E.D., Thaller, C., Smith, S.M. 1997. Temporally-regulated retinoic acid depletion produces specific neural crest, ocular and nervous system defects. Development 124, 3111–3121.

Diez del Corral, R., Storey, KG. 2004. Opposing FGF and retinoid pathways: A signalling switch that controls differentiation and patterning onset in the extending vertebrate body axis. Bioessays 26, 857–869.

Diez del Corral, R., Olivera-Martinez, I., Goriely, A., Gale, E., Maden, M., Storey, K. 2003. Opposing FGF and retinoid pathways control ventral neural pattern, neuronal differentiation, and segmentation during body axis extension. Neuron 40, 65–79.

Drysdale, T.A., Patterson, K.D., Saha, M., Krieg, P.A. 1997. Retinoic acid can block differentiation of the myocardium after heart specification. Dev. Biol. 188, 205–215.

Dubrulle, J., Pourquié, O. 2004a. Coupling segmentation to axis formation. Development 131, 5783–5793.

Dubrulle, J., Pourquié, O. 2004b. Fgf8 mRNA decay establishes a gradient that couples axial elongation to patterning in the vertebrate embryo. Nature 427, 419–422.

Dubrulle, J., McGrew, M.J., Pourquié, O. 2001. FGF signaling controls somite boundary position and regulates segmentation clock control of spatiotemporal Hox gene activation. Cell 106, 219–232.

Duester, G. 1991. A hypothetical mechanism for fetal alcohol syndrome involving ethanol inhibition of retinoic acid synthesis at the alcohol dehydrogenase step. Alcohol Clin. Exp. Res. 15, 568–572.

Duester, G. 2000. Families of retinoid dehydrogenases regulating vitamin A function: Production of visual pigment and retinoic acid. Eur. J. Biochem. 267, 4315–4324.

Duester, G., Mic, F.A., Molotkov, A. 2003. Cytosolic retinoid dehydrogenases govern ubiquitous metabolism of retinol to retinaldehyde followed by tissue-specific metabolism to retinoic acid. Chem. Biol. Interact. 143–144, 201–210.

Dupé, V., Davenne, M., Brocard, J., Dollé, P., Mark, M., Dierich, A., Chambon, P., Rijli, F.M. 1997. In vivo functional analysis of the Hoxa-1 3′ retinoic acid response element (3′ RARE). Development 124, 399–410.

Dupé, V., Ghyselinck, N.B., Wendling, O., Chambon, P., Mark, M. 1999. Key roles of retinoic acid receptors alpha and beta in the patterning of the caudal hindbrain, pharyngeal arches and otocyst in the mouse. Development 126, 5051–5059.

Dupé, V., Matt, N., Garnier, J.M., Chambon, P., Mark, M., Ghyselinck, N.B. 2003. A newborn lethal defect due to inactivation of retinaldehyde dehydrogenase type 3 is prevented by maternal retinoic acid treatment. Proc. Natl. Acad. Sci. USA 100, 14036–14041.

Durston, A.J., Timmermans, J.P., Hage, W.J., Hendriks, H.F., de Vries, N.J., Heideveld, M., Nieuwkoop, P.D. 1989. Retinoic acid causes an anteroposterior transformation in the developing central nervous system. Nature 340, 140–144.

Echevarria, D., Vieira, C., Gimeno, L., Martinez, S. 2003. Neuroepithelial secondary organizers and cell fate specification in the developing brain. Brain Res. Brain Res. Rev. 43, 179–191.

Fan, X., Molotkov, A., Manabe, S., Donmoyer, C.M., Deltour, L., Foglio, M.H., Cuenca, A.E., Blaner, W.S., Lipton, S.A., Duester, G. 2003. Targeted disruption of Aldh1a1 (Raldh1) provides evidence for a complex mechanism of retinoic acid synthesis in the developing retina. Mol. Cell. Biol. 23, 4637–4648.

Fernandez-Teran, M., Piedra, M.E., Kathiriya, I.S., Srivastava, D., Rodriguez-Rey, J.C., Ros, M.A. 2000. Role of dHAND in the anterior-posterior polarization of the limb bud: Implications for the Sonic hedgehog pathway. Development 127, 2133–2142.

Foley, A.C., Skromne, I., Stern, C.D. 2000. Reconciling different models of forebrain induction and patterning: A dual role for the hypoblast. Development 127, 3839–3854.

Gagnon, I., Duester, G., Bhat, P.V. 2002. Kinetic analysis of mouse retinal dehydrogenase type-2 (RALDH2) for retinal substrates. Biochim. Biophys. Acta 1596, 156–162.

Gale, E., Zile, M., Maden, M. 1999. Hindbrain respecification in the retinoid-deficient quail. Mech. Dev. 89, 43–54.

Gavalas, A. 2002. ArRAnging the hindbrain. Trends Neurosci. 25, 61–64.

Ghyselinck, N.B., Dupé, V., Dierich, A., Messaddeq, N., Garnier, J.M., Rochette-Egly, C., Chambon, P., Mark, M. 1997. Role of the retinoic acid receptor beta (RARbeta) during mouse development. Int. J. Dev. Biol. 41, 425–447.

Ghyselinck, N.B., Bavik, C., Sapin, V., Mark, M., Bonnier, D., Hindelang, C., Dierich, A., Nilsson, C.B., Hakansson, H., Sauvant, P., Azais-Braesco, V., Frasson, M., et al. 1999. Cellular retinol-binding protein I is essential for vitamin A homeostasis. EMBO J. 18, 4903–4914.

Gould, A., Itasaki, N., Krumlauf, R. 1998. Initiation of rhombomeric Hoxb4 expression requires induction by somites and a retinoid pathway. Neuron 21, 39–51.

Grun, F., Hirose, Y., Kawauchi, S., Ogura, T., Umesono, K. 2000. Aldehyde dehydrogenase 6, a cytosolic retinaldehyde dehydrogenase prominently expressed in sensory neuroepithelia during development. J. Biol. Chem. 275, 41210–41218.

Guris, D.L., Duester, G., Papaioannou, V.E., Imamoto, A. 2006. Dose-dependent interaction of Tbx1 and Crkl and locally aberrant RA signaling in a model of del22q11 syndrome. Dev. Cell 10, 81–92.

Halilagic, A., Zile, M.H., Studer, M. 2003. A novel role for retinoids in patterning the avian forebrain during presomite stages. Development 130, 2039–2050.

Haselbeck, R.J., Hoffmann, I., Duester, G. 1999. Distinct functions for Aldh1 and Raldh2 in the control of ligand production for embryonic retinoid signaling pathways. Dev. Genet. 25, 353–364.

Heine, U.I., Roberts, A.B., Munoz, E.F., Roche, N.S., Sporn, M.B. 1985. Effects of retinoid deficiency on the development of the heart and vascular system of the quail embryo. Virchows Arch. B Cell Pathol. Incl. Mol. Pathol. 50, 135–152.

Henderson, G.I., Chen, J.J., Schenker, S. 1999. Ethanol, oxidative stress, reactive aldehydes, and the fetus. Front. Biosci. 15, D541–D550.

Heyman, R.A., Mangelsdorf, D.J., Dyck, J.A., Stein, R.B., Eichele, G., Evans, R.M., Thaller, C. 1992. 9-Cis retinoic acid is a high affinity ligand for the retinoid X receptor. Cell 68, 397–406.

Hochgreb, T., Linhares, V.L., Menezes, D.C., Sampaio, A.C., Yan, C.Y., Cardoso, W.V., Rosenthal, N., Xavier-Neto, J. 2003. A caudorostral wave of RALDH2 conveys anteroposterior information to the cardiac field. Development 130, 5363–5374.

Hogan, B.L., Thaller, C., Eichele, G. 1992. Evidence that Hensen's node is a site of retinoic acid synthesis. Nature 359, 237–241.

Iulianella, A., Lohnes, D. 2002. Chimeric analysis of retinoic acid receptor function during cardiac looping. Dev. Biol. 247, 62–75.

Jepsen, K., Rosenfeld, M.G. 2002. Biological roles and mechanistic actions of co-repressor complexes. J. Cell Sci. 115, 689–698.

Jonk, L.J., de Jonge, M.E., Vervaart, J.M., Wissink, S., Kruijer, W. 1994. Isolation and developmental expression of retinoic-acid-induced genes. Dev. Biol. 161, 604–614.

Kastner, P., Grondona, J.M., Mark, M., Gansmuller, A., LeMeur, M., Decimo, D., Vonesch, J.L., Dollé, P., Chambon, P. 1994. Genetic analysis of RXR alpha developmental function: Convergence of RXR and RAR signaling pathways in heart and eye morphogenesis. Cell 78, 987–1003.

Kastner, P., Mark, M., Leid, M., Gansmuller, A., Chin, W., Grondona, J.M., Decimo, D., Krezel, W., Dierich, A., Chambon, P. 1996. Abnormal spermatogenesis in RXR beta mutant mice. Genes Dev. 10, 80–92.

Kastner, P., Messaddeq, N., Mark, M., Wendling, O., Grondona, J.M., Ward, S., Ghyselinck, N., Chambon, P. 1997a. Vitamin A deficiency and mutations of RXRalpha, RXRbeta and RAR-alpha lead to early differentiation of embryonic ventricular cardiomyocytes. Development 124, 4749–4758.

Kastner, P., Mark, M., Ghyselinck, N., Krezel, W., Dupe, V., Grondona, J.M., Chambon, P. 1997b. Genetic evidence that the retinoid signal is transduced by heterodimeric RXR/RAR functional units during mouse development. Development 124, 313–326.

Kawakami, Y., Raya, A., Raya, R.M., Rodriguez-Esteban, C., Belmonte, J.C. 2005. Retinoic acid signalling links left-right asymmetric patterning and bilaterally symmetric somitogen-esis in the zebrafish embryo. Nature 435, 165–171.

Keegan, B.R., Feldman, J.L., Begemann, G., Ingham, P.W., Yelon, D. 2005. Retinoic acid signaling restricts the cardiac progenitor pool. Science 307, 247–249.

Kelly, R.G., Buckingham, M.E. 2002. The anterior heart-forming field: Voyage to the arterial pole of the heart. Trends Genet. 18, 210–216.

Kessel, M., Gruss, P. 1991. Homeotic transformations of murine vertebrae and concomitant alteration of Hox codes induced by retinoic acid. Cell 67, 89–104.

Kostetskii, I., Yuan, S.Y., Kostetskaia, E., Linask, K.K., Blanchet, S., Seleiro, E., Michaille, J.J., Brickell, P., Zile, M. 1998. Initial retinoid requirement for early avian development coincides with retinoid receptor coexpression in the precardiac fields and induction of normal cardio-vascular development. Dev. Dyn. 213, 188–198.

Kostetskii, I., Jiang, Y., Kostetskaia, E., Yuan, S., Evans, T., Zile, M. 1999. Retinoid signaling required for normal heart development regulates GATA-4 in a pathway distinct from cardiomyocyte differentiation. Dev. Biol. 206, 206–218.

Lamb, A.L., Newcomer, M.E. 1999. The structure of retinal dehydrogenase type II at 2.7 A resolution: Implications for retinal specificity. Biochemistry 38, 6003–6011.

Lamb, T.D., Pugh, E.N., Jr. 2004. Dark adaptation and the retinoid cycle of vision. Prog. Retin. Eye. Res. 23, 307–380.

Lammer, E.J., Chen, D.T., Hoar, R.M., Agnish, N.D., Benke, P.J., Braun, J.T., Curry, C.J., Fernhoff, P.M., Grix, A.W., Jr., Lott, I.T., Richard, J.M., Sun, S.C., et al. 1985. Retinoic acid embryopathy. N. Engl. J. Med. 313, 837–841.

Lampron, C., Rochette-Egly, C., Gorry, P., Dollé, P., Mark, M., Lufkin, T., LeMeur, M., Chambon, P. 1995. Mice deficient in cellular retinoic acid binding protein II (CRABPII) or in both CRABPI and CRABPII are essentially normal. Development 121, 539–548.

Levin, A.A., Sturzenbecker, L.J., Kazmer, S., Bosakowski, T., Huselton, C., Allenby, G., Speck, J., Kratzeisen, C., Rosenberger, M., Lovey, A., Grippo, J.F. 1992. 9-Cis retinoic acid stereoisomer binds and activates the nuclear receptor RXR alpha. Nature 355, 359–361.

Li, H., Wagner, E., McCaffery, P., Smith, D., Andreadis, A., Dräger, U.C. 2000. A retinoic acid synthesizing enzyme in ventral retina and telencephalon of the embryonic mouse. Mech. Dev. 95, 283–289.

Lin, M., Zhang, M., Abraham, M., Smith, S.M., Napoli, J.L. 2003. Mouse retinal dehydro-genase 4 (RALDH4), molecular cloning, cellular expression, and activity in 9-cis-retinoic acid biosynthesis in intact cells. J. Biol. Chem. 278, 9856–9861.

Lohnes, D., Kastner, P., Dierich, A., Mark, M., LeMeur, M., Chambon, P. 1993. Function of retinoic acid receptor gamma in the mouse. Cell 73, 643–658.

Lohnes, D., Mark, M., Mendelsohn, C., Dollé, P., Dierich, A., Gorry, P., Gansmuller, A., Chambon, P. 1994. Function of the retinoic acid receptors (RARs) during development (I). Craniofacial and skeletal abnormalities in RAR double mutants. Development 120, 2723–2748.

Lufkin, T., Lohnes, D., Mark, M., Dierich, A., Gorry, P., Gaub, M.P., LeMeur, M., Chambon, P. 1993. High postnatal lethality and testis degeneration in retinoic acid receptor alpha mutant mice. Proc. Natl. Acad. Sci. USA 90(15), 7225–7229.

MacLean, G., Abu-Abed, S., Dollé, P., Tahayato, A., Chambon, P., Petkovich, M. 2001. Cloning of a novel retinoic-acid metabolizing cytochrome P450, Cyp26B1, and comparative expression analysis with Cyp26A1 during early murine development. Mech. Dev. 107, 195–201.

Maden, M. 2002. Retinoid signalling in the development of the central nervous system. Nat. Rev. Neurosci. 3, 843–853.

Maden, M., Sonneveld, E., van der Saag, P.T., Gale, E. 1998. The distribution of endogenous retinoic acid in the chick embryo: Implications for developmental mechanisms. Development 125, 4133–4144.

Marcucio, R.S., Cordero, D.R., Hu, D., Helms, J.A. 2005. Molecular interactions coordinating the development of the forebrain and face. Dev. Biol. 284, 48–61.

Mark, M., Ghyselinck, N.B., Wendling, O., Dupé, V., Mascrez, B., Kastner, P., Chambon, P. 1999. A genetic dissection of the retinoid signalling pathway in the mouse. Proc. Nutr. Soc. 58, 609–613.

Marshall, H., Studer, M., Popperl, H., Aparicio, S., Kuroiwa, A., Brenner, S., Krumlauf, R. 1994. A conserved retinoic acid response element required for early expression of the homeobox gene Hoxb-1. Nature 370, 567–571.

Matt, N., Dupé, V., Garnier, J.M., Dennefeld, C., Chambon, P., Mark, M., Ghyselinck, N.B. 2005a. Retinoic acid-dependent eye morphogenesis is orchestrated by neural crest cells. Development 132, 4789–4800.

Matt, N., Schmidt, C.K., Dupé, V., Dennefeld, C., Nau, H., Chambon, P., Mark, M., Ghyselinck, N.B. 2005b. Contribution of cellular retinol-binding protein type 1 to retinol metabolism during mouse development. Dev. Dyn. 233, 167–176.

McCaffery, P., Dräger, U.C. 1997. A sensitive bioassay for enzymes that synthesize retinoic acid. Brain Res. Brain Res. Protoc. 1, 232–236.

Mendelsohn, C., Lohnes, D., Décimo, D., Lufkin, T., LeMeur, M., Chambon, P., Mark, M. 1994. Function of the retinoic acid receptors (RARs) during development (II). Multiple abnormalities at various stages of organogenesis in RAR double mutants. Development 120, 2749–2771.

Merki, E., Zamora, M., Raya, A., Kawakami, Y., Wang, J., Zhang, X., Burch, J., Kubalak, S.W., Kaliman, P., Belmonte, J.C., Chien, K.R., Ruiz-Lozano, P. 2005. Epicardial retinoid X receptor alpha is required for myocardial growth and coronary artery formation. Proc. Natl. Acad. Sci. USA 102, 18455–18460.

Meyers, E.N., Lewandoski, M., Martin, G.R. 1998. An Fgf8 mutant allelic series generated by Cre- and Flp-mediated recombination. Nat. Genet. 18, 136–141.

Mic, F.A., Molotkov, A., Fan, X., Cuenca, A.E., Duester, G. 2000. RALDH3, a retinaldehyde dehydrogenase that generates retinoic acid, is expressed in the ventral retina, otic vesicle and olfactory pit during mouse development. Mech. Dev. 97, 227–230.

Mic, F.A., Haselbeck, R.J., Cuenca, A.E., Duester, G. 2002. Novel retinoic acid generating activities in the neural tube and heart identified by conditional rescue of Raldh2 null mutant mice. Development 129, 2271–2282.

Mic, F.A., Molotkov, A., Benbrook, D.M., Duester, G. 2003. Retinoid activation of retinoic acid receptor but not retinoid X receptor is sufficient to rescue lethal defect in retinoic acid synthesis. Proc. Natl. Acad. Sci. USA 100, 7135–7140.

Mic, F.A., Molotkov, A., Molotkova, N., Duester, G. 2004a. Raldh2 expression in optic vesicle generates a retinoic acid signal needed for invagination of retina during optic cup formation. Dev. Dyn. 231, 270–277.

Mic, F.A., Sirbu, I.O., Duester, G. 2004b. Retinoic acid synthesis controlled by Raldh2 is required early for limb bud initiation and then later as a proximodistal signal during apical ectodermal ridge formation. J. Biol. Chem. 279, 26698–26706.

Mollard, R., Viville, S., Ward, S.J., Décimo, D., Chambon, P., Dollé, P. 2000. Tissue-specific expression of retinoic acid receptor isoform transcripts in the mouse embryo. Mech. Dev. 94, 223–232.

Molotkov, A., Duester, G. 2002. Retinol/ethanol drug interaction during acute alcohol intoxication in mice involves inhibition of retinol metabolism to retinoic acid by alcohol dehydrogenase. J. Biol. Chem. 277, 22553–22557.

Molotkov, A., Fan, X., Deltour, L., Foglio, M.H., Martras, S., Farres, J., Pares, X., Duester, G. 2002a. Stimulation of retinoic acid production and growth by ubiquitously expressed alcohol dehydrogenase Adh3. Proc. Natl. Acad. Sci. USA 99, 5337–53342.

Molotkov, A., Deltour, L., Foglio, M.H., Cuenca, A.E., Duester, G. 2002b. Distinct retinoid metabolic functions for alcohol dehydrogenase genes Adh1 and Adh4 in protection against vitamin A toxicity or deficiency revealed in double null mutant mice. J. Biol. Chem. 277, 13804–13811.

Molotkov, A., Molotkova, N., Duester, G. 2006. Retinoic acid guides eye morphogenetic movements via paracrine signaling but is unnecessary for retinal dorsoventral patterning. Development 133, 1901–1910.

Moon, A.M., Boulet, A.M., Capecchi, M.R. 2000. Normal limb development in conditional mutants of Fgf4. Development 127, 989–996.

Moore, S.A., Baker, H.M., Blythe, T.J., Kitson, K.E., Kitson, T.M., Baker, E.N. 1998. Sheep liver cytosolic aldehyde dehydrogenase: The structure reveals the basis for the retinal specificity of class 1 aldehyde dehydrogenases. Structure 6, 1541–1551.

Moss, J.B., Xavier-Neto, J., Shapiro, M.D., Nayeem, S.M., McCaffery, P., Drager, U.C., Rosenthal, N. 1998. Dynamic patterns of retinoic acid synthesis and response in the developing mammalian heart. Dev. Biol. 199, 55–71.

Napoli, J.L. 1996. Retinoic acid biosynthesis and metabolism. FASEB J. 10, 993–1001.

Napoli, J.L. 1999. Interactions of retinoid binding proteins and enzymes in retinoid metabolism. Biochim. Biophys. Acta 1440, 139–162.

Napoli, J.L. 2001. 17beta-Hydroxysteroid dehydrogenase type 9 and other short-chain dehydrogenases/reductases that catalyze retinoid, 17beta- and 3alpha-hydroxysteroid metabolism. Mol. Cell. Endocrinol. 171, 103–109.

Niederreither, K., McCaffery, P., Dräger, U.C., Chambon, P., Dollé, P. 1997. Restricted expression and retinoic acid-induced downregulation of the retinaldehyde dehydrogenase type 2 (RALDH-2) gene during mouse development. Mech. Dev. 62, 67–78.

Niederreither, K., Subbarayan, V., Dollé, P., Chambon, P. 1999. Embryonic retinoic acid synthesis is essential for early mouse post-implantation development. Nat. Genet. 21, 444–448.

Niederreither, K., Vermot, J., Schuhbaur, B., Chambon, P., Dollé, P. 2000. Retinoic acid synthesis and hindbrain patterning in the mouse embryo. Development 127, 75–85.

Niederreither, K., Vermot, J., Messaddeq, N., Schuhbaur, B., Chambon, P., Dollé, P. 2001. Embryonic retinoic acid synthesis is essential for heart morphogenesis in the mouse. Development 128, 1019–1031.

Niederreither, K., Fraulob, V., Garnier, J.M., Chambon, P., Dollé, P. 2002a. Differential expression of retinoic acid-synthesizing (RALDH) enzymes during fetal development and organ differentiation in the mouse. Mech. Dev. 110, 165–171.

Niederreither, K., Vermot, J., Schuhbaur, B., Chambon, P., Dollé, P. 2002b. Embryonic retinoic acid synthesis is required for forelimb growth and anteroposterior patterning in the mouse. Development 129, 3563–3574.

Niederreither, K., Vermot, J., Fraulob, V., Chambon, P., Dollé, P. 2002c. Retinaldehyde dehydrogenase 2 (RALDH2)-independent patterns of retinoic acid synthesis in the mouse embryo. Proc. Natl. Acad. Sci. USA 99, 16111–16116.

Niederreither, K., Vermot, J., Le Roux, I., Schuhbaur, B., Chambon, P., Dollé, P. 2003. The regional pattern of retinoic acid synthesis by RALDH2 is essential for the development of posterior pharyngeal arches and the enteric nervous system. Development 130, 2525–2534.

Nieuwkoop, P.D., Botterenbrood, E.C., Kremer, A., Bloesma, F.F.S.N., Hoessels, E.M.L.J., Meyer, G., Verheyen, F.J. 1952. Activation and organization of the central nervous system in amphibians. J. Exp. Zool. 120, 1–108.

Nonaka, S., Yoshiba, S., Watanabe, D., Ikeuchi, S., Goto, T., Marshall, W.F., Hamada, H. 2005. De novo formation of left-right asymmetry by posterior tilt of nodal cilia. PLoS Biol. 3, e268.

Novitch, B.G., Wichterle, H., Jessell, T.M., Sockanathan, S. 2003. A requirement for retinoic acid-mediated transcriptional activation in ventral neural patterning and motor neuron specification. Neuron 40, 81–95.

Piantedosi, R., Ghyselinck, N., Blaner, W.S., Vogel, S. 2005. Cellular retinol-binding protein type III is needed for retinoid incorporation into milk. J. Biol. Chem. 280, 24286–24292.

Quadro, L., Blaner, W.S., Salchow, D.J., Vogel, S., Piantedosi, R., Gouras, P., Freeman, S., Cosma, M.P., Colantuoni, V., Gottesman, M.E. 1999. Impaired retinal function and vitamin A availability in mice lacking retinol-binding protein. EMBO J. 18, 4633–4644.

Quadro, L., Hamberger, L., Gottesman, M.E., Wang, F., Colantuoni, V., Blaner, W.S., Mendelsohn, C.L. 2005. Pathways of vitamin A delivery to the embryo: Insights from a new tunable model of embryonic vitamin A deficiency. Endocrinology 146, 4479–4490.

Raya, A., Izpisùa-Belmonte, J.C. 2006. Left-right asymmetry in the vertebrate embryo: From early information to higher-level integration. Nat. Rev. Genet. 7, 283–293.

Reijntjes, S., Gale, E., Maden, M. 2004. Generating gradients of retinoic acid in the chick embryo: Cyp26C1 expression and a comparative analysis of the Cyp26 enzymes. Dev. Dyn. 230, 509–517.

Ribes, V., Wang, Z., Dollé, P., Niederreither, K. 2006. Retinaldehyde dehydrogenase 2 (RALDH2)-mediated retinoic acid synthesis regulates early mouse embryonic forebrain development by controlling FGF and sonic hedgehog signaling. Development 133, 351–361.

Rossant, J., Zirngibl, R., Cado, D., Shago, M., Gigure, V. 1991. Expression of a retinoic acid response element-hsplacZ transgene defines specific domains of transcriptional activity during mouse embryogenesis. Genes Dev. 5, 1333–1344.

Ruberte, E., Friederich, V., Morriss-Kay, G., Chambon, P. 1992. Differential distribution patterns of CRABP I and CRABP II transcripts during mouse embryogenesis. Development 115, 973–987.

Saga, Y., Takeda, H. 2001. The making of the somite: Molecular events in vertebrate segmentation. Nat. Rev. Genet. 2, 835–845.

Sakai, Y., Meno, C., Fujii, H., Nishino, J., Shiratori, H., Saijoh, Y., Rossant, J., Hamada, H. 2001. The retinoic acid-inactivating enzyme CYP26 is essential for establishing an uneven distribution of retinoic acid along the anterio-posterior axis within the mouse embryo. Genes Dev. 15, 213–225.

Sakai, Y., Luo, T., McCaffery, P., Hamada, H., Drager, U.C. 2004. CYP26A1 and CYP26C1 cooperate in degrading retinoic acid within the equatorial retina during later eye development. Dev. Biol. 276, 143–157.

Sapin, V., Dollé, P., Hindelang, C., Kastner, P., Chambon, P. 1997. Defects of the chorioallantoic placenta in mouse RXRalpha null fetuses. Dev. Biol. 191, 29–41.

Schneider, R.A., Hu, D., Rubenstein, J.L., Maden, M., Helms, J.A. 2001. Local retinoid signaling coordinates forebrain and facial morphogenesis by maintaining FGF8 and SHH. Development 128, 2755–2767.

Serpente, P., Tumpel, S., Ghyselinck, N.B., Niederreither, K., Wiedemann, L.M., Dollé, P., Chambon, P., Krumlauf, R., Gould, A.P. 2005. Direct crossregulation between retinoic acid receptor {beta} and Hox genes during hindbrain segmentation. Development 132, 503–513.

Simeone, A., Avantaggiato, V., Moroni, M.C., Mavilio, F., Arra, C., Cotelli, F., Nigro, V., Acampora, D. 1995. Retinoic acid induces stage-specific antero-posterior transformation of rostral central nervous system. Mech. Dev. 51, 83–98.

Sirbu, I.O., Duester, G. 2006. Retinoic-acid signalling in node ectoderm and posterior neural plate directs left-right patterning of somitic mesoderm. Nat. Cell Biol. 8, 271–277.

Sirbu, I.O., Gresh, L., Barra, J., Duester, G. 2005. Shifting boundaries of retinoic acid activity control hindbrain segmental gene expression. Development 132, 2611–2622.

Sockanathan, S., Jessell, T.M. 1998. Motor neuron-derived retinoid signaling specifies the subtype identity of spinal motor neurons. Cell 94, 503–514.

Sockanathan, S., Perlmann, T., Jessell, T.M. 2003. Retinoid receptor signaling in postmitotic motor neurons regulates rostrocaudal positional identity and axonal projection pattern. Neuron 40, 97–111.

Stainier, D.Y., Fishman, M.C. 1992. Patterning the zebrafish heart tube: Acquisition of anteroposterior polarity. Dev. Biol. 153, 91–101.

Stuckmann, I., Evans, S., Lassar, A.B. 2003. Erythropoietin and retinoic acid, secreted from the epicardium, are required for cardiac myocyte proliferation. Dev. Biol. 255, 334–349.

Studer, M., Gavalas, A., Marshall, H., Ariza-McNaughton, L., Rijli, F.M., Chambon, P., Krumlauf, R. 1998. Genetic interactions between Hoxa1 and Hoxb1 reveal new roles in regulation of early hindbrain patterning. Development 125, 1025–1036.

Sun, S.Y., Lotan, R. 2002. Retinoids and their receptors in cancer development and chemoprevention. Crit. Rev. Oncol. Hematol. 41, 41–55.

Tahayato, A., Dollé, P., Petkovich, M. 2003. Cyp26C1 encodes a novel retinoic acid-metabolizing enzyme expressed in the hindbrain, inner ear, first branchial arch and tooth buds during murine development. Gene Exp. Patterns 3, 449–454.

Tanaka, Y., Okada, Y., Hirokawa, N. 2005. FGF-induced vesicular release of Sonic hedgehog and retinoic acid in leftward nodal flow is critical for left-right determination. Nature 435, 172–177.

Tickle, C. 2006. Making digit patterns in the vertebrate limb. Nat. Rev. Mol. Cell Biol. 7, 45–53.

Ulven, S.M., Gundersen, T.E., Weedon, M.S., Landaas, V.O., Sakhi, A.K., Fromm, S.H., Geronimo, B.A., Moskaug, J.O., Blomhoff, R. 2000. Identification of endogenous retinoids, enzymes, binding proteins, and receptors during early postimplantation development in mouse: Important role of retinal dehydrogenase type 2 in synthesis of all-*trans*-retinoic acid. Dev. Biol. 220, 379–391.

Vermot, J., Pourquié, O. 2005. Retinoic acid coordinates somitogenesis and left-right patterning in vertebrate embryos. Nature 435, 215–220.

Vermot, J., Niederreither, K., Garnier, J.M., Chambon, P., Dollé, P. 2003. Decreased embryonic retinoic acid synthesis results in a DiGeorge syndrome phenotype in newborn mice. Proc. Natl. Acad. Sci. USA 100, 1763–1768.

Vermot, J., Gallego Llamas, J., Fraulob, V., Niederreither, K., Chambon, P., Dollé, P. 2005a. Retinoic acid controls the bilateral symmetry of somite formation in the mouse embryo. Science 308, 563–566.

Vermot, J., Schuhbaur, B., Le Mouellic, H., McCaffery, P., Garnier, J.M., Hentsch, D., Brulet, P., Niederreither, K., Chambon, P., Dollé, P., Le Roux, I. 2005b. Retinaldehyde dehydrogenase 2 and Hoxc8 are required in the murine brachial spinal cord for the specification of Lim1+ motoneurons and the correct distribution of Islet1+ motoneurons. Development 132, 1611–1621.

Vermot, J., Garnier, J.M., Dierich, A., Niederreither, K., Harvey, R.P., Chambon, P., Dollé, P. 2006. Conditional (loxP-flanked) allele for the gene encoding the retinoic acid-synthesizing enzyme retinaldehyde dehydrogenase 2 (RALDH2). Genesis 44, 155–158.

Vogel, S., Piantedosi, R., O'Byrne, S.M., Kako, Y., Quadro, L., Gottesman, M.E., Goldberg, I.J., Blaner, W.S. 2002. Retinol-binding protein-deficient mice: Biochemical basis for impaired vision. Biochemistry 41, 15360–15368.

Wagner, E., McCaffery, P., Dräger, U.C. 2000. Retinoic acid in the formation of the dorsoventral retina and its central projections. Dev. Biol. 222, 460–470.

Waldo, K.L., Hutson, M.R., Stadt, H.A., Zdanowicz, M., Zdanowicz, J., Kirby, M.L. 2005. Cardiac neural crest is necessary for normal addition of the myocardium to the arterial pole from the secondary heart field. Dev. Biol. 281, 66–77.

Ward, S.J., Chambon, P., Ong, D.E., Bavik, C. 1997. A retinol-binding protein receptor-mediated mechanism for uptake of vitamin A to postimplantation rat embryos. Biol. Reprod. 57, 751–755.

Wendling, O., Chambon, P., Mark, M. 1999. Retinoid X receptors are essential for early mouse development and placentogenesis. Proc. Natl. Acad. Sci. USA 96, 547–551.

Wendling, O., Dennefeld, C., Chambon, P., Mark, M. 2000. Retinoid signaling is essential for patterning the endoderm of the third and fourth pharyngeal arches. Development 127, 1553–1562.

Wendling, O., Ghyselinck, N.B., Chambon, P., Mark, M. 2001. Roles of retinoic acid receptors in early embryonic morphogenesis and hindbrain patterning. Development 128, 2031–2038.

White, J.C., Highland, M., Kaiser, M., Clagett-Dame, M. 2000. Vitamin A deficiency results in the dose-dependent acquisition of anterior character and shortening of the caudal hindbrain of the rat embryo. Dev. Biol. 220, 263–284.

Wilson, J.G., Roth, C.B., Warkany, J. 1953. An analysis of the syndrome of malformations induced by maternal vitamin A deficiency. Effects of restoration of vitamin A at various times during gestation. Am. J. Anat. 92, 189–217.

Wilson, L., Maden, M. 2005. The mechanisms of dorsoventral patterning in the vertebrate neural tube. Dev. Biol. 282, 1–13.

Wilson, S.W., Houart, C. 2004. Early steps in the development of the forebrain. Dev. Cell 6, 167–181.

Xavier-Neto, J., Neville, C.M., Shapiro, M.D., Houghton, L., Wang, G.F., Nikovits, W., Jr., Stockdale, F.E., Rosenthal, N. 1999. A retinoic acid-inducible transgenic marker of sino-atrial development in the mouse heart. Development 126, 2677–2687.

Xavier-Neto, J., Rosenthal, N., Silva, F.A., Matos, T.G., Hochgreb, T., Linhares, V.L. 2001. Retinoid signaling and cardiac anteroposterior segmentation. Genesis 31, 97–104.

Yashiro, K., Zhao, X., Uehara, M., Yamashita, K., Nishijima, M., Nishino, J., Saijoh, Y., Sakai, Y., Hamada, H. 2004. Regulation of retinoic acid distribution is required for proximodistal patterning and outgrowth of the developing mouse limb. Dev. Cell 6, 411–422.

Yutzey, K.E., Rhee, J.T., Bader, D. 1994. Expression of the atrial-specific myosin heavy chain AMHC1 and the establishment of anteroposterior polarity in the developing chicken heart. Development 120, 871–883.

Zhao, D., McCaffery, P., Ivins, K.J., Neve, R.L., Hogan, P., Chin, W.W., Dräger, U.C. 1996. Molecular identification of a major retinoic-acid-synthesizing enzyme, a retinaldehyde-specific dehydrogenase. Eur. J. Biochem. 240, 15–22.

Zile, M.H. 2001. Function of vitamin A in vertebrate embryonic development. J. Nutr. 131, 705–708.

Zile, M.H., Kostetskii, I., Yuan, S., Kostetskaia, E., St Amand, T.R., Chen, Y., Jiang, W. 2000. Retinoid signaling is required to complete the vertebrate cardiac left/right asymmetry pathway. Dev. Biol. 223, 323–338.

Hindbrain development and retinoids

Joel C. Glover,[1] Jean-Sébastien Renaud,[1]
Xavier Lampe[2] and Filippo M. Rijli[2]

[1]Department of Physiology, University of Oslo, Norway
[2]Institut de Génétique et de Biologie Moléculaire et Cellulaire,
CNRS/INSERM/ULP, Illkirch Cedex, France

Contents

Early studies in the vertebrate embryo showed that both retinoid excess and deficiency were teratogenic, resulting in significant defects in the developmental patterning of the hindbrain. These results led to the suggestion that endogenous retinoids could also play a physiological role in anteroposterior hindbrain patterning. The spatially restricted expression of retinoid

Advances in Developmental Biology
Volume 16 ISSN 1574-3349
DOI: 10.1016/S1574-3349(06)16005-6

synthetic and degradative enzymes, binding proteins, and receptors in the hindbrain and neighboring regions of the neuroepithelium and mesoderm lend further support to this idea. Moreover, the molecular patterning of the hindbrain, including the regionalized expression of the *Hox* genes and other developmental regulatory genes, is profoundly influenced by retinoid signaling. In this chapter, we review the work leading up to current models of retinoid action in vertebrate hindbrain development and outline some of the major challenges that remain.

1. Introduction

One of the regions in the central nervous system in which the developmental role of retinoids has been most intensively studied is the segmented hindbrain. This is partly due to the discovery that the patterning along the rostrocaudal axis of the hindbrain involves a combinatorial code of *Hox* gene expression (Wilkinson et al., 1989; Kessel and Gruss, 1991; Maden et al., 1991; Gale et al., 1996; Rijli et al., 1998) and that *Hox* expression can be regulated by retinoic acid (RA) both in cell culture (Simeone et al., 1990) and in the embryo (Kessel and Gruss, 1991; Conlon and Rossant, 1992; Marshall et al., 1996; Dupe et al., 1997). Moreover, the localization of teratogenic effects to the hindbrain region suggested retinoids as potentially important molecular agents in hindbrain development (Morriss, 1972; Durston et al., 1989; Holder and Hill, 1991). The subsequent characterization of retinoid responses (Rossant et al., 1991; Balkan et al., 1992; Colbert et al., 1993), endogenous retinoids (Wagner et al., 1990; Chen et al., 1992; Zhao et al., 1996; Maden et al., 1998; Maden, 2002), and retinoic acid synthesis (Niederreither et al., 1997; Berggren et al., 1999) demonstrating a source of retinoic acid in the cervical region, as well as the localization of degradative enzymes to the rostral hindbrain creating a sink (Yamamoto et al., 1998; de Roos et al., 1999), led to the suggestion that endogenous retinoid levels were graded from posterior to anterior hindbrain segments. Recent work dissecting the regulatory pathways of *Hox* and other segmentally expressed genes has provided further support to the idea that a spatiotemporally controlled gradient of retinoids contributes to the antero-posterior patterning of the hindbrain (Grapin-Botton et al., 1995; Gale et al., 1996; Itasaki et al., 1996; Marshall et al., 1996; Blumberg et al., 1997; Gould et al., 1998; Grapin-Botton et al., 1998; Rijli et al., 1998; van der Wees et al., 1998; Gavalas and Krumlauf, 2000; Trainor and Krumlauf, 2000; Dupe and Lumsden, 2001; Maden, 2002; Shiotsugu et al., 2004; Maves and Kimmel, 2005; Sirbu et al., 2005). In this chapter, we review the work leading up to current models of retinoid action in the hindbrain and outline some of the major challenges that remain.

2. Teratogenic effects of retinoid excess and deficiency on hindbrain development

Already over 30 years ago, it was demonstrated that fetal exposure to excess vitamin A leads to teratogenic alterations in the hindbrain of the vertebrate fetus (Morriss, 1972). In particular, vitamin A administration to pregnant rats shortly before or during the early neurulation stages led to a shortening of the preotic region of the hindbrain in the fetus (Morriss, 1972; Morriss and Thorogood, 1978). In the mouse (Morriss-Kay et al., 1991; Wood et al., 1994), teratogenic effects on the hindbrain were shown to be stage dependent, varying from an apparent loss of the rostral (preotic) hindbrain following exposures just prior to somite formation, to a normal morphology accompanied by an anterior shift in segmental patterns of gene expression following exposures just after somite formation (Fig. 1A). Similar apparent loss of preotic hindbrain on retinoid administration was observed in *Xenopus* (Durston et al., 1989; Papalopulu et al., 1991) and zebrafish embryos (Holder and Hill, 1991). However, based on the analysis of the expression of anterior hindbrain genes and neural crest migration and differentiation, Morriss-Kay et al. (1991) and Wood et al. (1994) interpreted such a phenotype as a contraction and a lack of segmentation, rather than a loss, of the preotic hindbrain that could conceivably involve a redirection of cell fate from neural to other lineages (Agarwal and Sato, 1993). Moreover, the expression patterns of *Hox* and segmentation genes, such as *Krox-20*, in the anterior hindbrain were changed toward a more posterior segmental identity (Durston et al., 1989; Morriss-Kay et al., 1991; Conlon and Rossant, 1992; Marshall et al., 1992; Sundin and Eichele, 1992; Wood et al., 1994; Hill et al., 1995). In the most severe cases, the region normally encompassing rhombomeres (r) 1–4 had the appearance of a single unsegmented territory carrying an r4 expression pattern. This "posteriorizing" effect also appeared to depend on which retinoid was administered, with 9-*cis*-RA having a more pronounced effect than atRA in zebrafish (Zhang et al., 1996; see also Kraft et al., 1994; Kraft and Juchau, 1995). These studies showed that a retinoid excess administered within a specific time window at early developmental stages profoundly disrupts the patterning of the anterior portion of the hindbrain, while leaving the posterior part evidently unaffected. However, a loss of segmentation without overt changes in gene expression patterns could also be obtained in the chicken embryo following local exposure to exogenous retinoids at somewhat later stages, when hindbrain segmentation is beginning (Nittenberg et al., 1997).

Notably, retinoid exposure at early stages *increases* the size of the hindbrain, at the expense of other brain regions (Avantaggiato et al., 1996). This is due to a reprogramming of the subdivision of the neural tube into its main territories (forebrain, midbrain, hindbrain, spinal cord), as opposed to a disruption of patterning within the hindbrain as discussed earlier. Whether the

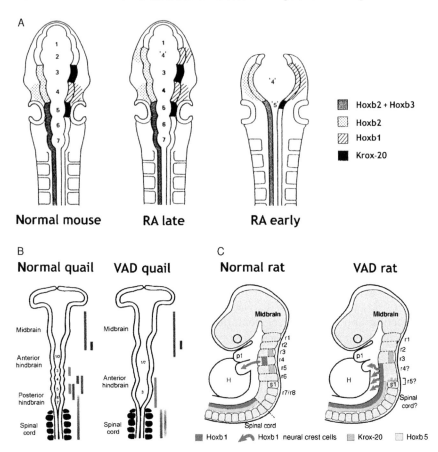

Fig. 1. The teratogenic effects of retinoid excess and deficiency on early patterning of the hindbrain. (A) Stage-dependent teratogenesis in the mouse embryo caused by exposure to exogenous retinoic acid (modified from Wood et al., 1994). (B) The effects of retinoid deficiency in the VAD quail embryo (modified from Gale et al., 1999). (C) The effects of retinoid deficiency in the VAD rat embryo (modified from White et al., 2000). (See Color Insert.)

increase in hindbrain territory is accompanied by abnormal hindbrain patterning has not been addressed.

Retinoid deficiency also has teratogenic effects (Kalter and Warkany, 1959) that have been most clearly elucidated in vitamin A deficient (VAD) animal models (Wilson et al., 1953; Thompson et al., 1969; Dersch and Zile, 1993). In the VAD quail, the anterior hindbrain appears normal, but the posterior hindbrain never differentiates the gene expression patterns characteristic of that region. Consequently, the cells that would normally give rise to the posterior rhombomeres (r4–r7) appear to contribute instead to an enlarged r3 that extends to the first somite, possibly due to a partial respecification of posterior rhombomeres to a more anterior fate (Gale et al., 1999; Fig. 1B). In a VAD rat model, the caudal part of the hindbrain is shortened and acquires an abnormal gene expression pattern, as if r5–r8 were transformed into r4, although the anterior hindbrain appears normal (White et al., 2000; Fig. 1C). Thus, the teratogenic effects induced by retinoid excess and deficiency appear to be complementary in the hindbrain, targeting predominantly anterior (r1–r4) and posterior (r4–r8) territories, respectively.

3. Endogenous sources of retinoids and sinks

3.1. Early patterning

The region-specific effects of retinoic acid excess and deficiency suggested that it could be acting as an endogenous signal to pattern the anteroposterior axis of the hindbrain, similar to its role in patterning the developing and regenerating limb (Thaller and Eichele, 1987; Scadding and Maden, 1994). Since retinoid excess disrupted the anterior hindbrain whereas retinoid deficiency impaired the posterior hindbrain, this implied that endogenous retinoid levels should be relatively high in the posterior hindbrain and relatively low in the anterior hindbrain. However, the detection of endogenous retinoids in small tissue samples has been a quite hard task, rendering a biochemical mapping of retinoid distribution within the neural tube a difficult challenge at the resolution of brain regions. Thus, several research groups turned to reporter mice and cell lines to detect endogenous retinoids in neural tissue. In mice carrying the *lacZ* reporter gene driven by retinoic acid response elements (RAREs), gene activity was initially reported in the spinal cord, eventually resolving at relatively late developmental stages to "hot spots" at cervical and lumbar levels, although no detectable activity was described in the hindbrain (Rossant et al., 1991; Balkan et al., 1992; Colbert et al., 1993). More recently, detailed temporal analyses at earlier developmental stages were carried out showing that, in E8.25–E9.25 *RARE-lacZ* embryos, retinoic acid activity is found in the posterior hindbrain up to r5 (Sakai et al., 2001; Mic et al., 2002). At even earlier stages, retinoic acid activity is transiently spreading from the presumptive posterior hindbrain up to a quite rostral limit corresponding to the future r2/r3 boundary (Sirbu et al., 2005; Fig. 2). The reporter gene approach has also been applied to the zebrafish by Perz-Edwards et al. (2001) who observed a dynamic retinoic acid signal in the neural tube that initially appeared in the region extending from somite 8 to just anterior to somite 1, then extended further into the posterior hindbrain over the next 12 hours before retreating back to thoracic spinal cord levels.

To assess whether such patterns of retinoic acid activity were due to spatially restricted production of retinoids and/or localization of retinoid

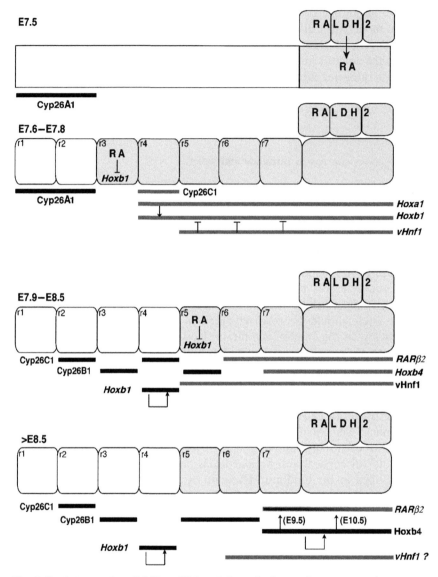

Fig. 2. Spatiotemporal availability of RA and dynamic changes in gene expression patterns in the mouse hindbrain (modified from Sirbu et al., 2005). The extent of RA activity detected in the developing hindbrain, as assessed by an RA-reporter transgene (*RARE-lacZ*), is depicted in light gray. Genes under the control of RA are depicted as gray boxes, and black boxes are for gene expression domains not controlled by RA. In the mouse embryo, rhombomeres become visible around E8.5–E9.0, although segmental gene expression could be appreciated already at earlier stages. For the sake of simplicity, we have drawn segment boundaries depicting presumptive rhombomeres starting from E7.6 to E7.8. E7.5: The expression of the RA-synthesizing enzyme RALDH2 starts around E7.0 in mesoderm lateral to the primitive streak. At this stage, there is no RA activity in the developing neural plate. At E7.5, the RARE-lacZ

receptors, Colbert et al. (1993) transfected a cell line with the same *RARE-lacZ* construct and placed selected neural tube regions onto these reporter cells. They found that reporter cells responded only to regions of the neural tube that showed strong reporter activity *in situ*, indicating that there are local sources of retinoid production that are restricted to spinal levels. Similar results using reporter cell lines were obtained by Chen et al. (1994) in *Xenopus*, Wagner et al. (1992) in the rat, and McCaffery and Drager (1994) in the mouse, who also used in parallel a zymography assay to assess the distribution of retinoid synthesis. Synthetic activity had the same anteroposterior distribution as the endogenous retinoids detected by the reporter cell line. The localization of a source of retinoic acid production at upper spinal and lower hindbrain levels was also supported by the analysis of the spatial distribution of a main retinoic acid synthetic enzyme, retinaldehyde dehydrogenase 2 (RALDH2; Zhao et al., 1996), using *in situ* hybridization and immunohistochemistry in mouse and chicken embryos (Niederreither et al., 1997; Berggren et al., 1999; Swindell et al., 1999). RALDH2 is expressed in paraxial (somitic) mesoderm along the spinal cord, although it is not present rostral to the first somite. Since the first 4–6 somites lie adjacent to the most posterior rhombomeres (r7–r8), this suggests that retinoic acid availability should be high in posterior hindbrain. RALDH2 expression was subsequently investigated by *in situ* hybridization in zebrafish (Begemann et al., 2001; Dobbs-McAuliffe et al., 2004) and *Xenopus* (Chen et al., 2001) embryos, and by immunohistochemistry in mouse (Haselbeck et al., 1999) embryos. The consistent finding is that RALDH2 is expressed in paraxial mesoderm up to and including the first somite (zebrafish, chicken, mouse) or in presomitic mesoderm and the spinal roof plate (*Xenopus*)

---

expression is detected near the node. The expression of the RA-degrading enzyme Cyp26A1 is already detected at the level of the presumptive r1 and r2. E7.6–E7.8: The RA produced from the somites diffuses anteriorly in the neural tube and reaches the r2/r3 border, next to the posterior border of expression of Cyp26A1. At the level of r4, RA triggers the expression of another RA-degrading enzyme, Cyp26C1. The initial phase of expression of *Hoxa1*, *Hoxb1*, and *vHnf1* requires RA. They progressively reach their anterior borders of expression around E8.0. E7.9–E8.5: Since E7.9, the anterior limit of RA activity retreats posteriorly and is now detected at the r4/r5 border where it is sustained at least until E9.5. This caudal withdrawal may be attributed to the expression of multiple Cyp26 enzymes in the anterior part of the hindbrain: Cyp26C1 in r2 and r4 and, from E8.0, Cyp26B1 in r3 and r5. At this stage, there is no more Cyp26A1 in r1 and r2. The expression of *Hoxb1* in r4 is at the center of a complex regulatory network (see Fig. 4 for legend). In addition to direct repression of *Hoxb1* in r3/r5 through its 5′ DR2 (see also Fig. 4), RA also negatively regulates *Hoxb1* indirectly through the activation of the transcription factor vHnf1, which helps repression of *Hoxb1* throughout the posterior hindbrain. It is not known whether RA is also required in the maintenance of *vHnf1* expression (*vHnf1*?). In the posterior hindbrain and in the neural tube, RA activates the expression of the retinoic acid receptor *RARβ2* and of the *Hoxb4* gene. E8.5 and later: *Hoxb1* maintenance is under the control of a strong autoregulatory loop and *vHnf1* boundary of expression retreats by about one rhombomere. From E9.5, Hoxb4 controls its own expression in the hindbrain and that of *RARβ2* in r7, and in r7 and anterior spinal cord from E10.5. The *RARβ2* expression shifts from an RA-induced pathway toward a Hox-regulated control (black–gray box).

during early developmental stages. Thus, in all four species, the source of retinoic acid synthesis is in a region neighboring the posterior part of the hindbrain.

Subsequently, targeted knockout of *RALDH2* in the mouse showed that the absence of this enzyme causes hindbrain defects very similar to (but not as severe as) those generated in VAD animal models: the posterior hindbrain is reduced in size and exhibits gene expression patterns characteristic of anterior rhombomeres (Niederreither et al., 2000). Similarly, the *neckless* zebrafish mutant, which carries a null mutation in *RALDH2*, causes a shortening in the size of the posterior hindbrain between r5 and the first somite, even though the effect appears to be restricted at the posteriormost part of this region (Begemann et al., 2001). By contrast, overexpression of RALDH2 in *Xenopus* caused a posteriorization of the hindbrain (Chen et al., 2001). Thus, these studies indicated that RALDH2 synthesizes retinoic acid at cervical levels, which then diffuses anteriorly into the hindbrain to exert effects essential for the patterning of the postotic and posterior hindbrain. These results may therefore be consistent with a gradient model in which posterior hindbrain experiences relatively high concentrations and anterior hindbrain relatively low concentrations of retinoic acid, the two regions then being patterned according to the concentration-dependent effects of retinoic acid on the transcription of genes such as the *Hox* genes. The localization of catabolic enzymes that break down retinoids in the rostral hindbrain (see later) appeared as a suitable complementary sink for maintaining such a retinoid gradient.

RALDH2, however, is not the only enzyme synthesizing retinoic acid. Three other retinal dehydrogenases, with different tissue distributions, substrate specificities, and enzymatic efficiencies, have been identified: RALDH1, RALDH3, and RALDH4. Of these, RALDH1 and RALDH3 are expressed predominantly in optic and nasal structures as well as some telencephalic and mesencephalic neuronal populations (Li et al., 2000; Mic et al., 2000; Niederreither et al., 2002; Blentic et al., 2003). RALDH4 is instead expressed predominantly in liver and kidney (Lin et al., 2003). Thus, none of these has been implicated in the synthesis of retinoic acid required for the early patterning of the hindbrain. Despite the significant evidence that RALDH2 is primarily responsible for generating the retinoic acid influencing the patterning of the early hindbrain, data indicate that additional, as yet unidentified synthetic sources may also be involved. This evidence comes from studies in which retinoic acid-sensitive reporter systems are expressed on a background of the *RALDH2* knockout mouse (Mic et al., 2002; Niederreither et al., 2002), as well as from pharmacological blockade of residual retinoic acid synthesis in the zebrafish *neckless* and *no-fin* mutants (Grandel et al., 2002; Begemann et al., 2004). *RALDH2* knockout mouse embryos normally die at E9.5–E10.5 but can be rescued from lethality by a transient maternal retinoic acid supplementation at E7.5–E9.5. In the absence of dietary supplementation, no retinoid response is detectable anywhere in the *RALDH2* knockout, as expected, but in the ocular region (where RALDH1 and RALDH3 are expressed). Two days after terminating the dietary supplementation, however, reporter activity is seen in the ventral spinal cord and hindbrain up to r5, despite the complete absence of RALDH2 and the lack of RALDH1 or RALDH3 expression. This suggests that some other, presumably enzymatic, source must be responsible. In the zebrafish *neckless* mutant (which lacks RALDH2), the phenotype more closely approaches a VAD phenotype when treated with pharmacological blockers of retinoic acid synthesis, again suggesting that RALDH2 is not the only enzyme involved. An additional source of retinoic acid synthesis in the posterior hindbrain could explain why the defects seen in the *RALDH2* knockout and the zebrafish *neckless* mutant are not as severe as those observed in VAD animals (Niederreither et al., 2000; Begemann et al., 2004). Whether the as yet unidentified source is active during normal development, or is only switched on as a consequence of large variations in retinoid availability (such as would occur with retinoic acid treatment in the absence of RALDH2), is unclear.

Reporter constructs and enzyme distributions indicate *where* retinoids are being produced, but they provide little information (aside from the substrate specificity profiles of enzymes) about *which* retinoids are being produced. A number of studies have addressed this issue by means of biochemical detection methods to identify and quantitate specific retinoids in neural tissue. This approach is technically challenging, due to the intrinsic instability of retinoids and their rapid conversion to alternative molecular forms by light and oxygen, the limits of detection of various methods, and the problem of discriminating molecular forms with similar chemical properties. Caution must therefore be exerted in interpreting the results of such studies, particularly when the conditions of tissue collection and sample processing are not described in detail.

In most species examined with acceptable methods, atRA has been detected in whole embryos or in the developing neural tube. These include zebrafish (Costaridis et al., 1996), chicken (Maden et al., 1998), and mouse (Horton and Maden, 1995; Ulven et al., 2001). In some species, other forms of retinoic acid have also been detected. In the neural tube of the chicken embryo, in particular, 3,4-didehydro-atRA is present in greater abundance than atRA (Maden et al., 1998; Hoover et al., 2001). In *Xenopus*, initial studies reported the presence of atRA and 9-*cis*-RA (Kraft et al., 1994). In contrast, later studies, using a more sensitive and specific detection system, showed that these retinoic acid isomers were not reproducibly detectable, whereas 4-oxo-ROL, 4-oxo-RAL, and 4-oxo-RA were present in abundance, and were bioactive (Blumberg et al., 1996). Since catabolic enzymes generate 4-oxo- and 4-hydroxy-retinoid derivatives (Pijnappel et al., 1993; see later), it is not clear whether the presence of these forms merely indicates that atRA and its precursors are much more rapidly catabolized in *Xenopus* than in other species.

Given the likely misidentification of 9-*cis*-RA in *Xenopus* (Kraft et al., 1994; Blumberg et al., 1996), and the general inability to detect 9-*cis*-RA in neural tissue in any other species, it is remarkable that two studies have observed strong reporter gene activation in the neural tube using ostensibly 9-*cis*-RA–specific reporter constructs. Solomin et al. (1998), using a transgenic mouse in which RXR activation drives a *Gal4* reporter, and Luria and Furlow (2004), using a similar *RXRα-Gal4* approach in *Xenopus*, have both observed reporter activity in the spinal neural tube. As with RAR-mediated reporter activity, the RXR-mediated reporter activity does not extend detectably into the hindbrain in either species. These results indicate that RXRs are being activated and since RXRs are generally assumed to be 9-*cis*-RA– selective receptors, the implication is that 9-*cis*-RA is present in cervical regions along with atRA and that both could be diffusing into the hindbrain. This could conceivably be reconciled with the lack of biochemical detection of 9-*cis*-RA if 9-*cis*-RA were produced by far fewer cells or in much lower concentrations per cell than atRA. However, two alternative explanations exist. First, RXRs can be activated by nonretinoid ligands (Goldstein et al., 2003) and by mechanisms other than ligand binding (Matkovits and Christakos, 1995). Second, RXRs, although they bind 9-*cis*-RA much more avidly than atRA, can bind and be activated by atRA with low affinity (Allenby et al., 1993), and this activity might be revealed if 9-*cis*-RA were completely absent. In any case, it seems most likely that atRA is the principal endogenous retinoic acid form acting on the hindbrain in most species, but there remains the possibility that 9-*cis*-RA and 4-oxo-RA, as well as 4-oxo-RAL, may also be involved and that there may be species differences in the relative levels of these retinoids.

Retinoic acid is catabolized by the cytochrome oxidase p450, subfamily 26 (CYP26) enzymes (White et al., 1996; Fujii et al., 1997; Ray et al., 1997), and substantial effort has been made to identify these and map their distributions in the hindbrain. In initial studies, Yamamoto et al. (1998) used a combined biochemical/reporter cell assay to demonstrate that the hindbrain of mouse embryos had particularly high levels of retinoic acid catabolic activity that was likely to be mediated by a cytochrome oxidase p450 enzyme. Subsequently, the first CYP26 enzymes identified in mouse and *Xenopus* (Fujii et al., 1997; Hollemann et al., 1998; de Roos et al., 1999) were found to be expressed in the anterior hindbrain, and the first identified in the chicken (Swindell et al., 1999) was found to be expressed in forebrain and midbrain. Thus, in all three species, the relative locations of RALDH2 and CYP26 indicated a posterior source and an anterior sink of retinoic acid, strengthening the notion of a posterior-to-anterior gradient. During the next few years, additional members of the family were identified and cloned, and their expression patterns assessed by *in situ* hybridization. There are now four identified members of the family, CYP26A1, CYP26B1, CYP26C1, and CYP26D1. Of these, CYP26A1 has been characterized in zebrafish, *Xenopus*, chick, and mouse, CYP26B1

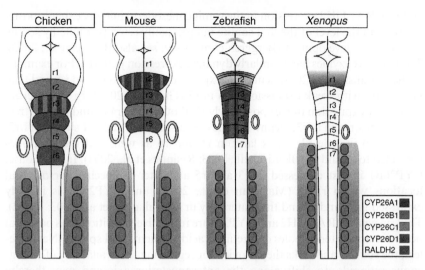

Fig. 3. Current knowledge of the expression patterns of RALDH2 and CYP26 enzymes in the developing zebrafish, *Xenopus*, chicken, and mouse hindbrains. Information obtained from the following sources: zebrafish (Hollemann et al., 1998; Nelson, 1999; Begemann et al., 2001; Kudoh et al., 2002; Dobbs-McAuliffe et al., 2004; Emoto et al., 2005; Gu et al., 2005); *Xenopus* (de Roos et al., 1999; Chen et al., 2001); chicken (Berggren et al., 1999; Blentic et al., 2003; Reijntjes et al., 2004); mouse (de Roos et al., 1999; Nelson, 1999; Abu-Abed et al., 2001; MacLean et al., 2001; Tahayato et al., 2003). (See Color Insert.)

and CYP26C1 in zebrafish, chick, and mouse, and CYP26D1 in zebrafish (Fig. 3). All four are expressed in the hindbrain in each species in which they are found, and the patterns of expression are generally conserved, with high levels of message particularly in r2 and r4. Knockout of *CYP26A1* in the mouse, mutation of *CYP26A1* in zebrafish, and knockdown of CYP26 in *Xenopus* causes misspecification of the anterior hindbrain associated with an expansion of posterior hindbrain-specific gene expression domains (Abu-Abed et al., 2001; Sakai et al., 2001; Kudoh et al., 2002; Emoto et al., 2005), similar to the effects of retinoid excess. By contrast, overexpression of CYP26 in *Xenopus* causes an anteriorization of the hindbrain similar to the effects of retinoid deficiency (Hollemann et al., 1998; Kudoh et al., 2002). The effect can be graded, with a progressive increase in CYP26 expression giving a stepwise anteriorization at the level of individual rhombomeres (Hollemann et al., 1998).

An important feature of the expression of RALDH2 and of the CYP26 enzymes is that the situation is dynamic. Static depictions, such as those shown in Fig. 2, therefore do not do justice to the changing scenario experienced by the hindbrain neural tube. As mentioned previously, both in zebrafish and in mouse retinoic acid reporter models, the retinoic acid signal extends into and retreats from the hindbrain during the course of development. Similarly, in the chicken

embryo, RALDH2 expression in the somitic mesoderm initially extends posteriorly from the level of the first somite but gradually retreats somite by somite
with further development (Berggren et al., 1999; see also Blentic et al., 2003).
The CYP26 enzymes have similarly dynamic expression patterns. For example,
in the zebrafish, CYP26D1 is initially expressed strongly in r2 and r4, and then
later in r6, with weak expression also in r3 (Gu et al., 2005). In the chicken,
CYP26A1 expression moves posteriorly to about the r3/r4 boundary and then
becomes restricted to r3; meanwhile the expression of CYP26C1 starts in r1/r2,
expands to include r3, arises later in r5 and r6 as well, and then becomes
restricted to r5 and r6 (Blentic et al., 2003; Reijntjes et al., 2004). In the mouse,
CYP26B1 is first expressed in r3 and r5 and later in specific dorsoventral
locations within r2–r6 (MacLean et al., 2001), and CYP26C1 is initially
expressed in r2 and r4 and then later only in r2 (Tahayato et al., 2003; Fig. 2).
Thus, in general RALDH2 and CYP26s are respectively positioned as potential
posterior and anterior poles of a retinoic acid gradient in all species. Nevertheless, the sources and sinks they provide for regulating the availability of endogenous retinoids do shift along the anteroposterior axis and may thereby
establish local maxima and minima that temporally modulate what might
otherwise be a smoothly graded pattern. In the hindbrain, such a mechanism
may underlie the segmental specification of rhombomere identity through the
positive and negative regulation of Hox genes (Sirbu et al., 2005; see later).

Potentially contributing to the dynamic nature of RALDH2 and CYP26
expression patterns is the fact that the expression of these enzymes is regulated
by retinoids in an apparently homeostatic fashion. RALDH2 is downregulated by retinoic acid (Niederreither et al., 1997), whereas CYP26 enzymes
are upregulated by retinoic acid (Wang et al., 2002). This has obvious implications for the interpretation of retinoid excess and deficiency but could
also play a role in defining the dynamic anteroposterior domains of enzymatic
activity.

3.2. Later synthetic sources

In addition to the posterior source represented by RALDH2 during the
early patterning of the hindbrain, sources of retinoic acid synthesis are also
found at later stages and play important roles in regulating the differentiation of specific hindbrain structures. RALDH2 is expressed abundantly in
the meninges surrounding the hindbrain by mid-gestational stages in the
chicken embryo (Zhang et al., 2003). The expression is nonuniform, and is
particularly high in areas overlying the rhombic lip, from which late-derived
neuron populations of the precerebellar system (pontine nuclei, inferior
olive) originate. This correlates with retinoid-dependent regulation of cell
proliferation and fate and later anatomical organization in these structures
(Yamamoto et al., 1999; Yamamoto et al., 2005), and with late teratogenic

effects of excess retinoids on the same structures in human embryos (Lammer and Armstrong, 1992). Another potential source of retinoic acid at late stages are the auditory afferents. These express high levels of RALDH3 in their axons and terminals in the hindbrain (Wagner et al., 2002), presumably as a temporal and anatomical extension of the RALDH3 expression present in the otic vesicle from which the inner ear develops (Blentic et al., 2003). Generation and release of retinoic acid by these axons could influence auditory target nuclei within the hindbrain.

4. Retinoid receptors and binding proteins

The effects of endogenous and exogenous retinoids on hindbrain patterning must be transduced at least in part by transcriptional regulatory mechanisms. Chief among these are the retinoic acid receptors (RARs) and the retinoid X receptors (RXRs) that mediate transcriptional regulation directly as ligand-bound heterodimers. Although evidence is appearing that retinoids and activated retinoid receptors may exert some effects through nontranscriptional mechanisms (Liao et al., 2004), retinoid receptors have been unequivocally implicated in the transcriptional regulation of *Hox* and other genes important for hindbrain patterning (see later). Thus, considerable effort has been made toward mapping the distributions of these receptors in the hindbrain. In addition, the availability of retinoic acid for signaling and metabolism is regulated by cellular retinoic acid-binding proteins (CRABPs), so these have also been prime targets for mapping studies. Here we summarize the distributions of RARs, RXRs, and CRABPs to the extent that they are currently characterized in the developing hindbrain of zebrafish, *Xenopus*, chicken, and mouse embryos (Figs. 4 and 5). It is important to realize that some of the information on which this summary is based comes from *in situ* hybridization studies in which either the spatial resolution of the signal is not optimal or the documentation does not clearly indicate rhombomere boundaries. Moreover, the expression of retinoid receptors may in some cases be dynamic. This is an area that deserves closer scrutiny in the future and Fig. 4 therefore should be considered a rough approximation only.

In the mouse, three RARs (RARα, RARβ, and RARγ) and three RXRs (RXRα, RXRβ, and RXRγ) have been identified (Chambon, 1996). In the zebrafish, only two of these RARs have been identified (RARα and RARγ), but four RXRs have been identified (RARα, RARγ, RARδ, and RARε; Jones et al., 1995). In each of *Xenopus* and chicken, only five of the receptors found in mouse have been identified; RARβ has not been detected in *Xenopus* (Blumberg et al., 1992; Marklew et al., 1994; Sharpe, 1994), whereas *RXRβ* evidently is not present in the chicken genome (Rowe and Brickell, 1993; Michaille et al., 1995). Based on current evidence, most if not all of the receptors present in each species are expressed in the developing hindbrain, but their

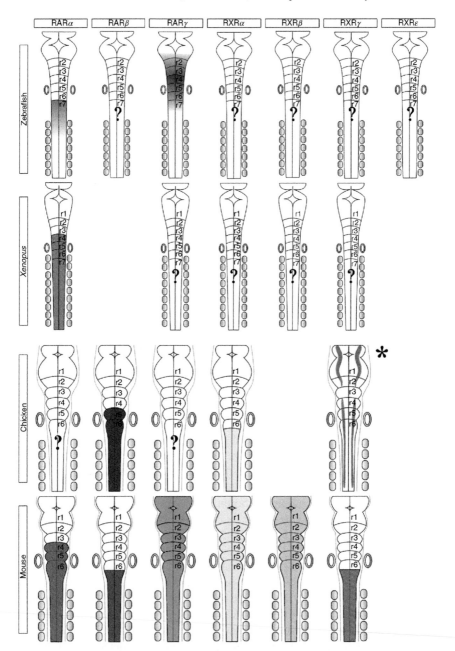

Fig. 4. Current knowledge of the expression patterns of retinoid receptors in the developing hindbrains of zebrafish, *Xenopus*, chicken, and mouse embryos. Information obtained from the following sources: zebrafish (Joore et al., 1994; Jones et al., 1995; Sharma et al., 2003); *Xenopus* (Sharpe, 1992; Dekker et al., 1994; Ho et al., 1994; Marklew et al., 1994; Sharpe, 1994; Crawford

individual distributions vary from specific rhombomeric domains to being ubiquitous. Thus, in the mouse embryo, RARα is expressed from r4 and posteriorly, and RARβ and RXRγ from r7 and posteriorly, whereas the other receptors are expressed throughout the hindbrain (Fig. 4). In the chicken embryo, RARβ2 is expressed from r5 and posteriorly, RXRα is expressed from r7 and posteriorly, and RXRγ is expressed only at later stages of hindbrain development after anteroposterior patterning is complete (Hoover et al., 2000). In *Xenopus*, RARα is expressed from r4 and posteriorly, as in mouse (Ruberte et al., 1991; Sharpe, 1992). In zebrafish, by contrast, RARα is expressed from r7 and posteriorly. This illustrates greater interspecies diversity than is the case for the expression of RALDH2 and the CYP26 enzymes (Fig. 3), and suggests that specific roles for retinoid receptors in hindbrain patterning are not highly conserved. Above all, however, the information presented in Fig. 4 shows that large holes exist in current knowledge about retinoid receptor distributions.

A number of studies have confirmed the role of retinoid receptors in transducing retinoid signals into gene expression patterns, either through genetic knockout by ectopically overexpressing receptors or by blocking their action using dominant-negative constructs. Single knockouts of each of the *RAR* and *RXR* genes in mouse generally cause no developmental defects, but specific double knockouts have noticeable effects on hindbrain patterning, indicating a great deal of functional redundancy among individual receptors. Double knockout of *RARα* and *RARβ* or of *RARα* and *RARγ* disrupts hindbrain patterning in a way that is reminiscent of the anteriorizing effects of retinoid deficiency (Dupe et al., 1999; Wendling et al., 2001). In the *RARα/β* double knockout, the boundary between r5 and r6 is abolished, the expression of genes normally restricted to r4 or r5 or r6 spreads into the more posterior regions, and genes normally expressed in the posterior region are silenced. In the *RARα/γ* double knockout, the anteriorization is more severe, as the posterior portion of the hindbrain takes on an r3/r4 phenotype, quite similar to that seen in the *RALDH2* knockout mouse. Wendling et al. (2001) also compared these double knockout phenotypes to the effects of blocking RAR receptors pharmacologically. Early exposure to a pan-RAR antagonist phenocopied the *RARα/γ* double knockout, whereas later exposure phenocopied the *RARα/β* double knockout, leading to the proposal that RARα or RARγ are critical for setting up the r5/r6 territory and that RARβ is involved in setting the posterior border of this territory. Curiously, given the prevailing view that RXRs are important heterodimer partners for RARs, double knockout of either *RXRα*

---

et al., 1995); chicken (Maden et al., 1991; Smith and Eichele, 1991; Smith et al., 1994; Hoover and Glover, 1998; *note that RXRγ is expressed in this pattern in the chicken only at later stages); mouse (Ruberte et al., 1990, 1991, 1992, 1993; Dolle et al., 1994; Mollard et al., 2000). (See Color Insert.)

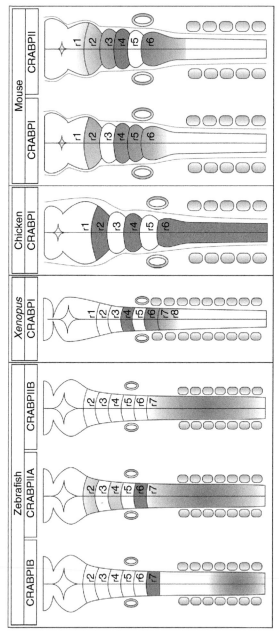

Fig. 5. Current knowledge of the expression patterns of CRABPI and -II in the developing hindbrains of zebrafish, *Xenopus*, chicken, and mouse embryos. Information obtained from the following sources: zebrafish (Sharma et al., 2003); *Xenopus* (Ho et al., 1994); chicken (Maden et al., 1991); mouse (Ruberte et al., 1991, 1992).

or *RXRβ* together with either *RARα* or *RARβ* does not lead to hindbrain defects (Kastner et al., 1997). Thus, it appears in the mouse hindbrain that RAR heterodimers are principally responsible for transducing the retinoid signal and that they may be differentially utilized to this end at different developmental time points (Wendling et al., 2001).

Expression of dominant-negative forms of RARβ has been shown to disrupt hindbrain patterning in both *Xenopus* (van der Wees et al., 1998) and chicken (Gould et al., 1998). In *Xenopus*, the effect is akin to the anteriorization seen in retinoid deficiency, with posterior expansion of r4 and r5 territories and serial multiplications of the r4-specific Mauthner neuron. Curiously, the rostral hindbrain appeared to acquire an r3 identity, suggesting a bidirectional defect not normally seen in retinoid deficiency (van der Wees et al., 1998). In the chicken embryo, dominant-negative RARβ expression blocks the repatterning of hindbrain neural tube transplanted to somitic levels, indicating that the retinoic acid synthesized in the somites exerts its effects on *Hox* gene expression and subsequent neuronal patterning through RARβ-containing heterodimers (Gould et al., 1998). Similarly, injection of a dominant-negative RARα into *Xenopus* embryos leads to an anteriorization, and conversely the injection of constitutively active RARα leads to posteriorization, of the hindbrain (Blumberg et al., 1997).

Pharmacological blockade of retinoid receptors has been facilitated by the development of a spectrum of specific synthetic agonists and antagonists. In addition to the studies mentioned earlier that have utilized them in the mouse embryo, such drugs have been used in *Xenopus* to show that antagonizing endogenous retinoid signaling has the opposite effect on *Hox* gene expression as application of exogenous retinoids (Kolm et al., 1997). In the chicken embryo, application of a pan-RAR antagonist at varying concentrations has been used to show that progressively more posterior rhombomeres require progressively higher concentrations of retinoic acid for their correct patterning (Dupé and Lumsden, 2001), evidence that is consistent with a posterior-to-anterior retinoid gradient.

Evidence has shown that retinoid receptors also have non-ligand-bound repressive roles in regulating gene expression (Weston et al., 2003) and may also have nontranscriptional functions. These activities have not been investigated with respect to hindbrain patterning, and this area should receive more focus in the future.

Another class of proteins that bind retinoids comprises the CRABPs. These have been postulated to bind retinoids intracellularly and have been ascribed various potential roles, including storage, removal from signaling pools, and protection from enzymatic degradation. CRABP expression is also present in the hindbrain of the four vertebrate species we have focused on in this chapter (Fig. 5). As is the case for the retinoid receptors, there is a great deal of interspecies diversity in CRABP expression patterns. However, despite interesting correlations with rhombomeric domains and with the expression of certain patterning genes, there are few indications of an important role for CRABP in patterning the hindbrain. Overexpression of CRABP in *Xenopus* disrupts hindbrain segmentation on a background of a general reduction of the brain (Dekker et al., 1994), thus resembling the effects of retinoid excess. On the other hand, neither single nor double knockout of *CRABPI* and *CRABPII* in the mouse produces any noticeable defects in the hindbrain or elsewhere (Gorry et al., 1994; Lampron et al., 1995; Romand et al., 2000). Thus, the significance of the evolutionary conservation and segmentally modulated expression patterns of CRABP proteins remains a mystery.

5. Nature of the gradient

The posterior location of RALDH2 and the anterior location of CYP26 enzymes is consistent with a spatial gradient of retinoic acid in the hindbrain, with high levels of retinoic acid being produced posteriorly and low levels being available anteriorly. Although initial models presupposed a static linear or curvilinear distribution of retinoic acid along the posterior-to-anterior axis, this simple situation is generally considered to be untenable on several counts.

First, many of the components of retinoid metabolism and signaling are dynamic. As noted earlier, both RALDH2 and the CYP26 enzymes have highly dynamic patterns of expression, with boundaries that shift and expression territories that appear and disappear during the stages when retinoids exert patterning effects. In the mouse embryo, the shifting expression patterns of these enzymes have been shown to be critical for the precise spatial pattern of *Hox* gene induction, suggesting that unequal temporal pulses of retinoic acid are produced in different rhombomeric domains (Sirbu et al., 2005). Such dynamics make sense with regard to *Hox* gene induction in general, as many *Hox* genes are also expressed in temporally dynamic patterns within the developing hindbrain (Sundin et al., 1990; Prince et al., 1998; Davenne et al., 1999). Existing descriptions of retinoid receptor expression patterns do not indicate equally dynamic expression, but on the other hand these descriptions do not offer particularly high spatial and temporal resolution and many dynamic features may have gone unnoticed.

Second, it is hard to imagine any point in time when a retinoid gradient could be linear, simply curvilinear, or even particularly smooth, given the many modulating influences located along the anteroposterior axis. Not only are the CYP26 enzymes segmentally patterned, but CRABP and some retinoid receptors are also segmentally patterned. This suggests substantial heterogeneity in retinoid availability along the anteroposterior axis. Moreover, synthetic sources in addition to RALDH2 clearly exist in the posterior hindbrain, in patterns that appear more complex than a simple gradient would suggest (Mic et al., 2002).

Third, rescue of retinoid deficiency by administration of retinoic acid or retinoid receptor agonists reveals an intriguing feature, namely that a surprisingly high degree of the normal pattern can be reinstated by systemic and/or temporally restricted application (Gale et al., 1999; Niederreither et al., 2000; White et al., 2000; Begemann et al., 2001; Grandel et al., 2002; Mic et al., 2002; Maves and Kimmel, 2005). How can a pulse of retinoic acid or a uniform exposure to retinoic acid recreate a gradient? In this respect, Mic et al. (2002) observed that a brief exposure to exogenous retinoic acid, although spatially uniform, triggers endogenous, nonuniform, non-RALDH2–dependent sources of retinoid synthesis in the hindbrain of RALDH2-negative mice. Thus, some sort of anteroposteriorly patterned synthetic activity might be stimulated by retinoid pulses as part of a homeostatic response to large swings in retinoid availability.

Fourth, several studies suggest that appropriate patterning of the hindbrain by retinoids requires both a spatial gradient and correctly timed availability. In addition to the study by Sirbu et al. (2005), which shows a critical spatiotemporal relationship between retinoic acid synthesis and *Hox* gene expression, Dupé and Lumsden (2001) have shown in the chicken embryo, through timed treatments with a pan-RAR antagonist, that successively more posterior rhombomeres require retinoic acid signaling at successively later stages for their proper establishment. Thus, not only is retinoic acid concentration important, but also the timing of exposure is important (Conlon and Rossant, 1992; Grapin-Botton et al., 1998; Dupé and Lumsden, 2001). In a study in the zebrafish, Maves and Kimmel (2005) show that blocking retinoid signaling or synthesis just before a given rhombomere-specific gene is induced is sufficient to block induction of that gene and that application of exogenous retinoic acid can induce the gene earlier than normal. Their results suggest that the critical feature of retinoid signaling in patterning hindbrain gene expression is the specific time at which retinoids reach a given concentration threshold. The model they propose for the sequential induction of more and more posterior rhombomere identities elegantly resolves the time and concentration dependence of the process, and goes a long way toward explaining the patterns of gene expression that are seen in many other studies of retinoid deficiency and excess. However, their model is still based on a linear gradient and will likely require modification to explain the specific segmental patterns that arise.

Another important issue that has not been resolved is exactly where the retinoid gradient resides. Several studies indicate that hindbrain patterning may be dependent as much on signals from adjacent mesoderm (somitic mesoderm with respect to the posterior hindbrain, head mesoderm with respect to anterior hindbrain) as on retinoic acid itself (Grapin-Botton et al., 1995; Itasaki et al., 1996; Gould et al., 1998; Grapin-Botton et al., 1998). In these studies, specific rhombomeric domains or somites have been transplanted to ectopic positions along the anteroposterior axis of the neural tube, thereby shifting the segmental relationship between the neural tube and the surrounding mesodermal environment. The results indicate that rhombomeres transplanted to more posterior positions are induced to express *Hox* genes characteristic of the new position and that the altered Hox code engenders a homeotic transformation with the differentiation of posterior neuronal groups characteristic of the new position (Grapin-Botton et al., 1995). This effect discriminates among individual rhombomeric domains within the posterior hindbrain, such that, for example, r5/r6 can be transformed into r7/r8 if transplanted posteriorly. Parallel changes in the expression of the *Hox* repressor gene *MafB/Kr/val* (Cordes and Barsh, 1994; Eichmann et al., 1997; Moens et al., 1998) suggest that a retinoid-dependent regulation of *MafB/Kr/val* may operate upstream or in parallel with a direct retinoid-mediated change in Hox code. The regulation of *MafB/Kr/val* is complex, however, and is dependent on protein signaling factors in addition to retinoids (Hernandez et al., 2004). In addition, experiments in which rhombomeres and somites are cocultured but separated by filters with specific molecular weight cutoffs show that the posterior transformation of rhombomeres by somites must be mediated by a factor of 10–200 kDa, that is, much larger than a retinoid molecule (Gould et al., 1998), even though retinoic acid itself can mimic the effect of the somites. This has led to the suggestion that the normal patterning effect exerted on the hindbrain by paraxial mesoderm is mediated by retinoids and a parallel retinoid-dependent protein signal (Gould et al., 1998).

The above information suggests two quite different scenarios that are likely to be acting in concert. In one, retinoids synthesized in somitic mesoderm diffuse into the cervical neural tube and then anteriorly within it, thereby setting up a posterior-to-anterior gradient of retinoid signaling in the hindbrain. In the other, retinoids synthesized in somitic mesoderm diffuse anteriorly within the head paraxial mesoderm, setting up a retinoid gradient there that can then impinge on the adjacent hindbrain as well as induce the production of non-retinoid factors that are important for regulating hindbrain gene expression. In support of a retinoid gradient within the mesoderm itself, Reijntjes et al. (2004) pointed out that, similar to the mesodermal expression of RALDH2, the CYP26 enzymes are also expressed in mesoderm lateral to the hindbrain, not just in the hindbrain. Thus, a source and sink for retinoic acid exists in the mesoderm alongside the hindbrain in addition to the sink within the hindbrain.

6. Control of *Hox* gene expression by retinoids

The link between retinoid activity and *Hox* gene expression was first determined in studies where cultured cells were exposed to distinct concentrations of retinoic acid. It was shown that posterior, 5′ *Hox* genes belonging to paralogue group 5–9 are activated later and at higher retinoic acid concentrations than anterior, 3′ *Hox* genes from paralogue group 1–4 (Simeone et al., 1990; Papalopulu et al., 1991; Simeone et al., 1991). Moreover, *Hox* genes were activated in a sequence colinear with their physical position along the cluster, from 3′ to 5′ (Simeone et al., 1990). The importance of *Hox* genes as potential mediators of retinoic acid-induced patterning in the embryo was subsequently suggested by several studies that showed that the administration of exogenous retinoids resulted in dramatic anterior shifts of *Hox* gene expression boundaries, correlating with anterior truncations or partial respecifications of vertebral and rhombomere identities (Kessel and Gruss, 1991; Morriss-Kay et al., 1991; Conlon and Rossant, 1992; Dekker et al., 1992; Marshall et al., 1992). Conversely, blocking or decreasing retinoic acid signaling in the hindbrain resulted in downregulation or impairment of *Hox* gene expression (Blumberg et al., 1997; Kolm et al., 1997; van der Wees et al., 1998; Gale et al., 1999).

Efforts to decipher whether the response of *Hox* genes to retinoic acid in the hindbrain is directly induced have revealed the presence of RAREs located 3′ and/or 5′ to the coding regions of several *Hox* genes. These elements consist of two direct repeats separated either by 2 or 5 nucleotide spacers (DR2 or DR5, respectively) that are bound by specific RXR/RAR heterodimers and display distinct receptor–DNA affinities (Fig. 6A). For paralogue group 1–4, at least four *Hox* genes have been shown to possess RAREs located 3′ and/or 5′ to the promoters, namely *Hoxa1* (Langston and Gudas, 1992; Frasch et al., 1995), *Hoxb1* (Marshall et al., 1994; Studer et al., 1994), *Hoxb4* (Gould et al., 1998), and *Hoxd4* (Morrison et al., 1996, 1997; Zhang et al., 1997, 2000; Nolte et al., 2003). The conservation of such elements in different vertebrate species further suggested their importance in *Hox* gene regulation. Analysis in transgenic mice carrying RARE-containing enhancer elements driving the *lacZ* reporter has shown that regulation via the RAREs is important for establishing normal *Hox* expression domains in the developing hindbrain and for inducing ectopic expression on exogenous retinoic acid administration. For instance, multiple RAREs contribute to the expression of *Hoxb1*. A 3′ DR2 RARE is required for early induction in the posterior neuroectoderm up to the presumptive r3/r4 boundary (Marshall et al., 1994), whereas a 3′ DR5 RARE is required for regulation in the foregut region (Huang et al., 1998, 2002). Moreover, an additional RARE located 5′ to the promoter is required for retinoic acid-mediated repression of *Hoxb1* in r3 and r5 (Studer et al., 1994; Fig. 6B and see also Fig. 2). Swapping of the *Hoxb4* and *Hoxd4* DR5 elements with the neuroectoderm-specific 3′ *Hoxb1* DR2 RARE results in distinct changes in the anteroposterior specificity of reporter expression in the neural tube, suggesting that the establishment of anterior expression border of individual *Hox* genes is not a simple function of the affinity of retinoid receptors for their cognate element (Gould et al., 1998; Nolte et al., 2003). The setting and maintenance of anterior borders of *Hox* gene expression in the hindbrain requires the additional contribution of distinct conserved enhancer elements acting at

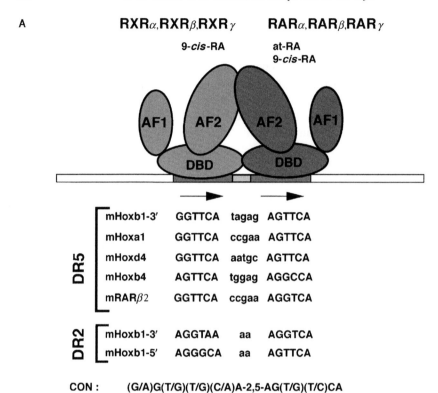

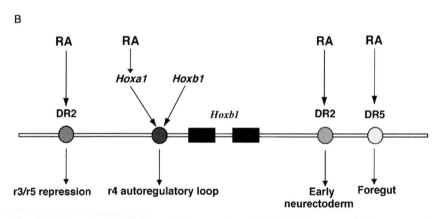

Fig. 6. (A) Organization and sequence comparison of RAREs from several *Hox* genes and mouse RARβ2 (mRARβ2) gene. The RAREs are composed of two hexameric direct repeats, separated by a spacer of variable length, normally, one, two, or five nucleotides (DR1, DR2, and DR5, respectively). RAR and RXR receptor heterodimers contain two transcriptional activating domains (AF1/AF2) and a zinc finger DNA-binding domain (DBD) per receptor. On the DR5 and DR2 elements, the RXR sits down on the 5′ hexanucleotide, whereas the RAR sits down on the 3′ core (reviewed in Bastien and Rochette-Egly, 2004). Among *Hox* genes from

later stages and mediating auto- and cross-regulation between Hox factors (e.g., Popperl et al., 1995; Maconochie et al., 1997; Gould et al., 1998; Manzanares et al., 2001). The *in vivo* contribution of single RAREs to Hox-mediated patterning was further demonstrated by targeted mutations of specific RAREs in the mouse. For example, the single knockouts of the 3' RAREs of *Hoxa1* and *Hoxb1* resulted in molecular and morphological defects that, although milder, were reminiscent of the hindbrain phenotypes obtained with full inactivation of the cognate genes (Dupé et al., 1997; Gavalas et al., 1998; Studer et al., 1998). Analysis of compound mutations resulted in stronger phenotypes in r4, indicating genetic interactions between *Hoxa1* and *Hoxb1*, and retinoid signaling in this domain (Studer et al., 1998). Based on these results, Studer et al. (1998) proposed an elegant model integrating *Hox* paralogue group 1 regulation and retinoid signaling in r4. Retinoid receptors activate the early expression of both *Hoxa1* and *Hoxb1* through their 3' RAREs and, in turn, Hoxa1 initiates the expression of *Hoxb1* in r4 through the binding of a *Hoxb1* r4-specific enhancer (Popperl et al., 1995). At later stages, a direct positive feedback circuit represented by an autoregulatory loop maintains high levels of *Hoxb1* expression selectively in r4, while *Hoxa1* is downregulated (Popperl et al., 1995; Studer et al., 1998). In r3 and r5, *Hoxb1* is directly repressed by retinoic acid through the above-mentioned 5' RAREs (Studer et al., 1994), while retinoic acid-mediated repression of *Hoxb1* in the posterior hindbrain is indirectly achieved through induction of vHnf1 (Hernandez et al., 2004; Sirbu et al., 2005). In the caudal hindbrain, retinoic acid positively regulates the early establishment of *Hox* paralogue group 4 expression domains up to the presumptive r6/r7 boundary (Gould et al., 1998), and late neural enhancers (LNE) containing Hox-responsive elements ensure their maintenance (Fig. 2). A study has provided evidence for a direct feedback transcriptional regulation of *RARβ* by Hoxb4 and Hoxd4 proteins that occurs in parallel to the maintenance of Hoxb4 and Hoxd4 expression and is involved in aligning the expression of multiple *Hox* and *RAR* genes at a single segment boundary (Serpente et al., 2005).

---

paralogous group 1–4, at least *Hoxa1, Hoxb1, Hoxb4*, and *Hoxd4* are known to possess one or several RAREs involved in the regulation of their expression in the hindbrain. (B) RA and Hox cross- and autoregulation at the *Hoxb1* locus (modified from Studer et al., 1998). RA initiates the expression of *Hoxb1* in the neuroectoderm through the binding of retinoid receptors on its 3' DR2 (gray circle) and by activating the expression of *Hoxa1*. Hoxa1 participate in the initiation of the expression of *Hoxb1* at the level of presumptive r4, while maintenance of Hoxb1 expression relies instead on an autoregulatory loop (black circle). Retinoids are also involved in the direct repression of *Hoxb1* in r3/r5 through the 5' DR2 RARE (dark gray circle). Thus, precise restriction of *Hoxb1* expression in r4 requires the concerted activity of both 3' DR2 and 5' DR2 and an autoregulatory loop. Finally, a 3' DR5 (light gray circle) is involved in regulating *Hoxb1* in the gut.

Whereas retinoid signaling is directly involved in the early establishment of *Hox* paralogue group 1–4 expression in the hindbrain, a role for retinoic acid in the regulation of caudal 5' *Hox* genes and patterning of the spinal cord has been less clear. Recent work shows that colinear activation of caudal *Hox* genes requires fibroblast growth factor (FGF) signaling, although some *Hox* genes can be induced both by FGF and by retinoic acid in the developing CNS, indicating a series of mutually interactive feedback loops among FGFs, RARs, and *Hox* genes (Liu et al., 2001; Bel-Vialar et al., 2002; Oosterveen et al., 2003; Shiotsugu et al., 2004). To integrate these and other data, Diez del Corral and Storey (2004) proposed an "opposing signal" model for colinear activation of caudal *Hox* genes. In this scenario, the mutual inhibition between retinoic acid and FGF secreted by the somites and the caudal region of the embryo, respectively, may prevent the expression of progressively more 5' *Hox* genes in cells leaving the posterior zone (the so-called caudal stem zone) during caudal extension of the body axis, thus assigning progressive rostrocaudal identity to newly formed spinal cord cells. While cells in the caudal stem zone experience high levels of FGF and express progressively more 5' *Hox* genes, retinoic acid limits the influence of FGF signaling just rostral to the stem zone, preventing the expression of additional caudal *Hox* genes. Furthermore, retinoic acid could be directly involved in the activation of later expression domains of certain 5' *Hox* genes. This has been shown for the late rostral extensions of *Hoxb5*, *Hoxb6*, and *Hoxb8* into the hindbrain that are mediated through a specific RARE located downstream to *Hoxb5* (Oosterveen et al., 2003). Similar opposing activities of FGF and retinoic acid may operate in the rostral hindbrain, where FGF signaling from the isthmus represses *Hox* genes in rhombomere 1 (Irving and Mason, 2000), thus counteracting retinoic acid activity from the posterior hindbrain.

7. Specification of individual neuron types

Thus far, we have focused on the way retinoid signaling regulates the expression of patterning genes and how this contributes to establishing domains of gene expression within the hindbrain neuroepithelium. An important course of inquiry is the role retinoid signaling plays in the specification of different neuron types and the circuits they participate in.

On the one hand, it is to be expected that patterning, repatterning, or disruption of the anteroposterior patterning of the hindbrain at the gene expression level will lead to changes in the numbers of types of neurons present. For example, the Mauthner neuron, a large easily identifiable reticulospinal neuron located in r4 of anamniotes, is serially multiplicated in *Xenopus* after either exogenous retinoid application (Manns and Fritzsch, 1992) or in the case of deficient retinoid signaling in the posterior hindbrain (van der Wees et al., 1998). Similarly, somatic motoneurons, which normally differentiate only in the posterior rhombomeres r5–r8 of the hindbrain (although they also differentiate near the hindbrain/midbrain transition), differentiate in all rhombomeres following application of retinoic acid to the chicken embryo prior to segmentation into rhombomeres (Guidato et al., 2003; see also Linville et al., 2004; Begemann et al., 2004). This parallels the hindbrain anteriorization of Hox gene expression and rhombomere identity induced by excess retinoids described earlier.

On the other hand, although serial multiplications of neurons could be automatic sequelae of an anteroposterior repatterning, it should be remembered that retinoid signaling may be more protracted developmentally and therefore exert effects beyond the stages of anteroposterior patterning. This opens the possibility that neuronal differentiation may be affected by alterations in retinoid signaling independently of anteroposterior patterning. For example, the effects described earlier could conceivably arise from a later dorsoventral patterning influence of retinoic acid (Renaud et al., 2006), with retinoic acid exposure leading to a ventralization of anterior rhombomeres and thus the production of ventral neuron types such as somatic motoneurons there, and retinoic acid deficiency leading to a dorsalization of posterior rhombomeres and thus the production of dorsal neuron types such as Mauthner neurons there. In the zebrafish, the *neckless* mutant (which lacks RALDH2) affects the differentiation of secondary, branchiomotor neurons without affecting the segmentally disposed primary reticulospinal neurons, suggesting an effect that is temporally targeted to stages after anteroposterior patterning (Begemann et al., 2004; Linville et al., 2004; see also Sharpe and Goldstone, 2000). Retinoid signaling might therefore have direct effects on neuronal differentiation independent of its role in anteroposterior patterning. This question deserves closer attention in the future, and the effects of retinoid signaling perturbations on neuron differentiation at later stages need to be assessed more comprehensively.

8. Final remarks

Despite the impressive array of information that has already been obtained about how retinoid signaling patterns the developing hindbrain, there are still large gaps in our knowledge. Major challenges that remain include the following:

Determining the identity and dynamic availability of endogenous retinoids at high spatial resolution. This will require a combination of state-of-the-art chemical assays at at least the resolution of individual rhombomeres as well as assiduous analysis of high-sensitivity reporter systems.

Obtaining more complete assessments of the expression patterns of metabolic enzymes, binding proteins, and retinoid receptors, with a focus on spatiotemporal dynamics. Both mRNA and protein expression need to be assessed, since most of the information at hand is of mRNA expression only.

Obtaining more complete assessments of the regulation of enzyme, binding protein, and receptor expression by retinoids. Elucidating the nature and role of feedback loops in setting up the expression patterns of these retinoid signaling components will be pivotal in understanding the inherent dynamics of retinoid signaling.

Obtaining information about the regulatory mechanisms setting up the expression domains of retinoid signaling components. Surprisingly little is currently known about the molecules and transcription factors that contribute to set up the spatiotemporal patterns of expression of retinoid signaling components in the developing hindbrain and adjacent mesenchyme. What regulates the regulator? This should be an active domain of research for the future.

Obtaining a better understanding of how retinoid signaling regulates the expression of the many developmental regulatory genes that are involved in hindbrain patterning. The primary focus to date in this area has been on the *Hox* genes. We need to determine whether other families of patterning genes are directly or indirectly regulated by retinoids. One important aspect is to elucidate the interactions between anteroposterior and dorsoventral patterning as regulatory targets of retinoid signaling. Are dorsoventral patterning effects downstream of *Hox* gene expression, or can the patterning of anteroposterior and dorsoventral axes be regulated independently by retinoids?

Determining the dependencies of different neuron types on retinoid signaling. Indications that not all neuron types are equally sensitive to perturbation of retinoid signaling suggest heterogeneity in either the temporal or the cell type-specific dependence of neuronal specification on retinoids. More neuron types need to be studied over a broader range of stages.

In summary, tremendous advances have been made in recent years in understanding hindbrain patterning in general and its regulation by retinoids in particular. Given the scope of the remaining challenges, however, it is probably safe to say that research into the way retinoid signaling patterns the hindbrain will continue for quite some time to come.

Acknowledgments

Work in J.C.G's laboratory is supported by grants from the EEC (Brainstem Genetics—QLG2-CT01-01467), the Human Frontiers Science Program, the Norwegian Research Council, the Norwegian Center for Stem Cell Research, the University of Oslo, and the Odd Fellow Foundation. Work in F.M.R's laboratory is supported by grants from the EEC Brainstem Genetics Program (QLG2-CT01-01467), the Agence Nationale pour la Recherche (ANR), the Fondation pour la Recherche Medicale (FRM), the Association pour la

Recherche contre le Cancer (ARC), the Association Française contre les Myopathies (AFM), the Ministère pour le Recherche (ACI program), and by institutional funds from CNRS and INSERM.

References

Abu-Abed, S., Dolle, P., Metzger, D., Beckett, B., Chambon, P., Petkovich, M. 2001. The retinoic acid-metabolizing enzyme, CYP26A1, is essential for normal hindbrain patterning, vertebral identity, and development of posterior structures. Genes Dev. 15, 226–240.

Agarwal, V.R., Sato, S.M. 1993. Retinoic acid affects central nervous system development of Xenopus by changing cell fate. Mech. Dev. 44, 167–173.

Allenby, G., Bocquel, M.T., Saunders, M., Kazmer, S., Speck, J., Rosenberger, M., Lovey, A., Kastner, P., Grippo, J.F., Chambon, P., Levin, A.A. 1993. Retinoic acid receptors and retinoid X receptors: Interactions with endogenous retinoic acids. Proc. Natl. Acad. Sci. USA 90, 30–34.

Avantaggiato, V., Acampora, D., Tuorto, F., Simeone, A. 1996. Retinoic acid induces stage-specific repatterning of the rostral central nervous system. Dev. Biol. 175, 347–357.

Balkan, W., Colbert, M., Bock, C., Linney, E. 1992. Transgenic indicator mice for studying activated retinoic acid receptors during development. Proc. Natl. Acad. Sci. USA 89, 3347–3351.

Bastien, J., Rochette-Egly, C. 2004. Nuclear retinoid receptors and the transcription of retinoid-target genes. Gene 328, 1–16.

Begemann, G., Schilling, T.F., Rauch, G.J., Geisler, R., Ingham, P.W. 2001. The zebrafish neckless mutation reveals a requirement for raldh2 in mesodermal signals that pattern the hindbrain. Development 128, 3081–3094.

Begemann, G., Marx, M., Mebus, K., Meyer, A., Bastmeyer, M. 2004. Beyond the neckless phenotype: Influence of reduced retinoic acid signaling on motor neuron development in the zebrafish hindbrain. Dev. Biol. 271, 119–129.

Bel-Vialar, S., Itasaki, N., Krumlauf, R. 2002. Initiating Hox gene expression: In the early chick neural tube differential sensitivity to FGF and RA signaling subdivides the HoxB genes in two distinct groups. Development 129, 5103–5115.

Berggren, K., McCaffery, P., Drager, U., Forehand, C.J. 1999. Differential distribution of retinoic acid synthesis in the chicken embryo as determined by immunolocalization of the retinoic acid synthetic enzyme, RALDH-2. Dev. Biol. 210, 288–304.

Blentic, A., Gale, E., Maden, M. 2003. Retinoic acid signalling centres in the avian embryo identified by sites of expression of synthesising and catabolising enzymes. Dev. Dyn. 227, 114–127.

Blumberg, B., Mangelsdorf, D.J., Dyck, J.A., Bittner, D.A., Evans, R.M., De Robertis, E.M. 1992. Multiple retinoid-responsive receptors in a single cell: Families of retinoid "X" receptors and retinoic acid receptors in the Xenopus egg Proc. Natl. Acad. Sci. USA 89, 2321–2325.

Blumberg, B., Bolado, J., Jr., Derguini, F., Craig, A.G., Moreno, T.A., Chakravarti, D., Heyman, R.A., Buck, J., Evans, R.M. 1996. Novel retinoic acid receptor ligands in Xenopus embryos. Proc. Natl. Acad. Sci. USA 93, 4873–4878.

Blumberg, B., Bolado, J., Jr., Moreno, T.A., Kintner, C., Evans, R.M., Papalopulu, N. 1997. An essential role for retinoid signaling in anteroposterior neural patterning. Development 124, 373–379.

Chambon, P. 1996. A decade of molecular biology of retinoic acid receptors. FASEB J. 10, 940–954.

Chen, Y., Huang, L., Russo, A.F., Solursh, M. 1992. Retinoic acid is enriched in Hensen's node and is developmentally regulated in the early chicken embryo. Proc. Natl. Acad. Sci. USA 89, 10056–10059.

Chen, Y., Huang, L., Solursh, M. 1994. A concentration gradient of retinoids in the early *Xenopus laevis* embryo. Dev. Biol. 161, 70–76.

Chen, Y., Pollet, N., Niehrs, C., Pieler, T. 2001. Increased XRALDH2 activity has a posteriorizing effect on the central nervous system of Xenopus embryos. Mech. Dev. 101, 91–103.

Colbert, M.C., Linney, E., LaMantia, A.S. 1993. Local sources of retinoic acid coincide with retinoid-mediated transgene activity during embryonic development. Proc. Natl. Acad. Sci. USA 90, 6572–6576.

Conlon, R.A., Rossant, J. 1992. Exogenous retinoic acid rapidly induces anterior ectopic expression of murine Hox-2 genes *in vivo*. Development 116, 357–368.

Cordes, S.P., Barsh, G.S. 1994. The mouse segmentation gene kr encodes a novel basic domain-leucine zipper transcription factor. Cell 79, 1025–1034.

Costaridis, P., Horton, C., Zeitlinger, J., Holder, N., Maden, M. 1996. Endogenous retinoids in the zebrafish embryo and adult. Dev. Dyn. 205, 41–51.

Crawford, M.J., Liversage, R.A., Varmuza, S.L. 1995. Two isoforms of Xenopus retinoic acid receptor gamma 2 (B) exhibit differential expression and sensitivity to retinoic acid during embryogenesis. Dev. Genet. 17, 291–302.

Davenne, M., Maconochie, M.K., Neun, R., Pattyn, A., Chambon, P., Krumlauf, R., Rijli, F.M. 1999. Hoxa2 and Hoxb2 control dorsoventral patterns of neuronal development in the rostral hindbrain. Neuron 22, 677–691.

de Roos, K., Sonneveld, E., Compaan, B., ten Berge, D., Durston, A.J., van der Saag, P.T. 1999. Expression of retinoic acid 4-hydroxylase (CYP26) during mouse and *Xenopus laevis* embryogenesis. Mech. Dev. 82, 205–211.

Dekker, E.J., Pannese, M., Houtzager, E., Timmermans, A., Boncinelli, E., Durston, A. 1992. Xenopus Hox-2 genes are expressed sequentially after the onset of gastrulation and are differentially inducible by retinoic acid. Dev. Suppl. 195–202.

Dekker, E.J., Vaessen, M.J., van den Berg, C., Timmermans, A., Godsave, S., Holling, T., Nieuwkoop, P., Geurts, van Kessel,A., Durston, A. 1994. Overexpression of a cellular retinoic acid binding protein (xCRABP) causes anteroposterior defects in developing Xenopus embryos. Development 120, 973–985.

Dersch, H., Zile, M.H. 1993. Induction of normal cardiovascular development in the vitamin A-deprived quail embryo by natural retinoids. Dev. Biol. 160, 424–433.

Diez del Corral, R., Storey, K.G. 2004. Opposing FGF and retinoid pathways: A signalling switch that controls differentiation and patterning onset in the extending vertebrate body axis. Bioessays 26, 857–869.

Dobbs-McAuliffe, B., Zhao, Q., Linney, E. 2004. Feedback mechanisms regulate retinoic acid production and degradation in the zebrafish embryo. Mech. Dev. 121, 339–350.

Dolle, P., Fraulob, V., Kastner, P., Chambon, P. 1994. Developmental expression of murine retinoid X receptor (RXR) genes. Mech. Dev. 45, 91–104.

Dupe, V., Lumsden, A. 2001. Hindbrain patterning involves graded responses to retinoic acid signalling. Development 128, 2199–2208.

Dupe, V., Davenne, M., Brocard, J., Dolle, P., Mark, M., Dierich, A., Chambon, P., Rijli, F.M. 1997. In vivo functional analysis of the Hoxa-1,3′ retinoic acid response element (3′RARE). Development 124, 399–410.

Dupe, V., Ghyselinck, N.B., Wendling, O., Chambon, P., Mark, M. 1999. Key roles of retinoic acid receptors alpha and beta in the patterning of the caudal hindbrain, pharyngeal arches and otocyst in the mouse. Development 126, 5051–5059.

Durston, A.J., Timmermans, J.P., Hage, W.J., Hendriks, H.F., de Vries, N.J., Heideveld, M., Nieuwkoop, P.D. 1989. Retinoic acid causes an anteroposterior transformation in the developing central nervous system. Nature 340, 140–144.

Eichmann, A., Grapin-Botton, A., Kelly, L., Graf, T., Le Douarin, N.M., Sieweke, M. 1997. The expression pattern of the mafB/kr gene in birds and mice reveals that the kreisler phenotype does not represent a null mutant. Mech. Dev. 65, 111–122.

Emoto, Y., Wada, H., Okamoto, H., Kudo, A., Imai, Y. 2005. Retinoic acid-metabolizing enzyme Cyp26a1 is essential for determining territories of hindbrain and spinal cord in zebrafish. Dev. Biol. 278, 415–427.

Frasch, M., Chen, X., Lufkin, T. 1995. Evolutionary-conserved enhancers direct region specific expression of the murine Hoxa-1 and Hoxa-2 loci in both mice and Drosophila. Development 121, 957–974.

Fujii, H., Sato, T., Kaneko, S., Gotoh, O., Fujii-Kuriyama, Y., Osawa, K., Kato, S., Hamada, H. 1997. Metabolic inactivation of retinoic acid by a novel P450 differentially expressed in developing mouse embryos. EMBO J. 16, 4163–4173.

Gale, E., Prince, V., Lumsden, A., Clarke, J., Holder, N., Maden, M. 1996. Late effects of retinoic acid on neural crest and aspects of rhombomere. Development 122, 783–793.

Gale, E., Zile, M., Maden, M. 1999. Hindbrain respecification in the retinoid-deficient quail. Mech. Dev. 89, 43–54.

Gavalas, A., Krumlauf, R. 2000. Retinoid signalling and hindbrain patterning. Curr. Opin. Genet. Dev. 10, 380–386.

Gavalas, A., Studer, M., Lumsden, A., Rijli, F.M., Krumlauf, R., Chambon, P. 1998. Hoxa1 and Hoxb1 synergize in patterning the hindbrain, cranial nerves and second pharyngeal arch. Development 125, 1123–1136.

Goldstein, J.T., Dobrzyn, A., Clagett-Dame, M., Pike, J.W., DeLuca, H.F. 2003. Isolation and characterization of unsaturated fatty acids as natural ligands for the retinoid-X receptor. Arch. Biochem. Biophys. 420, 185–193.

Gorry, P., Lufkin, T., Dierich, A., Rochette-Egly, C., Decimo, D., Dolle, P., Mark, M., Durand, B., Chambon, P. 1994. The cellular retinoic acid binding protein I is dispensable. Proc. Natl. Acad. Sci. USA 91, 9032–9036.

Gould, A., Itasaki, N., Krumlauf, R. 1998. Initiation of rhombomeric Hoxb4 expression requires induction by somites and a retinoid pathway. Neuron 21, 39–51.

Grandel, H., Lun, K., Rauch, G.J., Rhinn, M., Piotrowski, T., Houart, C., Sordino, P., Kuchler, A.M., Schulte-Merker, S., Geisler, R., Holder, N., Wilson, S.W., et al. 2002. Retinoic acid signalling in the zebrafish embryo is necessary during pre-segmentation stages to pattern the anterior-posterior axis of the CNS and to induce a pectoral fin bud. Development 129, 2851–2865.

Grapin-Botton, A., Bonnin, M.A., McNaughton, L.A., Krumlauf, R., Le Douarin, N.M. 1995. Plasticity of transposed rhombomeres: Hox gene induction is correlated with phenotypic modifications. Development 121, 2707–2721.

Grapin-Botton, A., Bonnin, M.A., Sieweke, M., Le Douarin, N.M. 1998. Defined concentrations of a posteriorizing signal are critical for MafB/Kreisler segmental expression in the hindbrain. Development 125, 1173–1181.

Gu, X., Xu, F., Wang, X., Gao, X., Zhao, Q. 2005. Molecular cloning and expression of a novel CYP26 gene (cyp26d1) during zebrafish early development. Gene Expr. Patterns 5, 733–739.

Guidato, S., Barrett, C., Guthrie, S. 2003. Patterning of motor neurons by retinoic acid in the chick embryo hindbrain *in vitro*. Mol. Cell. Neurosci. 23, 81–95.

Haselbeck, R.J., Hoffmann, I., Duester, G. 1999. Distinct functions for Aldh1 and Raldh2 in the control of ligand production for embryonic retinoid signaling pathways. Dev. Genet. 25, 353–364.

Hernandez, R.E., Rikhof, H.A., Bachmann, R., Moens, C.B. 2004. vhnf1 integrates global RA patterning and local FGF signals to direct posterior hindbrain development in zebrafish. Development 131, 4511–4520.

Hill, J., Clarke, J.D., Vargesson, N., Jowett, T., Holder, N. 1995. Exogenous retinoic acid causes specific alterations in the development of the midbrain and hindbrain of the zebrafish embryo including positional respecification of the Mauthner neuron. Mech. Dev. 50, 3–16.

Ho, L., Mercola, M., Gudas, L.J. 1994. *Xenopus laevis* cellular retinoic acid-binding protein: Temporal and spatial expression pattern during early embryogenesis. Mech. Dev. 47, 53–64.

Holder, N., Hill, J. 1991. Retinoic acid modifies development of the midbrain-hindbrain border and affects cranial ganglion formation in zebrafish embryos. Development 113, 1159–1170.

Hollemann, T., Chen, Y., Grunz, H., Pieler, T. 1998. Regionalized metabolic activity establishes boundaries of retinoic acid signalling. EMBO J. 17, 7361–7372.

Hoover, F., Glover, J.C. 1998. Regional pattern of retinoid X receptor-alpha gene expression in the central nervous system of the chicken embryo and its up-regulation by exposure to 9-cis retinoic acid. J. Comp. Neurol. 398, 575–586.

Hoover, F., Kielland, A., Glover, J.C. 2000. RXR gamma gene is expressed by discrete cell columns within the alar plate of the brainstem of the chicken embryo. J. Comp. Neurol. 416, 417–428.

Hoover, F., Gundersen, T.E., Ulven, S.M., Michaille, J.J., Blanchet, S., Blomhoff, R., Glover, J.C. 2001. Quantitative assessment of retinoid signaling pathways in the developing eye and retina of the chicken embryo. J. Comp. Neurol. 436, 324–335.

Horton, C., Maden, M. 1995. Endogenous distribution of retinoids during normal development and teratogenesis in the mouse embryo. Dev. Dyn. 202, 312–323.

Huang, D., Chen, S.W., Langston, A.W., Gudas, L.J. 1998. A conserved retinoic acid responsive element in the murine Hoxb-1 gene is required for expression in the developing gut. Development 125, 3235–3246.

Huang, D., Chen, S.W., Gudas, L.J. 2002. Analysis of two distinct retinoic acid response elements in the homeobox gene Hoxb1 in transgenic mice. Dev. Dyn. 223, 353–370.

Irving, C., Mason, I. 2000. Signalling by FGF8 from the isthmus patterns anterior hindbrain and establishes the anterior limit of Hox gene expression. Development 127, 177–186.

Itasaki, N., Sharpe, J., Morrison, A., Krumlauf, R. 1996. Reprogramming Hox expression in the vertebrate hindbrain: Influence of paraxial mesoderm and rhombomere transposition. Neuron 16, 487–500.

Jones, B.B., Ohno, C.K., Allenby, G., Boffa, M.B., Levin, A.A., Grippo, J.F., Petkovich, M. 1995. New retinoid X receptor subtypes in zebra fish (Danio rerio) differentially modulate transcription and do not bind 9-cis retinoic acid. Mol. Cell. Biol. 15, 5226–5234.

Joore, J., van der Lans, G.B., Lanser, P.H., Vervaart, J.M., Zivkovic, D., Speksnijder, J.E., Kruijer, W. 1994. Effects of retinoic acid on the expression of retinoic acid receptors during zebrafish embryogenesis. Mech. Dev. 46, 137–150.

Kalter, H., Warkany, J. 1959. Experimental production of congenital malformations in mammals by metabolic procedure. Physiol. Rev. 39, 69–115.

Kastner, P., Messaddeq, M., Mark, N., Wendling, O., Grondona, J.M., Ward, S., Ghyselinck, N., Chambon, P. 1997. Vitamin A deficiency and mutations of RXRalpha, RXRbeta and RARalpha lead to early differentiation of embryonic ventricular cardiomyocytes. Development 124, 4749–4758.

Kessel, M., Gruss, P. 1991. Homeotic transformations of murine vertebrae and concomitant alteration of Hox codes induced by retinoic acid. Cell 67, 89–104.

Kolm, P.J., Apekin, V., Sive, H. 1997. Xenopus hindbrain patterning requires retinoid signaling. Dev. Biol. 192, 1–16.

Kraft, J.C., Juchau, M.R. 1995. *Xenopus laevis*: A model system for the study of embryonic retinoid metabolism. III. Isomerization and metabolism of all-trans-retinoic acid and 9-cis-retinoic acid and their dysmorphogenic effects in embryos during neurulation. Drug Metab. Dispos. 23, 1058–1071.

Kraft, J.C., Schuh, T., Juchau, M., Kimelman, D. 1994. The retinoid X receptor ligand, 9-cis-retinoic acid, is a potential regulator of early Xenopus development. Proc. Natl. Acad. Sci. USA 91, 3067–3071.

Kudoh, T., Wilson, S.W., Dawid, I.B. 2002. Distinct roles for Fgf, Wnt and retinoic acid in posteriorizing the neural ectoderm. Development 129, 4335–4346.

Lammer, E.J., Armstrong, D.L. 1992. Malformations of hindbrain structures among humans exposed to isotretinoin 13-*cis*-retinoic acid during early embryogenesis. In: *Retinoids in Normal Development and Teratogenesis* (G.M. Morris-Kay, Ed.), Oxford: Oxford University Press, pp. 281–295.

Lampron, C., Rochette-Egly, C., Gorry, P., Dolle, P., Mark, M., Lufkin, T., LeMeur, M., Chambon, P. 1995. Mice deficient in cellular retinoic acid binding protein II (CRAB-PII) or in both CRABPI and CRABPII are essentially normal. Development 121, 539–548.

Langston, A.W., Gudas, L.J. 1992. Identification of a retinoic acid responsive enhancer 3' of the murine homeobox gene Hox-1.6. Mech. Dev. 38, 217–227.

Li, H., Wagner, E., McCaffery, P., Smith, D., Andreadis, A., Drager, U.C. 2000. A retinoic acid synthesizing enzyme in ventral retina and telencephalon of the embryonic mouse. Mech. Dev. 95, 283–289.

Liao, Y.P., Ho, S.Y., Liou, J.C. 2004. Non-genomic regulation of transmitter release by retinoic acid at developing motoneurons in Xenopus cell culture. J. Cell Sci. 117, 2917–2924.

Lin, M., Zhang, M., Abraham, M., Smith, S.M., Napoli, J.L. 2003. Mouse retinal dehydrogenase 4 (RALDH4), molecular cloning, cellular expression, and activity in 9-cis-retinoic acid biosynthesis in intact cells. J. Biol. Chem. 278, 9856–9861.

Linville, A., Gumusaneli, E., Chandraratna, R.A., Schilling, T.F. 2004. Independent roles for retinoic acid in segmentation and neuronal differentiation in the zebrafish hindbrain. Dev. Biol. 270, 186–199.

Liu, J.P., Laufer, E., Jessel, T.M. 2001. Assigning the positional identity of spinal motor neurons: Rostrocaudal patterning of Hox-C expression by FGFs, Gdfl 1, and retinoids. Neuron 32, 997–1012.

Luria, A., Furlow, J.D. 2004. Spatiotemporal retinoid-X receptor activation detected in live vertebrate embryos. Proc. Natl. Acad. Sci. USA 101, 8987–8992.

MacLean, G., Abu-Abed, S., Dolle, P., Tahayato, A., Chambon, P., Petkovich, M. 2001. Cloning of a novel retinoic-acid metabolizing cytochrome P450, Cyp26B1, and comparative expression analysis with Cyp26A1 during early murine development. Mech. Dev. 107, 195–201.

Maconochie, M.K., Nonchev, S., Studer, M., Chan, S.K., Popperl, H., Sham, M.H., Mann, R.S., Krumlauf, R. 1997. Cross-regulation in the mouse HoxB complex: The expression of Hoxb2 in rhombomere 4 is regulated by Hoxb1. Genes Dev. 11, 1885–1895.

Maden, M. 2002. Retinoid signalling in the development of the central nervous system. Nat. Rev. Neurosci. 3, 843–853.

Maden, M., Hunt, P., Eriksson, U., Kuroiwa, A., Krumlauf, R., Summerbell, D. 1991. Retinoic acid-binding protein, rhombomeres and the neural crest. Development 111, 35–43.

Maden, M., Sonneveld, E., van der Saag, P.T., Gale, E. 1998. The distribution of endogenous retinoic acid in the chick embryo: Implications for developmental mechanisms. Development 125, 4133–4144.

Manns, M., Fritzsch, B. 1992. Retinoic acid affects the organization of reticulospinal neurons in developing Xenopus. Neurosci. Lett. 139, 253–256.

Manzanares, M., Bel-Vialar, S., Ariza-McNaughton, L., Ferretti, E., Marshall, H., Maconochie, M.M., Blasi, F., Krumlauf, R. 2001. Independent regulation of initiation and maintenance phases of Hoxa3 expression in the vertebrate hindbrain involve auto- and cross-regulatory mechanisms. Development 128, 3595–3607.

Marklew, S., Smith, D.P., Mason, C.S., Old, R.W. 1994. Isolation of a novel RXR from Xenopus that most closely resembles mammalian RXR beta and is expressed throughout early development. Biochim. Biophys. Acta 1218, 267–272.

Marshall, H., Nonchev, S., Sham, M.H., Muchamore, I., Lumsden, A., Krumlauf, R. 1992. Retinoic acid alters hindbrain Hox code and induces transformation of rhombomeres 2/3 into a 4/5 identity. Nature 360, 737–741.

Marshall, H., Studer, M., Popperl, H., Aparicio, S., Kuroiwa, A., Brenner, S., Krumlauf, R. 1994. A conserved retinoic acid response element required for early expression of the homeobox gene Hoxb-1. Nature 370, 567–571.

Marshall, H., Morrison, A., Studer, M., Popperl, H., Krumlauf, R. 1996. Retinoids and Hox genes. FASEB J. 10, 969–978.

Matkovits, T., Christakos, S. 1995. Ligand occupancy is not required for vitamin D receptor and retinoid receptor-mediated transcriptional activation. Mol. Endocrinol. 9, 232–242.

Maves, L., Kimmel, C.B. 2005. Dynamic and sequential patterning of the zebrafish posterior hindbrain by retinoic acid. Dev. Biol. 285, 593–605.

McCaffery, P., Drager, U.C. 1994. Hot spots of retinoic acid synthesis in the developing spinal cord. Proc. Natl. Acad. Sci. USA 91, 7194–7197.

Mic, F.A., Molotkov, A., Fan, X., Cuenca, A.E., Duester, G. 2000. RALDH3, a retinaldehyde dehydrogenase that generates retinoic acid, is expressed in the ventral retina, otic vesicle and olfactory pit during mouse development. Mech. Dev. 97, 227–230.

Mic, F.A., Haselbeck, R.J., Cuenca, A.E., Duester, G. 2002. Novel retinoic acid generating activities in the neural tube and heart identified by conditional rescue of Raldh2 null mutant mice. Development 129, 2271–2282.

Michaille, J.J., Kanzler, B., Blanchet, S., Garnier, J.M., Dhouailly, D. 1995. Characterization of cDNAs encoding two chick retinoic acid receptor alpha isoforms and distribution of retinoic acid receptor alpha, beta and gamma transcripts during chick skin development. Int. J. Dev. Biol. 39, 587–596.

Moens, C.B., Cordes, S.P., Giorgianni, M.W., Barsh, G.S., Kimmel, C.B. 1998. Equivalence in the genetic control of hindbrain segmentation in fish and mouse. Development 125, 381–391.

Mollard, R., Viville, S., Ward, S.J., Decimo, D., Chambon, P., Dolle, P. 2000. Tissue-specific expression of retinoic acid receptor isoform transcripts in the mouse embryo. Mech. Dev. 94, 223–232.

Morriss, G.M. 1972. Morphogenesis of the malformations induced in rat embryos by maternal hypervitaminosis A. J. Anat. 113, 241–250.

Morriss, G.M., Thorogood, P.V.T. 1978. In: *Development in Mammals* (M.H. Johnson, Ed.), Amsterdam: Elsevier North-Holland, vol. 3, pp. 363–412.

Morrison, A., Moroni, M.C., Ariza-McNaughton, L., Krumlauf, R., Mavilio, F. 1996. *In vitro* and transgenic analysis of a human HOXD4 retinoid-responsive enhancer. Development 122, 1895–1907.

Morrison, A., Ariza-McNaughton, L., Gould, A., Featherstone, M., Krumlauf, R. 1997. HOXD4 and regulation of the group 4 paralog genes. Development 124, 3135–3146.

Morriss-Kay, G.M., Murphy, P., Hill, R.E., Davidson, D.R. 1991. Effects of retinoic acid excess on expression of Hox-2.9 and Krox-20 and on morphological segmentation in the hindbrain of mouse embryos. EMBO J. 10, 2985–2995.

Nelson, D.R. 1999. A second CYP26 P450 in humans and zebrafish: CYP26B1. Arch. Biochem. Biophys. 371, 345–347.

Niederreither, K., McCaffery, P., Drager, U.C., Chambon, P., Dolle, P. 1997. Restricted expression and retinoic acid-induced downregulation of the retinaldehyde dehydrogenase type 2 (RALDH-2) gene during mouse development. Mech. Dev. 62, 67–78.

Niederreither, K., Vermot, J., Schuhbaur, B., Chambon, P., Dolle, P. 2000. Retinoic acid synthesis and hindbrain patterning in the mouse embryo. Development 127, 75–85.

Niederreither, K., Fraulob, V., Garnier, J.M., Chambon, P., Dolle, P. 2002. Differential expression of retinoic acid-synthesizing (RALDH) enzymes during fetal development and organ differentiation in the mouse. Mech. Dev. 110, 165–171.

Nittenberg, R., Patel, K., Joshi, Y., Krumlauf, R., Wilkinson, D.G., Brickell, P.M., Tickle, C., Clarke, J.D. 1997. Cell movements, neuronal organisation and gene expression in hindbrains lacking morphological boundaries. Development 124, 2297–2306.

Nolte, C., Amores, A., Nagy Kovacs, E., Postlethwait, J., Featherstone, M. 2003. The role of a retinoic acid response element in establishing the anterior neural expression border of Hoxd4 transgenes. Mech. Dev. 120, 325–335.

Oosterveen, T., Niederreither, K., Dolle, P., Chambon, P., Meijlink, F., Deschamps, J. 2003. Retinoids regulate the anterior expression boundaries of 5' Hoxb genes in posterior hindbrain. EMBO J. 22, 262–269.

Papalopulu, N., Clarke, J.D., Bradley, L., Wilkinson, D., Krumlauf, R., Holder, N. 1991. Retinoic acid causes abnormal development and segmental patterning of the anterior hindbrain in Xenopus embryos. Development 113, 1145–1158.

Perz-Edwards, A., Hardison, N.L., Linney, E. 2001. Retinoic acid-mediated gene expression in transgenic reporter zebrafish. Dev. Biol. 229, 89–101.

Pijnappel, W.W., Hendriks, H.F., Folkers, G.E., van den Brink, C.E., Dekker, E.J., Edelenbosch, C., van der Saag, P.T., Durston, A.J. 1993. The retinoid ligand 4-oxo-retinoic acid is a highly active modulator of positional specification. Nature 366, 340–344.

Popperl, H., Bienz, M., Studer, M., Chan, S.K., Aparicio, S., Brenner, S., Mann, R.S., Krumlauf, R. 1995. Segmental expression of Hoxb-1 is controlled by a highly conserved autoregulatory loop dependent upon exd/pbx. Cell 81, 1031–1042.

Prince, V.E., Moens, C.B., Kimmel, C.B., Ho, R.K. 1998. Zebrafish hox genes: Expression in the hindbrain region of wild-type and mutants of the segmentation gene, valentino. Development 125, 393–406.

Ray, W.J., Bain, G., Yao, M., Gottlieb, D.I. 1997. CYP26, a novel mammalian cytochrome P450, is induced by retinoic acid and defines a new family. J. Biol. Chem. 272, 18702–18708.

Reijntjes, S., Gale, E., Maden, M. 2004. Generating gradients of retinoic acid in the chick embryo: Cyp26C1 expression and a comparative analysis of the Cyp26 enzymes. Dev. Dyn. 230, 509–517.

Renaud, J.S., Nordheim, S., Glover, J.C. 2006. Retinoic acid integrates anteroposterior and dorsoventral patterning in the developing hindbrian. (Accepted pending revision, Development).

Rijli, F.M., Gavalas, A., Chambon, P. 1998. Segmentation and specification in the branchial region of the head: The role of the Hox selector genes. Int. J. Dev. Biol. 42, 393–401.

Romand, R., Sapin, V., Ghyselinck, N.B., Avan, P., Le Calvez, S., Dolle, P., Chambon, P., Mark, M. 2000. Spatio-temporal distribution of cellular retinoid binding protein gene transcripts in the developing and the adult cochlea. Morphological and functional consequences in CRABP- and CRBPI-null mutant mice. Eur. J. Neurosci. 12, 2793–2804.

Rossant, J., Zirngibl, R., Cado, D., Shago, M., Giguere, V. 1991. Expression of a retinoic acid response element-hsplacZ transgene defines specific domains of transcriptional activity during mouse embryogenesis. Genes Dev. 5, 1333–1344.

Rowe, A., Brickell, P.M. 1993. The nuclear retinoid receptors. Int. J. Exp. Pathol. 74, 117–126.

Ruberte, E., Dolle, P., Krust, A., Zelent, A., Morriss-Kay, G., Chambon, P. 1990. Specific spatial and temporal distribution of retinoic acid receptor gamma transcripts during mouse embryogenesis. Development 108, 213–222.

Ruberte, E., Dolle, P., Chambon, P., Morriss-Kay, G. 1991. Retinoic acid receptors and cellular retinoid binding proteins. II. Their differential pattern of transcription during early morphogenesis in mouse embryos. Development 111, 45–60.

Ruberte, E., Friederich, V., Morriss-Kay, G., Chambon, P. 1992. Differential distribution patterns of CRABP I and CRABP II transcripts during mouse embryogenesis. Development 115, 973–987.

Ruberte, E., Friederich, V., Chambon, P., Morriss-Kay, G. 1993. Retinoic acid receptors and cellular retinoid binding proteins. III. Their differential transcript distribution during mouse nervous system development. Development 118, 267–282.

Sakai, Y., Meno, C., Fujii, H., Nishino, J., Shiratori, H., Saijoh, Y., Rossant, J., Hamada, H. 2001. The retinoic acid-inactivating enzyme CYP26 is essential for establishing an uneven distribution of retinoic acid along the anterio-posterior axis within the mouse embryo. Genes Dev. 15, 213–225.

Scadding, S.R., Maden, M. 1994. Retinoic acid gradients during limb regeneration. Dev. Biol. 162, 608–617.

Serpente, P., Tumpel, S., Ghyselinck, N.B., Niederreither, K., Wiedemann, L.M., Dolle, P., Chambon, P., Krumlauf, R., Gould, A.P. 2005. Direct crossregulation between retinoic acid receptor {beta} and Hox genes during hindbrain segmentation. Development 132, 503–513.

Sharma, M.K., Denovan-Wright, E.M., Boudreau, M.E., Wright, J.M. 2003. A cellular retinoic acid-binding protein from zebrafish (Danio rerio): cDNA sequence, phylogenetic analysis, mRNA expression, and gene linkage mapping. Gene 311, 119–128.

Sharpe, C., Goldstone, K. 2000. The control of Xenopus embryonic primary neurogenesis is mediated by retinoid signalling in the neurectoderm. Mech. Dev. 91, 69–80.

Sharpe, C.R. 1992. Two isoforms of retinoic acid receptor alpha expressed during Xenopus development respond to retinoic acid. Mech. Dev. 39, 81–93.

Sharpe, C.R. 1994. The expression of retinoic acid receptors in Xenopus development. Biochem. Soc. Trans. 22, 575–579.

Shiotsugu, J., Katsuyama, Y., Arima, K., Baxter, A., Koide, T., Song, J., Chandraratna, R.A., Blumberg, B. 2004. Multiple points of interaction between retinoic acid and FGF signaling during embryonic axis formation. Development 131, 2653–2667.

Simeone, A., Acampora, D., Arcioni, L., Andrews, P.W., Boncinelli, E., Mavilio, F. 1990. Sequential activation of HOX2 homeobox genes by retinoic acid in human embryonal carcinoma cells. Nature 346, 763–766.

Simeone, A., Acampora, D., Nigro, V., Faiella, A., D'Esposito, M., Stornaiuolo, A., Mavilio, F., Boncinelli, E. 1991. Differential regulation by retinoic acid of the homeobox genes of the four HOX loci in human embryonal carcinoma cells. Mech. Dev. 33, 215–227.

Sirbu, I.O., Gresh, L., Barra, J., Duester, G. 2005. Shifting boundaries of retinoic acid activity control hindbrain segmental gene expression. Development 132, 2611–2622.

Smith, D.P., Mason, C.S., Jones, E.A., Old, R.W. 1994. A novel nuclear receptor superfamily member in Xenopus that associates with RXR, and shares extensive sequence similarity to the mammalian vitamin D3 receptor. Nucleic Acids Res. 22, 66–71.

Smith, S.M., Eichele, G. 1991. Temporal and regional differences in the expression pattern of distinct retinoic acid receptor-beta transcripts in the chick embryo. Development 111, 245–252.

Solomin, L., Johansson, C.B., Zetterstrom, R.H., Bissonnette, R.P., Heyman, R.A., Olson, L., Lendahl, U., Frisen, J., Perlmann, T. 1998. Retinoid-X receptor signalling in the developing spinal cord. Nature 395, 398–402.

Studer, M., Popperl, H., Marshall, H., Kuroiwa, A., Krumlauf, R. 1994. Role of a conserved retinoic acid response element in rhombomere restriction of Hoxb-1. Science 265, 1728–1732.

Studer, M., Gavalas, A., Marshall, H., Ariza-McNaughton, L., Rijli, F.M., Chambon, P., Krumlauf, R. 1998. Genetic interactions between Hoxa1 and Hoxb1 reveal new roles in regulation of early hindbrain patterning. Development 125, 1025–1036.

Sundin, O., Eichele, G. 1992. An early marker of axial pattern in the chick embryo and its respecification by retinoic acid. Development 114, 841–852.

Sundin, O.H., Busse, H.G., Rogers, M.B., Gudas, L.J., Eichele, G. 1990. Region-specific expression in early chick and mouse embryos of Ghox-lab and Hox 1.6, vertebrate homeobox-containing genes related to Drosophila labial. Development 108, 47–58.

Swindell, E.C., Thaller, C., Sockanathan, S., Petkovich, M., Jessell, T.M., Eichele, G. 1999. Complementary domains of retinoic acid production and degradation in the early chick embryo. Dev. Biol. 216, 282–296.

Tahayato, A., Dolle, P., Petkovich, M. 2003. Cyp26C1 encodes a novel retinoic acid-metabolizing enzyme expressed in the hindbrain, inner ear, first branchial arch and tooth buds during murine development. Gene Expr. Patterns 3, 449–454.

Thaller, C., Eichele, G. 1987. Identification and spatial distribution of retinoids in the developing chick limb bud. Nature 327, 625–628.

Thompson, J.N., Howell, J.M., Pitt, G.A., McLaughlin, C.I. 1969. The biological activity of retinoic acid in the domestic fowl and the effects of vitamin A deficiency on the chick embryo. Br. J. Nutr. 23, 471–490.

Trainor, P.A., Krumlauf, R. 2000. Patterning the cranial neural crest: Hindbrain segmentation and Hox gene plasticity. Nat. Rev. Neurosci. 1, 116–124.

Ulven, S.M., Gundersen, T.E., Sakhi, A.K., Glover, J.C., Blomhoff, R. 2001. Quantitative axial profiles of retinoic acid in the embryonic mouse spinal cord: 9-cis retinoic acid only detected after all-trans-retinoic acid levels are super-elevated experimentally. Dev. Dyn. 222, 341–353.

van der Wees, J., Schilthuis, J.G., Koster, C.H., Diesveld-Schipper, H., Folkers, G.E., van der Saag, P.T., Dawson, M.I., Shudo, K., van der Burg, B., Durston, A.J. 1998. Inhibition of retinoic acid receptor-mediated signalling alters positional identity in the developing hindbrain. Development 125, 545–556.

Wagner, E., Luo, T., Drager, U.C. 2002. Retinoic acid synthesis in the postnatal mouse brain marks distinct developmental stages and functional systems. Cereb. Cortex 12, 1244–1253.

Wagner, M., Thaller, C., Jessell, T., Eichele, G. 1990. Polarizing activity and retinoid synthesis in the floor plate of the neural tube. Nature 345, 819–822.

Wagner, M., Han, B., Jessell, T.M. 1992. Regional differences in retinoid release from embryonic neural tissue detected by an *in vitro* reporter assay. Development 116, 55–66.

Wang, Y., Zolfaghari, R., Ross, A.C. 2002. Cloning of rat cytochrome P450RAI (CYP26) cDNA and regulation of its gene expression by all-trans-retinoic acid *in vivo*. Arch Biochem. Biophys. 401, 235–243.

Wendling, O., Ghyselinck, N.B., Chambon, P., Mark, M. 2001. Roles of retinoic acid receptors in early embryonic morphogenesis and hindbrain patterning. Development 128, 2031–2038.

Weston, A.D., Blumberg, B., Underhill, T.M. 2003. Active repression by unliganded retinoid receptors in development: Less is sometimes more. J. Cell Biol. 161, 223–228.

White, J.A., Guo, Y.D., Baetz, K., Beckett-Jones, B., Bonasoro, J., Hsu, K.E., Dilworth, F.J., Petkovich, M., Jones, G. 1996. Identification of the retinoic acid-inducible all-trans-retinoic acid 4-hydroxylase. J. Biol. Chem. 271, 29922–29927.

White, J.C., Highland, M., Kaiser, M., Clagett-Dame, M. 2000. Vitamin A deficiency results in the dose-dependent acquisition of anterior character and shortening of the caudal hindbrain of the rat embryo. Dev. Biol. 220, 263–284.

Wilkinson, D.G., Bhatt, S., Cook, M., Boncinelli, E., Krumlauf, R. 1989. Segmental expression of Hox-2 homoeobox-containing genes in the developing mouse hindbrain. Nature 341, 405–409.

Wilson, J.G., Roth, C.B., Warkany, J. 1953. An analysis of the syndrome of malformations induced by maternal vitamin A deficiency. Effects of restoration of vitamin A at various times during gestation. Am. J. Anat. 92, 189–217.

Wood, H., Pall, G., Morriss-Kay, G. 1994. Exposure to retinoic acid before or after the onset of somitogenesis reveals separate effects on rhombomeric segmentation and 3′ HoxB gene expression domains. Development 120, 2279–2285.

Yamamoto, M., Drager, U.C., McCaffery, P. 1998. A novel assay for retinoic acid catabolic enzymes shows high expression in the developing hindbrain. Brain Res. Dev. Brain Res. 107, 103–111.

Yamamoto, M., Ullman, D., Drager, U.C., McCaffery, P. 1999. Postnatal effects of retinoic acid on cerebellar development. Neurotoxicol. Teratol. 21, 141–146.

Yamamoto, M., Fujinuma, M., Hirano, S., Hayakawa, Y., Clagett-Dame, M., Zhang, J., McCaffery, P. 2005. Retinoic acid influences the development of the inferior olivary nucleus in the rodent. Dev. Biol. 280, 421–433.

Zhang, F., Popperl, H., Morrison, A., Kovacs, E.N., Prideaux, V., Schwarz, L., Krumlauf, R., Rossant, J., Featherstone, M.S. 1997. Elements both 5′ and 3′ to the murine Hoxd4 gene establish anterior borders of expression in mesoderm and neurectoderm. Mech. Dev. 67, 49–58.

Zhang, F., Nagy Kovacs, E., Featherstone, M.S. 2000. Murine hoxd4 expression in the CNS requires multiple elements including a retinoic acid response element. Mech. Dev. 96, 79–89.

Zhang, J., Smith, D., Yamamoto, M., Ma, L., McCaffery, P. 2003. The meninges is a source of retinoic acid for the late-developing hindbrain. J. Neurosci. 23, 7610–7620.

Zhang, Z., Balmer, J.E., Lovlie, A., Fromm, S.H., Blomhoff, R. 1996. Specific teratogenic effects of different retinoic acid isomers and analogs in the developing anterior central nervous system of zebrafish. Dev. Dyn. 206, 73–86.

Zhao, D., McCaffery, P., Ivins, K.J., Neve, R.L., Hogan, P., Chin, W.W., Drager, U.C. 1996. Molecular identification of a major retinoic-acid-synthesizing enzyme, a retinaldehyde-specific dehydrogenase. Eur. J. Biochem. 240, 15–22.

Retinoid receptors in vertebral patterning

Charlotte Rhodes and David Lohnes

Department of Cellular and Molecular Medicine,
University of Ottawa, Ottawa, Canada K1H 8M5

Contents

Advances in Developmental Biology
Volume 16 ISSN 1574-3349
DOI: 10.1016/S1574-3349(06)16006-8

Retinoic acid (RA) is critical for growth and development. The RA signal is mediated by heterodimeric retinoid receptors, RXR–RAR, which have been implicated in vertebral patterning through regulation of *Hox* gene expression. Data suggest that RA may regulate vertebral patterning and *Hox* expression indirectly through the homeodomain transcription factor *Cdx1*. Several additional signaling molecules, notably members of the Wnt (Wingless) and fibroblast growth factor (FGF) families are also implicated in patterning of the posterior vertebrate embryo and evidence suggests that Wnt and RA signaling synergize to regulate Cdx1 expression in the tail bud. Moreover, axial extension is intrinsically linked to somitogenesis and studies have led to a model whereby opposing gradients of FGF and RA link somitogenesis to elongation of the anteroposterior (AP) axis and to left–right patterning.

1. Introduction

The process of gastrulation generates the three primary tissue layers (ectoderm, endoderm, and mesoderm). After the initiation of gastrulation, the principle axis of the embryo is established with the subdivision of the embryo along the AP axis. During this process, patterning information is conveyed to the embryonic axis through the activation of patterning genes. The *Hox* gene products are key players in this process as pertains to derivatives of the caudal embryo, and are essential for normal patterning of the central nervous system (CNS) caudal to the anterior hindbrain, as well as derivatives of the somitic mesoderm and endoderm.

The activation–transformation model, defined by Pietr Nieuwkoop (Stern, 2001), initially proposed a general paradigm for the neurectodermal patterning processes in vertebrates. In this model, the neurectoderm is initially "activated" by signals emanating from the underlying mesoderm. This neurectoderm, however, exhibits a default anterior (forebrain-like) state. Subsequent patterning along the AP axis is imparted by a "transformation" step, again mediated by signals derived from the underlying mesoderm. A number of studies have implicated three major players in this latter process: the vitamin A metabolite, RA; certain members of the Wnt/wingless family; and fibroblast growth factors (FGFs). These signaling molecules can suppress markers of anterior identity and induce the expression of genes associated with posterior fate, such as the *Hox* genes.

In addition to patterning the neuraxis, FGFs, Wnts, and RA can also impact on AP patterning of mesodermal derivatives, including paraxial mesoderm. Paraxial mesoderm can be subdivided into two major domains. The head, or cephalic, paraxial mesoderm extends from the anterior tip of the embryonic axis to the otic vesicle and gives rise to several muscles and certain bones of the head (Noden, 1991; Couly et al., 1992, 1993). Caudal to this region, the paraxial mesoderm generates somites, paired epithelial structures juxtaposed on either side of the prospective neural tube, which extend along the axis to the end of the tail. Somites are produced in a rostral–caudal sequence initially from cells ingressing from the epiblast of the primitive streak during gastrulation. Following the regression of the primitive streak, paraxial mesoderm that will give rise to the more posterior somites is derived from the tail bud, which is located at the caudal extremity of the axis. The tail bud acts as a terminal growth zone and contains a pool of pluripotent mesenchymal stem cells that are considered to represent the remnant of the primitive streak (Tam and Tan, 1992).

Following ingression of epiblast cells through the streak (or generated by the tail bud), paraxial mesoderm forms as a contiguous population of presomitic mesoderm (PSM) which subsequently undergoes segmentation in a rostral–caudal manner, resulting in the formation of somites along the AP axis in the process of somitogenesis. The first somite forms directly posterior to the otic vesicle (Hamilton and Hinsch, 1956; Huang et al., 1997; Vermot and Pourquie, 2005), with subsequent somites generated in a sequential manner along the AP axis from a reiteration of the segmentation process, resulting in formation of somites at regular intervals. This segmentation occurs at a uniform pace in a manner that is species dependent, requiring ~90 min in the chick, 120 min in the mouse, and 30 min in the zebrafish, under optimal conditions (Tam, 1981). The segmental process lasts until the number of somites characteristic of a given species is reached. This termination is believed to be determined by a change in the tissue environment in the tail bud rather than a loss of somitogenetic potential (Tam and Tan, 1992).

By its very nature, somitogenesis is tightly coupled to extension of the body axis as PSM, derivatives, initially produced caudally, are displaced rostrally following which they differentiate to form somites (Ruberte et al., 1990; Fujii et al., 1997; Dubrulle and Pourquie, 2004a). This periodic production of somites along the AP axis involves a molecular oscillator termed the "segmentation clock." Evidence for such mechanism acting within the PSM comes from the dynamic and periodic expression of genes belonging to the Notch pathway, collectively referred to as cyclic genes, which appear to be downstream of a fluctuating Wnt signal. The periodicity of alteration in expression of such cyclic genes varies once during the formation of each somite, and is believed to impart a wave of information necessary for somitogenesis that sweeps along the PSM in a caudal–rostral wave. This rhythmic expression begins during late gastrulation and is maintained throughout somitogenesis. The details underlying the process of somitogenesis and our understanding of the molecular basis of this event are beyond the scope of this chapter, and have been discussed elsewhere (Pourquie, 2001; Dubrulle and Pourquie, 2004a).

The translation of cyclic gene expression along the PSM into the spatial periodicity of somite formation is proposed to be under the control of a "determination front," a region of cell competence which dictates the differentiation events leading to segmentation of PSM and the generation of

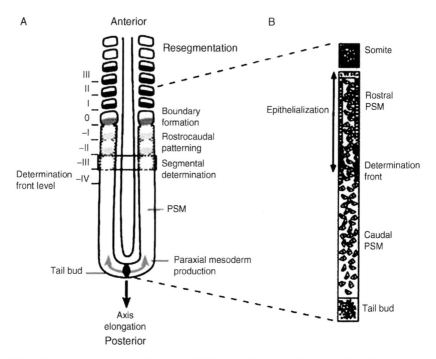

Fig. 1. Successive steps in somitogenesis. (A) The caudal region of a 2-day-old chicken embryo, representing the events taking place in the presomitic mesoderm (PSM) leading to somite formation. Somite nomenclature, as defined by Pourquie and Tam (2001) is indicated on the left. Paraxial mesoderm production is indicated by gray arrows. (B) A parasagittal section through the PSM illustrating the cellular organization of cells along the AP axis. Caudal PSM cells are of a loose mesenchymal character with epithelialization beginning anterior to level of the determination front. Reproduced with permission from Dubrulle and Pourquie (2004a).

somites (Fig. 1). As discussed later, studies suggest that opposing gradients of FGF and RA are integral to the establishment of the determination front (Diez del Corral et al., 2003; Moreno and Kintner, 2004; Dubrulle and Pourquie, 2004a).

Once formed, somites differentiate in accordance with environmental cues from the surrounding tissues, resulting in the generation of dermomyotome from the dorsal–lateral somite and sclerotome from the ventral somitic aspect. The dermomyotome gives rise to both dermis and skeletal muscles of the trunk and limbs, while the sclerotome is the anlage of the vertebrae and ribs. In addition, each somite also undergoes compartmentalization into anterior and posterior halves, a process which appears to be established prior to somite formation and is dependent on information imparted by signaling through the Notch pathway in the PSM (Stern et al., 1991; Dubrulle and Pourquie, 2004a). This subdivision of the somite has important ramifications. The definitive vertebrae are formed from the fusion of the caudal half of one somite with the rostral half of the following somite in a process called resegmentation; therefore, the vertebrae are offset by a half-segment relative to the somites themselves (Bagnall et al., 1988; Christ et al., 1998). In contrast to the vertebrae, each axial muscle derives from the myotome of a single somite, which results in the muscles of the back being attached to two successive vertebrae. This is critical to mobility as muscle contraction is now integrated to two consecutive vertebral units (Huang et al., 2000).

Ultimately, the sclerotomal derivative of each somite is patterned into cervical, thoracic, lumbar, sacral, and caudal vertebral identities according to their position along the AP axis (Meinhardt, 1986; Tam and Trainor, 1994; Gossler and Hrabe, 1998; Pourquie, 2000). In the mouse, the somites give rise to occipital bones (formed by condensation of the anterior-most somites) 7 cervical (C1–C7), 13 thoracic (T1–T13), 6 lumbar (L1–L6), 3 or 4 sacral (S1–S4), and 31 caudal vertebrae. Morphological differences are evident between both vertebral regions, and frequently between vertebrae within a given axial region (Meinhardt, 1986; Christ et al., 1998, 2000).

Prior to overt differentiation, the somites themselves are morphologically indistinguishable. The distinct vertebral morphologies, however, necessitate that patterning along the AP axis must be imparted at some stage of their ontogenesis. Transplantation experiments in the chick (Kieny et al., 1972; Burke and Nowicki, 2003) and mouse (Beddington et al., 1992) indicate that such positional information is established in the PSM prior to even the formation of somites themselves. In the chick embryo, heterotopic grafting of unsegmented PSM destined to generate cervical vertebrae results in the development of vertebrae which retain this cervical morphology (Kieny et al., 1972; Burke and Nowicki, 2003). Furthermore, the transplants continue to express patterning genes (e.g., *Hox*; see in a later section) characteristic of their presumptive cervical origin. Therefore, the transcriptional program necessary to establish gene expression relevant to vertebral AP patterning operates early, likely shortly after gastrulation in the PSM.

2. Retinoids and development

The importance of retinoids (vitamin A and its derivatives) has been known for nearly a century. The physiological functions of retinoids were initially inferred from studies on vitamin A-deficient (VAD) animals which demonstrated that retinoids play essential roles in differentiation and in the maintenance of numerous tissues in the adult. Rodents fed on a VAD diet display a well-defined spectrum of symptoms during postnatal life. These typically include weight loss, sterility, and squamous metaplasia of various epithelia, including the respiratory, alimentary, and genitourinary tracts (Lohnes et al., 1995; Niederreither et al., 1999; Clagett-Dame and DeLuca, 2002).

Offspring from VAD females also exhibit a plethora of congenital defects involving the genitourinary tract, the eye, and the respiratory tract among others (Wilson et al., 1953), demonstrating critical roles for retinoid signaling in a number of ontogenic programs. With the exception of vision, RA (the carboxylic acid derivative of vitamin A) can prevent or reverse most of the defects generated by postpartum or embryonic VAD, suggesting that RA is the principal biological form of vitamin A.

3. Regulation of retinoid biodistribution

Vitamin A must be obtained from dietary sources, such as carotenoids or retinyl esters, which are converted into retinol in the intestine, and subsequently stored in hepatocytes as retinyl esters (Clagett-Dame and DeLuca, 2002; Harrison, 2005). When required, retinol is released from these stores and is biologically activated by a two-step oxidation process. The first reaction, conversion of retinol to retinaldehyde, appears to be ubiquitous and can be catalyzed by any one of a number of alcohol dehydrogenases (Meinhardt, 1986; Napoli et al., 1995; Duester, 2000; Reijntjes et al., 2005). The subsequent, irreversible, oxidation of retinal to RA can be catalyzed by any one of several retinaldehyde dehydrogenases (RALDH) of which RALDH2 is critical for production of RA in most of the early embryo (McCaffery and Drager, 1993, 1995; Drager and McCaffery, 1995). In this regard, although the stereoisomer 9-*cis*-RA has been implicated as an important bioactive retinoid, for the sake of brevity, its potential roles in retinoid signaling will not be elaborated in this chapter.

In the developing mouse, *Raldh2* transcripts are found in the primitive streak region at embryonic day 7.5 (E7.5) and subsequently shift to the embryonic somitic mesoderm around E8.5. There is a strong concordance between *Raldh2* expression and activity of an RA-responsive transgenic reporter at these stages (Mendelsohn et al., 1991; Rossant et al., 1991; Fujii et al., 1997; Niederreither et al., 1997; Maden et al., 1998), suggesting that this enzyme plays a critical role in delimiting retinoid activity during development. This role is further underscored by gene knockout studies. *Raldh2*-null mutant embryos die prior to E10.5 and exhibit a number of defects, including a dilated midline heart, hindbrain-patterning defects, a foreshortened axis, and small somites, and failure to activate the RA reporter at early somite stages (with the exception of the developing eye which expresses other *Raldh* types Napoli et al., 1995; Zhao et al., 1996; Niederreither et al., 1997; Berggren et al., 1999; Maeda et al., 2005; Matt et al., 2005; Reijntjes et al., 2005; Salvador-Silva et al., 2005). Administration of maternal RA can rescue the defects inherent to the *Raldh2*-null mutant, demonstrating a critical role for retinoid signaling in a number of early developmental programs, including axial extension and somitogenesis.

Prior work has shown that catabolic inactivation of RA also plays a critical role in delimiting fields of active retinoid signaling during development. Members of the *Cyp26* family of P450 cytochrome oxidases (*Cyp26A1,*

Cyp26B1, and *Cyp26C1*) have been identified in several vertebrate species, all of which catalyze the conversion of RA to more oxidized derivatives (Fujii et al., 1997; Ray et al., 1997; Tahayato et al., 2003; Reijntjes et al., 2004; Taimi et al., 2004; Gu et al., 2006).

A critical role for *Cyp26A1* in limiting retinoid signaling in the tail bud (among other tissues) is evidenced by gene-targeting studies. Among other defects, *Cyp26A1*-null mutant embryos exhibit caudal regression associated with spina bifida, posterior homeotic transformations of cervical vertebrae, and malformed lumbosacral skeletal elements, all of which are also common outcomes of embryonic retinoid excess (Abu-Abed et al., 2001; Sakai et al., 2001). *Cyp26A1* is also a direct RA target gene (Loudig et al., 2005) and is profoundly induced in the tail bud by exogenous RA, suggesting a biofeedback mechanism to limit retinoid signaling tissue(s) (Iulianella et al., 1999; Yashiro et al., 2004; Sirbu et al., 2005). As discussed later, *Cyp26A1* has also been implicated as an important player in somitogenesis.

4. Retinoid receptors

The RA signal is mediated by two families of nuclear receptors, the RA receptors (RARs) and the retinoid X receptors (RXRs). Both RARs and RXRs are expressed as three subtypes (α, β, and γ), with each subtype also expressed as multiple N-terminal variant isoforms (Brand et al., 1988, 1990; Krust et al., 1989; Kastner et al., 1990, 1997; Leroy et al., 1991; Mangelsdorf and Evans, 1995; Chambon, 1996; Mark et al., 1999). The retinoid receptors mediate the RA signal through RXR–RAR heterodimers which impact on transcription through association with RA response elements (RAREs) in the promoter/enhancer regions of target genes. RAREs are typically composed of two direct repeats with the consensus half-site 5'-PuGGTCA-3' spaced by two or five intervening nucleotides (DR2 and DR5, respectively), although a number of variant motifs have been identified (Altucci and Gronemeyer, 2001). RXRs are also requisite heterodimeric partners for a number of additional nuclear receptors (Mangelsdorf and Evans, 1995), and RXR homodimers have been shown to respond to 9-*cis*-RA on DR1 elements (Ahuja et al., 2003).

In the absence of ligand, RXR–RAR heterodimers act as repressors of at least some target genes through recruitment of histone deacetylase (HDAC)-containing corepressor complexes. HDAC activity contributes to transcriptional repression via deacetylation of N-terminal histone lysine residues, resulting in chromatin compaction and rendering the template less accessible to the basal transcriptional machinery and/or other transcription factors. Target gene transcription, initiated on ligand association of the RAR moiety of the heterodimer, induces a conformational change in the activation function 2 domain of the RAR, resulting in destabilization of corepressor association and subsequent recruitment of coactivator complexes. Histone acetyltransferase (HAT) activity, associated with such coactivator complexes, leads to chromatin decondensation over the target gene promoter region. Ligand-dependent recruitment of additional regulatory proteins, such as components of the mediator complex, also contributes to RA-dependent gene regulation (Giguere et al., 1987; Petkovich et al., 1987; Dolle et al., 1990; Mangelsdorf et al., 1990; Ruberte et al., 1991; Leid et al., 1992; Chambon, 1994; Altucci and Gronemeyer, 2001).

Targeted mutagenesis has shown that loss of a single RAR isoform typically results in negligible phenotypic consequence, while mice lacking all *RARα* or *RARγ* isoforms exhibit a number of defects, some of which are reminiscent of the outcome of postpartum VAD. *RARγ*-null mice also exhibit vertebral homeotic transformations (Fig. 2) consistent with a role for retinoid signaling in regulation of *Hox* gene expression (Lohnes et al., 1993), as described in greater detail later (Kastner et al., 1995; Sucov and Evans, 1995). In contrast to single null mutants, mice lacking multiple RARs exhibit a multitude of congenital defects and recapitulate essentially all aspects of fetal VAD syndrome. In addition, *RARα1/RARγ* and *RARα/RARγ* double null mutants exhibit an exacerbation of the vertebral homeosis

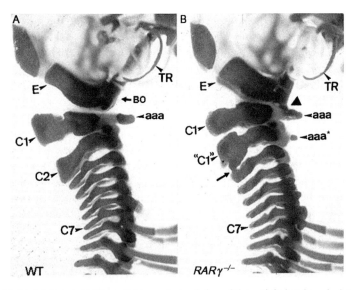

Fig. 2. Skeletal defects in *RARγ*-null fetuses. Lateral view of the occipital and cervical region of wild type (A) and *RARγ*$^{-/-}$ (B) skeletons. Note the homeotic transformation of the second cervical vertebra ("C1" in B) to a first cervical (C1)—like identity as evidenced by wider, sculpted, neural arches and the presence of an ectopic anterior arch of the atlas (aaa* in B). The *RARγ*$^{-/-}$ skeleton also exhibits fusion between the basoccipital bone (BO) and the anterior arch of the atlas (aaa) (arrowhead in B, compare with A). E, exoccipital; TR, tympanic ring. Reprinted from Cell, Vol. 73, Lohnes et al., Function of retinoic acid receptor gamma in the mouse, pp. 643–658, Copyright May 21, 1993, with permission from Elsevier.

seen in *RAR*γ single null mice (Lohnes et al., 1994; Mendelsohn et al., 1994; Luo et al., 1996). *RXRα/RAR* compound mutants also exhibit increased penetrance and expressivity of the abnormalities observed in RAR single null offspring, consistent with a role for RXR/RAR heterodimers in transducing the retinoid signal *in vivo* (Kastner et al., 1997). These studies underscore a critical role for retinoid signaling in vertebral AP patterning and demonstrate a considerable degree of functional redundancy between these receptors.

5. *Hox* genes and vertebral patterning

A wealth of gain- and loss-of-function experiments has shown that *Hox* gene products are critical determinants of neurectoderm and mesoderm patterning along the AP axis. The 39 mammalian *Hox* genes are arranged in four clusters, *Hoxa* to *Hoxd*, that have likely evolved by duplication of an ancestral complex related to the *HOM-C* genes of *Drosophila melanogaster* (Duboule and Dolle, 1989; Duboule, 1998; Ferrier and Holland, 2001). *Hox* genes can be further categorized into paralogous groups, 1–13, based on their physical location within a given *Hox* cluster and the sequence relatedness of their homeodomains.

Hox genes within a given cluster are expressed in a temporally controlled manner along the developing axial and appendicular axes. As regards expression along the AP axis, *Hox* genes within a given paralogue group typically exhibit similar temporal onset of expression in the primitive streak region, with expression subsequently spreading to achieve comparable rostral boundaries (Krust et al., 1989; Deschamps and van Nes, 2005). This pattern of expression, termed colinearity, was first documented in *Drosophila* (Duboule and Dolle, 1989; Duboule, 1998; Ferrier and Holland, 2001) and subsequently characterized in the developing mouse (Duboule and Dolle, 1989). Briefly, *Hox* genes situated at the 3′ end of a given cluster are typically expressed at earlier stages, and achieve a more rostral limit of expression, than more 5′ *Hox* genes for a given locus (Fig. 3). In the mouse, this culminates in a nested set of expression domains, with distinct axial levels exhibiting a unique combination of *Hox* transcripts which has been referred to as the "Hox code" (Kessel and Gruss, 1991; Conlon and Rossant, 1992; Krumlauf, 1994; Conlon, 1995). Inherent to this model is the implication that the particular combination of *Hox* gene products impacts on the developmental fate of that particular structure. This model is supported by the phenotypes evoked by a number of gain- and loss-of-function experiments, which frequently manifest as homeotic transformation(s) of elements derived from the rhombomeres or of the somites (Popperl and Featherstone, 1993; Marshall et al., 1994; Studer et al., 1994; Conlon, 1995; Frasch et al., 1995; Morrison et al., 1996; Dupe et al., 1997; Zhang et al., 1997; Gould et al., 1998; Huang et al., 1998; Packer et al., 1998). Implicit to these findings is that

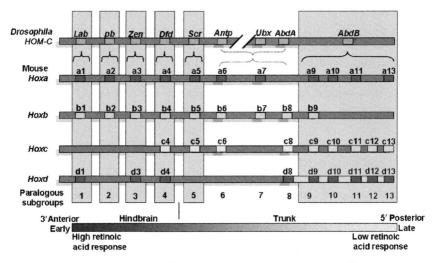

Fig. 3. *Hox* paralogous groups. Evolutionary conservation of the *Drosophila HOM-C* genes with the four mouse paralogous clusters, *Hoxa* to *Hoxd*. The order of the genes on the chromosome has been conserved between fly and mouse. The genes located at the 3' end of the chromosome are expressed first, respond to high doses of retinoic acid and have a more rostral limit of expression compared to the 5' *Hox* genes which respond to low doses of retinoic acid, are expressed later in development, and at sequentially more caudal levels. Figure adapted from Gilbert, S.F., Dev. Biol., 6th ed., p. 366. (See Color Insert.)

Hox expression must be strictly controlled, both spatially and temporally, for correct embryonic patterning.

6. Regulation of *Hox* expression by RA

As discussed earlier, RA is essential for numerous developmental programs. Of particular relevance to this chapter is the role of RA in vertebral AP patterning through regulation of *Hox* gene expression. One of the early observations suggesting a relationship between retinoid signaling and regulation of *Hox* expression came from the finding that RA is able to activate *Hox* genes in embryocarcinoma cells in culture (Simeone et al., 1990; Boncinelli et al., 1991; Bottero et al., 1991; Perantoni et al., 2005). Remarkably, this induction closely reflected the normal colinear onset of *Hox* gene activation typically seen during development, with 3' genes induced by RA more acutely, and more strongly, than 5' genes. Subsequent work suggested that RA also played a critical role in regulating *Hox* expression *in vivo*. In the mouse, *Hox* expression is initiated at E7.5 in the primitive streak and subsequently expands anteriorly in the neural tube and PSM. At E6.5, prior to initiation of *Hox* expression, exposure to exogenous RA has no effect on axial patterning, while RA treatment at E7.0–E8.0 results in vertebral homeotic transformations throughout the axial skeleton which occur concomitant with rostral shifts in the expression of several *Hox* genes (Kessel and Gruss, 1991; Conlon and Rossant, 1992). These homeoses consist largely of posterior vertebral transformations, consistent with the anticipated impact of anteriorized *Hox* expression on interpretation of the Hox code. Similar effects of RA treatment are also seen as regards *Hox* expression and patterning of the rhombomeres of the hindbrain (Gavalas and Krumlauf, 2000; Niederreither et al., 2000; Begemann and Meyer, 2001; Maden, 2001; Gavalas, 2002; Melton et al., 2004).

The above-mentioned observations indicate a key role for RA in vertebral AP patterning through regulation of the Hox code. Consistent with this, mice null for *RARγ* or *RARα* and *RARγ* exhibit anterior vertebral transformations, suggestive of posteriorized *Hox* expression. Notably, these transformations are largely confined to elements comprising the cervical and anterior thoracic skeleton, in agreement with the biodistribution of RA in the primitive streak region at E7.5 when the PSM precursors of these elements are likely undergoing AP patterning (Kieny et al., 1972). Although more posterior vertebrae do not exhibit a high frequency of homeosis in any RAR mutants background, this could be due to functional redundancy among these receptors. However, the restriction of affected vertebral elements to anterior somite derivatives is in agreement with exclusion of RA from the primitive streak/tail bud from E8.5 on, suggesting that retinoid signaling impacts on vertebral patterning during a discrete temporal window corresponding to the late primitive streak stage (see later).

The finding of functional RAREs within several *Hox* loci suggests that RA plays a direct role in regulating the expression of at least some retinoid-responsive *Hox* genes. The most rigorously characterized of these include *Hoxa1* and *Hoxa2* (Frasch et al., 1995), *Hoxb1* (Marshall et al., 1994; Huang et al., 1998), *Hoxa4* (Zhang et al., 1997; Packer et al., 1998), *Hoxd4* (Popperl and Featherstone, 1993; Frasch et al., 1995; Morrison et al., 1996; Zhang et al., 1997), and *Hoxb5* (Oosterveen et al., 2003). A role for several of these RAREs in directing spatial expression *in vivo* has been well established. For example, transgenic reporter analysis of a *Hoxa1* enhancer has demonstrated that the RARE within these sequences is essential for directing reporter expression in the neuroepithelium caudal to rhombomere (r) 4 (Frasch et al., 1995). Moreover, deletion of this RARE by gene-targeting results in hindbrain and cranial nerve abnormalities which represent a partial phenocopy of the *Hoxa1*-null mutant (Dupe et al., 1997). *Hoxb1* expression has been shown to be influenced by two RAREs 3' to the coding sequences of this gene. The 5'-most RARE, a DR2 element, restricts transgenic reporter expression to r4 in the hindbrain (Studer et al., 1994), while the more 3' element, a DR5 motif, is essential for transgenic expression in the gut (Huang et al., 1998).

Additional work has suggested that the fundamental composition of an RARE may provide an additional level of regulation. By replacing the *Hoxb4* DR5 element with the *Hoxb1* DR2 element, Gould et al. (1998) found that the latter motif directed anteriorized transgene expression, resulting in a shift of the *Hoxb4* expression domain from its normal anterior boundary in presumptive r6/7 to r3/4. This suggests that retinoid signaling may participate in establishing the segment-restricted expression of *Hox* genes in the rhombomeres through differential interpretation of the retinoid signal by utilization of RAREs of variable efficiency. Alternatively, or in addition, either a temporal or spatial retinoid gradient may be a key determinant in establishing *Hox* expression patterns in the hindbrain and other retinoid-target tissues (Dupe et al., 1997; Yashiro et al., 2004).

While a role for direct RA regulation of *Hox* genes relevant to patterning the neurectoderm has been well established, the relationship between RA and regulation of mesodermal *Hox* expression as pertains to vertebral patterning is less clear. An RARE critical for *Hox* function in vertebral AP patterning has not been documented, suggesting that RA may impact on vertebral patterning through an intermediary. Based on their function in a number of vertebrate model systems, members of the *Cdx* gene family have emerged as potential candidates for such a role.

7. *Cdx* genes and axial patterning

The *Drosophila caudal* (*cad*) gene was the first *Cdx* member to be identified. *Cad* encodes a homeodomain transcription factor and functions in a manner analogous to a *Hox* gene product in specifying the identity of the posterior abdominal segment (Mlodzik and Gehring, 1987; Moreno and Morata, 1999). This feature may represent a vestigial function given that *Cad* is a member of the *Protohox* cluster which is believed to be an ancestral antecedent of the *Hox* genes (Brooke et al., 1998; Ferrier and Holland, 2001). *Cad* homologues have been identified in a number of vertebrate species (Duprey et al., 1988; Burglin et al., 1989; Blumberg et al., 1991; Frumkin et al., 1991; Joly et al., 1992; Serrano et al., 1993; Northrop and Kimelman, 1994), and all appear to be involved in specification of the posterior embryo and/or patterning the AP axis in a manner similar to *Cad*. However, unlike *Cad*, which appears to function in a *Hox*-like manner, vertebrate Cdx proteins appear to act upstream of the *Hox* genes, as discussed later.

Analysis of *Cdx* function and expression in vertebrates has focused largely on the mouse, chick, zebrafish, and frog *Xenopus*. The following description of expression patterns is focused on studies using the mouse relevant to vertebral patterning, although recent work suggests there may be important species-specific differences in *Cdx* expression and/or function.

In the murine embryo proper, *Cdx1* transcripts are first detected during midgastrulation (E7.5) in the ectoderm and mesoderm in the primitive streak region (Gamer and Wright, 1993; Meyer and Gruss, 1993) (Fig. 4). At E7.75, *Cdx1* achieves an anterior-most limit of expression, with transcripts observed in the neurectoderm at the level of the preotic sulcus in the presumptive hindbrain and with a slightly more posterior boundary of expression in the mesoderm. This anterior boundary in the neural tube subsequently recedes to the level of the presumptive spinal cord and becomes restricted to the dorsal region of the neural folds and in some populations of newly migrating neural crest cells. At late streak stages, *Cdx1* is also detected in unsegmented paraxial mesoderm and, transiently, in the newly formed somites. From E9 to E10, as the somites differentiate, *Cdx1* expression is restricted to a dorsal region corresponding to presumptive dermamyotome. In the tail bud, *Cdx1* expression continues until E12, when transcripts are no longer detected. While *Cdx1* (and *Cdx2*) is subsequently expressed in the hindgut endoderm, a role for Cdx function in intestinal patterning is parenthetical to the present chapter, and has been discussed elsewhere (Silberg et al., 2000; Guo et al., 2004).

Cdx2 is first expressed in the trophectoderm at E3.5, where it plays critical roles in this lineage (Niwa et al., 2005; Strumpf et al., 2005; Deb et al., 2006).

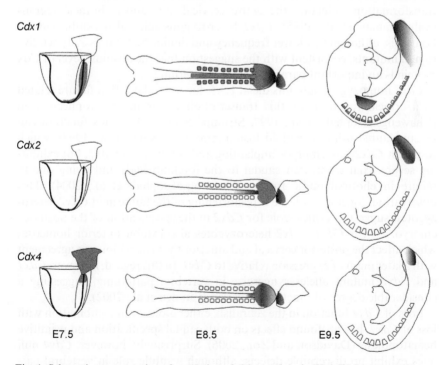

Cdx1

Cdx2

Cdx4

E7.5   E8.5   E9.5

Fig. 4. Schematic representation of expression of *Cdx* members in E7.5–E9.5 mouse embryos. Relative level of expression is denoted by shading. Extraembryonic expression is not indicated.

Cdx2 transcripts in the embryo are first observed at E8.5 in the tail bud, the base of the allantois as well as the posterior neural plate, notochord, hindgut endoderm, and unsegmented paraxial mesoderm. Expression continues in the tail bud until E12.5 (Boncinelli et al., 1991; Beck et al., 1995).

Cdx4 expression is first detected in the allantois and the posterior primitive streak at E7.5, and subsequently expressed throughout the primitive streak region with transcripts in the paraxial mesoderm reaching a rostral limit posterior to the last formed somite at E8.5. Expression is also observed in the posterior lateral plate mesoderm (LPM), intermediate mesoderm, and hindgut endoderm until E10.5 (Gamer and Wright, 1993).

The above-mentioned data, together with studies in the developing chick, have shown that *Cdx* members are expressed in nested domains with a caudal-high distribution along the developing AP axis from midgastrulation to tail bud stages. These observations have led to the suggestion that relative combinations of Cdx levels in the primitive streak/tail bud may be an important determinant of their function in axial patterning (Charite et al., 1995, 1998; Marom et al., 1997; Gaunt et al., 2004).

Loss-of-function studies in mice have shown that both Cdx1 and Cdx2 play roles in AP patterning. *Cdx1*-null mice are viable and survive to adulthood. However, they exhibit vertebral homeosis which typically consist of anterior transformations affecting the entire cervical and upper thoracic regions (Subramanian et al., 1995). *Cdx1* heterozygous mice also exhibit similar homeosis, although at a lower frequency and diminished expressivity relative to null mutants, consistent with the suggestion that the absolute level of Cdx protein is an important parameter in AP patterning.

Cdx2 homozygous mutants die at E3.5 due to implantation failure related to a critical function for this transcription factor in the trophectoderm (Chawengsaksophak et al., 1997; Strumpf et al., 2005). This function can be supplanted using tetraploid fusion strategies (Nagy et al., 1993), which results in *Cdx2*-null embryos implanting and surviving to E10.5, but exhibiting severe axial truncation caudal to the forelimb buds and dying due to failure in chorioallantoic fusion (Chawengsaksophak et al., 2004). This observation, together with the foreshortening of the tail seen in *Cdx2* heterozygotes, suggests a critical role for *Cdx2* in the specification of the posterior embryo. As for *Cdx1, Cdx2* heterozygotes also exhibit anterior homeoses which affect the posterior cervical and anterior thoracic regions, in agreement with a later onset of expression relative to *Cdx1*. In this regard, however, *Cdx1* and *Cdx2* mutant alleles synergize in vertebral patterning, suggesting a considerable degree of functional overlap (van den et al., 2002).

Loss of *Cdx4* function in the zebrafish, either alone or in combination with loss of *Cdx1*, has profound effects on both caudal specification and primitive hematopoiesis (Davidson and Zon, 2006). Surprisingly, however, *Cdx4*-null mice exhibit no discernible defects, although a subtle role in vertebral patterning and placentogenesis are seen with subsequent loss of *Cdx1* or *Cdx2*, respectively (van Nes et al., 2006). A complete understanding of the roles of Cdx members likely necessitates ablation of all three genes.

8. Regulation of *Hox* expression by Cdx

The vertebral transformations seen in *Cdx1*-null and *Cdx2* heterozygous offspring recapitulate the defects seen in a number of *Hox*-null mutants, suggesting that Cdx proteins function upstream of *Hox*. The vertebral transformations in *Cdx* mutants are accompanied by posterior shifts in the expression of the same *Hox* gene(s) implicated in patterning a given vertebral element. For example, both *Cdx1*- and *Hoxd3*-null offspring exhibit a broadening of the basioccipital bone and fusion of this element with the anterior arch of the atlas (Gruss and Kessel, 1991; Condie and Capecchi, 1993). In this regard, disruption of a single *Hox* typically impacts on patterning of a limited number of vertebrae, while *Cdx* mutants exhibit a series of anterior transformations affecting a broad region of the cervical and anterior thoracic regions. *Cdx* loss of function correlates with posteriorization of somitic expression of multiple *Hox* genes, consistent with a pleiotropic patterning affect and suggesting that Cdx proteins are required to establish normal patterns of expression of multiple *Hox* genes (Subramanian et al., 1995; Isaacs et al., 1998; van den et al., 2002).

Cdx members function as transcription factors by binding to *cis*-acting regulatory elements in target genes, the first of which was described for the *Drosophila* segmentation gene *fushi tarazu* (Dearolf et al., 1989). Subsequent work has identified a conserved TTTATG *Cdx*-binding motif, which has been found in the regulatory region of several *Hox* loci (Knittel et al., 1995; Shashikant et al., 1995; Subramanian et al., 1995; Charite et al., 1998; Tabaries et al., 2005). For example, the *Hoxa7* 5' promoter region contains two such motifs which have been shown to respond to Cdx1 in transfection assays (Subramanian et al., 1995), and which have been implicated in establishing proper expression of *Hoxa7 in vivo* (Knittel et al., 1995). A similar element has been documented in a *Hoxb8* enhancer which is essential for the proper spatial expression in transgenic reporter assays (Charite et al., 1998). Together with the finding of altered *Hox* expression in Cdx loss-of-function models, these findings are consistent with *Cdx* members impacting on AP patterning by direct regulation of at least some *Hox* genes.

As discussed earlier, it has been suggested that the levels of Cdx proteins are important in establishing normal patterns of target genes along the AP axis. In agreement with this, multimerization of *Cdx* response elements, or anteriorization of Cdx expression, results in anteriorization of reporter expression in transgenic models (Charite et al., 1995, 1998; Gaunt et al., 2004). These findings imply that *Hox* expression may be dictated by a gradient of Cdx expression along the AP axis which may be differentially interpreted by the number and/or affinity of *Cdx* response elements in target genes. Factors governing *Cdx* expression would therefore be predicted to be important modulators of AP patterning through such a paradigm.

9. *Cdx1* functions as an intermediary in RA-dependent vertebral patterning

Retinoid signaling is critical to normal AP vertebral patterning, presumably through regulation of mesodermal *Hox* gene expression. Considerable evidence now suggests that RA exerts its effects on *Hox* gene expression, and subsequent vertebral patterning, in part through direct regulation of *Cdx1*.

The concerted actions of *Raldh2* and members of the *Cyp26* gene family serve to limit retinoid signaling to discrete embryonic fields during development. At E7.5, RA bioactivity is found in the primitive streak region in a manner which correlates with the onset of *Cdx1* expression. This period also corresponds with the window when exogenous RA can affect both *Hox* expression and vertebral patterning (Kessel and Gruss, 1991), and with the temporal window during which the cervical vertebrae are patterned. It is these elements which are typically affected by either RAR or Cdx1 loss of function.

A relationship between retinoid signaling and *Cdx1* was first suggested by the observation that *Cdx1* responds rapidly to exogenous RA treatment at E7.5, while expression is reduced in *RARα1/γ*-null mutant mice at this stage (Houle et al., 2000). In agreement with a direct relationship, the *Cdx1* promoter harbors an atypical RARE which mediates retinoid response in tissue culture models (Houle et al., 2000), and contributes to spatial–temporal expression of a transgenic reporter gene under the control of *Cdx1* promoter sequences *in vivo* (Lickert and Kemler, 2002). Ablation of this RARE by gene-targeting results in reduced *Cdx1* expression with concomitant posteriorization of a number of *Hox* genes and elicits a subset of the vertebral defects seen in both *Cdx1*- and *RAR*-null mutants (Houle et al., 2003). Together, these data strongly support a role for *Cdx1* as an intermediary which serves to relay the retinoid signal to regulate mesodermal *Hox* gene expression required for normal vertebral patterning.

Although targeted ablation of the *Cdx1* RARE has phenotypic consequence, *Cdx1* still responds to exogenous RA in this knockout background. This induction could be mediated by potential RAREs identified in either the first intron (Gaunt et al., 2004) or a more distal 5′ element (Houle et al., 2003). In this regard, the atypical RARE is conserved between mouse and human, while the intronic motif appears more ancestral. Given this, it is tempting to speculate that the murine and human *Cdx1* promoters may have co-opted additional regulatory elements, such as this RARE, which may reflect evolutionary differences in vertebral patterning.

Cdx1/RARγ compound null mutants exhibit a synergistic interaction in vertebral patterning. However, the effects of RA on vertebral patterning and *Hox* gene expression are only partially attenuated in the absence of *Cdx1*

(Allan et al., 2001) suggesting that RA impacts on *Hox* gene expression and vertebral patterning through other means in addition to direct regulation of *Cdx1*.

10. Regulation of *Cdx1* expression by Wnt signaling

The spatial and temporal expression of retinoid bioactivity correlates with the onset of *Cdx1* and *Hox* gene expression in the PSM during primitive streak stages. However, RA is not detected in the tail bud from E8.5 on, while *Cdx1* expression persists. Moreover, *Cdx1* is affected in *RARα1/γ* double null mutants transiently (at E7.5), while expression appears unaffected at E8.5. This is consistent with a role for RA in a discrete, early, phase of *Cdx* expression relevant to patterning cervical vertebral elements, and suggests that the later phase of expression involves other factors; given their documented role as caudalizers, members of the *Wnt* gene family are potential candidates for such a function.

Wnts are a large family of secreted glycoprotein signaling molecules and play important roles in diverse ontogenic processes including cell proliferation, differentiation, cell polarity, and migration. Two signaling cascades mediate the Wnt signal: the canonical and noncanonical pathways. In the canonical pathway, Wnt signaling, transduced via receptors of the Frizzled family, induces nuclear accumulation of β-catenin which translocates to the nucleus, forms a complex with transcription factors of the LEF/TCF family, and induces transcription of target genes (Nusse, 1999; Brantjes et al., 2002; Katoh, 2005; Li et al., 2006).

Among other family members, *Wnt3a* is expressed in a domain which overlaps with that of *Cdx1* in the caudal embryo. Both *Wnt3a* and *vestigial tail* (Greco et al., 1996) mutant offspring exhibit vertebral transformations that correlate with altered *Hox* gene expression and reduced expression of *Cdx1* (Ikeya and Takada, 2001). Consistent with a direct relationship, LEF/TCF-binding sites have been identified in the *Cdx1* promoter that mediate a response to Wnt signaling in transfection assays, and are essential for expression of a transgenic reporter gene *in vivo* (Prinos et al., 2001; Lickert and Kemler, 2002). A critical requirement for Wnt signaling in directing *Cdx1* expression, through these particular binding sites, has also been demonstrated by targeted mutagenesis, which results in essentially complete loss of *Cdx1* expression and a close phenocopy of the *Cdx1*-null phenotype (Rhodes, C. and Lohnes, D., unpublished data).

Both Wnt and retinoid signaling pathways exhibit caudalizing activity and, in *Xenopus*, can synergize in inducing genes implicated in posterior development (Cho and De Robertis, 1990; Shiotsugu et al., 2004). Both pathways are also direct effectors of *Cdx1* expression, and it is conceivable that some of their synergistic interactions could manifest through convergent regulation of *Cdx1*. In agreement with this, RA and Wnt induce stable *Cdx1* reporter constructs in embryocarcinoma cells, with simultaneous treatment resulting in a strong, synergistic, induction of expression (Prinos et al., 2001). This interaction is likely direct as it requires the presence of both the RARE and the LEF/TCF-binding motifs. This combinatorial regulation is also likely physiologically relevant, as endogenous *Cdx1*, which is typically induced by exogenous RA at late primitive streak stages, responds poorly on ablation of the LEF/TCF-binding sites (Rhodes, C. and Lohnes, D., unpublished observation). These findings suggest that Wnt and retinoid signaling pathways synergize on the *Cdx1* promoter *in vivo* and that such interactions are essential for *Cdx1* expression and for subsequent establishment of the normal pattern of *Hox* expression.

11. FGF, RA, and somitogenesis

As discussed earlier, the generation of somites from the unsegmented PSM occurs at a uniform, albeit species-dependent, rate which depends on information from a molecular clock which is integrated into a spatial determination front. As axial extension is intrinsically linked to the process of somitogenesis, this determination front must remain in a fixed position relative to the PSM as new somites are added. The work suggests that opposing gradients of FGF and RA serve important roles in setting this determination front.

FGFs are extracellular signaling molecules which signal through transmembrane tyrosine kinase receptors. FGF receptor activation can trigger a number of downstream effectors, including mitogen-activated protein kinase (MAPK)/extracellular signal-regulated kinase (ERK), MAPK/p38, phosphatidyl-inositol 3-kinase (PI3K), or phospholipase Cγ (PLCγ) (Bottcher and Niehrs, 2005).

A number of FGFs are expressed in the primitive streak and tail bud, including *Fgf8*. *Fgf8* transcripts are distributed in the posterior embryo in a caudal-high manner through an mRNA decay mechanism (Dubrulle and Pourquie, 2004b). This is converted into a graded distribution of FGF8 protein which correlates with a similar gradient of activated MAPK/ERK (Dubrulle et al., 2001; Sawada et al., 2001). The significance of this gradient is underscored by a number of observations. Inversion of the caudal-most PSM (which exhibits relatively high FGF8 levels) along the AP axis in chick embryos has no effect on somite segmentation or on AP patterning. However, inversion of the rostral (FGF8-low) PSM leads to somites with inverted polarities and segmentation defects, suggesting that somite polarity has been fixed in these cells and that attenuation of FGF function is an important step in somite maturation (Dubrulle et al., 2001; Dubrulle and Pourquie, 2004a). This is supported by the finding that overexpression of FGF8 in the anterior PSM leads to retention of early presomitic markers, such as *brachyury*, and a delay in the expression of genes characteristic of more mature somites, such as *Pax3* and *MyoD* (Dubrulle et al., 2001). Moreover, such treatments result in somites that are typically reduced in size. Conversely, exposure of embryos to SU5402, a drug that effectively blocks FGF receptor signaling (Mohammadi et al., 1997), results in precocious somite differentiation and larger somites (Dubrulle et al., 2001; Sawada et al., 2001; Diez del Corral et al., 2003). Taken together these, and other, results suggest that the PSM can be divided into an FGF-low rostral domain, where the segmentation program has been established and is irreversible, and a caudal domain in which PSM cells are maintained in a more primitive state by FGF signaling. The boundary between these two regions defines the "determination front" where FGF levels reach a critical threshold permissive for the execution of the segmentation program (Fig. 5). It is also notable that altered FGF signaling also impacts on *Hox* gene expression in affected somites, suggesting that spatiotemporal *Hox* expression is also under segmental control (Dubrulle et al., 2001; Dubrulle and Pourquie, 2004b).

A role for RA in affecting somitogenesis, and in particular establishment of the determination front, comes from a number of studies. As discussed

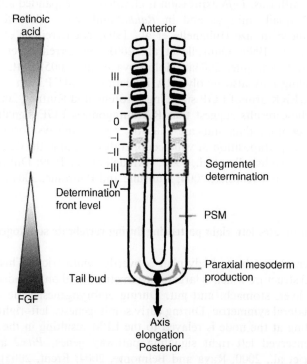

Fig. 5. RA/FGF gradients and the determination front. Paraxial mesoderm, produced caudally, is sequentially segmented into somites at the determination front. Mutually opposing gradients of fibroblast growth factor (FGF) produced caudally by the presomitic mesoderm (PSM)/tail bud and retinoic acid, produced rostrally by the somitic mesoderm specify the location of the determination front.

earlier, at E8.5 and later, RA bioactivity is found in a region immediately anterior to the tail bud, concordant with expression of *Raldh2* in the somites. By contrast, *Cyp26A1* is strongly expressed in the retinoid-deficient tail bud in a manner complementary to that of *Raldh2*, a relationship that is maintained throughout somitogenesis (Fujii et al., 1997; Niederreither et al., 1997; Ray et al., 1997; Maden, 1998; Sakai et al., 2001; Abu-Abed et al., 2003; Blentic et al., 2003). Retinoid excess, or targeted disruption of *Cyp26A1*, results in an increase in RA levels in the tail bud leading to caudal dysmorphogenesis and a number of vertebral malformations (Kessel, 1992; Abu-Abed et al., 2001; Sakai et al., 2001; Diez del Corral et al., 2003). These effects correlate with loss of several markers of nascent mesoderm in the tail bud, including *Fgf8* (Iulianella et al., 1999; Dubrulle et al., 2001). Moreover, in *Xenopus*, treatment with RA results in the formation of enlarged somites (Moreno and Kintner, 2004), suggesting that some of the effects of retinoid excess manifest through reduced FGF signaling.

The above-mentioned data are consistent with a role for retinoid signaling in influencing FGF-dependent mechanisms such as the determination front. Consistent with this, *Fgf8* expression is elevated and expanded anteriorly in both VAD quail embryos and in *Raldh2*-null mice, while the somites appear reduced in size (Iulianella et al., 1999; Niederreither et al., 1999; Swindell et al., 1999; Dubrulle et al., 2001; Niederreither et al., 2002; Dubrulle and Pourquie, 2004a; Molotkova et al., 2005). Attenuation of FGF signaling may also involve RA induction of MKP3, an inhibitor of the MAPK/ERK arm of FGF signaling (Moreno and Kintner, 2004). Taken together, these results suggest that RA antagonizes FGF signaling in the PSM, while other data indicate that the two pathways form opposing gradients, the juxtaposition of which cooperates to define the determination front (Fig. 5) (Iulianella et al., 1999; Swindell et al., 1999; Dubrulle et al., 2001; Moreno and Kintner, 2004; Dubrulle and Pourquie, 2004a).

12. RA coordinates left–right patterning during vertebrate somitogenesis

The external vertebrate body plan is overtly symmetrical. This contrasts the internal structure which is largely asymmetric based on the positioning of the heart, liver, stomach, and gut. During embryogenesis, the embryo is initially bilateral symmetric. During early somitogenesis, left–right information initiating at the node is relayed to the LPM resulting in the induction of two conserved left–right signaling pathway genes, *Pitx2* and *Nodal* (Capdevila et al., 2000; Raya and Belmonte, 2004; Brent, 2005). A direct interaction between the left–right patterning and somitogenesis was previously unknown. However, studies in mouse, chick, and zebrafish have revealed that inhibition of RA signaling during somitogenesis results in retarded somite formation on one side of the embryo (Kawakami et al., 2005; Vermot and Pourquie, 2005; Vermot et al., 2005). The side in which the delay occurs appears to be species specific, being on the right in zebrafish and mouse and on the left in chick. This may be related to anteriorization of expression of *Fgf8* in the RA-deficient models, as FGF8 acts as a right determinant in chick and a left in mouse (Meyers and Martin, 1999). Irrespective, a consistent observation in all three species is a desynchronization of the segmentation clock suggesting that the periodicity of the clock is no longer coordinated in the left and right PSM.

Retinoid deficiency leads to asymmetric somitogenesis during a specific developmental window which corresponds to the stage when the left–right information is transferred from the node to the LPM. This revelation prompted the suggestion that the left–right signaling pathway may impact on somitogenesis. Experimentally manipulating the *situs* of RA-deficient chick embryos with Sonic Hedgehog-soaked beads results in a reciprocal shift of somite asymmetry (Vermot and Pourquie, 2005). Similar results were also obtained in *Raldh2 - iv* double mutant mice, which exhibit an inversion of the somitogenesis asymmetry concomitant with an alteration in their *situs* determination (Vermot et al., 2005). These results are consistent with an interaction between the left–right signaling pathway and RA-dependent somitogenesis, leading to the proposal that RA acts as a buffer to compensate for the asymmetrical influence of the left–right pathway, thus preventing disruption of synchronous somite formation. More recent work also suggests that the impact of retinoid (and FGF) signaling on somite laterality may actually manifest through the neurectoderm (Sirbu and Duester, 2006), rather than somitic mesoderm, a finding that may relate to the discovery that ablation of *Fgf8* in nascent mesoderm has no effect on somitogenesis (Perantoni et al., 2005), although functional redundancy cannot be ruled out at present.

13. Conclusions

Paradigms suggest that RA performs two major functions during vertebral patterning: as a signal to induce the expression of genes associated with vertebral patterning (e.g., *Cdx, Hox*) and as a gradient that antagonizes FGF function to control the progression of the wavefront and the periodic segmentation of the PSM into somites during somitogenesis. Despite these advances, there remains a considerable number of unanswered questions. For example, although *Cdx1* may act to relay RA signals to mesodermal *Hox* expression, there are clearly other unknown pathways at work. Likewise, the relationship between RA-regulated vertebral patterning and the role of retinoid signaling in establishing the determination front is unclear: does RA act in two windows, with early function in the primitive streak effecting patterning of the cervical vertebrae, and later in the somites impacting on the determination front? The involvement of RA in positioning the determination front and its antagonizing effect on FGF signaling is supported by a number of compelling observations, but the molecular mechanisms influenced by this front are unresolved. Finally, the finding that RA impacts on left–right signaling as pertains to somitogenesis opens up a new perspective on the roles of retinoid signaling and the promise of future revelations.

References

Abu-Abed, S., Dolle, P., Metzger, D., Beckett, B., Chambon, P., Petkovich, M. 2001. The retinoic acid-metabolizing enzyme, CYP26A1, is essential for normal hindbrain patterning, vertebral identity, and development of posterior structures. Genes Dev. 15, 226–240.

Abu-Abed, S., Dolle, P., Metzger, D., Wood, C., MacLean, G., Chambon, P., Petkovich, M. 2003. Developing with lethal RA levels: Genetic ablation of Rarg can restore the viability of mice lacking Cyp26a1. Development 130, 1449–1459.

Ahuja, H.S., Szanto, A., Nagy, L., Davies, P.J. 2003. The retinoid X receptor and its ligands: Versatile regulators of metabolic function, cell differentiation and cell death. J. Biol. Regul. Homeost. Agents 17, 29–45.

Allan, D., Houle, M., Bouchard, N., Meyer, B.I., Gruss, P., Lohnes, D. 2001. RARgamma and Cdx1 interactions in vertebral patterning. Dev. Biol. 240, 46–60.

Altucci, L., Gronemeyer, H. 2001. The promise of retinoids to fight against cancer. Nat. Rev. Cancer 1, 181–193.

Bagnall, K.M., Higgins, S.J., Sanders, E.J. 1988. The contribution made by a single somite to the vertebral column: Experimental evidence in support of resegmentation using the chick-quail chimaera model. Development 103, 69–85.

Beck, F., Erler, T., Russell, A., James, R. 1995. Expression of Cdx-2 in the mouse embryo and placenta: Possible role in patterning of the extra-embryonic membranes. Dev. Dyn. 204, 219–227.

Beddington, R.S., Puschel, A.W., Rashbass, P. 1992. Use of chimeras to study gene function in mesodermal tissues during gastrulation and early organogenesis. Ciba Found Symp. 165, 61–74.

Begemann, G., Meyer, A. 2001. Hindbrain patterning revisited: Timing and effects of retinoic acid signalling. Bioessays 23, 981–986.

Berggren, K., McCaffery, P., Drager, U., Forehand, C.J. 1999. Differential distribution of retinoic acid synthesis in the chicken embryo as determined by immunolocalization of the retinoic acid synthetic enzyme, RALDH-2. Dev. Biol. 210, 288–304.

Blentic, A., Gale, E., Maden, M. 2003. Retinoic acid signalling centres in the avian embryo identified by sites of expression of synthesising and catabolising enzymes. Dev. Dyn. 227, 114–127.

Blumberg, B., Wright, C.V., De Robertis, E.M., Cho, K.W. 1991. Organizer-specific homeobox genes in *Xenopus laevis* embryos. Science 253, 194–196.

Boncinelli, E., Simeone, A., Acampora, D., Mavilio, F. 1991. HOX gene activation by retinoic acid. Trends Genet. 7, 329–334.

Bottcher, R.T., Niehrs, C. 2005. Fibroblast growth factor signaling during early vertebrate development. Endocr. Rev. 26, 63–77.

Bottero, L., Simeone, A., Arcioni, L., Acampora, D., Andrews, P.W., Boncinelli, E., Mavilio, F. 1991. Differential activation of homeobox genes by retinoic acid in human embryonal carcinoma cells. Recent Results Cancer Res. 123, 133–143.

Brand, N., Petkovich, M., Krust, A., Chambon, P., de The, H., Marchio, A., Tiollais, P., Dejean, A. 1988. Identification of a second human retinoic acid receptor. Nature 332, 850–853.

Brand, N.J., Petkovich, M., Chambon, P. 1990. Characterization of a functional promoter for the human retinoic acid receptor-alpha (hRAR-alpha). Nucleic Acids Res. 18, 6799–6806.

Brantjes, H., Barker, N., van Es, J., Clevers, H. 2002. TCF: Lady Justice casting the final verdict on the outcome of Wnt signalling. Biol. Chem. 383, 255–261.

Brent, A.E. 2005. Somite formation: Where left meets right. Curr. Biol. 15, R468–R470.

Brooke, N.M., Garcia-Fernandez, J., Holland, P.W. 1998. The ParaHox gene cluster is an evolutionary sister of the Hox gene cluster. Nature 392, 920–922.

Burglin, T.R., Finney, M., Coulson, A., Ruvkun, G. 1989. *Caenorhabditis elegans* has scores of homoeobox-containing genes. Nature 341, 239–243.

Burke, A.C., Nowicki, J.L. 2003. A new view of patterning domains in the vertebrate mesoderm. Dev. Cell 4, 159–165.

Capdevila, J., Vogan, K.J., Tabin, C.J., Izpisua Belmonte, J.C. 2000. Mechanisms of left-right determination in vertebrates. Cell 101, 9–21.

Chambon, P. 1994. The retinoid signaling pathway: Molecular and genetic analyses. Semin. Cell Biol. 5, 115–125.

Chambon, P. 1996. A decade of molecular biology of retinoic acid receptors. FASEB J. 10, 940–954.

Charite, J., de Graaff, W., Vogels, R., Meijlink, F., Deschamps, J. 1995. Regulation of the Hoxb-8 gene: Synergism between multimerized cis-acting elements increases responsiveness to positional information. Dev. Biol. 171, 294–305.

Charite, J., de Graaff, W., Consten, D., Reijnen, M.J., Korving, J., Deschamps, J. 1998. Transducing positional information to the Hox genes: Critical interaction of cdx gene products with position-sensitive regulatory elements. Development 125, 4349–4358.

Chawengsaksophak, K., James, R., Hammond, V.E., Kontgen, F., Beck, F. 1997. Homeosis and intestinal tumours in Cdx2 mutant mice. Nature 386, 84–87.

Chawengsaksophak, K., de Graaff, W., Rossant, J., Deschamps, J., Beck, F. 2004. Cdx2 is essential for axial elongation in mouse development. Proc. Natl. Acad. Sci. USA 101, 7641–7645.

Cho, K.W., De Robertis, E.M. 1990. Differential activation of *Xenopus* homeo box genes by mesoderm-inducing growth factors and retinoic acid. Genes Dev. 4, 1910–1916.

Christ, B., Schmidt, C., Huang, R., Wilting, J., Brand-Saberi, B. 1998. Segmentation of the vertebrate body. Anat. Embryol. (Berl.) 197, 1–8.

Christ, B., Huang, R., Wilting, J. 2000. The development of the avian vertebral column. Anat. Embryol. (Berl.) 202, 179–194.

Clagett-Dame, M., DeLuca, H.F. 2002. The role of vitamin A in mammalian reproduction and embryonic development. Annu. Rev. Nutr. 22, 347–381.

Condie, B.G., Capecchi, M.R. 1993. Mice homozygous for a targeted disruption of Hoxd-3 (Hox-4.1) exhibit anterior transformations of the first and second cervical vertebrae, the atlas and the axis. Development 119, 579–595.

Conlon, R.A. 1995. Retinoic acid and pattern formation in vertebrates. Trends Genet. 11, 314–319.

Conlon, R.A., Rossant, J. 1992. Exogenous retinoic acid rapidly induces anterior ectopic expression of murine Hox-2 genes *in vivo*. Development 116, 357–368.

Couly, G.F., Coltey, P.M., Le Douarin, N.M. 1992. The developmental fate of the cephalic mesoderm in quail-chick chimeras. Development 114, 1–15.

Couly, G.F., Coltey, P.M., Le Douarin, N.M. 1993. The triple origin of skull in higher vertebrates: A study in quail-chick chimeras. Development 117, 409–429.

Davidson, A.J., Zon, L.I. 2006. The caudal-related homeobox genes cdx1a and cdx4 act redundantly to regulate hox gene expression and the formation of putative hematopoietic stem cells during zebrafish embryogenesis. Dev. Biol. 292(2), 506–518.

Dearolf, C.R., Topol, J., Parker, C.S. 1989. The caudal gene product is a direct activator of *fushi tarazu* transcription during *Drosophila* embryogenesis. Nature 341, 340–343.

Deb, K., Sivaguru, M., Yong, H.Y., Roberts, R.M. 2006. Cdx2 gene expression and trophecto-derm lineage specification in mouse embryos. Science 311, 992–996.

Deschamps, J., van Nes, J. 2005. Developmental regulation of the Hox genes during axial morphogenesis in the mouse. Development 132, 2931–2942.

Diez del Corral, R., Olivera-Martinez, I., Goriely, A., Gale, E., Maden, M., Storey, K. 2003. Opposing FGF and retinoid pathways control ventral neural pattern, neuronal differentiation, and segmentation during body axis extension. Neuron 40, 65–79.

Dolle, P., Ruberte, E., Leroy, P., Morriss-Kay, G., Chambon, P. 1990. Retinoic acid receptors and cellular retinoid binding proteins. I. A systematic study of their differential pattern of transcription during mouse organogenesis. Development 110, 1133–1151.

Drager, U.C., McCaffery, P. 1995. Retinoic acid synthesis in the developing spinal cord. Adv. Exp. Med. Biol. 372, 185–192.

Duboule, D. 1998. Vertebrate hox gene regulation: Clustering and/or colinearity? Curr. Opin. Genet. Dev. 8, 514–518.

Duboule, D., Dolle, P. 1989. The structural and functional organization of the murine HOX gene family resembles that of *Drosophila* homeotic genes. EMBO J. 8, 1497–1505.

Dubrulle, J., Pourquie, O. 2004a. Coupling segmentation to axis formation. Development 131, 5783–5793.

Dubrulle, J., Pourquie, O. 2004b. fgf8 mRNA decay establishes a gradient that couples axial elongation to patterning in the vertebrate embryo. Nature 427, 419–422.

Dubrulle, J., McGrew, M.J., Pourquie, O. 2001. FGF signaling controls somite boundary position and regulates segmentation clock control of spatiotemporal Hox gene activation. Cell 106, 219–232.

Duester, G. 2000. Families of retinoid dehydrogenases regulating vitamin A function: Production of visual pigment and retinoic acid. Eur. J. Biochem. 267, 4315–4324.

Dupe, V., Davenne, M., Brocard, J., Dolle, P., Mark, M., Dierich, A., Chambon, P., Rijli, F.M. 1997. *In vivo* functional analysis of the Hoxa-1 3' retinoic acid response element (3'RARE). Development 124, 399–410.

Duprey, P., Chowdhury, K., Dressler, G.R., Balling, R., Simon, D., Guenet, J.L., Gruss, P. 1988. A mouse gene homologous to the *Drosophila* gene caudal is expressed in epithelial cells from the embryonic intestine. Genes Dev. 2, 1647–1654.

Ferrier, D.E., Holland, P.W. 2001. Ancient origin of the Hox gene cluster. Nat. Rev. Genet. 2, 33–38.

Frasch, M., Chen, X., Lufkin, T. 1995. Evolutionary-conserved enhancers direct region-specific expression of the murine Hoxa-1 and Hoxa-2 loci in both mice and *Drosophila*. Development 121, 957–974.

Frumkin, A., Rangini, Z., Ben Yehuda, A., Gruenbaum, Y., Fainsod, A. 1991. A chicken caudal homologue, CHox-cad, is expressed in the epiblast with posterior localization and in the early endodermal lineage. Development 112, 207–219.

Fujii, H., Sato, T., Kaneko, S., Gotoh, O., Fujii-Kuriyama, Y., Osawa, K., Kato, S., Hamada, H. 1997. Metabolic inactivation of retinoic acid by a novel P450 differentially expressed in developing mouse embryos. EMBO J. 16, 4163–4173.

Gamer, L.W., Wright, C.V. 1993. Murine Cdx-4 bears striking similarities to the *Drosophila* caudal gene in its homeodomain sequence and early expression pattern. Mech. Dev. 43, 71–81.

Gaunt, S.J., Cockley, A., Drage, D. 2004. Additional enhancer copies, with intact cdx binding sites, anteriorize Hoxa-7/lacZ expression in mouse embryos: Evidence in keeping with an instructional cdx gradient. Int. J. Dev. Biol. 48, 613–622.

Gavalas, A. 2002. Arranging the hindbrain. Trends Neurosci. 25, 61–64.

Gavalas, A., Krumlauf, R. 2000. Retinoid signalling and hindbrain patterning. Curr. Opin. Genet. Dev. 10, 380–386.

Giguere, V., Ong, E.S., Segui, P., Evans, R.M. 1987. Identification of a receptor for the morphogen retinoic acid. Nature 330, 624–629.

Gossler, A., Hrabé de Angelis, M. 1998. Somitogenesis. Curr. Top. Dev. Biol. 38, 225–287.

Gould, A., Itasaki, N., Krumlauf, R. 1998. Initiation of rhombomeric Hoxb4 expression requires induction by somites and a retinoid pathway. Neuron 21, 39–51.

Greco, T.L., Takada, S., Newhouse, M.M., McMahon, J.A., McMahon, A.P., Camper, S.A. 1996. Analysis of the vestigial tail mutation demonstrates that Wnt-3a gene dosage regulates mouse axial development. Genes Dev. 10, 313–324.

Gruss, P., Kessel, M. 1991. Axial specification in higher vertebrates. Curr. Opin. Genet. Dev. 1, 204–210.

Gu, X., Xu, F., Song, W., Wang, X., Hu, P., Yang, Y., Gao, X., Zhao, Q. 2006. A Novel Cytochrome P450, Zebrafish Cyp26D1, is involved in metabolism of all-trans retinoic acid. Mol. Endocrinol. 20(7), 1661–1672.

Guo, R.J., Suh, E.R., Lynch, J.P. 2004. The role of cdx proteins in intestinal development and cancer. Cancer Biol. Ther. 3, 593–601.

Hamilton, H.L., Hinsch, G.W. 1956. The developmental fate of the first somite of the chick. Anat. Rec. 125, 225–245.

Harrison, E.H. 2005. Mechanisms of digestion and absorption of dietary vitamin A. Annu. Rev. Nutr. 25, 87–103.

Houle, M., Prinos, P., Iulianella, A., Bouchard, N., Lohnes, D. 2000. Retinoic acid regulation of Cdx1: An indirect mechanism for retinoids and vertebral specification. Mol. Cell. Biol. 20, 6579–6586.

Houle, M., Sylvestre, J.R., Lohnes, D. 2003. Retinoic acid regulates a subset of Cdx1 function *in vivo*. Development 130, 6555–6567.

Huang, D., Chen, S.W., Langston, A.W., Gudas, L.J. 1998. A conserved retinoic acid responsive element in the murine Hoxb-1 gene is required for expression in the developing gut. Development 125, 3235–3246.

Huang, R., Zhi, Q., Ordahl, C.P., Christ, B. 1997. The fate of the first avian somite. Anat. Embryol. (Berl.) 195, 435–449.

Huang, R., Zhi, Q., Brand-Saberi, B., Christ, B. 2000. New experimental evidence for somite resegmentation. Anat. Embryol. (Berl.) 202, 195–200.

Ikeya, M., Takada, S. 2001. Wnt-3a is required for somite specification along the anteroposterior axis of the mouse embryo and for regulation of cdx-1 expression. Mech. Dev. 103, 27–33.

Isaacs, H.V., Pownall, M.E., Slack, J.M. 1998. Regulation of Hox gene expression and posterior development by the *Xenopus* caudal homologue Xcad3. EMBO J. 17, 3413–3427.

Iulianella, A., Beckett, B., Petkovich, M., Lohnes, D. 1999. A molecular basis for retinoic acid-induced axial truncation. Dev. Biol. 205, 33–48.

Joly, J.S., Maury, M., Joly, C., Duprey, P., Boulekbache, H., Condamine, H. 1992. Expression of a zebrafish caudal homeobox gene correlates with the establishment of posterior cell lineages at gastrulation. Differentiation 50, 75–87.

Kastner, P., Krust, A., Mendelsohn, C., Garnier, J.M., Zelent, A., Leroy, P., Staub, A., Chambon, P. 1990. Murine isoforms of retinoic acid receptor gamma with specific patterns of expression. Proc. Natl. Acad. Sci. USA 87, 2700–2704.

Kastner, P., Mark, M., Chambon, P. 1995. Nonsteroid nuclear receptors: What are genetic studies telling us about their role in real life? Cell 83, 859–869.

Kastner, P., Mark, M., Ghyselinck, N., Krezel, W., Dupe, V., Grondona, J.M., Chambon, P. 1997. Genetic evidence that the retinoid signal is transduced by heterodimeric RXR/RAR functional units during mouse development. Development 124, 313–326.

Katoh, M. 2005. WNT/PCP signaling pathway and human cancer (review). Oncol. Rep. 14, 1583–1588.

Kawakami, Y., Raya, A., Raya, R.M., Rodriguez-Esteban, C., Belmonte, J.C. 2005. Retinoic acid signalling links left-right asymmetric patterning and bilaterally symmetric somitogenesis in the zebrafish embryo. Nature 435, 165–171.

Kessel, M. 1992. Respecification of vertebral identities by retinoic acid. Development 115, 487–501.

Kessel, M., Gruss, P. 1991. Homeotic transformations of murine vertebrae and concomitant alteration of Hox codes induced by retinoic acid. Cell 67, 89–104.

Kieny, M., Mauger, A., Sengel, P. 1972. Early regionalization of somitic mesoderm as studied by the development of axial skeleton of the chick embryo. Dev. Biol. 28, 142–161.

Knittel, T., Kessel, M., Kim, M.H., Gruss, P. 1995. A conserved enhancer of the human and murine Hoxa-7 gene specifies the anterior boundary of expression during embryonal development. Development 121, 1077–1088.

Krumlauf, R. 1994. Hox genes in vertebrate development. Cell 78, 191–201.

Krust, A., Kastner, P., Petkovich, M., Zelent, A., Chambon, P. 1989. A third human retinoic acid receptor, hRAR-gamma. Proc. Natl. Acad. Sci. USA 86, 5310–5314.

Leid, M., Kastner, P., Chambon, P. 1992. Multiplicity generates diversity in the retinoic acid signalling pathways. Trends Biochem. Sci. 17, 427–433.

Leroy, P., Krust, A., Zelent, A., Mendelsohn, C., Garnier, J.M., Kastner, P., Dierich, A., Chambon, P. 1991. Multiple isoforms of the mouse retinoic acid receptor alpha are generated by alternative splicing and differential induction by retinoic acid. EMBO J. 10, 59–69.

Li, F., Chong, Z.Z., Maiese, K. 2006. Winding through the WNT pathway during cellular development and demise. Histol. Histopathol. 21, 103–124.

Lickert, H., Kemler, R. 2002. Functional analysis of cis-regulatory elements controlling initiation and maintenance of early Cdx1 gene expression in the mouse. Dev. Dyn. 225, 216–220.

Lohnes, D., Kastner, P., Dierich, A., Mark, M., LeMeur, M., Chambon, P. 1993. Function of retinoic acid receptor gamma in the mouse. Cell 73, 643–658.

Lohnes, D., Mark, M., Mendelsohn, C., Dolle, P., Dierich, A., Gorry, P., Gansmuller, A., Chambon, P. 1994. Function of the retinoic acid receptors (RARs) during development (I). Craniofacial and skeletal abnormalities in RAR double mutants. Development 120, 2723–2748.

Lohnes, D., Mark, M., Mendelsohn, C., Dolle, P., Decimo, D., LeMeur, M., Dierich, A., Gorry, P., Chambon, P. 1995. Developmental roles of the retinoic acid receptors. J. Steroid Biochem. Mol. Biol. 53, 475–486.

Loudig, O., Maclean, G.A., Dore, N.L., Luu, L., Petkovich, M. 2005. Transcriptional co-operativity between distant retinoic acid response elements in regulation of Cyp26A1 inducibility. Biochem. J. 392, 241–248.

Luo, J., Sucov, H.M., Bader, J.A., Evans, R.M., Giguere, V. 1996. Compound mutants for retinoic acid receptor (RAR) beta and RAR alpha 1 reveal developmental functions for multiple RAR beta isoforms. Mech. Dev. 55, 33–44.

Maden, M. 1998. The role of retinoids in developmental mechanisms in embryos. Subcell. Biochem. 30, 81–111.

Maden, M. 2001. Role and distribution of retinoic acid during CNS development. Int. Rev. Cytol. 209, 1–77.

Maden, M., Sonneveld, E., van der Saag, P.T., Gale, E. 1998. The distribution of endogenous retinoic acid in the chick embryo: Implications for developmental mechanisms. Development 125, 4133–4144.

Maeda, A., Maeda, T., Imanishi, Y., Kuksa, V., Alekseev, A., Bronson, J.D., Zhang, H., Zhu, L., Sun, W., Saperstein, D.A., Rieke, F., Baehr, W., et al. 2005. Role of photoreceptor-specific retinol dehydrogenase in the retinoid cycle in vivo. J. Biol. Chem. 280, 18822–18832.

Mangelsdorf, D.J., Evans, R.M. 1995. The RXR heterodimers and orphan receptors. Cell 83, 841–850.

Mangelsdorf, D.J., Ong, E.S., Dyck, J.A., Evans, R.M. 1990. Nuclear receptor that identifies a novel retinoic acid response pathway. Nature 345, 224–229.

Mark, M., Ghyselinck, N.B., Wendling, O., Dupe, V., Mascrez, B., Kastner, P., Chambon, P. 1999. A genetic dissection of the retinoid signalling pathway in the mouse. Proc. Nutr. Soc. 58, 609–613.

Marom, K., Shapira, E., Fainsod, A. 1997. The chicken caudal genes establish an anterior-posterior gradient by partially overlapping temporal and spatial patterns of expression. Mech. Dev. 64, 41–52.

Marshall, H., Studer, M., Popperl, H., Aparicio, S., Kuroiwa, A., Brenner, S., Krumlauf, R. 1994. A conserved retinoic acid response element required for early expression of the homeobox gene Hoxb-1. Nature 370, 567–571.

Matt, N., Dupe, V., Garnier, J.M., Dennefeld, C., Chambon, P., Mark, M., Ghyselinck, N.B. 2005. Retinoic acid-dependent eye morphogenesis is orchestrated by neural crest cells. Development 132, 4789–4800.

McCaffery, P., Drager, U.C. 1993. Retinoic acid synthesis in the developing retina. Adv. Exp. Med. Biol. 328, 181–190.

McCaffery, P., Drager, U.C. 1995. Retinoic acid synthesizing enzymes in the embryonic and adult vertebrate. Adv. Exp. Med. Biol. 372, 173–183.

Meinhardt, H. 1986. Hierarchical inductions of cell states: A model for segmentation in *Drosophila*. J. Cell Sci. (Suppl. 4), 357–381.

Melton, K.R., Iulianella, A., Trainor, P.A. 2004. Gene expression and regulation of hindbrain and spinal cord development. Front. Biosci. 9, 117–138.

Mendelsohn, C., Ruberte, E., LeMeur, M., Morriss-Kay, G., Chambon, P. 1991. Developmental analysis of the retinoic acid-inducible RAR-beta 2 promoter in transgenic animals. Development 113, 723–734.

Mendelsohn, C., Mark, M., Dolle, P., Dierich, A., Gaub, M.P., Krust, A., Lampron, C., Chambon, P. 1994. Retinoic acid receptor beta 2 (RAR beta 2) null mutant mice appear normal. Dev. Biol. 166, 246–258.

Meyer, B.I., Gruss, P. 1993. Mouse Cdx-1 expression during gastrulation. Development 117, 191–203.

Meyers, E.N., Martin, G.R. 1999. Differences in left-right axis pathways in mouse and chick: Functions of FGF8 and SHH. Science 285, 403–406.

Mlodzik, M., Gehring, W.J. 1987. Expression of the caudal gene in the germ line of *Drosophila*: Formation of an RNA and protein gradient during early embryogenesis. Cell 48, 465–478.

Mohammadi, M., McMahon, G., Sun, L., Tang, C., Hirth, P., Yeh, B.K., Hubbard, S.R., Schlessinger, J. 1997. Structures of the tyrosine kinase domain of fibroblast growth factor receptor in complex with inhibitors. Science 276, 955–960.

Molotkova, N., Molotkov, A., Sirbu, I.O., Duester, G. 2005. Requirement of mesodermal retinoic acid generated by Raldh2 for posterior neural transformation. Mech. Dev. 122, 145–155.

Moreno, E., Morata, G. 1999. Caudal is the Hox gene that specifies the most posterior Drosophile segment. Nature 400, 873–877.

Moreno, T.A., Kintner, C. 2004. Regulation of segmental patterning by retinoic acid signaling during *Xenopus* somitogenesis. Dev. Cell 6, 205–218.

Morrison, A., Moroni, M.C., Ariza-McNaughton, L., Krumlauf, R., Mavilio, F. 1996. *In vitro* and transgenic analysis of a human HOXD4 retinoid-responsive enhancer. Development 122, 1895–1907.

Nagy, A., Rossant, J., Nagy, R., Abramow-Newerly, W., Roder, J.C. 1993. Derivation of completely cell culture-derived mice from early-passage embryonic stem cells. Proc. Natl. Acad. Sci. USA 90, 8424–8428.

Napoli, J.L., Boerman, M.H., Chai, X., Zhai, Y., Fiorella, P.D. 1995. Enzymes and binding proteins affecting retinoic acid concentrations. J. Steroid Biochem. Mol. Biol. 53, 497–502.

Niederreither, K., McCaffery, P., Drager, U.C., Chambon, P., Dolle, P. 1997. Restricted expression and retinoic acid-induced downregulation of the retinaldehyde dehydrogenase type 2 (RALDH-2) gene during mouse development. Mech. Dev. 62, 67–78.

Niederreither, K., Subbarayan, V., Dolle, P., Chambon, P. 1999. Embryonic retinoic acid synthesis is essential for early mouse post-implantation development. Nat. Genet. 21, 444–448.

Niederreither, K., Vermot, J., Schuhbaur, B., Chambon, P., Dolle, P. 2000. Retinoic acid synthesis and hindbrain patterning in the mouse embryo. Development 127, 75–85.

Niederreither, K., Vermot, J., Fraulob, V., Chambon, P., Dolle, P. 2002. Retinaldehyde dehydrogenase 2 (RALDH2)-independent patterns of retinoic acid synthesis in the mouse embryo. Proc. Natl. Acad. Sci. USA 99, 16111–16116.

Niwa, H., Toyooka, Y., Shimosato, D., Strumpf, D., Takahashi, K., Yagi, R., Rossant, J. 2005. Interaction between Oct3/4 and Cdx2 determines trophectoderm differentiation. Cell 123, 917–929.

Noden, D.M. 1991. Cell movements and control of patterned tissue assembly during craniofacial development. J. Craniofac. Genet. Dev. Biol. 11, 192–213.

Northrop, J.L., Kimelman, D. 1994. Dorsal-ventral differences in Xcad-3 expression in response to FGF-mediated induction in Xenopus. Dev. Biol. 161, 490–503.

Nusse, R. 1999. WNT targets. Repression and activation. Trends Genet. 15, 1–3.

Oosterveen, T., Niederreither, K., Dolle, P., Chambon, P., Meijlink, F., Deschamps, J. 2003. Retinoids regulate the anterior expression boundaries of 5' Hoxb genes in posterior hindbrain. EMBO J. 22, 262–269.

Packer, A.I., Crotty, D.A., Elwell, V.A., Wolgemuth, D.J. 1998. Expression of the murine Hoxa4 gene requires both autoregulation and a conserved retinoic acid response element. Development 125, 1991–1998.

Perantoni, A.O., Timofeeva, O., Naillat, F., Richman, C., Pajni-Underwood, S., Wilson, C., Vainio, S., Dove, L.F., Lewandoski, M. 2005. Inactivation of FGF8 in early mesoderm reveals an essential role in kidney development. Development 132, 3859–3871.

Petkovich, M., Brand, N.J., Krust, A., Chambon, P. 1987. A human retinoic acid receptor which belongs to the family of nuclear receptors. Nature 330, 444–450.

Popperl, H., Featherstone, M.S. 1993. Identification of a retinoic acid response element upstream of the murine Hox-4.2 gene. Mol. Cell. Biol. 13, 257–265.

Pourquie, O. 2000. Segmentation of the paraxial mesoderm and vertebrate somitogenesis. Curr. Top. Dev. Biol. 47, 81–105.

Pourquie, O. 2001. Vertebrate somitogenesis. Annu. Rev. Cell Dev. Biol. 17, 311–350.

Pourquie, O., Tam, P.P. 2001. A nomenclature for prospective somites and phases of cyclic gene expression in the presomitic mesoderm. Dev. cell 1, 619–620.

Prinos, P., Joseph, S., Oh, K., Meyer, B.I., Gruss, P., Lohnes, D. 2001. Multiple pathways governing Cdx1 expression during murine development. Dev. Biol. 239, 257–269.

Ray, W.J., Bain, G., Yao, M., Gottlieb, D.I. 1997. CYP26, a novel mammalian cytochrome P450, is induced by retinoic acid and defines a new family. J. Biol. Chem. 272, 18702–18708.

Raya, A., Belmonte, J.C. 2004. Sequential transfer of left-right information during vertebrate embryo development. Curr. Opin. Genet. Dev. 14, 575–581.

Reijntjes, S., Gale, E., Maden, M. 2004. Generating gradients of retinoic acid in the chick embryo: Cyp26C1 expression and a comparative analysis of the Cyp26 enzymes. Dev. Dyn. 230, 509–517.

Reijntjes, S., Blentic, A., Gale, E., Maden, M. 2005. The control of morphogen signalling: Regulation of the synthesis and catabolism of retinoic acid in the developing embryo. Dev. Biol. 285, 224–237.

Rossant, J., Zirngibl, R., Cado, D., Shago, M., Giguere, V. 1991. Expression of a retinoic acid response element-hsplacZ transgene defines specific domains of transcriptional activity during mouse embryogenesis. Genes Dev. 5, 1333–1344.

Ruberte, E., Dolle, P., Krust, A., Zelent, A., Morriss-Kay, G., Chambon, P. 1990. Specific spatial and temporal distribution of retinoic acid receptor gamma transcripts during mouse embryogenesis. Development 108, 213–222.

Ruberte, E., Dolle, P., Chambon, P., Morriss-Kay, G. 1991. Retinoic acid receptors and cellular retinoid binding proteins. II. Their differential pattern of transcription during early morphogenesis in mouse embryos. Development 111, 45–60.

Sakai, Y., Meno, C., Fujii, H., Nishino, J., Shiratori, H., Saijoh, Y., Rossant, J., Hamada, H. 2001. The retinoic acid-inactivating enzyme CYP26 is essential for establishing an uneven distribution of retinoic acid along the anterio-posterior axis within the mouse embryo. Genes Dev. 15, 213–225.

Salvador-Silva, M., Ghosh, S., Bertazolli-Filho, R., Boatright, J.H., Nickerson, J.M., Garwin, G.G., Saari, J.C., Coca-Prados, M. 2005. Retinoid processing proteins in the ocular ciliary epithelium. Mol. Vis. 11, 356–365.

Sawada, A., Shinya, M., Jiang, Y.J., Kawakami, A., Kuroiwa, A., Takeda, H. 2001. Fgf/MAPK signalling is a crucial positional cue in somite boundary formation. Development 128, 4873–4880.

Serrano, J., Scavo, L., Roth, J., de la Rosa, E.J., de Pablo, F. 1993. A novel chicken homeobox-containing gene expressed in neurulating embryos. Biochem. Biophys. Res. Commun. 190, 270–276.

Shashikant, C.S., Bieberich, C.J., Belting, H.G., Wang, J.C., Borbely, M.A., Ruddle, F.H. 1995. Regulation of Hoxc-8 during mouse embryonic development: Identification and characterization of critical elements involved in early neural tube expression. Development 121, 4339–4347.

Shiotsugu, J., Katsuyama, Y., Arima, K., Baxter, A., Koide, T., Song, J., Chandraratna, R.A., Blumberg, B. 2004. Multiple points of interaction between retinoic acid and FGF signaling during embryonic axis formation. Development 131, 2653–2667.

Silberg, D.G., Swain, G.P., Suh, E.R., Traber, P.G. 2000. Cdx1 and cdx2 expression during intestinal development. Gastroenterology 119, 961–971.

Simeone, A., Acampora, D., Arcioni, L., Andrews, P.W., Boncinelli, E., Mavilio, F. 1990. Sequential activation of HOX2 homeobox genes by retinoic acid in human embryonal carcinoma cells. Nature 346, 763–766.

Sirbu, I.O., Duester, G. 2006. Retinoic-acid signalling in node ectoderm and posterior neural plate directs left-right patterning of somitic mesoderm. Nat. Cell Biol. 8, 271–277.

Sirbu, I.O., Gresh, L., Barra, J., Duester, G. 2005. Shifting boundaries of retinoic acid activity control hindbrain segmental gene expression. Development 132, 2611–2622.

Stern, C.D. 2001. Initial patterning of the central nervous system: How many organizers? Nat. Rev. Neurosci. 2, 92–98.

Stern, C.D., Jaques, K.F., Lim, T.M., Fraser, S.E., Keynes, R.J. 1991. Segmental lineage restrictions in the chick embryo spinal cord depend on the adjacent somites. Development 113, 239–244.

Strumpf, D., Mao, C.A., Yamanaka, Y., Ralston, A., Chawengsaksophak, K., Beck, F., Rossant, J. 2005. Cdx2 is required for correct cell fate specification and differentiation of trophectoderm in the mouse blastocyst. Development 132, 2093–2102.

Studer, M., Popperl, H., Marshall, H., Kuroiwa, A., Krumlauf, R. 1994. Role of a conserved retinoic acid response element in rhombomere restriction of Hoxb-1. Science 265, 1728–1732.

Subramanian, V., Meyer, B.I., Gruss, P. 1995. Disruption of the murine homeobox gene Cdx1 affects axial skeletal identities by altering the mesodermal expression domains of Hox genes. Cell 83, 641–653.

Sucov, H.M., Evans, R.M. 1995. Retinoic acid and retinoic acid receptors in development. Mol. Neurobiol. 10, 169–184.

Swindell, E.C., Thaller, C., Sockanathan, S., Petkovich, M., Jessell, T.M., Eichele, G. 1999. Complementary domains of retinoic acid production and degradation in the early chick embryo. Dev. Biol. 216, 282–296.

Tabaries, S., Lapointe, J., Besch, T., Carter, M., Woollard, J., Tuggle, C.K., Jeannotte, L. 2005. Cdx protein interaction with Hoxa5 regulatory sequences contributes to Hoxa5 regional expression along the axial skeleton. Mol. Cell Biol. 25, 1389–1401.

Tahayato, A., Dolle, P., Petkovich, M. 2003. Cyp26C1 encodes a novel retinoic acid-metabolizing enzyme expressed in the hindbrain, inner ear, first branchial arch and tooth buds during murine development. Gene Expr. Patterns 3, 449–454.

Taimi, M., Helvig, C., Wisniewski, J., Ramshaw, H., White, J., Amad, M., Korczak, B., Petkovich, M. 2004. A novel human cytochrome P450, CYP26C1, involved in metabolism of 9-cis and all-trans isomers of retinoic acid. J. Biol. Chem. 279, 77–85.

Tam, P.P. 1981. The control of somitogenesis in mouse embryos. J. Embryol. Exp. Morphol. 65 (Suppl.), 103–128.

Tam, P.P., Tan, S.S. 1992. The somitogenetic potential of cells in the primitive streak and the tail bud of the organogenesis-stage mouse embryo. Development 115, 703–715.

Tam, P.P., Trainor, P.A. 1994. Specification and segmentation of the paraxial mesoderm. Anat. Embryol. (Berl.) 189, 275–305.

van den, A.E., Forlani, S., Chawengsaksophak, K., de Graaff, W., Beck, F., Meyer, B.I., Deschamps, J. 2002. Cdx1 and Cdx2 have overlapping functions in anteroposterior patterning and posterior axis elongation. Development 129, 2181–2193.

van Nes, J., de Graaff, W., Lebrin, F., Gerhard, M., Beck, F., Deschamps, J. 2006. The Cdx4 mutation affects axial development and reveals an essential role of Cdx genes in the ontogenesis of the placental labyrinth in mice. Development 133, 419–428.

Vermot, J., Pourquie, O. 2005. Retinoic acid coordinates somitogenesis and left-right patterning in vertebrate embryos. Nature 435, 215–220.

Vermot, J., Gallego, L.J., Fraulob, V., Niederreither, K., Chambon, P., Dolle, P. 2005. Retinoic acid controls the bilateral symmetry of somite formation in the mouse embryo. Science 308, 563–566.

Wilson, J.G., Roth, C.B., Warkany, J. 1953. An analysis of the syndrome of malformations induced by maternal vitamin A deficiency. Effects of restoration of vitamin A at various times during gestation. Am. J. Anat. 92, 189–217.

Yashiro, K., Zhao, X., Uehara, M., Yamashita, K., Nishijima, M., Nishino, J., Saijoh, Y., Sakai, Y., Hamada, H. 2004. Regulation of retinoic acid distribution is required for proximodistal patterning and outgrowth of the developing mouse limb. Dev. Cell 6, 411–422.

Zhang, F., Popperl, H., Morrison, A., Kovacs, E.N., Prideaux, V., Schwarz, L., Krumlauf, R., Rossant, J., Featherstone, M.S. 1997. Elements both 5′ and 3′ to the murine Hoxd4 gene establish anterior borders of expression in mesoderm and neurectoderm. Mech. Dev. 67, 49–58.

Zhao, D., McCaffery, P., Ivins, K.J., Neve, R.L., Hogan, P., Chin, W.W., Drager, U.C. 1996. Molecular identification of a major retinoic-acid-synthesizing enzyme, a retinaldehyde-specific dehydrogenase. Eur. J. Biochem. 240, 15–22.

Mouse embryocarcinoma F9 cells and retinoic acid: A model to study the molecular mechanisms of endodermal differentiation

Gaétan Bour,[1] Reshma Taneja[2] and Cécile Rochette-Egly[1]

[1]*Institut de Génétique et de Biologie Moléculaire et Cellulaire,*
CNRS/INSERM/Université Louis Pasteur,
UMR 7104, BP 10142, 67404 Illkirch Cedex, France
[2]*Brookdale Department of Molecular, Cell and Developmental Biology,*
Mount Sinai School of Medicine, New York

Contents

Advances in Developmental Biology
Volume 16 ISSN 1574-3349
DOI: 10.1016/S1574-3349(06)16007-X

F9 cells are derived from mouse embryocarcinoma and markedly resemble embryonic cells from the blastocyst. They differentiate into primitive, parietal or visceral endoderm-like cells in response to retinoic acid (RA) which is known to play a crucial role in development. Therefore, they provided useful tools to study the critical events involved in early endodermal differentiation. Studies using new high throughput technologies (DNA microarrays, inducible expression of proteins, siRNA-mediated gene silencing, and targeted gene disruption) demonstrated that the expression of the components characteristic of the endodermal differentiation state (the basement membrane components, the formation of the apical junctions, and the reorganization of the cellular cytoskeleton) involves the activation of an extremely complex network of gene activation cascades. While the immediate early genes are direct RA targets in line with the presence at their promoter of response elements which bind nuclear RA receptors, the late genes are regulated by other transcriptional activators that are themselves induced by RA. During RA-induced endodermal differentiation, several signaling pathways are also activated leading to cascades of kinase activation and the subsequent phosphorylation of several substrates located either at the membrane or in the nucleus. *In fine*, these phosphorylation cascades coordinate and fine-tune the differentiation events (reorganization of the cell and gene activation cascades), resulting in a network with increasing complexity.

1. Introduction

During normal murine embryogenesis, at the fourth day of gestation, primitive endodermal (PrE) cells arise from the irreversible differentiation of the blastocoelic outer layer of the inner cell mass (ICM) which is made of embryonic stem (ES) cells. This primitive endoderm rapidly gives rise to two distinct populations of endoderm (Fig. 1): visceral endoderm (VE) which

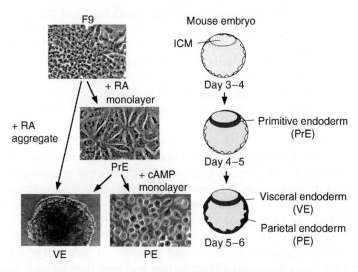

Fig. 1. Comparison of F9 embryocarcinoma cells differentiation (left panels) to normal mouse embryogenesis at the blastocyst stage (right panels). At the fourth day of gestation, before implantation, some ES cells in the ICM differentiate into either extraembryonic PE or VE, which remains associated with the ICM. The remaining ES cells will form the embryo. The left panels show the morphology of the F9 cells during their differentiation into primitive or VE when grown in monolayer and aggregates, respectively, in the presence of RA. Cyclic AMP enhances parietal differentiation.

remains associated with the ICM and parietal endoderm (PE) which is migratory, secretes large amounts of extracellular matrix forming the Reichert's membrane, and colonizes the inner surface of the trophectoderm. The formation of the endoderm is one of the earliest steps in the differentiation of pluripotent cells of the ICM during the early stages of embryonic development (Hogan et al., 1983). Evidence is mounting that an orderly succession of cell interactions between one another and with the extracellular matrix, together with a complex network of gene activations, plays a pivotal role during these developmental events.

Certain types of tumors, called teratocarcinoma (Andrews, 1988), contain pluripotential stem cells, named embryonal carcinoma cells, which are similar in morphology and biochemical characteristics to the totipotential ES cells of the mouse blastocyst. The ability of these cells to undergo normal cell differentiation pathways that resemble certain stages of embryonic development (Hogan et al., 1983) led many investigators to use these cells with the goal of defining the signaling pathways and molecular mechanisms involved in endodermal differentiation.

Among the many mouse embryocarcinoma cell lines that have been generated, the F9 cell line has been widely utilized due to its ability to differentiate into primitive, parietal or VE in response to retinoic acid (RA)

(Fig. 1), depending on the culture conditions (Strickland and Mahdavi, 1978; Strickland et al., 1980; Hogan et al., 1983, 1986; Gudas et al., 1994). RA is a vitamin A derivative which functions as a key regulatory signaling molecule for cell growth and differentiation during embryogenesis (Ross et al., 2000). The action of RA is mediated by specific nuclear receptors (Chambon, 1996) which act as ligand-dependent transcription factors and regulate the transcription of a battery of genes that play an important role in determining the differentiation program.

Although the question of whether RA actually is the differentiation inducing signal for the ES cells in the mouse blastocyst has not yet been answered, F9 cells provided an important tool to study the critical events involved in endodermal differentiation. Moreover, complementary strategies based on the generation of F9 cells ablated for specific nuclear RA receptors (Rochette-Egly and Chambon, 2001) highlighted the key role of the RA signaling pathway in the formation of endodermal cells. This chapter will describe which signaling pathways and cascades of genes expression are triggered during endodermal differentiation and how they interact with one another and with other signaling pathways.

2. F9 cell differentiation in response to retinoic acid

2.1. Origin of F9 cells

Teratocarcinoma is generally induced experimentally by the transplantation in many mouse strains of early embryos (up to 7 days of gestation) to ectopic sites. The F9 teratocarcinoma cell line was established in 1973 by Berstine et al. (1973) and derives from a subline of tumor OTT6050 obtained after transplantation of a 6-day-old embryo in the testis of a 129/Sv mouse.

Karyotype analysis has revealed that F9 cells are pseudodiploid with 39 chromosomes and 1 X chromosome and are characterized by a Robertsian translocation (Hogan et al., 1986; Alonso et al., 1991). Despite discrepancies concerning chromosomes 4 and 8 (Hogan et al., 1986; Alonso et al., 1991), comparative genomic hybridization on DNA microarray (CGH array) revealed deletions in chromosomes 4D3-4E1 and 6B3-6C3 and gains of chromosomes 8, 12A2-12ter, and 15 (Blelloch et al., 2004). Since their isolation, F9 cells have been subjected to an evolution which might be at the origin of some discrepancies between laboratories. However, they always retained their capacity to differentiate in response to RA.

2.2. RA-induced morphological changes

In vitro, F9 cells are characterized by poor spontaneous differentiation and adherence to cell culture dishes (Hogan et al., 1981). However, when cultured on gelatin coated dishes, F9 cells grow at relatively high density as a monolayer with a polygonal shape (Fig. 1). This population of undifferentiated cells depicts a poorly differentiated endoplasmic reticulum and no intercellular junctional complexes (Hogan et al., 1986). After RA (10^{-6} to 10^{-7} M) addition, the same cells acquire a triangular and flat morphology resembling PrE cells (Hogan et al., 1983) (Fig. 1). One distinguishing feature is their more extensive rough endoplasmic reticulum and the appearance of junctional complexes. In addition, when differentiation becomes completed, the growth rate of the cells decreases significantly and they accumulate in the G0/G1 phase of the cell cycle (Clifford et al., 1996; Chiba et al., 1997b).

Further differentiation into PE cells occurs after addition of cyclic AMP (Strickland et al., 1980), usually dibutyryl cyclic AMP or agents known to increase intracellular cyclic AMP levels (forskolin or phosphodiesterase inhibitors). Parietal endodermal cells are more rounded and refractile with numerous long filopodia (Fig. 1). Such changes in cell shape point to a reorganization of the cytoskeleton networks, massive synthesis of basement membrane components, and the loss of intercellular contacts (Hogan et al., 1983).

A very different response to RA is seen when F9 cells are cultured as small floating aggregates in bacteriological petri dishes rather than as monolayers. After RA treatment, these aggregates differentiate into embryoid bodies with an outer layer resembling VE cells and surrounding an inner layer of undifferentiated cells (Fig. 1). Later, many of these embryoid bodies develop a central cavity and become cystic (Hogan et al., 1983). When examined by electron microscopy, VE cells on the outside of the aggregates are clearly polarized with an apical surface depicting microvilli and a basal surface resting on a thin basement membrane. In addition, they are joined by apical junctional complexes made up of tight junctions and desmosomes. When cyclic AMP is combined with RA, the aggregates differentiate into PE-like cells without desmosome or apical junctional complex (Hogan et al., 1981).

2.3. Biochemical and molecular changes induced by RA

Many biochemical and molecular changes accompany differentiation into primitive, PE and VE. They reflect a complex activation of several networks and are to a certain point similar to what is observed in embryo (reviewed in Hogan et al., 1983). Some of these changes are common to all three cell types such as the expression of cytokeratins, the secretion of large amount of basement membrane components including type IV procollagen and laminin 1, and the synthesis and secretion of the plasminogen activator. Other changes are specific to a differentiation state. As an example, increased expression of α-fetoprotein accompanies VE differentiation (Casanova and Grabel, 1988), while the induction into PE is associated with an increase in the synthesis of thrombomodulin (Weiler-Guettler et al., 1992) and a glycoprotein named

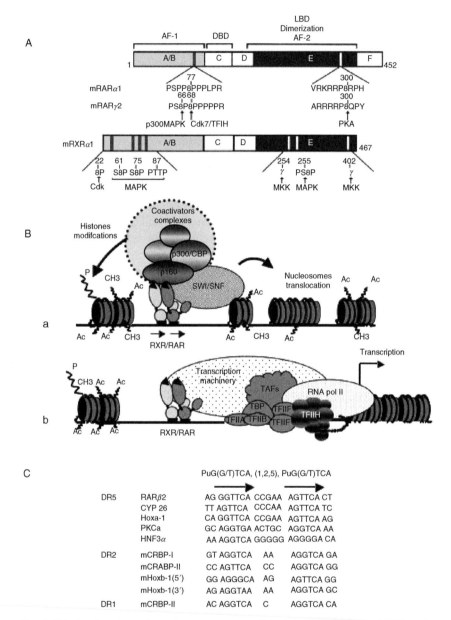

Fig. 2. RA signaling. (A) Schematic representation of the functional domains and the major phosphorylation sites of RARα, RARγ, and RXRα. The target sequences for phosphorylation by cdks, MAPKs, MKKs, and PKA are also shown. (B) Model of nuclear RA receptors action. On ligand binding (a) the RAR/RXR heterodimers bound to response elements located in the promoters of target genes undergo conformational changes, allowing the recruitment of coactivators associated with complexes displaying histone acetyltransferase, methyltransferase, kinase or ATP-dependent remodeling (SWI/SNF) activities that decompact repressive

SPARC (secreted protein acidic and rich in cysteine) (Mason et al., 1986). The expression of these components was extensively used as "markers," in combination with the morphological changes to identify a defined differentiated state. On the contrary, the expression of other markers, such as SSEA-1 or Rex-1 (Hosler et al., 1989), which are specific to the undifferentiated state, decreases during RA-induced differentiation of F9 cells.

The variation in expression of the different marker proteins has been reported to be related to changes in the levels of the corresponding mRNAs and thus to be regulated, at least in part, at the transcriptional level. However, the expression of these differentiation specific genes occurs relatively late after the addition of RA, indicating that it is indirect and depends either on the synthesis of upstream positive regulatory proteins or on the loss of repressors whose expression is direcly controlled by RA. Now there is increasing evidence that the expression of these markers involves increasingly complex cascades of gene activations resulting from the induction of upstream genes by RA nuclear receptors.

3. Role of nuclear retinoid receptors in F9 cell differentiation

3.1. Nuclear retinoic acid receptors and the basis of retinoic acid signaling

The RA signal is transduced by two families of nuclear receptors: the retinoic acid receptors (RARs) and the retinoid X receptors (RXRs) which work as RAR/RXR heterodimers. Each family is composed of three isotypes (α,β,γ) encoded by separate genes and for each isotype there are at least two main isoforms which are generated by differential promoter usage and alternative splicing and differ only in their N-terminal regions (for review see Chambon, 1996).

The modular structural organization of RARs and RXRs (Fig. 2A) has been extensively described elsewhere as well as their three-dimensional structure (Chambon, 1996; Renaud and Moras, 2000; Laudet and Gronemeyer, 2001). Briefly, the basic mechanism of RARs and RXRs activity relies on three specific domains: a central DNA-binding domain (DBD), the activation function 2 (AF-2) domain located in the C-terminal ligand-binding domain (LBD), and the N-terminal activation function 1 (AF-1) domain. On ligand binding, the LBD which is composed of 12 α helices organized as an antiparallel sandwich undergoes major structural rearrangements (Moras

---

chromatin. Such events pave the way for the recruitment (b) of the transcription machinery including the RNA Pol II and the general transcription factors, resulting in transcription initiation. (C) Description of the retinoic acid response elements (RAREs). The classical RARE is a direct repeat of the motif 5'-PuG(G/T)TCA spaced by 1 (DR1), 2 (DR2), or 5 (DR5) base pairs. The natural RARE from the promoters of some RA-target genes are shown.

and Gronemeyer, 1998; Egea et al., 2001; Bastien and Rochette-Egly, 2004) and cooperates with the AF-1 domain to recruit coactivators and multiple complexes to alter the chromatin structure surrounding the promoter of target genes (Fig. 2B). *In fine*, these events pave the way for the recruitment of the transcription machinery, including RNA polymerase II and the general transcription factors (Fig. 2B) (Bastien and Rochette-Egly, 2004; Rochette-Egly, 2005).

RARs and RXRs have been shown to be subjected to rapid modifications such as phosphorylations (for review see Rochette-Egly, 2003). In line with this, both the AF-1 and AF-2 domains contain serine residues which can be phosphorylated by several types of kinases (Fig. 2A). It emerged that phosphorylation, in cooperation with RA, fine-tunes the exchanges between RAR/RXR heterodimers and their coregulators and thereby transcription (Rochette-Egly, 2005).

3.2. RAR/RXR heterodimers in F9 cell differentiation

F9 cells express all RAR and RXR isotypes with RARα1 and RARγ2 being the main RAR isoforms. Two strategies have been used to investigate the role played by the various RARs and RXRs in the response of F9 cells to RA. First, the genes for either RARα, RARγ, or RXRα were disrupted by homologous recombination (Boylan et al., 1993, 1995; Clifford et al., 1996). Cells lacking both RARγ and RXRα or both RARα and RXRα were also engineered (Chiba et al., 1997a,b). Second, wild-type and mutant F9 cells were treated with pan-RXR and RAR-isotype selective retinoids (Roy et al., 1995; Taneja et al., 1996; Chiba et al., 1997a). Such strategies allowed to establish which RAR/RXR heterodimers are involved in each differentiation pathway. The conclusion of this work was that RARγ/RXRα heterodimers are the functional subunit mediating the RA-induced differentiation of F9 cells into primitive and VE cells (reviewed in Rochette-Egly and Chambon, 2001). They also represent an essential receptor pair for the concomitant decrease in cell growth rate. In contrast, differentiation into PE requires RARα/RXRα heterodimers, with the RARγ/RXRα-dependent formation of primitive endoderm as a prerequisite (Taneja et al., 1997).

The contribution of the different activating domains of RARγ, RARα, and RXRα in the RA response has been dissected by generating "rescue" cell lines reexpressing the receptors lacking the AF-1 or AF-2 domains starting from a null background. This "rescue" strategy highlighted that both RARγ and RARα need the integrity of both their AF-1 and AF-2 domains to efficiently transduce the RA signal (Taneja et al., 1997; Plassat et al., 2000; Rochette-Egly et al., 2000).

3.3. RAR/RXR transcriptional coregulators

According to the classical model of nuclear receptors activation, in the absence of ligand, RAR/RXR heterodimers are associated via their AF-2 domain with corepressors, resulting in chromatin compaction and silencing of the promoters of the target genes (Glass and Rosenfeld, 2000). Subsequent to the conformational changes induced by ligand binding, the corepressor-binding surface is destabilized leading to their dissociation, and a new surface is created for recruiting coactivators. Among the large number of coactivators that have been identified, the steroid receptor SRC/p160 family and p300/CBP which are endowed with intrinsic histone acetyltransferase activity and are involved in the recruitment of several other proteins and complexes involved in chromatin modifications and decondensation, stand out (McKenna and O'Malley, 2002). There are also the SWI/SNF and ASWI-based families of chromatin remodeling complexes which displace impeding nucleosomes through their ATP-dependent disruption activity (Narlikar et al., 2002).

F9 cells express the p160 family of coactivators (SRC-1/NcoA1, SRC-2/TIF2/GRIP-1, and SRC-3/pCIP/RAC3/ACTR/AIB-1/TRAM-1) (our unpublished observations), but there are no data reporting whether and how they contribute to RA signaling. In contrast, the role of CBP and/or p300 has been extensively studied. Experiments performed after knockdown of p300 or CBP with hammerhead ribozymes capable of cleaving p300 or CBP mRNAs (Kawasaki et al., 1998a) demonstrated that, despite their similarities, p300 and CBP have distinct functions, with p300 being required for RA-induced differentiation of F9 cells whereas CBP is involved in growth arrest. Finally, the brahma-related SNF2β/BRG-1 protein within SWI/SNF complexes has been shown to play an essential role in the survival of F9 cells, as its inactivation by homologous recombination resulted in nonviable cells (Sumi-Ichinose et al., 1997).

Other coactivators can interact with the AF-2 domain of RARs such as the TIF1 family (TIF1α, TIF1β, and TIF1γ). F9 cells express TIF1β which is a corepressor for KRAB-domain-containing zinc finger proteins and exerts essential functions for early embryonic development by regulating chromatin organization and recruitment of heterochromatin protein 1 (HP1) (Cammas et al., 2000). During F9 cell differentiation, TIF1β relocates from euchromatin to heterochromatin through interaction with HP1 (Cammas et al., 2002). This relocalization does not occur in F9 cells lacking RARα and RARγ, in line with the ability of RARs to recruit TIF1β in response to RA. By engineering a modified F9 cell line expressing a mutated TIF1β protein unable to interact with HP1 proteins, it has been shown that the TIF1β–HP1 interaction is not required for the formation of the PrE but is essential for further differentiation into PE and for differentiation of F9 cell aggregates into VE (Cammas et al., 2004).

Finally, studies from our group led to the discovery of a protein interacting with the N-terminal domain of RARγ (Bour et al., 2005). This protein, vinexin β, is a recently identified cytoskeletal protein which belongs to the vinexin/CAP/ponsin/ArgBP2 family of proteins and is also involved in signal transduction pathways. In F9 cells, vinexin β has been shown to be an inhibitor of RA signaling (Bour et al., 2005) through a mechanism which remains to be determined.

4. Genes activated in response to retinoic acid

At the beginning, the RA-induced differentiation of F9 cells was correlated to the expression of a few "marker" genes, including essentially the genes involved in the expression of basement components such as laminin 1 and collagen IV (Hogan et al., 1983; Wang et al., 1985; Gudas, 1992; Gudas et al., 1994) and in the formation of functional junctional complexes (Kubota et al., 2001). However, the analysis of the expression levels of up to 40,000 murine genes using new high throughput technologies, such as DNA microarrays (Harris and Childs, 2002; Futaki et al., 2003; Sangster-Guity et al., 2004), revealed a huge and extremely complex network of upstream genes whose expression is altered by RA. Almost 300 genes (Harris and Childs, 2002) are induced or repressed during the transition from undifferentiated F9 cells to primitive and finally parietal endoderm. Such strategies also demonstrated that these modifications exhibit a biphasic pattern with immediate response genes which are upregulated very quickly in response to RA, followed by an extremely complex set of late gene clusters (Harris and Childs, 2002). Further analysis of some of the selected RA-target genes by quantitative RT-PCR confirmed such orderly and successive activation processes (our unpublished observations).

The induction of most of the early genes occurs in the absence of de novo protein synthesis, suggesting that they are primary targets for RA action (Harris and Childs, 2002). In line with this, their promoter contains authentic RA response elements (RAREs) (Fig. 2C) which are able to bind RAR/RXR heterodimers. In contrast, the other genes that are induced later are regulated through the recruitment at their promoter of other transcription factors that are themselves induced by RA. In turn, the products of these genes activate other genes, resulting into cascades with increasing complexity. Now efforts are going on in order to decipher such cascades and how they lead to the final expression of the differentiation markers. With that aim, new technologies based on inducible overexpression of proteins (Shen et al., 2000; Chiba et al., 2003; Satohisa et al., 2005) and loss of function screening systems using vector-based short interfering RNA (siRNA)-mediating gene silencing (Futaki et al., 2004), or targeted gene disruption (Zhuang et al., 2003), are providing increasing information concerning such cascades.

4.1. Early genes with a retinoic acid response element

RAR/RXR heterodimers bind to the promoters of RA-target genes at cognate DNA response elements (RAREs) which are typically composed of two direct repeats of a core hexameric motif PuG(G/T)TCA (Fig. 2C). The classical RARE (Laudet and Gronemeyer, 2001; Germain et al., 2003) is a 5-bp-spaced direct repeat (referred to as a DR5). However, RAR/RXR heterodimers also bind to direct repeats separated by 2 bp (DR2) or 1 bp (DR1). The DR1 elements can recruit either RXR/RXR homodimers or RAR/RXR heterodimers which then switch from activators to repressors of the corresponding RA-responsive genes. RAREs have been identified in the promoters of a large number of RA-target genes implicated in a wide variety of functions. The classical DR5 elements are found in the promoters of genes which encode transcription factors such as the *RARβ2* gene itself (de The et al., 1990; Hoffmann et al., 1990), *Hox* genes, notably *Hoxa-1*, which encode transcription factors with a highly conserved DBD, the homeodomain (Langston and Gudas, 1992), and the *Hepatocyte Nuclear Factor 3α* (*HNF3α*) gene (Jacob et al., 1999). They are also present in the promoter of genes involved in RA metabolism (*Cyp26*; Loudig et al., 2000, 2005) or in cell signaling pathways (*PKCα*; Desai et al., 1999). DR2 elements were identified in the *CRBPI* (Smith et al., 1991), *Hoxb-1* (Huang et al., 1998) and *cdx-1* promoters (Gaunt et al., 2003; Houle et al., 2003), and DR1 elements in the *CRBPII* promoter (Mangelsdorf et al., 1991). Finally the promoter of the *CRABPII* gene was found to contain two RAREs, DR1 and DR2 (Durand et al., 1992).

In F9 cells, some of these RARE-driven genes, including at least *Cyp26, RARβ2, HNF3α*, and *Hoxa-1*, are activated rapidly in response to RA (Table 1) in the absence of *de novo* protein synthesis (Abu-Abed et al., 1998; Jacob et al., 1999; Shen et al., 2000). By using RAR isotype-specific agonists and F9 cells disrupted for different RARs, it has been demonstrated that the activation of these target genes results from the activation of RARγ/RXRα heterodimers (Rochette-Egly and Chambon, 2001).

The *PKCα* (Khuri et al., 1996), *CRABPII* (Chiba et al., 1997a; Plassat et al., 2000), and *cdx-1* (Chiba et al., 1997a) genes are also rapidly activated, but there are no data mentioning whether their induction results from the direct activation of RARγ/RXR heterodimers or from the RA-induced expression of RARβ2. Additional RA-target genes have been identified, but all of them did not exhibit a canonical RARE, as some have mutation of one or more bases. Among such genes are the gene encoding the p85α subunit of the phosphatidylinositol-3-kinase (PI3K) (Bastien et al., 2006).

Finally, it must be highlighted that the regulation of expression of some of these genes is more complex than expected due to the composite structure of their promoter. As an example, the promoter of the *Cyp26* gene contains in addition to the canonical RARE, other regulatory sequences involved in

Table 1
Genes regulated by retinoic acid in F9 cells[a]

Function / Official symbol (NCBI)	Official name (NCBI)	Alternate name	Induction/ repression	References	RARE in promoter	References	Other regulators	References
Transcription factor								
ATF3	Activating transcription factor 3	LRG-21	+	Futaki et al., 2004				
Cdx1	Caudal type homeo box 1	cdx1	+	Taneja et al., 1995; Houle et al., 2000	Atypical RARE/DR2	Gaunt et al., 2003; Houle et al., 2003		
Foxa1	Forkhead box A1	HNF-3α	+	Jacob et al., 1999	DR5	Jacob et al., 1999	HNF	Tan et al., 2001
Foxa2	Forkhead box A2	HNF-3β	+	Reichel et al., 1994			HNF3	Tan et al., 2001
GATA4	GATA-binding protein 4	GATA-4	+	Arceci et al., 1993				
GATA6	GATA-binding protein 6	GATA-6	+	Futaki et al., 2003				
Hnf4a	Hepatic nuclear factor 4, α	HNF-4α	+	Nakhei et al., 1998	DR1	Hatzis and Talianidis, 2001	HNF-1α, HNF-1β HNF-6 Sp1 GATA6	Hatzis and Talianidis, 2001
Hoxa1	Homeo box A1	Hox 1.6	+	Iwai et al., 1995	DR5	Langston and Gudas, 1992	Other regulators	Thompson et al., 1998
Hoxa5	Homeo box A5	Hox 1.3	+	Murphy et al., 1988				
Hoxa7	Homeo box A7	Hox 1.1	+	Schulze et al., 1987	DR5, DR3	Kim et al., 2002	cdx1	Subramanian et al., 1995

Hoxb1	Homeo box B1	Hox 2.9	+	Boylan et al., 1993	DR5	Huang et al., 1998	DRE	Kawasaki et al., 1998b
Jun	Jun oncogene	c-Jun	+	Yang-Yen et al., 1990				
Myc	Myelocytomatosis oncogene	c-myc	−	Dean et al., 1986; Zhuang et al., 2003				
NR2C2	Nuclear receptor subfamily 2, group C, member 2	TR4	+	Lee et al., 1998				
nr2f1	Nuclear receptor subfamily 2, group F, member 1	coup-tf1/ear3	+	Futaki et al., 2003				
nr2f2	Nuclear receptor subfamily 2, group F, member 2	coup-tf2/arp1	+	Futaki et al., 2003				
RARB2	Retinoic acid receptor, β	RARB2	+	Hu and Gudas, 1990	DR5	de Thé et al., 1990	TRE CRE	Shen et al., 1991
SOX17	SRY-box containing gene 17	SRY box 17	+	Futaki et al., 2003				
SOX7	SRY-box containing gene 7	SRY box 7	+	Futaki et al., 2003				
Tcf1	Transcription factor 1	HNF-1α	+	Kuo et al., 1991			HNF3	Tan et al., 2001
Tcf2	Transcription factor 2	HNF-1β	+	Kuo et al., 1991	DR1		HNF3	Tan et al., 2001
Zfp42	Zinc finger protein 42	Rex-1	−	Hosler et al., 1989			Oct3/4, Rox1	Ben-Shushan et al., 1998

(continued)

Table 1 (*continued*)

Function	Official symbol (NCBI)	Official name (NCBI)	Alternate name	Induction/ repression	References	RARE in promoter	References	Other regulators	References
Transcription regulator									
	Cand1	Cullin associated and neddylation disassociated 1	TIP120A	+	Yogosawa et al., 1999				
	Dab2	Disabled homologue 2 (*Drosophila*)	Dab2	+	Cho and Park, 2000				
	Phc1	Polyhomeotic-like 1	Rae28/Mph1	+	Motaleb et al., 1999			DRS	Motaleb et al., 1999
Membrane protein									
Intracellular protein									
	krt1–18	Keratin complex 1, acidic, gene 18	Endo C	+	Tabor and Oshima, 1982				
	krt2–8	Keratin complex 2, basic, gene 8	Endo A	+	Tabor and Oshima, 1982			Ets	Seth et al., 1994
	lmna	Lamin A	Lamin A	+	Lebel et al., 1987				
	lmnb	Lamin B	Lamin B	+	Lebel et al., 1987				
	sec61a1	Sec61 α1 subunit	Sec61 α	+	Ferreira et al., 2003				
Extracellular protein									
	B3GALT5	UDP-Gal:βGlcNAc β1,3-galactosyl transferase, polypeptide 5	β3GalT-V	+	Zhou et al., 2000				
	Col4a	Procollagen, type IV, α	Col4a	+	Marotti et al., 1985			Not identified	Tanaka et al., 1993

Gene	Name	+/−	Reference	RARE	Reference	Factors	Reference
Lama1	Laminin, α 1	+	Futaki et al., 2003			Sp1/Sp3, NF-Y, Sox7/Sox17	Niimi et al., 2004
Lamb1	Laminin B1 subunit 1	+	Vasios et al., 1989	Complex RARE	Vasios et al., 1991	Sox-5, HNF-1, GATA-1, …	Sharif et al., 2001
Lamc1	Laminin, γ1	+	Gudas et al., 1990	Putative RARE	Chan et al., 1996		Chang et al., 1996
Lgals1	Lectin, galactose binding, soluble 1	+	Lu et al., 1998				
Thbd	Thrombomodulin	+	Weiler-Guettler et al., 1992	Putative DR	Niforas et al., 1996	CRE, Sp1, AP1, AP2	Yao et al., 1999
Itgb1	Integrin β1	+	Morini et al., 1999				
Pdia3	Protein disulfide isomerase associated 3	+	Miyaishi et al., 1998				

Cytoskeleton

Gene	Name	+/−	Reference	RARE	Reference	Factors	Reference
Vim	Vimentin	+	Benazzouz and Duprev, 1999			Hox-a5 + other regulators	Benazzouz and Duprev, 1999

Link to RA metabolism, transport

Gene	Name	+/−	Reference	RARE	Reference	Factors	Reference
Crabp1	Cellular retinoic acid-binding protein I	−	Means et al., 2000	Not identified	Means et al., 2000		
Crabp2	Cellular retinoic acid-binding protein II	+	Giguere et al., 1990	DR1, DR2	Durand et al., 1992	Sp1, AP-1, AP-2	MacGregor et al., 1992
Cyp26A1	Cytochrome P450, family 26, subfamily a, polypeptide 1	+	Abu-Abed et al., 1998	DR5	Loudig et al., 2000, 2005	Sp1/Sp3	Loudig et al., 2000

(continued)

Table 1 (*continued*)

Function / Official symbol (NCBI)	Official name (NCBI)	Alternate name	Induction/ repression	References	RARE in promoter	References	Other regulators	References
Rbp1	Retinol-binding protein 1, cellular	CRBPI	+	Chiba et al., 1997a	DR3	Smith et al., 1991		
Cell cycle regulators								
Ccnd1	Cyclin D1	Cyclin D1	−	Li et al., 1999				
Ccnd2	Cyclin D2	Cyclin D2	+	Li et al., 1999				
Ccnd3	Cyclin D3	Cyclin D3	−	Faria et al., 1998; Li et al., 1999				
Cdkn1a	Cyclin-dependent kinase inhibitor 1A (P21)	p21	+	Drdova and Vachtenheim, 2005	DR5	Liu et al., 1996		
Cdkn1b	Cyclin-dependent kinase inhibitor 1B (P27)	p27	+	Li et al., 1999, 2004b				
Signaling pathway								
Secreted protein								
AFP	α fetoprotein	AFP	+	Casanova and Grabel, 1988			DAS, HNF-3	Chen et al., 1999; Costa et al., 2001
Bmp2	Bone morphogenetic protein 2	BMP2	+	Grabel et al., 1998			Sp1	Heller et al., 1999; Abrams et al., 2004
fgf3	Fibroblast growth factor 3	Int-2	+	Smith et al., 1988			GATA4, SOX7	Murakami et al., 2004
Plat	Plasminogen activator, tissue	tPA	+	Cheng and Grabel, 1997				
Sparc	Secreted acidic cysteine rich glycoprotein	Sparc	+	Mason et al., 1986				

Symbol	Name		+/−					
Tgfb2	Transforming growth factor, β2	Tgf-b2	+	Mummery et al., 1990			CREB	Kingsley-Kallesen et al., 1999
Ttr	Transthyretin	TTR	+	Soprano et al., 1988			HNF	Tan et al., 2001
	Dickkopf-1	Dkk-1	+	Shibamoto et al., 2004				
Membrane receptor								
adrb1	Adrenergic receptor, β1	Adrb1	+	Bahouth et al., 1998	DR5	Bahouth et al., 1998	TR	Bahouth et al., 1998
Pdgfra	Platelet derived growth factor receptor, α polypeptide	PDGFRA	+	Wang et al., 1990			GATA	Wang and Song, 1996
Intracellular								
Pik3r1	Phosphatidylinositol 3-kinase, regulatory subunit, polypeptide 1	p85a	+	Bastien et al., 2006	Noncanonical Bastien et al., 2006			
Plcg	Phospholipase C, γ	PLCγ	−	Lee et al., 1993				
Prkca	Protein kinase C, α	PKCα	+	Kindregan et al., 1994	Desai et al., 1999			
Prkcb1	Protein kinase C, β1	PKCβ	−	Kindregan et al., 1994				
Prkcc	Protein kinase C, γ	PKCγ	−	Kindregan et al., 1994				
Ywhab	Tyrosine 3-monooxygenase/ tryptophan 5-monooxygenase activation protein, β polypeptide	14-3-3	+	Takihara et al., 2000				

[a]Selected RA-target genes were classified according to their major functions. Official names from the NCBI as well as usual names are given. If reported in the literature, the presence of binding sequences for RAR/RXR (RARE) or other regulators in the promoter of the target gene is mentioned. Some genes require the combination of RA with cyclic AMP to be induced.

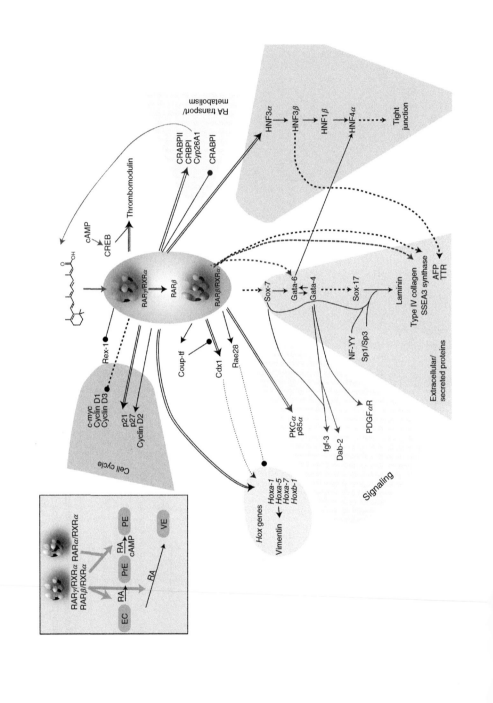

the recruitment of additional transcriptional factors, such as Sp1, which cooperate with RAR/RXR heterodimers for optimal transcription (Loudig et al., 2000). Similarly, cyclic AMP response elements (CREs) cooperate with RAREs for the activation of *RARβ2* and *Thrombomodulin* genes (Niforas et al., 1996) during parietal endodermal differentiation that occurs on combination of cyclic AMP to RA.

The height in the complexity is that RA can activate very rapidly some genes which do not contain any RARE in their promoter. A typical example is that of *c-jun* whose promoter contains a differentiation response element (DRE) involved in the recruitment of activation transcription factor 2 (ATF-2). In fact, the rapid activation of this gene would rely on the RA-induced phosphorylation of ATF-2 (see later) and/or expression of the Rcd1 factor which subsequently assembles with RAR and ATF-2 at the DRE (Hiroi et al., 2002).

4.2. Downstream genes involved in differentiation

Making an extensive list of all the RA-induced genes is out of the scope of this chapter. We will focus on some groups or families of genes which have been more extensively studied than the others (Table 1) and shown to be involved in cascades leading to the formation of the basement membrane and the junctional complexes that are characteristic of the differentiated state (Fig. 3).

As *RARβ2* is one of the earliest genes induced in response to RA through a RARE, it has been hypothesized that the corresponding protein regulates the expression of a unique set of downstream genes. Some of the RARβ2-target genes have been identified through DNA array analysis and substractive hybridization of F9 cells that have been knocked out for RARβ2 (Zhuang et al., 2003). That they are specific RARβ2-target genes has been confirmed by a comparison of WT, RARβ2, RARα, and RARγ null cells. Using such strategies, three main groups of genes have been shown to be regulated by RARβ2 (Zhuang et al., 2003; Futaki et al., 2004): (1) genes encoding transcription factors such as the Gata factors (Gata-6 and Gata-4), hepatocyte nuclear factors (HNF3β, HNF1β, and HNF4α), and the SRY-box containing genes 7 and 17 (*Sox7* and *Sox17*); (2) genes encoding intracellular signaling factors such as CRABPII and Disabled-2 (Dab-2); and (3) genes encoding

Fig. 3. Schematic representation of the cascades of RA-target genes. Genes that are induced in response to RA were arbitrarily classified in six categories to highlight the cascades and cross-regulation concepts. Among these genes, some are direct targets for RAR/RXR heterodimers that bind to RAREs located in their promoter ($=\!\!=$). Other genes are indirectly activated via the recruitment at their promoter of other transcriptional activators that are themselves induced by RA, subsequently to cascades of gene activations which have been characterized (——) or not (-----). Similarly, sets of genes are downregulated either directly or indirectly (——•). Note that the cdx/Rae28-mediated regulation of Hox genes (·······) has been described in other cell lines.

secreted factors including the fibroblast growth factor 3 (FGF3) and Midkine, or cell surface receptors such as the platelet-derived growth factor α-receptor (PDGFα-R).

However, additional studies based on overexpression or siRNA-mediated silencing experiments revealed that many of these genes are not regulated directly by RARβ2, in line with the lack of a known RARE in their promoter, but rather through the upregulation of a cascade of activations (Fig. 3). As an example, a study (Futaki et al., 2004) demonstrated that RARβ2 action would involve first the upregulation of Sox7 that is required for the induction of Gata-6 which in turn positively regulates Gata-4 and Sox17. Finally, the Sox factors converge toward the activation of the *laminin* genes and the formation of the basement membrane in line with the presence of binding sites for these factors in the *Lama1* enhancer (Niimi et al., 2004).

Similar strategies have been used to analyze the putative downstream targets of Hoxa-1 (Shen et al., 2000) and HNF3α (Tan et al., 2001). They converged toward the conclusion that HNF3α is the initiator of a network of cascades within the HNF family (Fig. 3) leading to the expression of the AFP marker or to the formation of functional tight junctions. According to studies performed with mouse ES cells, HNF3α would activate the *HNF3β* and *HNF1β* genes (Levinson-Dushnik and Benvenisty, 1997; Kaestner, 2000) which in turn direct the activation of *HNF4α* (Coffinier et al., 1999; Bailly et al., 2001). In line with this, elegant experiments using F9 cells expressing doxycycline-inducible HNF4α demonstrated that this factor can provoke *de novo* the formation of tight junctions (Chiba et al., 2003; Satohisa et al., 2005).

These cascades are even more complex as they can interfere with each other and with other pathways (Fig. 3). In such cross talks, the Gata-4/Gata-6 and Sox7 factors appear to play central roles as they are part of a network directing not only the expression of *Sox17* and *laminin* but also the *HNF* family (Morrisey et al., 1998). They also control the expression of genes encoding growth factors, including *fgf-3* (Murakami et al., 1999) and trans-membrane receptors such as the *PDGFα-R* (Wang and Song, 1996), in line with the presence of binding site for Gata-4/Gata-6 and Sox7 in their promoter. Finally, the complexity of these genes regulations is further increased due to the requirement of additional factors (such as Sp1) and the fact that the Gata factors are redundant, and induce themselves through positive feedback loops (Futaki et al., 2004).

4.3. Other genes involved in cell proliferation

F9 cells lacking RARγ/RXRα heterodimers were not only deficient in the expression of several genes involved in differentiation but also exhibited no growth arrest in response to RA (Chiba et al., 1997b). The same observations were made with RARβ2-null cells (Faria et al., 1999), indicating that the cascades described earlier for endodermal differentiation also play a role in the concomitant decrease in cell proliferation induced by RA. In line with this, RA induces the upregulation of p27$^{CIP/KIP}$, a CDK inhibitor (Li et al., 2004b), and downregulates the expression of cyclin D3 via the action of RARβ2 (Faria et al., 1998). It must be noted, however, that these transcriptional effects are amplified by increases in the stability of the p27 protein and in the turnover of cyclin D3 (Li et al., 2004b). Finally, another CDK inhibitor, p21$^{CIP/KIP}$, was found to be increased via the action of HNF4α (Chiba et al., 2005). Altogether, these observations indicate that the different elements of the cascades activated in response to RA via the action of RAR/RXR heterodimers can modulate either the differentiation or the proliferation of F9 cells.

5. Signaling pathways activated during RA-induced F9 cells endodermal differentiation

The formation of the endodermal cells is accompanied by a multitude of other cellular effects, involving contacts between cells and/or their communication with each other by the transduction of autocrine or paracrine signals. Differentiating cells secrete growth factors which interact with high-affinity cell surface receptors with tyrosine kinase activity (RTKs) or with members of a large superfamily of receptors with seven membrane-spanning regions (Fig. 4) linked to heterodimeric guanine nucleotide-binding proteins (G-proteins) which engage well-defined effector pathways, including adenylate cyclase, phospholipase C-β, mitogen-activated protein kinases (MAPKs), and phosphatidylinositol-3-kinase (PI3K). These receptors route the signal to several distinct intracellular pathways, through well-conserved kinase phosphorylation cascades (Fig. 4). The activated kinases lead to the phosphorylation of several substrates located in the cytosol or at the membrane which will regulate the organization of the cell. They also transduce the signals toward or within the nucleus through their translocation to the nucleus or through phosphorylating (1) kinases involved in a nucleus-directed kinase cascade, (2) proteins that are subsequently translocated to the nucleus, (3) anchoring proteins leading to the release and translocation to the nucleus of another protein (Buchner, 1995; Brivanlou and Darnell, 2002). As all cell surface receptors engage multiple signaling pathways which interact with one another and can be integrated with signals from other receptors, all these interconnections result in a network with increasing complexity (Pawson and Nash, 2000; Toker, 2000; Neves et al., 2002).

Evidence is accumulating that the activation of signaling pathways and protein kinase cascades are important self-perpetuating driving forces for endodermal differentiation. As F9 cells exhibit many features of normal cell differentiation pathways and cell–cell interactions, they constitute a suitable model to study the molecular mechanisms involved in the cascades of kinases and the nature of the phosphorylated substrates. Numerous studies reported that, in F9 cells, RA has been shown to induce variations in the expression of

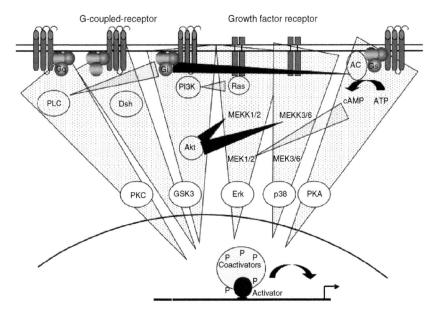

Fig. 4. Schematic representation of the signaling pathways activating MAPKs (Erks and p38), Akt, PKA, PKC, and Dsh. The classical signaling cascades are represented within dotted triangles. Some of these signaling pathways interact with one another: the interactions with a positive influence are represented with gray triangles and those which are inhibitory with black ones. PI3K, phosphatidylinositol-3-kinase; PLC, phospholipase C; MEK, MAPK kinase; MEKK, MEK kinase; AC, adenylate cyclase; Dsh, Dishevelled.

G-proteins (Watkins et al., 1992). RA also increases markedly the expression of cell surface RTKs (PDGFα-R and EGF-R) (Wang et al., 1990; Joh et al., 1992) and of growth factors (Grotendorst et al., 1996; Kingsley-Kallesen et al., 1999), subsequently to the activation of the corresponding genes via the recruitment at their promoters of transcription factors which are themselves upregulated in response to RA through complex cascades (Wang and Song, 1996; Murakami et al., 1999) (Figs. 3 and 5).

During the last decade, the understanding for RA cross talks with signaling pathways increased rapidly despite the complexity of the signaling networks. It emerged that in F9 cells, RA induces the activation of numerous protein kinases pathways, including the MAPK, Akt, and PKC pathways. It also synergizes with the PKA pathway when activated by high cyclic AMP levels.

5.1. MAPK signaling pathways

A common mechanism of signal transduction consists of the sequential activation of protein kinases within the MAPK cascades. At least four families have been identified: extracellular-regulated protein kinases 1 and 2 (ERK1/2), ERK5, c-Jun NH2-terminal protein kinases (JNKs), and p38MAPKs.

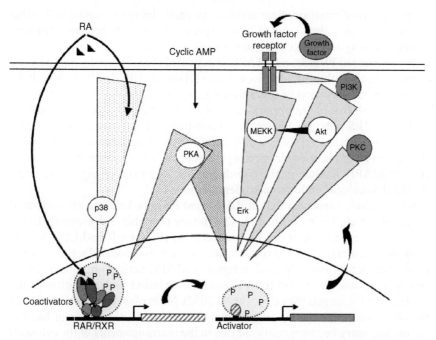

Fig. 5. Recapitulation of the signaling pathways activated by RA. Some kinases (PKC, PI3K) are upregulated at the transcriptional level, subsequently to the activation of the corresponding genes (dark gray) through the recruitment at their promoter of transcription activators (hatched) which are themselves upregulated in response to RA through gene activation cascades. Cell surface receptors for growth factors and growth factors are similarly transcriptionally upregulated. Consequently, the PI3K/Akt, MAPK, and PKC signaling pathways are activated (gray triangles). The PI3K/Akt pathway negatively influences the MAPK pathway (black triangles). Overall, these processes which are not immediate, contribute to the changes observed during the formation of endodermal cells. RA also activates rapidly p38MAPK (dotted triangles) through a mechanism which has not been elucidated yet, leading to the phosphorylation of transcriptional activators, such as RARs and RXRs, and of their coregulators and thereby to the induction of the corresponding target genes. The activation of PKA by cyclic AMP is also shown (hatched triangles).

How MAPKs are activated by phosphorylation cascades has been extensively reviewed (Martin-Blanco, 2000; Pearson et al., 2001; Weston and Davis, 2002; Weston et al., 2002). In brief, MAPKs activation results from the activation of RTKs or G-protein–coupled receptors (Fig. 4), followed by the sequential activation of GTPases, MEK kinases (MEKK), and MAPK kinases (MEK or MKK). The MAPK kinases include MKK1 and MKK2 for ERK1/2, MKK5 for ERK5, MKK4 and MKK7 for JNKs, and MKK3 and MKK6 for p38MAPK. Activated MAPKs phosphorylate a number of substrates located at the cell membrane or in the cytosol. They can also enter the nucleus and phosphorylate numerous transcription factors, leading to changes affecting either their conformation, their stability, or their ability to interact with other partners. *In fine*, they control gene expression and cell behavior through communicating an extracellular signal to the nucleus.

In RA-treated F9 cells, p38MAPK is activated (Fig. 5) rapidly and plays an important role in the differentiation process (Gianni et al., 2002). Although the mechanism of activation of this kinase has not been elucidated yet, it does not seem to result from a transcriptional process. According to studies performed with mouse ES cells, this kinase would constitute an early switch, committing the embryonic cells into specific differentiation pathways (Aouadi et al., 2006). Work is in progress in our laboratory in order to determine the importance of this p38MAPK module in the coordination of the early changes that occur during RA-induced endodermal differentiation.

Several studies reported that in RA-treated F9 cells, Erks are also activated (Verheijen et al., 1999; Smith et al., 2001), very probably, subsequently to the RA-induced expression of cell surface receptors (PDGFα-R and EGF-R) and of growth factors (Fig. 5). Consistent with this, in RA-treated cells, the response to growth factors was enhanced in terms of MAPK activation. Unexpectedly this process was uncoupled from the phosphorylation and activation of the nuclear MAPK targets (Smedberg et al., 2002). No explanation was found until the observation that during endodermal differentiation, the access of MAPK for nuclear entry becomes restricted due to the rearrangements of the cytoskeleton and of the basement membrane (Smith et al., 2004). It has been proposed that the subsequent restriction of the phosphorylation of the nuclear MAPK targets (transcription factors) would be the causative mechanism of the growth suppression of F9 cells treated with RA, while cytosolic MAPK activity would allow the differentiation processes to proceed.

5.2. PI3K/Akt pathway

During the last decade, a new pathway, the PI3K/Akt pathway, came onto the scene as a key signaling pathway that participates in a myriad of cellular processes, including cell growth and differentiation, insulin action, and cell survival to name a few (Datta et al., 1999; Brazil et al., 2004). There are three Akt kinases (Akt1, Akt2, and Akt3) and the biochemical mechanisms leading to their activation are well defined and have been reviewed in detail elsewhere (Scheid and Woodgett, 2001; Vivanco and Sawyers, 2002; Cantley, 2002; Neves et al., 2002; Hanada et al., 2004). Phosphoinositide-3-kinase (PI3K) is activated on stimulation of RTKs or G-protein–coupled receptors (Fig. 4) and catalyzes the formation of phosphatidylinositol 3,4,5-triphosphate (PIP3), which mediates the translocation of Akt at the plasma membrane, rendering the protein accessible for kinases (PDK1 and PDK2) and subsequent phosphorylation. This activation of Akt is antagonized by PTEN, a lipid phosphatase which dephosphorylates PIP3 (Leslie and

Downes, 2002). Akt is the primary mediator of PI3K signaling and has a number of downstream substrates at the cell membrane, in the cytosol and in the nucleus, which contribute to cell proliferation and survival. Note that the PI3K/Akt pathway interferes with the MAPK pathway through the phosphoinactivation of MEKKs by Akt (Fig. 4).

The PI3K/Akt pathway has been found to be activated in RA-treated F9 cells (Bastien et al., 2006). PI3K is activated subsequent to the RA-dependent increase in the expression of its p85α regulatory subunit (Fig. 5). The enhanced PI3K activity has been correlated to an increase in PIP3 levels, resulting in activation of the downstream Akt kinase and subsequently to the phosphoinactivation of its targets such as GSK3. By using an siRNA approach, the activation of this pathway has been found to be a key determinant for endodermal differentiation of F9 cells, in line with the emerging concept that PI3K/Akt signaling has diversified capacities and is apparently capable not only to promote cell growth but also to contribute to cell differentiation. It has been suggested that the PI3K/Akt pathway contributes to F9 cell differentiation through regulating the transcription of some genes involved in the differentiation process and/or the growth and the survival of the newly differentiated cells. RA-induced activation of the PI3K/Akt pathway is followed by an inhibition of Akt (Bastien et al., 2006) which has been correlated to the RA-induced decrease in cell proliferation.

5.3. PKC and PKA pathways

Not only the MAPK and PI3K/Akt pathways but also other signaling pathways are transduced on activation of cell surface G-protein–coupled receptors. Among these pathways are the classical adenylyl cyclase and phospholipase C-β (PLC-β) pathways (Fig. 4). The former transmits the signal to protein kinase A (PKA) through the formation of cyclic AMP, while the latter leads to the activation of protein kinase C (PKC) through the formation of diacylglycerol. The mechanisms of activation of PKA and PKC have been well established (Newton, 1997; Parekh et al., 2000; Fimia and Sassone-Corsi, 2001). Once activated, PKA and PKC transduce the signals toward or within the nucleus through their translocation to the nucleus or through phosphorylating proteins involved in nucleus-directed cascades (Buchner, 1995). Both PKC and PKA pathways are activated and involved in the endodermal differentiation of F9 cells.

The PKC family consists of several subfamilies (classic: α, β, γ; novel: δ, ε, η, θ; and atypical λ, ζ, ι) classified according to their requirement for calcium, phospholipids, or phorbol esters for activation. Undifferentiated F9 cells express the conventional calcium, diacyl-glycerol, and phospholipid-dependent PKCβ but not PKCα (Khuri et al., 1996). RA-induced differentiation of F9 cells leads to a transition from PKCβ to PKCα expression (Kindregan et al.,

1994; Cho et al., 1998) (Fig. 5) in line with the presence of an RA response element in the promoter of the *PKCα* gene (Desai et al., 1999). By establishing F9-derived cell lines constitutively overexpressing PKCα or antisense PKCα (Cho et al., 1998), it has been demonstrated that PKCα is critical for establishing and maintaining the endoderm phenotype. However, there are no data indicating whether, in F9 cells, RA also regulates PKCα activity through direct binding as demonstrated *in vitro* (Lopez-Andreo et al., 2005). Whether RA also activates PLCδ and induces its translocation to the nucleus, as in several other RA-responsive cell lines (Kambhampati et al., 2003) would require further investigations.

Concerning PKA, there are at least two major isoforms named I and II, each one being composed of a dimeric regulatory subunit (R2) and two monomeric catalytic subunits (C). PKA type I is predominantly cytoplasmic while PKA type II is associated with membranes and the cytoskeleton. This intracellular localization is controlled by anchoring proteins (Malbon et al., 2004; Wong and Scott, 2004). Binding of cyclic AMP to regulatory subunits induces conformational changes that activate PKA by exposing the substrate binding site of the C subunit with or without subunit dissociation.

An increase in PKA activity associated with the plasma membranes of F9 cells has been observed in response to RA (Plet et al., 1982). However, there are no data mentioning whether it is associated to an increase in cyclic AMP levels and/or to the subsequent translocation of the catalytic subunit to the nucleus as in APL cells (Parrella et al., 2004; Zhao et al., 2004). Nevertheless, conceivably, these observations suggest that alterations in PKA activity and distribution may contribute to the phosphorylation of substrates involved in the dramatic changes observed during the formation of endodermal cells.

In fact, a key role for PKA in F9 cell differentiation arises from the observation that, when combined to RA, cyclic AMP or adenylate cyclase activators promote the transition from primitive to PE-like cells, which is associated to the increased expression of basement membrane components (collagen IV and laminin) and of several RA-target genes (Rochette-Egly et al., 2000; Futaki et al., 2003).

5.4. Wnt–β-catenin signaling pathway

The classical Wnt–*β*-catenin signaling paradigm is based on the ability of Wnt signaling proteins to activate frizzled receptors via G-proteins, resulting in the activation of Dishevelled (Dsh) and the subsequent phosphorylation and inhibition of GSK3 (Fig. 4). Subsequently, *β*-catenin, a key component of adherens junctions, becomes stabilized, accumulates in the nucleus, and interacts with the Lef/Tcf transcriptional complex to activate genes involved in early development (Bejsovec, 2005). Several studies demonstrated that the ability of RA to induce the differentiation of F9 cells (Liu et al., 2002) and several other cell lines (Otero et al., 2004) is absolutely dependent on the integrity of β-catenin signaling, suggesting that the two pathways interact with each other. Although the mechanism of this cross talk is still unclear, most studies converge toward the conclusion that RA stabilizes the β-catenin protein and that its subsequent accumulation contributes to the expression of genes involved in the formation of primitive endoderm by F9 cells. As GSK3 resides at the junction of the PI3K/Akt and Wnt/β-catenin pathways (Fig. 4), an attractive hypothesis would be that RA promotes the dephosphorylation and stabilization of β-catenin, through the phosphoinactivation of GSK3 (Mulholland et al., 2005). Differentiated F9 cells are no longer able to respond to Wnt signaling (Shibamoto et al., 2004), most probably due to the RA-induced expression of Wnt inhibitors resulting in a negative feedback loop.

6. Phosphorylation processes during RA-induced F9 cell differentiation

6.1. Phosphorylated substrates

In line with the mounting evidence that several kinases are induced or activated in response to RA (Fig. 5), the extent of phosphorylation of several proteins is significantly altered during the RA-induced differentiation of F9 cells (Kitabayashi et al., 1994).

The first candidates for phosphorylations are the RARs and RXRs themselves, in line with their implication in the activation of the early RA-target genes. RARs and RXRs depict in their N-terminal AF-1 domain several consensus phosphorylation sites for proline-dependent kinases which include cyclin-dependent kinases (cdks) and MAPKs (Fig. 2A). Our laboratory demonstrated that the RARα_1 and RARγ2 isotypes are phosphorylated at one residue (S77 and S68, respectively) by the cyclin H-dependent kinase cdk7 associated to the general transcription factor TFIIH (Rochette-Egly et al., 1997; Bastien et al., 2000). RARγ2 can be also phosphorylated at an additional nearby serine residue (S66) by p38MAPK (Gianni et al., 2002). Concerning RXRα, this receptor is phosphorylated at S22 by a cdk that is distinct from cdk7 and at three additional residues (S61, S75, and T87) by MAPKs (Adam-Stitah et al., 1999).

RARs and RXRs can be phosphorylated not only in their N-terminal AF-1 domain but also in their C-terminal LBD. RARα and RARγ contain a conserved serine residue located in loop 9–10 (S369 for RARα1 and S362 for RARγ2) which is a target for PKA (Rochette-Egly et al., 1995), while RXRα contains several residues which are targets for MAPKs and MEKs in response to stress agents (Adam-Stitah et al., 1999; Lee et al., 2000).

In RA-treated F9 cells, the phosphorylation of the N-terminal domain of both partners within RARγ/RXRα heterodimers is increased: RARγ becomes rapidly phosphorylated at S66 subsequently to the activation of p38MAPK (Gianni et al., 2002) and RXRα at the three MAPK sites (Ser61, Ser75, and Thr87) (Gianni et al., 2003; Bruck et al., 2005). RARα and RARγ become phosphorylated at their PKA site when cyclic AMP is combined with RA (Rochette-Egly et al., 1995). It must be noted that RARα and RXRα can be also phosphorylated at additional residues by JNKs in response to stress agents (Bruck et al., 2005; Srinivas et al., 2005; Tarrade et al., 2005), but whether these phosphorylation processes are relevant to F9 cell differentiation and/or proliferation has not been investigated yet.

Not only RARs and RXRs but also their coactivators including the p160 family of coactivators and p300 are targets for phosphorylations as they can be phosphorylated in response to several signals and in many cell types (Wu et al., 2004; Poizat et al., 2005; Gianni et al., 2006). In F9 cells, there are no data indicating that the p160 coactivators are phosphorylated in response to RA. However, p300 is phosphorylated and its phosphorylation state changes during F9 cell differentiation (Kitabayashi et al., 1995; Brouillard and Cremisi, 2003).

Finally, other transcription factors involved in the cascades of genes activated in response to RA are also targets for phosphorylations. In line with this, ATF-2 which mediates the activation of c-jun is a well-known target for MAPKs (Morton et al., 2004). In F9 cells, it is phosphorylated by PKCα at a serine residue (S121) located in the region involved in the interaction with p300 (Kawasaki et al., 1998b). This PKC-dependent phosphorylation of ATF-2 is determinant for the RA-induced activation of c-jun (Ugai et al., 1999). In addition, the CRE/ATF transcription factors ATF-1 and CREB which are involved in the RA-induced expression of the TGFβ gene are targets for phosphorylation by PKA (Kingsley-Kallesen et al., 1999).

In conclusion, in F9 cells, phosphorylation of specific transcription factors or coregulators appears to be a key event participating to the activation of RA-target genes and therefore to endodermal differentiation. Whether other transcription factors and/or coactivators involved in the activation of the cascade of RA-target genes are phosphorylated would represent a new interesting field of investigations. Interesting potential candidates would be the Gata factors as some studies reported that they are targets for as many kinases as MAPKs, PKA, and PKC (Kerkela et al., 2002; Tremblay and Viger, 2003; Ishida et al., 2005). Similarly, future work should examine whether other kinases, such as Akt or PDK1, shuttle within the nucleus and phosphorylate new nuclear substrates involved in F9 cell differentiation (Srinivas et al., 2006).

6.2. Role of MAPK-mediated RARγ/RXRα phosphorylation in primitive endodermal differentiation of F9 cells

F9 cells which are obviouly less complex than an intact animal provide an attractive system to study at the molecular level that how phosphorylations of nuclear activators of transcription are involved in the mechanism of endodermal differentiation.

The role of RARγ and RXRα phosphorylation in the RA response of F9 cells has been dissected in our laboratory, by using "rescue" cell lines stably expressing RARγ or RXRα mutated at the phosphorylation sites located in their N-terminal domain (RARγS66/68A and RXRαS61A/S75A/T87A), in an RARγ- or RXRα-null background (Rochette-Egly and Chambon, 2001). Taking advantage of this strategy, it has been demonstrated that the integrity of the two RARγ phosphorylation sites is indispensable to the activation of a subset of RA-target genes and to the formation of primitive endoderm-like cells (Taneja et al., 1997). It has been suggested that during endodermal differentiation, phosphorylation of the N-terminal domain of RARγ participates in the expression of several early RA-target genes, most probably through the regulation of the association–dissociation of RAR/RXR heterodimers with coregulators (Bastien and Rochette-Egly, 2004; Rochette-Egly, 2005; Bour et al., in press). Then phosphorylation negatively controls the transcription of RA-target genes through signaling the degradation of the receptor by the ubiquitin-proteasome pathway (Kopf et al., 2000; Gianni et al., 2002).

The same "rescue" strategy demonstrated that phosphorylation of the N-terminal domain of RXRα at the three MAPK sites participates to the cooperation between both partners for optimal activity of the RARγ/RXRα heterodimers (Gianni et al., 2003; Bruck et al., 2005). In contrast, phosphorylation of RXRα at S22 was not required for endodermal differentiation but instead implicated in the inhibition of proliferation induced by RA (Bastien et al., 2002).

In conclusion, this "rescue" strategy provided useful tools to demonstrate the importance of the phosphorylation of both partners within the RARγ/RXRα heterodimers in the RA-induced differentiation of F9 cells. However, it did not allow to discriminate the role of p38MAPK per se which is activated very rapidly in response to RA. Although it is increasingly clear that p38MAPK targets RARγ, one cannot exclude that it also phosphorylates the p160 family of coactivators and/or p300 (Poizat et al., 2005; Gianni et al., 2006), therefore influencing their turnover and/or exchanges with RAR/RXR heterodimers. It is not excluded either that such a scenario might be applicable to other activators of transcription, thereby increasing the complexity of the regulatory network involved in the regulation of the cascade of RA-target genes.

6.3. Role of PKA-mediated phosphorylation in parietal endodermal differentiation of F9 cells

The analysis of the F9 rescue lines, reexpressing RARγ or RARα mutated at the PKA site in the corresponding null background (Taneja et al., 1997;

Rochette-Egly et al., 2000), demonstrated that the phosphorylation of RARα (and not of RARγ) by PKA is implicated together with cyclic AMP for PE differentiation, most probably through the regulation of specific RARα-target genes.

Parietal differentiation required to be preceded by the RARγ-dependent formation of PrE (Taneja et al., 1997). As RARγ is degraded when primitive differentiation is completed (Gianni et al., 2002), it has been assumed that this degradation process is a prerequisite for allowing RARα to activate a series of genes involved specifically in the formation of parietal endodermal cells. However, such a hypothesis has been ruled out subsequently to the observation that parietal differentiation could occur in the absence of RARγ degradation and that complete endodermal differentiation is not required (Perletti et al., 2001). In fact, a number of studies converged toward the conclusion that some RARγ-mediated events that occur during primitive differentiation, that is, the expression of transcription factors, such as the Sox or Gata factors, are mandatory to pave the way for the action of PKA (Futaki et al., 2004). As Sox and Gata factors are targets for PKA (Tremblay and Viger, 2003; Ishida et al., 2005) as well as p300 (Brouillard and Cremisi, 2003), one can propose that PKA regulates a wide variety of genes involved in parietal differentiation through the phosphorylation of these factors. In line with this, in F9 cells, several genes which are up- or downregulated in response to the RA/cyclic AMP combination (Kingsley-Kallesen et al., 1999; Futaki et al., 2003) involve transcription factors that are targets for PKA. Finally, one should not rule out that PKA can also phosphorylate several substrates located at the membrane or in the cytosol and involved in the organization of the basement membrane components.

In conclusion, although the overall mechanisms are still only partially understood, it appears that during endodermal differentiation, the expression of specific genes is coordinated through a well-organized regulatory network. Therefore, it will be important to unveil which factors can be regulated by phosphorylation mediated by MAPKs, PKCα, and PKA and how phosphorylation by these kinases orchestrates and/or optimizes the gene regulatory network involved in endoderm differentiation.

7. Conclusions

F9 cells which are derived from mouse embryocarcinoma markedly resemble embryonic cells from the blastocyst. Due to their ability to differentiate into primitive, PE- or VE-like cells in response to RA which is known to play a crucial role in development, they provided useful tools to study the critical events involved in these early differentiation processes. Studies performed with F9 cells in our laboratory and others, demonstrated the importance of the nuclear RARs in line with their ability to induce the activation of cascades of genes resulting in the expression of the components characteristic of the endodermal differentiation state (the basement membrane components, the formation of the apical junctions, and the reorganization of the cellular cytoskeleton). New technologies based on the analysis of DNA microarrays, inducible expression of proteins and siRNA-mediated gene silencing, have revealed an extremely complex network of gene activation and its fine-tuning by phosphorylation cascades. Current and future investigations will undoubtedly improve our understanding of such a network.

ES cell technology has progressed impressively during the past few years, offering the possibility of studying a broader spectrum of cell types than F9 cells (Mummery et al., 1990; Keller and Snodgrass, 1999). ES cells which are derived directly from the ICM of blastocysts can spontaneously differentiate to form embryoid bodies that contain elements of all three embryonic germ layers: ectoderm, mesoderm, and endoderm. As differentiation continues, a wide range of cell types including hematopoietic, endothelial, muscle, and neuronal develop within the EBs, as in the normal embryo gene. Studies performed with ES cells have highlighted the importance of Gata and HNF factors (Morrisey et al., 1998; Fujikura et al., 2002; Yoshida-Koide et al., 2004; Li et al., 2004a; Capo-Chichi et al., 2005) during endodermal differentiation. In addition, quantitative screening technologies have allowed the identification of growth factors that act to enhance endoderm formation (Chang and Zandstra, 2004). However, up to now, F9 cells proved to be unique tools to correlate the activity of nuclear RA receptors with early endodermal differentiation, highlighting the interest of such cells.

References

Abrams, K.L., Xu, J., Nativelle-Serpentini, C., Dabirshahsahebi, S., Rogers, M.B. 2004. An evolutionary and molecular analysis of Bmp2 expression. J. Biol. Chem. 279, 15916–15928.

Abu-Abed, S.S., Beckett, B.R., Chiba, H., Chithalen, J.V., Jones, G., Metzger, D., Chambon, P., Petkovich, M. 1998. Mouse P450RAI (CYP26) expression and retinoic acid-inducible retinoic acid metabolism in F9 cells are regulated by retinoic acid receptor gamma and retinoid X receptor alpha. J. Biol. Chem. 273, 2409–2415.

Adam-Stitah, S., Penna, L., Chambon, P., Rochette-Egly, C. 1999. Hyperphosphorylation of the retinoid X receptor alpha (RXRα) by activated c-Jun N-terminal Kinases (JNKs). J. Biol. Chem. 274, 18932–18941.

Alonso, A., Breuer, B., Steuer, B., Fischer, J. 1991. The F9-EC cell line as a model for the analysis of differentiation. Int. J. Dev. Biol. 35, 389–397.

Andrews, P.W. 1988. Human teratocarcinomas. Biochem. Biophys. Acta 948, 17–36.

Aouadi, M., Bost, F., Caron, L., Laurent, K., Le Marchand Brustel, Y., Binetruy, B. 2006. P38MAPK activity commits embryonic stem cells to either neurogenesis or cardiomyogenesis. Stem Cells 24(5), 1399–1406.

Arceci, R.J., King, A.A., Simon, M.C., Orkin, S.H., Wilson, D.B. 1993. Mouse GATA-4: A retinoic acid-inducible GATA-binding transcription factor expressed in endodermally derived tissues and heart. Mol. Cell. Biol. 13, 2235–2246.

Bahouth, S.W., Beauchamp, M.J., Park, E.A. 1998. Identification of a retinoic acid response domain involved in the activation of the beta 1-adrenergic receptor gene by retinoic acid in F9 teratocarcinoma cells. Biochem. Pharmacol. 55, 215–225.

Bailly, A., Torres-Padilla, M.E., Tinel, A.P., Weiss, M.C. 2001. An enhancer element 6 kb upstream of the mouse HNF4alpha1 promoter is activated by glucocorticoids and liver-enriched transcription factors. Nucleic Acids Res. 29, 3495–3505.

Bastien, J., Rochette-Egly, C. 2004. Nuclear retinoid receptors and the transcription of retinoid-target genes. Gene 328, 1–16.

Bastien, J., Adam-Stitah, S., Riedl, T., Egly, J.M., Chambon, P., Rochette-Egly, C. 2000. TFIIH interacts with the retinoic acid receptor gamma and phosphorylates its AF-1-activating domain through cdk7. J. Biol. Chem. 275, 21896–21904.

Bastien, J., Adam-Stitah, S., Plassat, J.L., Chambon, P., Rochette-Egly, C. 2002. The phosphorylation site located in the A region of RXRα is required for the anti-proliferative effect of retinoic acid and the activation of RA-target genes in F9 cells. J. Biol. Chem. 24, 24.

Bastien, J., Plassat, J.L., Payrastre, B., Rochette-Egly, C. 2006. The phosphoinositide 3-kinase/Akt pathway is essential for the retinoic acid-induced differentiation of F9 cells. Oncogene 25, 2040–2047.

Bejsovec, A. 2005. Wnt pathway activation: New relations and locations. Cell 120, 11–14.

Benazzouz, A., Duprev, P. 1999. The vimentin promoter as a tool to analyze the early events of retinoic acid-induced differentiation of cultured embryonal carcinoma cells. Differentiation 65, 171–180.

Ben-Shushan, E., Thompson, J.R., Gudas, L.J., Bergman, Y. 1998. Rex-1, a gene encoding a transcription factor expressed in the early embryo, is regulated via Oct-3/4 and Oct-6 binding to an octamer site and a novel protein, Rox-1, binding to an adjacent site. Mol. Cell. Biol. 18, 1866–1878.

Berstine, E.G., Hooper, M.L., Grandchamp, S., Ephrussi, B. 1973. Alkaline phosphatase activity in mouse teratoma. Proc. Natl. Acad. Sci. USA 70, 3899–3903.

Blelloch, R.H., Hochedlinger, K., Yamada, Y., Brennan, C., Kim, M., Mintz, B., Chin, L., Jaenisch, R. 2004. Nuclear cloning of embryonal carcinoma cells. Proc. Natl. Acad. Sci. USA 101, 13985–13990.

Bour, G., Plassat, J.L., Bauer, A., Lalevee, S., Rochette-Egly, C. 2005. Vinexin beta interacts with the non-phosphorylated AF-1 domain of retinoid receptor gamma (RAR gamma) and represses RAR gamma-mediated transcription. J. Biol. Chem. 280, 17027–17037.

Boylan, J.F., Lohnes, D., Taneja, R., Chambon, P., Gudas, L.J. 1993. Loss of retinoic acid receptor gamma function in F9 cells by gene disruption results in aberrant Hoxa-1 expression and differentiation upon retinoic acid treatment. Proc. Natl. Acad. Sci. USA 90, 9601–9605.

Boylan, J.F., Lufkin, T., Achkar, C.C., Taneja, R., Chambon, P., Gudas, L.J. 1995. Targeted disruption of retinoic acid receptor alpha (RAR alpha) and RAR gamma results in receptor-specific alterations in retinoic acid-mediated differentiation and retinoic acid metabolism. Mol. Cell. Biol. 15, 843–851.

Brazil, D.P., Yang, Z.Z., Hemmings, B.A. 2004. Advances in protein kinase B signalling: AKT ion on multiple fronts. Trends Biochem. Sci. 29, 233–242.

Brivanlou, A.H., Darnell, J.E., Jr. 2002. Signal transduction and the control of gene expression. Science 295, 813–818.

Brouillard, F., Cremisi, C.E. 2003. Concomitant increase of histone acetyltransferase activity and degradation of p300 during retinoic acid-induced differentiation of F9 cells. J. Biol. Chem. 278, 39509–39516.

Bruck, N., Bastien, J., Bour, G., Tarrade, A., Plassat, J.L., Bauer, A., Adam-Stitah, S., Rochette-Egly, C. 2005. Phosphorylation of the retinoid x receptor at the omega loop, modulates the expression of retinoic-acid-target genes with a promoter context specificity. Cell. Signal. 17, 1229–1239.

Buchner, K. 1995. Protein kinase C in the transduction of signals toward and within the cell nucleus. Eur. J. Biochem. 228, 211–221.

Cammas, F., Mark, M., Dolle, P., Dierich, A., Chambon, P., Losson, R. 2000. Mice lacking the transcriptional corepressor TIF1beta are defective in early postimplantation development. Development 127, 2955–2963.

Cammas, F., Oulad-Abdelghani, M., Vonesch, J.L., Huss-Garcia, Y., Chambon, P., Losson, R. 2002. Cell differentiation induces TIF1beta association with centromeric heterochromatin via an HP1 interaction. J. Cell Sci. 115, 3439–3448.

Cammas, F., Herzog, M., Lerouge, T., Chambon, P., Losson, R. 2004. Association of the transcriptional corepressor TIF1beta with heterochromatin protein 1 (HP1): An essential role for progression through differentiation. Genes Dev. 18, 2147–2160.

Cantley, L.C. 2002. The phosphoinositide 3-kinase pathway. Science 296, 1655–1657.

Capo-Chichi, C.D., Rula, M.E., Smedberg, J.L., Vanderveer, L., Parmacek, M.S., Morrisey, E. E., Godwin, A.K., Xu, X.X. 2005. Perception of differentiation cues by GATA factors in primitive endoderm lineage determination of mouse embryonic stem cells. Dev. Biol. 286, 574–586.

Casanova, J.E., Grabel, L.B. 1988. The role of cell interactions in the differentiation of teratocarcinoma-derived parietal and visceral endoderm. Dev. Biol. 129, 124–139.

Chambon, P. 1996. A decade of molecular biology of retinoic acid receptors. FASEB J 10, 940–954.

Chang, H.S., Kim, N.B., Phillips, S.L. 1996. Positive elements in the laminin gamma 1 gene synergize to activate high level transcription during cellular differentiation. Nucleic Acids Res. 24, 1360–1368.

Chang, K.H., Zandstra, P.W. 2004. Quantitative screening of embryonic stem cell differentiation: Endoderm formation as a model. Biotechnol. Bioeng. 88, 287–298.

Chen, H., Dong, J.M., Liu, Y., Chiu, J.F. 1999. Identification of a cis-acting element in the rat alpha-fetoprotein gene and its specific binding proteins in F9 cells during retinoic acid-induced differentiation. J. Cell Biochem. 72, 25–34.

Cheng, L., Grabel, L.B. 1997. The involvement of tissue-type plasminogen activator in parietal endoderm outgrowth. Exp. Cell Res. 230, 187–196.

Chiba, H., Clifford, J., Metzger, D., Chambon, P. 1997a. Distinct retinoid X receptor-retinoic acid receptor heterodimers are differentially involved in the control of expression of retinoid target genes in F9 embryonal carcinoma cells. Mol. Cell. Biol. 17, 3013–3020.

Chiba, H., Clifford, J., Metzger, D., Chambon, P. 1997b. Specific and redundant functions of retinoid X receptor/retinoic acid receptor heterodimers in differentiation, proliferation, and apoptosis of F9 embryonal carcinoma cells. J. Cell Biol. 139, 735–747.

Chiba, H., Gotoh, T., Kojima, T., Satohisa, S., Kikuchi, K., Osanai, M., Sawada, N. 2003. Hepatocyte nuclear factor (HNF)-4alpha triggers formation of functional tight junctions and establishment of polarized epithelial morphology in F9 embryonal carcinoma cells. Exp. Cell Res. 286, 288–297.

Chiba, H., Itoh, T., Satohisa, S., Sakai, N., Noguchi, H., Osanai, M., Kojima, T., Sawada, N. 2005. Activation of p21CIP1/WAF1 gene expression and inhibition of cell proliferation by overexpression of hepatocyte nuclear factor-4alpha. Exp. Cell Res. 302, 11–21.

Cho, S.Y., Park, S.S. 2000. Genomic organization and promoter analysis of mouse disabled 2 gene. Biochem. Biophys. Res. Commun. 275, 189–194.

Cho, Y., Klein, M.G., Talmage, D.A. 1998. Distinct functions of protein kinase C alpha and protein kinase C beta during retinoic acid-induced differentiation of F9 cells. Cell Growth Differ. 9, 147–154.

Clifford, J., Chiba, H., Sobieszczuk, D., Metzger, D., Chambon, P. 1996. RXRalpha-null F9 embryonal carcinoma cells are resistant to the differentiation, anti-proliferative and apoptotic effects of retinoids. EMBO J. 15, 4142–4155.

Coffinier, C., Thepot, D., Babinet, C., Yaniv, M., Barra, J. 1999. Essential role for the homeoprotein vHNF1/HNF1beta in visceral endoderm differentiation. Development 126, 4785–4794.

Costa, R.H., Kalinichenko, V.V., Lim, L. 2001. Transcription factors in mouse lung development and function. Am. J. Physiol. Lung Cell Mol. Physiol. 280, L823–L838.

Datta, S.R., Brunet, A., Greenberg, M.E. 1999. Cellular survival: A play in three Akts. Genes Dev. 13, 2905–2927.

de The, H., Vivanco-Ruiz, M.M., Tiollais, P., Stunnenberg, H., Dejean, A. 1990. Identification of a retinoic acid responsive element in the retinoic acid receptor beta gene. Nature 343, 177–180.

Dean, M., Levine, R.A., Campisi, J. 1986. c-myc regulation during retinoic acid-induced differentiation of F9 cells is posttranscriptional and associated with growth arrest. Mol. Cell. Biol. 6, 518–524.

Desai, D.S., Hirai, S., Karnes, W.E., Jr., Niles, R.M., Ohno, S. 1999. Cloning and characterization of the murine PKC alpha promoter: Identification of a retinoic acid response element. Biochem. Biophys. Res. Commun. 263, 28–34.

Drdova, B., Vachtenheim, J. 2005. A role for p21 (WAF1) in the cAMP-dependent differentiation of F9 teratocarcinoma cells into parietal endoderm. Exp. Cell Res. 304, 293–304.

Durand, B., Saunders, M., Leroy, P., Leid, M., Chambon, P. 1992. All-trans and 9-cis retinoic acid induction of CRABPII transcription is mediated by RAR-RXR heterodimers bound to DR1 and DR2 repeated motifs. Cell 71, 73–85.

Egea, P.F., Rochel, N., Birck, C., Vachette, P., Timmins, P.A., Moras, D. 2001. Effects of ligand binding on the association properties and conformation in solution of retinoic acid receptors RXR and RAR. J. Mol. Biol. 307, 557–576.

Faria, T.N., LaRosa, G.J., Wilen, E., Liao, J., Gudas, L.J. 1998. Characterization of genes which exhibit reduced expression during the retinoic acid-induced differentiation of F9 teratocarcinoma cells: Involvement of cyclin D3 in RA-mediated growth arrest. Mol. Cell. Endocrinol. 143, 155–166.

Faria, T.N., Mendelsohn, C., Chambon, P., Gudas, L.J. 1999. The targeted disruption of both alleles of RARbeta(2) in F9 cells results in the loss of retinoic acid-associated growth arrest. J. Biol. Chem. 274, 26783–26788.

Ferreira, L.R., Velano, C.E., Braga, E.C., Paula, C.C., Martelli-Junior, H., Sauk, J.J. 2003. Expression of Sec61 alpha in F9 and P19 teratocarcinoma cells after retinoic acid treatment. Braz. J. Biol. 63, 245–252.

Fimia, G.M., Sassone-Corsi, P. 2001. Cyclic AMP signalling. J. Cell Sci. 114, 1971–1972.

Fujikura, J., Yamato, E., Yonemura, S., Hosoda, K., Masui, S., Nakao, K., Miyazaki Ji, J., Niwa, H. 2002. Differentiation of embryonic stem cells is induced by GATA factors. Genes Dev. 16, 784–789.

Futaki, S., Hayashi, Y., Yamashita, M., Yagi, K., Bono, H., Hayashizaki, Y., Okazaki, Y., Sekiguchi, K. 2003. Molecular basis of constitutive production of basement membrane components. Gene expression profiles of Engelbreth-Holm-Swarm tumor and F9 embryonal carcinoma cells. J. Biol. Chem. 278, 50691–50701.

Futaki, S., Hayashi, Y., Emoto, T., Weber, C.N., Sekiguchi, K. 2004. Sox7 plays crucial roles in parietal endoderm differentiation in F9 embryonal carcinoma cells through regulating Gata-4 and Gata-6 expression. Mol. Cell. Biol. 24, 10492–10503.

Gaunt, S.J., Drage, D., Cockley, A. 2003. Vertebrate caudal gene expression gradients investigated by use of chick cdx-A/lacZ and mouse cdx-1/lacZ reporters in transgenic mouse embryos: Evidence for an intron enhancer. Mech. Dev. 120, 573–586.

Germain, P., Altucci, L., Bourguet, W., Rochette-Egly, C., Gronemeyer, H. 2003. Nuclear receptor superfamily: Principles of signalling. Pure Appl. Chem. 75, 1619–1664.

Gianni, M., Bauer, A., Garattini, E., Chambon, P., Rochette-Egly, C. 2002. Phosphorylation by p38MAPK and recruitment of SUG-1 are required for RA-indced RARγ degradation and transactivation. EMBO J. 21, 3760–3769.

Gianni, M., Tarrade, A., Nigro, E.A., Garattini, E., Rochette-Egly, C. 2003. The AF-1 and AF-2 domains of RAR gamma 2 and RXR alpha cooperate for triggering the transactivation and the degradation of RAR gamma 2/RXR alpha heterodimers. J. Biol. Chem. 278, 34458–34466.

Gianni, M., Parrella, E., Raska, I., Gaillard, E., Nigro, E.A., Gaudon, C., Garattini, E., Rochette-Egly, C. 2006. P38MAPK-dependent phosphorylation and degradation of SRC-3/AIB1 and RARalpha-mediated transcription. EMBO J. 25, 739–751.

Giguere, V., Lyn, S., Yip, P., Siu, C.H., Amin, S. 1990. Molecular cloning of cDNA encoding a second cellular retinoic acid-binding protein. Proc. Natl. Acad. Sci. USA 87, 6233–6237.

Glass, C.K., Rosenfeld, M.G. 2000. The coregulator exchange in transcriptional functions of nuclear receptors. Genes Dev. 14, 121–141.

Grabel, L., Becker, S., Lock, L., Maye, P., Zanders, T. 1998. Using EC and ES cell culture to study early development: Recent observations on Indian hedgehog and Bmps. Int. J. Dev. Biol. 42, 917–925.

Grotendorst, G.R., Okochi, H., Hayashi, N. 1996. A novel transforming growth factor beta response element controls the expression of the connective tissue growth factor gene. Cell Growth Differ. 7, 469–480.

Gudas, L.J. 1992. Retinoids, retinoid-responsive genes, cell differentiation, and cancer. Cell Growth Differ. 3, 655–662.

Gudas, L.J., Grippo, J.F., Kim, K.W., Larosa, G.J., Stoner, C.M. 1990. The regulation of the expression of genes encoding basement membrane proteins during the retinoic acid-associated differentiation of murine teratocarcinoma cells. Ann. NY Acad. Sci. 580, 245–251.

Gudas, L.J., Sporn, M.B., Roberts, A.B. 1994. Cellular biology and biochemistry of the retinoids. In: *The Retinoids: Biology, Chemistry and Medecine* (M.B. Sporn, A.B. Roberts, D.S. Goodman, Eds.), New York: Raven Press, Ltd., pp. 443–520.

Hanada, M., Feng, J., Hemmings, B.A. 2004. Structure, regulation and function of PKB/AKT—a major therapeutic target. Biochim. Biophys. Acta 1697, 3–16.

Harris, T.M., Childs, G. 2002. Global gene expression patterns during differentiation of F9 embryonal carcinoma cells into parietal endoderm. Funct. Integr. Genomics 2, 105–119.

Hatzis, P., Talianidis, I. 2001. Regulatory mechanisms controlling human hepatocyte nuclear factor 4alpha gene expression. Mol. Cell. Biol. 21, 7320–7330.

Heller, L.C., Li, Y., Abrams, K.L., Rogers, M.B. 1999. Transcriptional regulation of the Bmp2 gene. Retinoic acid induction in F9 embryonal carcinoma cells and Saccharomyces cerevisiae. J. Biol. Chem. 274, 1394–1400.

Hiroi, N., Ito, T., Yamamoto, H., Ochiya, T., Jinno, S., Okayama, H. 2002. Mammalian Rcd1 is a novel transcriptional cofactor that mediates retinoic acid-induced cell differentiation. EMBO J. 21, 5235–5244.

Hoffmann, B., Lehmann, J.M., Zhang, X.K., Hermann, T., Husmann, M., Graupner, G., Pfahl, M. 1990. A retinoic acid receptor-specific element controls the retinoic acid receptor-beta promoter. Mol. Endocrinol. 4, 1727–1736.

Hogan, B.L., Taylor, A., Adamson, E. 1981. Cell interactions modulate embryonal carcinoma cell differentiation into parietal or visceral endoderm. Nature 291, 235–237.

Hogan, B.L., Barlow, D.P., Tilly, R. 1983. F9 teratocarcinoma cells as a model for the differentiation of parietal and visceral endoderm in the mouse embryo. Cancer Surveys 2, 115–140.

Hogan, B.L., Constantini, F., Lacy, E. 1986. *Manipulating the Mouse Embryo: A Laboratory Manual.* New York: C. S. H. Laboratory.

Hosler, B.A., LaRosa, G.J., Grippo, J.F., Gudas, L.J. 1989. Expression of REX-1, a gene containing zinc finger motifs, is rapidly reduced by retinoic acid in F9 teratocarcinoma cells. Mol. Cell. Biol. 9, 5623–5629.

Houle, M., Prinos, P., Iulianella, A., Bouchard, N., Lohnes, D. 2000. Retinoic acid regulation of Cdx1: An indirect mechanism for retinoids and vertebral specification. Mol. Cell. Biol. 20, 6579–6586.

Houle, M., Sylvestre, J.R., Lohnes, D. 2003. Retinoic acid regulates a subset of Cdx1 function *in vivo*. Development 130, 6555–6567.

Hu, L., Gudas, L.J. 1990. Cyclic AMP analogs and retinoic acid influence the expression of retinoic acid receptor alpha, beta, and gamma mRNAs in F9 teratocarcinoma cells. Mol. Cell. Biol. 10, 391–396.

Huang, D., Chen, S.W., Langston, A.W., Gudas, L.J. 1998. A conserved retinoic acid responsive element in the murine Hoxb-1 gene is required for expression in the developing gut. Development 125, 3235–3246.

Ishida, A., Iijima, R., Kobayashi, A., Maeda, M. 2005. Characterization of cAMP-dependent proteolysis of GATA-6. Biochem. Biophys. Res. Commun. 332, 976–981.

Iwai, S.A., Nishina, Y., Kosaka, M., Sumi, T., Doi, T., Sakuda, M., Nishimune, Y. 1995. The kinetics of induction of Hox1.6 and C-jun mRNA during three different ways of inducing differentiation in teratocarcinoma F9 cells. *In Vitro* Cell Dev. Biol. Anim. 31, 462–466.

Jacob, A., Budhiraja, S., Reichel, R.R. 1999. The HNF-3alpha transcription factor is a primary target for retinoic acid action. Exp. Cell Res. 250, 1–9.

Joh, T., Darland, T., Samuels, M., Wu, J.X., Adamson, E.D. 1992. Regulation of epidermal growth factor receptor gene expression in murine embryonal carcinoma cells. Cell Growth Differ. 3, 315–325.

Kaestner, K.H. 2000. The hepatocyte nuclear factor 3 (HNF3 or FOXA) family in metabolism. Trends Endocrinol. Metab. 11, 281–285.

Kambhampati, S., Li, Y., Verma, A., Sassano, A., Majchrzak, B., Deb, D.K., Parmar, S., Giafis, N., Kalvakolanu, D.V., Rahman, A., Uddin, S., Minucci, S., et al. 2003. Activation of protein kinase C delta by all-trans-retinoic acid. J. Biol. Chem. 278, 32544–32551.

Kawasaki, H., Eckner, R., Yao, T.P., Taira, K., Chiu, R., Livingston, D.M., Yokoyama, K.K. 1998a. Distinct roles of the co-activators p300 and CBP in retinoic-acid-induced F9-cell differentiation. Nature 393, 284–289.

Kawasaki, H., Song, J., Eckner, R., Ugai, H., Chiu, R., Taira, K., Shi, Y., Jones, N., Yokoyama, K.K. 1998b. p300 and ATF-2 are components of the DRF complex, which regulates retinoic acid- and E1A-mediated transcription of the c-jun gene in F9 cells. Genes Dev. 12, 233–245.

Keller, G., Snodgrass, H.R. 1999. Human embryonic stem cells: The future is now. Nat. Med. 5, 151–152.

Kerkela, R., Pikkarainen, S., Majalahti-Palviainen, T., Tokola, H., Ruskoaho, H. 2002. Distinct roles of mitogen-activated protein kinase pathways in GATA-4 transcription factor-mediated regulation of B-type natriuretic peptide gene. J. Biol. Chem. 277, 13752–13760.

Khuri, F.R., Cho, Y., Talmage, D.A. 1996. Retinoic acid-induced transition from protein kinase C beta to protein kinase C alpha in differentiated F9 cells: Correlation with altered regulation of proto-oncogene expression by phorbol esters. Cell Growth Differ. 7, 595–602.

Kim, M.H., Shin, J.S., Park, S., Hur, M.W., Lee, M.O., Park, H., Lee, C.S. 2002. Retinoic acid response element in HOXA-7 regulatory region affects the rate, not the formation of anterior boundary expression. Int. J. Dev. Biol. 46, 325–328.

Kindregan, H.C., Rosenbaum, S.E., Ohno, S., Niles, R.M. 1994. Characterization of conventional protein kinase C (PKC) isotype expression during F9 teratocarcinoma differentiation. Overexpression of PKC alpha alters the expression of some differentiation-dependent genes. J. Biol. Chem. 269, 27756–27761.

Kingsley-Kallesen, M.L., Kelly, D., Rizzino, A. 1999. Transcriptional regulation of the transforming growth factor-beta2 promoter by cAMP-responsive element-binding protein (CREB) and activating transcription factor-1 (ATF-1) is modulated by protein kinases and the coactivators p300 and CREB-binding protein. J. Biol. Chem. 274, 34020–34028.

Kitabayashi, I., Chiu, R., Umesono, K., Evans, R.M., Gachelin, G., Yokoyama, K. 1994. A novel pathway for retinoic acid-induced differentiation of F9 cells that is distinct from receptor-mediated trans-activation. *In Vitro* Cell Dev. Biol. Anim. 30A, 761–768.

Kitabayashi, I., Eckner, R., Arany, Z., Chiu, R., Gachelin, G., Livingston, D.M., Yokoyama, K.K. 1995. Phosphorylation of the adenovirus E1A-associated 300 kDa protein in response to retinoic acid and E1A during the differentiation of F9 cells. EMBO J. 14, 3496–3509.

Kopf, E., Plassat, J.L., Vivat, V., de The, H., Chambon, P., Rochette-Egly, C. 2000. Dimerization with retinoid X receptors and phosphorylation modulate the retinoic acid-induced degradation of retinoic acid receptors alpha and gamma through the ubiquitin-proteasome pathway. J. Biol. Chem. 275, 33280–33288.

Kubota, H., Chiba, H., Takakuwa, Y., Osanai, M., Tobioka, H., Kohama, G., Mori, M., Sawada, N. 2001. Retinoid X receptor alpha and retinoic acid receptor gamma mediate expression of genes encoding tight-junction proteins and barrier function in F9 cells during visceral endodermal differentiation. Exp. Cell Res. 263, 163–172.

Kuo, C.J., Mendel, D.B., Hansen, L.P., Crabtree, G.R. 1991. Independent regulation of HNF-1 alpha and HNF-1 beta by retinoic acid in F9 teratocarcinoma cells. EMBO J. 10, 2231–2236.

Langston, A.W., Gudas, L.J. 1992. Identification of a retinoic acid responsive enhancer 3' of the murine homeobox gene Hox-1.6. Mech. Dev. 38, 217–227.

Laudet, V., Gronemeyer, H. 2001. *Nuclear Receptor Factsbook,* London: Academic Press.

Lebel, S., Lampron, C., Royal, A., Raymond, Y. 1987. Lamins A and C appear during retinoic acid-induced differentiation of mouse embryonal carcinoma cells. J. Cell Biol. 105, 1099–1104.

Lee, H.Y., Suh, Y.A., Robinson, M.J., Clifford, J.L., Hong, W.K., Woodgett, J.R., Cobb, M. H., Mangelsdorf, D.J., Kurie, J.M. 2000. Stress pathway activation induces phosphorylation of retinoid X receptor. J. Biol. Chem. 275, 32193–32199.

Lee, Y.F., Young, W.J., Burbach, J.P., Chang, C. 1998. Negative feedback control of the retinoid-retinoic acid/retinoid X receptor pathway by the human TR4 orphan receptor, a member of the steroid receptor superfamily. J. Biol. Chem. 273, 13437–13443.

Lee, Y.H., Lee, H.Y., Ryu, S.H., Suh, P.G., Kim, K.W. 1993. Reduced expression of PLC-gamma during the differentiation of mouse F9 teratocarcinoma cells. Cancer Lett. 68, 237–242.

Leslie, N.R., Downes, C.P. 2002. PTEN: The down side of PI 3-kinase signalling. Cell Signal. 14, 285–295.

Levinson-Dushnik, M., Benvenisty, N. 1997. Involvement of hepatocyte nuclear factor 3 in endoderm differentiation of embryonic stem cells. Mol. Cell. Biol. 17, 3817–3822.

Li, L., Arman, E., Ekblom, P., Edgar, D., Murray, P., Lonai, P. 2004a. Distinct GATA6- and laminin-dependent mechanisms regulate endodermal and ectodermal embryonic stem cell fates. Development 131, 5277–5286.

Li, R., Faria, T.N., Boehm, M., Nabel, E.G., Gudas, L.J. 2004b. Retinoic acid causes cell growth arrest and an increase in p27 in F9 wild type but not in F9 retinoic acid receptor beta2 knockout cells. Exp. Cell Res. 294, 290–300.

Li, Y., Glozak, M.A., Smith, S.M., Rogers, M.B. 1999. The expression and activity of D-type cyclins in F9 embryonal carcinoma cells: Modulation of growth by RXR-selective retinoids. Exp. Cell Res. 253, 372–384.

Liu, M., Iavarone, A., Freedman, L.P. 1996. Transcriptional activation of the human p21 (WAF1/CIP1) gene by retinoic acid receptor. Correlation with retinoid induction of U937 cell differentiation. J. Biol. Chem. 271, 31723–31728.

Liu, T., Lee, Y.N., Malbon, C.C., Wang, H.Y. 2002. Activation of the beta-catenin/Lef-Tcf pathway is obligate for formation of primitive endoderm by mouse F9 totipotent teratocarcinoma cells in response to retinoic acid. J. Biol. Chem. 277, 30887–30891.

Lopez-Andreo, M.J., Torrecillas, A., Conesa-Zamora, P., Corbalan-Garcia, S., Gomez-Fernandez, J.C. 2005. Retinoic acid as a modulator of the activity of protein kinase C alpha. Biochemistry 44, 11353–11360.

Loudig, O., Babichuk, C., White, J., Abu-Abed, S., Mueller, C., Petkovich, M. 2000. Cytochrome P450RAI(CYP26) promoter: A distinct composite retinoic acid response element underlies the complex regulation of retinoic acid metabolism. Mol. Endocrinol. 14, 1483–1497.

Loudig, O., Maclean, G.A., Dore, N.L., Luu, L., Petkovich, M. 2005. Transcriptional co-operativity between distant retinoic acid response elements in regulation of Cyp26A1 inducibility. Biochem. J. 392, 241–248.

Lu, Y., Amos, B., Cruise, E., Lotan, D., Lotan, R. 1998. A parallel association between differentiation and induction of galectin-1, and inhibition of galectin-3 by retinoic acid in mouse embryonal carcinoma F9 cells. Biol. Chem. 379, 1323–1331.

MacGregor, T.M., Copeland, N.G., Jenkins, N.A., Giguere, V. 1992. The murine gene for cellular retinoic acid-binding protein type II. Genomic organization, chromosomal localization, and post-transcriptional regulation by retinoic acid. J. Biol. Chem. 267, 7777–7783.

Malbon, C.C., Tao, J., Wang, H.Y. 2004. AKAPs (A-kinase anchoring proteins) and molecules that compose their G-protein-coupled receptor signalling complexes. Biochem. J. 379, 1–9.

Mangelsdorf, D.J., Umesono, K., Kliewer, S.A., Borgmeyer, U., Ong, E.S., Evans, R.M. 1991. A direct repeat in the cellular retinol-binding protein type II gene confers differential regulation by RXR and RAR. Cell 66, 555–561.

Marotti, K.R., Brown, G.D., Strickland, S. 1985. Two-stage hormonal control of type IV collagen mRNA levels during differentiation of F9 teratocarcinoma cells. Dev. Biol. 108, 26–31.

Martin-Blanco, E. 2000. p38 MAPK signalling cascades: Ancient roles and new functions. Bioessays 22, 637–645.

Mason, I.J., Murphy, D., Munke, M., Francke, U., Elliott, R.W., Hogan, B.L. 1986. Developmental and transformation-sensitive expression of the Sparc gene on mouse chromosome 11. EMBO J. 5, 1831–1837.

McKenna, N.J., O'Malley, B.W. 2002. Combinatorial control of gene expression by nuclear receptors and coregulators. Cell 108, 465–474.

Means, A.L., Thompson, J.R., Gudas, L.J. 2000. Transcriptional regulation of the cellular retinoic acid binding protein I gene in F9 teratocarcinoma cells. Cell Growth Differ. 11, 71–82.

Miyaishi, O., Kozaki, K., Iida, K., Isobe, K., Hashizume, Y., Saga, S. 1998. Elevated expression of PDI family proteins during differentiation of mouse F9 teratocarcinoma cells. J. Cell. Biochem. 68, 436–445.

Moras, D., Gronemeyer, H. 1998. The nuclear receptor ligand-binding domain: Structure and function. Curr. Opin. Cell Biol. 10, 384–391.

Morini, M., Piccini, D., De Santanna, A., Levi, G., Barbieri, O., Astigiano, S. 1999. Localization and expression of integrin subunits in the embryoid bodies of F9 teratocarcinoma cells. Exp. Cell Res. 247, 114–122.

Morrisey, E.E., Tang, Z., Sigrist, K., Lu, M.M., Jiang, F., Ip, H.S., Parmacek, M.S. 1998. GATA6 regulates HNF4 and is required for differentiation of visceral endoderm in the mouse embryo. Genes Dev. 12, 3579–3590.

Morton, S., Davis, R.J., Cohen, P. 2004. Signalling pathways involved in multisite phosphorylation of the transcription factor ATF-2. FEBS Lett. 572, 177–183.

Motaleb, M.A., Takihara, Y., Ohta, H., Shimada, K. 1999. Characterization of cis-elements required for the transcriptional activation of the rae28/mph1 gene in F9 cells. Biochem. Biophys. Res. Commun. 262, 509–515.

Mulholland, D.J., Dedhar, S., Coetzee, G.A., Nelson, C.C. 2005. Interaction of nuclear receptors with the Wnt/beta-catenin/Tcf signaling axis: Wnt you like to know? Endocr. Rev. 26, 898–915.

Mummery, C.L., Feyen, A., Freund, E., Shen, S. 1990. Characteristics of embryonic stem cell differentiation: A comparison with two embryonal carcinoma cell lines. Cell Differ. Dev. 30, 195–206.

Murakami, A., Thurlow, J., Dickson, C. 1999. Retinoic acid-regulated expression of fibroblast growth factor 3 requires the interaction between a novel transcription factor and GATA-4. J. Biol. Chem. 274, 17242–17248.

Murakami, A., Shen, H., Ishida, S., Dickson, C. 2004. SOX7 and GATA-4 are competitive activators of Fgf-3 transcription. J. Biol. Chem. 279, 28564–28573.

Murphy, S.P., Garbern, J., Odenwald, W.F., Lazzarini, R.A., Linney, E. 1988. Differential expression of the homeobox gene Hox-1.3 in F9 embryonal carcinoma cells. Proc. Natl. Acad. Sci. USA 85, 5587–5591.

Nakhei, H., Lingott, A., Lemm, I., Ryffel, G.U. 1998. An alternative splice variant of the tissue specific transcription factor HNF4alpha predominates in undifferentiated murine cell types. Nucleic Acids Res. 26, 497–504.

Narlikar, G.J., Fan, H.Y., Kingston, R.E. 2002. Cooperation between complexes that regulate chromatin structure and transcription. Cell 108, 475–487.

Neves, S.R., Ram, P.T., Iyengar, R. 2002. G protein pathways. Science 296, 1636–1639.

Newton, A.C. 1997. Regulation of protein kinase C. Curr. Opin. Cell Biol. 9, 161–167.

Niforas, P., Chu, M.D., Bird, P. 1996. A retinoic acid/cAMP-responsive enhancer containing a cAMP responsive element is required for the activation of the mouse thrombomodulin-encoding gene in differentiating F9 cells. Gene 176, 139–147.

Niimi, T., Hayashi, Y., Futaki, S., Sekiguchi, K. 2004. SOX7 and SOX17 regulate the parietal endoderm-specific enhancer activity of mouse laminin alpha1 gene. J. Biol. Chem. 279, 38055–38061.

Otero, J.J., Fu, W., Kan, L., Cuadra, A.E., Kessler, J.A. 2004. Beta-catenin signaling is required for neural differentiation of embryonic stem cells. Development 131, 3545–3557.

Parekh, D.B., Ziegler, W., Parker, P.J. 2000. Multiple pathways control protein kinase C phosphorylation. EMBO J. 19, 496–503.

Parrella, E., Gianni, M., Cecconi, V., Nigro, E., Barzago, M.M., Rambaldi, A., Rochette-Egly, C., Terao, M., Garattini, E. 2004. Phosphodiesterase IV inhibition by piclamilast potentiates the cytodifferentiating action of retinoids in myeloid leukemia cells. Cross-talk between the cAMP and the retinoic acid signaling pathways. J. Biol. Chem. 279, 42026–42040.

Pawson, T., Nash, P. 2000. Protein-protein interactions define specificity in signal transduction. Genes Dev. 14, 1027–1047.

Pearson, G., Robinson, F., Beers Gibson, T., Xu, B.E., Karandikar, M., Berman, K., Cobb, M.H. 2001. Mitogen-activated protein (MAP) kinase pathways: Regulation and physiological functions. Endocr. Rev. 22, 153–183.

Perletti, L., Kopf, E., Carre, L., Davidson, I. 2001. Coordinate regulation of RARgamma2, TBP, and TAFII135 by targeted proteolysis during retinoic acid-induced differentiation of F9 embryonal carcinoma cells. BMC Mol. Biol. 2, 4.

Plassat, J., Penna, L., Chambon, P., Rochette-Egly, C. 2000. The conserved amphipatic alpha-helical core motif of RARgamma and RARalpha activating domains is indispensable for RA-induced differentiation of F9 cells. J. Cell Sci. 113, 2887–2895.

Plet, A., Evain, D., Anderson, W.B. 1982. Effect of retinoic acid treatment of F9 embryonal carcinoma cells on the activity and distribution of cyclic AMP-dependent protein kinase. J. Biol. Chem. 257, 889–893.

Poizat, C., Puri, P.L., Bai, Y., Kedes, L. 2005. Phosphorylation-dependent degradation of p300 by doxorubicin-activated p38 mitogen-activated protein kinase in cardiac cells. Mol. Cell. Biol. 25, 2673–2687.

Reichel, R.R., Budhiraja, S., Jacob, A. 1994. Delayed activation of HNF-3 beta upon retinoic acid-induced teratocarcinoma cell differentiation. Exp. Cell Res. 214, 634–641.

Renaud, J.P., Moras, D. 2000. Structural studies on nuclear receptors. Cell. Mol. Life Sci. 57, 1748–1769.

Rochette-Egly, C. 2003. Nuclear receptors: Integration of multiple signalling pathways through phosphorylation. Cell. Signal. 15, 355–366.

Rochette-Egly, C. 2005. Dynamic combinatorial networks in nuclear receptor-mediated transcription. J. Biol. Chem. 280, 32565–32568.

Rochette-Egly, C., Chambon, P. 2001. F9 embryocarcinoma cells: A cell autonomous model to study the functional selectivity of RARs and RXRs in retinoid signaling. Histol. Histopathol. 16, 909–922.

Rochette-Egly, C., Oulad-Abdelghani, M., Staub, A., Pfister, V., Scheuer, I., Chambon, P., Gaub, M.P. 1995. Phosphorylation of the retinoic acid receptor-alpha by protein kinase A. Mol. Endocrinol. 9, 860–871.

Rochette-Egly, C., Adam, S., Rossignol, M., Egly, J.M., Chambon, P. 1997. Stimulation of RAR alpha activation function AF-1 through binding to the general transcription factor TFIIH and phosphorylation by CDK7. Cell 90, 97–107.

Rochette-Egly, C., Plassat, J.L., Taneja, R., Chambon, P. 2000. The AF-1 and AF-2 activating domains of retinoic acid receptor-alpha (RARalpha) and their phosphorylation are differentially involved in parietal endodermal differentiation of F9 cells and retinoid-induced expression of target genes. Mol. Endocrinol. 14, 1398–1410.

Ross, S.A., McCaffery, P.J., Drager, U.C., De Luca, L.M. 2000. Retinoids in embryonal development. Physiol. Rev. 80, 1021–1054.

Roy, B., Taneja, R., Chambon, P. 1995. Synergistic activation of retinoic acid (RA)-responsive genes and induction of embryonal carcinoma cell differentiation by an RA receptor alpha (RAR alpha)-, RAR beta-, or RAR gamma-selective ligand in combination with a retinoid X receptor-specific ligand. Mol. Cell. Biol. 15, 6481–6487.

Sangster-Guity, N., Yu, L.M., McCormick, P. 2004. Molecular profiling of embryonal carcinoma cells following retinoic acid or histone deacetylase inhibitor treatment. Cancer Biol. Ther. 3, 1109–1120.

Satohisa, S., Chiba, H., Osanai, M., Ohno, S., Kojima, T., Saito, T., Sawada, N. 2005. Behavior of tight-junction, adherens-junction and cell polarity proteins during HNF-4alpha-induced epithelial polarization. Exp. Cell Res. 310, 66–78.

Scheid, M.P., Woodgett, J.R. 2001. PKB/AKT: Functional insights from genetic models. Nat. Rev. Mol. Cell. Biol. 2, 760–768.

Schulze, F., Chowdhury, K., Zimmer, A., Drescher, U., Gruss, P. 1987. The murine homeo box gene product, Hox 1.1 protein, is growth-controlled and associated with chromatin. Differentiation 36, 130–137.

Seth, A., Robinson, L., Panayiotakis, A., Thompson, D.M., Hodge, D.R., Zhang, X.K., Watson, D.K., Ozato, K., Papas, T.S. 1994. The EndoA enhancer contains multiple ETS binding site repeats and is regulated by ETS proteins. Oncogene 9, 469–477.

Sharif, K.A., Li, C., Gudas, L.J. 2001. cis-acting DNA regulatory elements, including the retinoic acid response element, are required for tissue specific laminin B1 promoter/lacZ expression in transgenic mice. Mech. Dev. 103, 13–25.

Shen, J., Wu, H., Gudas, L.J. 2000. Molecular cloning and analysis of a group of genes differentially expressed in cells which overexpress the Hoxa-1 homeobox gene. Exp. Cell Res. 259, 274–283.

Shen, S., Kruyt, F.A., den Hertog, J., van der Saag, P.T., Kruijer, W. 1991. Mouse and human retinoic acid receptor beta 2 promoters: Sequence comparison and localization of retinoic acid responsiveness. DNA Seq. 2, 111–119.

Shibamoto, S., Winer, J., Williams, M., Polakis, P. 2004. A blockade in Wnt signaling is activated following the differentiation of F9 teratocarcinoma cells. Exp. Cell Res. 292, 11–20.

Smedberg, J.L., Smith, E.R., Capo-Chichi, C.D., Frolov, A., Yang, D.H., Godwin, A.K., Xu, X.X. 2002. Ras/MAPK pathway confers basement membrane dependence upon endoderm differentiation of embryonic carcinoma cells. J. Biol. Chem. 277, 40911–40918.

Smith, E.R., Smedberg, J.L., Rula, M.E., Hamilton, T.C., Xu, X.X. 2001. Disassociation of MAPK activation and c-Fos expression in F9 embryonic carcinoma cells following retinoic acid-induced endoderm differentiation. J. Biol. Chem. 276, 32094–32100.

Smith, E.R., Smedberg, J.L., Rula, M.E., Xu, X.X. 2004. Regulation of Ras-MAPK pathway mitogenic activity by restricting nuclear entry of activated MAPK in endoderm differentiation of embryonic carcinoma and stem cells. J. Cell. Biol. 164, 689–699.

Smith, R., Peters, G., Dickson, C. 1988. Multiple RNAs expressed from the int-2 gene in mouse embryonal carcinoma cell lines encode a protein with homology to fibroblast growth factors. EMBO J. 7, 1013–1022.

Smith, W.C., Nakshatri, H., Leroy, P., Rees, J., Chambon, P. 1991. A retinoic acid response element is present in the mouse cellular retinol binding protein I (mCRBPI) promoter. EMBO J. 10, 2223–2230.

Soprano, D.R., Soprano, K.J., Wyatt, M.L., Goodman, D.S. 1988. Induction of the expression of retinol-binding protein and transthyretin in F9 embryonal carcinoma cells differentiated to embryoid bodies. J. Biol. Chem. 263, 17897–17900.

Srinivas, H., Juroske, D.M., Kalyankrishna, S., Cody, D.D., Price, R.E., Xu, X.C., Narayanan, R., Weigel, N.L., Kurie, J.M. 2005. c-Jun N-terminal kinase contributes to aberrant retinoid signaling in lung cancer cells by phosphorylating and inducing proteasomal degradation of retinoic acid receptor alpha. Mol. Cell. Biol. 25, 1054–1069.

Srinivas, H., Xia, D., Moore, N.L., Uray, I.P., Kim, H., Ma, L., Weigel, N.L., Brown, P.H., Kurie, J.M. 2006. Akt phosphorylates and suppresses the transactivation of retinoic acid receptor alpha. Biochem. J. 395, 653–662.

Strickland, S., Mahdavi, V. 1978. The induction of differentiation in teratocarcinoma cells by retinoic acid. Cell 15, 393–403.

Strickland, S., Smith, K.K., Marotti, K.R. 1980. Hormonal induction of differentiation in teratocarcinoma stem cells: Generation of parietal endoderm by retinoic acid and dibutyryl cAMP. Cell 21, 347–355.

Subramanian, V., Meyer, B.I., Gruss, P. 1995. Disruption of the murine homeobox gene Cdx1 affects axial skeletal identities by altering the mesodermal expression domains of Hox genes. Cell 83, 641–653.

Sumi-Ichinose, C., Ichinose, H., Metzger, D., Chambon, P. 1997. SNF2beta-BRG1 is essential for the viability of F9 murine embryonal carcinoma cells. Mol. Cell. Biol. 17, 5976–5986.

Tabor, J.M., Oshima, R.G. 1982. Identification of mRNA species that code for extra-embryonic endodermal cytoskeletal proteins in differentiated derivatives of murine embryonal carcinoma cells. J. Biol. Chem. 257, 8771–8774.

Takihara, Y., Matsuda, Y., Irie, K., Matsumoto, K., Hara, J. 2000. 14-3-3 protein family members have a regulatory role in retinoic acid-mediated induction of cytokeratins in F9 cells. Exp. Cell Res. 260, 96–104.

Tan, Y., Costa, R.H., Kovesdi, I., Reichel, R.R. 2001. Adenovirus-mediated increase of HNF-3 levels stimulates expression of transthyretin and sonic hedgehog, which is associated with F9 cell differentiation toward the visceral endoderm lineage. Gene Expr. 9, 237–248.

Tanaka, S., Kaytes, P., Kurkinen, M. 1993. An enhancer for transcription of collagen IV genes is activated by F9 cell differentiation. J. Biol. Chem. 268, 8862–8870.

Taneja, R., Bouillet, P., Boylan, J.F., Gaub, M.P., Roy, B., Gudas, L.J., Chambon, P. 1995. Reexpression of retinoic acid receptor (RAR) gamma or overexpression of RAR alpha or RAR beta in RAR gamma-null F9 cells reveals a partial functional redundancy between the three RAR types. Proc. Natl. Acad. Sci. USA 92, 7854–7858.

Taneja, R., Roy, B., Plassat, J.L., Zusi, C.F., Ostrowski, J., Reczek, P.R., Chambon, P. 1996. Cell-type and promoter-context dependent retinoic acid receptor (RAR) redundancies for RAR beta 2 and Hoxa-1 activation in F9 and P19 cells can be artefactually generated by gene knockouts. Proc. Natl. Acad. Sci. USA 93, 6197–6202.

Taneja, R., Rochette-Egly, C., Plassat, J.L., Penna, L., Gaub, M.P., Chambon, P. 1997. Phosphorylation of activation functions AF-1 and AF-2 of RAR alpha and RAR gamma is indispensable for differentiation of F9 cells upon retinoic acid and cAMP treatment. EMBO J. 16, 6452–6465.

Tarrade, A., Bastien, J., Bruck, N., Bauer, A., Gianni, M., Rochette-Egly, C. 2005. Retinoic acid and arsenic trioxide cooperate for apoptosis through phosphorylated RXR alpha. Oncogene 24, 2277–2288.

Thompson, J.R., Chen, S.W., Ho, L., Langston, A.W., Gudas, L.J. 1998. An evolutionary conserved element is essential for somite and adjacent mesenchymal expression of the Hoxa1 gene. Dev. Dyn. 211, 97–108.

Toker, A. 2000. Protein kinases as mediators of phosphoinositide 3-kinase signaling. Mol. Pharmacol. 57, 652–658.

Tremblay, J.J., Viger, R.S. 2003. Novel roles for GATA transcription factors in the regulation of steroidogenesis. J. Steroid Biochem. Mol. Biol. 85, 291–298.

Ugai, H., Uchida, K., Kawasaki, H., Yokoyama, K.K. 1999. The coactivators p300 and CBP have different functions during the differentiation of F9 cells. J. Mol. Med. 77, 481–494.

Vasios, G., Mader, S., Gold, J.D., Leid, M., Lutz, Y., Gaub, M.P., Chambon, P., Gudas, L. 1991. The late retinoic acid induction of laminin B1 gene transcription involves RAR binding to the responsive element. EMBO J. 10, 1149–1158.

Vasios, G.W., Gold, J.D., Petkovich, M., Chambon, P., Gudas, L.J. 1989. A retinoic acid-responsive element is present in the 5′ flanking region of the laminin B1 gene. Proc. Natl. Acad. Sci. USA 86, 9099–9103.

Verheijen, M.H., Wolthuis, R.M., Bos, J.L., Defize, L.H. 1999. The Ras/Erk pathway induces primitive endoderm but prevents parietal endoderm differentiation of F9 embryonal carcinoma cells. J. Biol. Chem. 274, 1487–1494.

Vivanco, I., Sawyers, C.L. 2002. The phosphatidylinositol 3-Kinase AKT pathway in human cancer. Nat. Rev. Cancer. 2, 489–501.

Wang, C., Song, B. 1996. Cell-type-specific expression of the platelet-derived growth factor alpha receptor: A role for GATA-binding protein. Mol. Cell. Biol. 16, 712–723.

Wang, C., Kelly, J., Bowen-Pope, D.F., Stiles, C.D. 1990. Retinoic acid promotes transcription of the platelet-derived growth factor alpha-receptor gene. Mol. Cell. Biol. 10, 6781–6784.

Wang, S.Y., LaRosa, G.J., Gudas, L.J. 1985. Molecular cloning of gene sequences transcriptionally regulated by retinoic acid and dibutyryl cyclic AMP in cultured mouse teratocarcinoma cells. Dev. Biol. 107, 75–86.

Watkins, D.C., Johnson, G.L., Malbon, C.C. 1992. Regulation of the differentiation of teratocarcinoma cells into primitive endoderm by G alpha i2. Science 258, 1373–1375.

Weiler-Guettler, H., Yu, K., Soff, G., Gudas, L.J., Rosenberg, R.D. 1992. Thrombomodulin gene regulation by cAMP and retinoic acid in F9 embryonal carcinoma cells. Proc. Natl. Acad. Sci. USA 89, 2155–2159.

Weston, C.R., Davis, R.J. 2002. The JNK signal transduction pathway. Curr. Opin. Genet. Dev. 12, 14–21.

Weston, C.R., Lambright, D.G., Davis, R.J. 2002. Signal transduction. MAP kinase signaling specificity. Science 296, 2345–2347.

Wong, W., Scott, J.D. 2004. AKAP signalling complexes: Focal points in space and time. Nat. Rev. Mol. Cell. Biol. 5, 959–970.

Wu, R.C., Qin, J., Yi, P., Wong, J., Tsai, S.Y., Tsai, M.J., O'Malley, B.W. 2004. Selective phosphorylations of the SRC-3/AIB1 coactivator integrate genomic reponses to multiple cellular signaling pathways. Mol. Cell 15, 937–949.

Yang-Yen, H.F., Chiu, R., Karin, M. 1990. Elevation of AP1 activity during F9 cell differentiation is due to increased c-jun transcription. New Biol. 2, 351–361.

Yao, A., Wang, J., Fink, L.M., Hardin, J.W., Hauer-Jensen, M. 1999. Molecular cloning and sequence analysis of the 5′-flanking region of the Sprague-Dawley rat thrombomodulin gene. DNA Seq. 10, 55–60.

Yogosawa, S., Kayukawa, K., Kawata, T., Makino, Y., Inoue, S., Okuda, A., Muramatsu, M., Tamura, T. 1999. Induced expression, localization, and chromosome mapping of a gene for the TBP-interacting protein 120A. Biochem. Biophys. Res. Commun. 266, 123–128.

Yoshida-Koide, U., Matsuda, T., Saikawa, K., Nakanuma, Y., Yokota, T., Asashima, M., Koide, H. 2004. Involvement of Ras in extraembryonic endoderm differentiation of embryonic stem cells. Biochem. Biophys. Res. Commun. 313, 475–481.

Zhao, Q., Tao, J., Zhu, Q., Jia, P.M., Dou, A.X., Li, X., Cheng, F., Waxman, S., Chen, G.Q., Chen, S.J., Lanotte, M., Chen, Z., et al. 2004. Rapid induction of cAMP/PKA pathway during retinoic acid-induced acute promyelocytic leukemia cell differentiation. Leukemia 18, 285–292.

Zhou, D., Henion, T.R., Jungalwala, F.B., Berger, E.G., Hennet, T. 2000. The beta 1, 3-galactosyltransferase beta 3GalT-V is a stage-specific embryonic antigen-3 (SSEA-3) synthase. J. Biol. Chem. 275, 22631–22634.

Zhuang, Y., Faria, T.N., Chambon, P., Gudas, L.J. 2003. Identification and characterization of retinoic acid receptor beta2 target genes in F9 teratocarcinoma cells. Mol. Cancer Res. 1, 619–630.

The Ftz-F1 family: Orphan nuclear receptors regulated by novel protein–protein interactions

Leslie Pick,[1,2] W. Ray Anderson,[2]
Jeffrey Shultz[1] and Craig T. Woodard[3]

[1]*Department of Entomology, 4112 Plant Sciences,*
University of Maryland, College Park, Maryland
[2]*Department of Cell Biology and Molecular Genetics,*
University of Maryland, College Park, Maryland
[3]*Department of Biological Sciences,*
Mount Holyoke College, South Hadley, Massachusetts

Contents

Advances in Developmental Biology
Volume 16 ISSN 1574-3349
DOI: 10.1016/S1574-3349(06)16008-1

Summary

The Ftz-F1 (NR5A) family of nuclear receptors (NRs) is a large group of proteins involved in a wide range of biological processes. This is an ancient family that is currently represented broadly throughout the animal kingdom. Ftz-F1 family NRs can bind to DNA as monomers; in addition to the zinc finger DNA-binding motifs, these proteins contain a C-terminal extension of their DBDs—the Ftz-F1 box—that contributes to DNA-binding specificity. However, direct interactions with other DNA-binding proteins, including homeodomain-containing transcription factors, influence the function of Ftz-F1 family members in animals as diverse as flies and mice. *Drosophila* Ftz-F1, the founding member of this family, is required for embryonic patterning at early stages of development and for ecdysone-triggered metamorphosis at later stages of development. Studies in *Drosophila* revealed a novel protein–protein interaction between Ftz-F1 and the Hox protein Ftz. The Ftz/Ftz-F1 protein–protein interaction mimics the interaction between NRs and standard coactivators as Ftz contains an LXXLL (NR-box) motif. The interaction of Ftz with Ftz-F1 also promoted the evolution of *ftz* from a homeotic gene to a segmentation gene in modern day *Drosophila*. Other noncanonical interactions between Ftz-F1 family members and DNA-binding transcription factors may function similarly to restrict the cell-specific function of other Ftz-F1 family NRs during development. These findings are discussed in light of evidence that Ftz-F1 family NRs bind phospholipid ligands.

1. Introduction: *Drosophila* nuclear receptors

The genome of *Drosophila melanogaster* contains 18 nuclear receptor (NR) genes. Most of these NRs participate in regulatory cascades involved in maturation. These cascades are controlled by two hormones in *Drosophila*: juvenile hormone and 20-hydroxyecdysone, referred to here as ecdysone. Ecdysone, the major regulator of larval molting, and of metamorphosis to produce the adult animal, functions by binding a heterodimer of the ecdysone receptor (EcR) and ultraspiracle (USP), which are orthologues of vertebrate farnesoid X receptor (FXR) and liver X receptor (LXR). Binding of ecdysone to its receptor initiates a cascade of gene activation; many of the direct and indirect targets are themselves NRs, which in turn regulate genes directly involved in metamorphosis. One of these targets is βFtz-F1, which serves as a competence factor for metamorphosis (see Section 5). The involvement of cascades of NRs in molting is also conserved in other invertebrates, such as *Caenorhabditis elegans*, where *nhr-23*, the orthologue of *Drosophila* DHR3, has been linked to molting (Kostrouchova et al., 1998). Tremendous insight into the role and function of the ecdysone cascade has been gained in recent years and is discussed in an excellent review (King-Jones and Thummel, 2005). Here, we focus on the orphan receptor Ftz-F1, which is involved in metamorphosis and is critical for embryonic development in *Drosophila*. The Ftz-F1 family of NRs appears to be ancient (Escriva et al., 2004), and orthologues have been identified in most metazoans examined to date. Ftz-F1 family proteins have been implicated in development in a range of species, including nematodes, flies, fish, and mammals. Here we discuss its initial discovery and its role in development in flies and other model systems, focusing on novel protein–protein interactions that appear to delimit the tissue-specific functions of Ftz-F1 in metazoans.

2. Identification, cloning, and structure of *Drosophila ftz-f1*

2.1. Ftz-F1 was identified by virtue of its binding to ftz regulatory elements

Drosophila Ftz-F1 (NR5A3; Fig. 1) was first identified in Carl Wu's laboratory over 15 years ago (Ueda et al., 1990) in the course of studies of the transcriptional regulation of the pair-rule segmentation gene *fushi tarazu* (*ftz*). *ftz* mutant embryos die before hatching, lacking alternate body segments; the

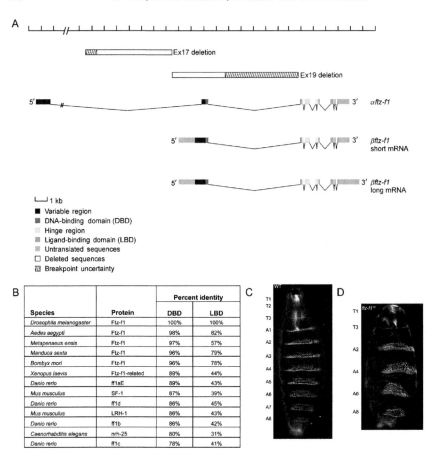

Fig. 1. Structure and function of the *Drosophila ftz-f1* gene. (A) Map of the *ftz-f1* gene (75D8-E1, Chromosome arm 3L). The first exon of α*ftz-f1* (Exon 1) is separated from Exon 2 by an intron of 33,674 bp. The transcription start site of α*ftz-f1* has not been mapped. Analysis of cDNAs suggests that the longer form of the β*ftz-f1* mRNA results from the use of an alternative polyadenylation signal about 800 bp downstream from that used for the short β*ftz-f1* mRNA and for the α*ftz-f1* mRNA. *ftz-f1*[19] was generated by the imprecise excision of a viable P element insertion, P[(w, ry)H] 2-2 (Levis et al., 1985; Broadus et al., 1999), which maps within 1 kb upstream of the β*ftz-f1* transcription start site. *ftz-f1*[19] results from deletion Ex19, which extends at least 4.5 kb downstream from the original insertion site of P[(w, ry)H]2-2, breaking in the first intron of β*ftz-f1* (Intron 2). This excision completely removes the transcription start site and first exon of β*ftz-f1* and is genetically an amorphic allele of β*ftz-f1*. Sequencing of α*ftz-f1* mRNAs revealed that Ex19 removes the second exon of α*ftz-f1* (Exon 2). In the resulting *ftz-f1*[19] mRNA, Exon 1 is spliced in-frame to Exon 3, and all exons other than Exon 2 are intact and in-frame. The *ftz-f1*[19] mRNA thus encodes an αFtz-F1 protein that is missing the DBD. *ftz-f1*[17], identified in the same screen, affects β*ftz-f1* specifically (Broadus et al., 1999), and results from deletion Ex17, which extends 6.5–7.5 kb upstream from the original insertion site of P[(w, ry)H] 2-2 (Cruz, 2000). *ftz-f1*[17] is a hypomorphic allele of β*ftz-f1* that encodes reduced levels of full-length β*ftz-f1* transcripts (Broadus et al., 1999). (B) Sequence conservation in the DBDs and LBDs of Ftz-F1 family members. Orthologues of Ftz-F1 were aligned by blastp at NCBI (Altschul et al., 1990) sorted by percent identity to the *Drosophila* Ftz-F1 DBD (aa 508–603)

name *ftz* means "not enough segments" in Japanese (Wakimoto et al., 1984). *ftz* is expressed in a seven-stripe pattern in *Drosophila* embryos (see Fig. 2, yellow) that correlates with its function; that is, the cells that express *ftz* in the blastoderm embryo are the primordia of the body regions missing in *ftz* mutant embryos. Thus, correct expression of *ftz* in its seven-stripe pattern is critical for the development of the embryo. Because of the importance of this striped expression pattern for function, considerable effort was focused on elucidating the mechanisms underlying *ftz* stripe establishment and mainte-nance. Two large regulatory elements, the zebra element and the upstream element, were shown to be required for expression of *ftz* in seven stripes (Hiromi et al., 1985; Hiromi and Gehring, 1987). To identify *trans*-acting factors that regulate *ftz* expression in stripes, Ueda et al. (1990) looked for sequence-specific DNA-binding proteins in *Drosophila* embryonic nuclear extracts that interacted with the *ftz* zebra element. They then purified one of these DNA-binding factors to homogeneity and named it Ftz-Factor 1 (Ftz-F1). Mutations in the Ftz-F1-binding sites in the zebra element were generated, and expression of zebra-element-*lacZ* reporter genes was exam-ined in transgenic *Drosophila* embryos. This revealed a requirement for these binding sites for high levels of expression of zebra-element-*lacZ* reporter genes, with possible differential effects on different sets of stripes (Ueda et al., 1990). Cloning of the *ftz-f1* cDNA by the same group using expression library screening revealed it to encode a member of the nuclear hormone receptor family (Lavorgna et al., 1991). This study also confirmed DNA-binding results that had suggested the existence of early (α*ftz-f1*) and late (β*ftz-f1*) forms of Ftz-F1, here shown to correspond to transcripts of different size, with a maternal transcript of 5.2 kb present in the early embryo when *ftz* is also expressed. Later, 4.8- and 5.6-kb transcripts are present when *ftz* is no longer expressed. This later expression of *ftz-f1* RNA, as well as DNA-binding activity, suggested an additional function(s) of Ftz-F1 (Lavorgna et al., 1991). Such functions were further supported by the finding that Ftz-F1 binds many sites on polytene chromosomes during the period of ecdysone-stimulated onset of metamorphosis (Lavorgna et al., 1993) (see Section 5 for further discussion on Ftz-F1 function in metamorphosis).

---

and LBD (aa 831–1027). *Drosophila melanogaster* Ftz-F1 (P33244), *Aedes aegypti* Ftz-f1 (AAF82307), *Metapenaeus ensis* Ftz-F1 (AAD41899), *Manduca Sexta* Ftz-F1 (AAL50351), *Bombyx mori* Ftz-F1 (AAL30663), *Xenopus laevis* Ftz-F1-related (AAA18357), *Danio rerio* ff1aE (AAK54449), ff1b (AAF43283), ff1c (AAK19303), ff1d (AAO59489), *Mus musculus* SF-1 (P33242), LRH-1 (NP_109601), *C. elegans* nrh-25 (NP_001024550). (C) *ftz-f1[19]* germ line clone embryos display an *ftz*-like pair-rule phenotype. *ftz-f1[19]* germ line clones were generated with FLP-DFS technique (Chou and Perrimon, 1996). Cuticle preparations of (C) a wild-type (*w[1118]*) larvae, displaying three thoracic and eight abdominal segments is compared to (D) the *ftz-f1* mutant, which has half the number of segments. The missing segments are the same segments that are missing in *ftz* mutant embryos. Cuticles were photographed with dark field optics on a Leica DM-RB microscope.

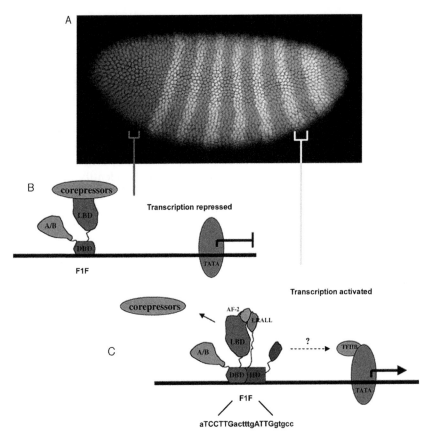

Fig. 2. Ftz mimics an NR coactivator for Ftz-F1. (A) Ftz and Ftz-F1 expression overlaps in the seven Ftz stripes. Ftz-F1 (green) is expressed and nuclear localized in all somatic cells of the embryo, shown here at the blastoderm stage. The embryo at this stage consists of approximately 5000 cells located around the circumference of the egg. Ftz is expressed in seven stripes, each 3–4 cells wide, in the primordia of the even-numbered parasegments that are missing in *ftz* mutants. Ftz protein was detected with a rhodamine-coupled secondary antibody (red); because expression overlaps precisely with Ftz-F1, stripes appear yellow. (B) Ftz-F1 is inactive in cells that do not express Ftz. When Ftz-F1 binds DNA in Ftz-nonexpressing (green) cells, corepressors are recruited to inhibit transcription. (C) Ftz-F1 and Ftz cooperate to activate target gene expression. Ftz and Ftz-F1 bind cooperatively to composite Ftz-F1/Ftz-binding sites, exemplified by F1F. Interaction of the Ftz LRALL sequence with the Ftz-F1 AF-2 domain displaces corepressors, allowing productive transcription complexes to form. (See Color Insert.)

2.2. Domain structure of Ftz-F1

Western blot analysis revealed that the early and late mRNAs encode distinct protein isoforms of Ftz-F1: a 130-kDa form, αFtz-F1, present in early embryos, and a 97-kDa form, βFtz-F1, present in late stage embryos (Lavorgna et al., 1993).

βFtz-F1 is also expressed preceding each larval molt and in the mid-prepupal period (Lavorgna et al., 1993; Woodard et al., 1994; White et al., 1997; Yamada et al., 2000). Both isoforms share a typical NR structure, with a highly conserved zinc finger DNA-binding domain (DBD) and a ligand-binding domain (LBD) that terminates in a canonical AF-2 domain.[1] αFtz-F1 has a long N-terminal arm relative to βFtz-F1, but no function has as yet been assigned to this portion of the protein. The distinct N termini of αFtz-F1 and βFtz-F1 are encoded by different first exons, which have distinct transcription start sites separated by over 30 kb of genomic DNA. The protein-coding portion of *αftz-f1* Exon 1 is 1195-bp long and is separated from Exon 2 by 33,674 bp (Consortium, 2003). The transcription start site of *βftz-f1* has been mapped (Lam et al., 1997), and it lies within the first intron of *αftz-f1* (Intron 1). A 3′ splice junction for the second exon of *αftz-f1* (Exon 2) resides within the first exon of *βftz-f1* (Exon 1a). Thus, the second exon of *αftz-f1* is identical to the 534 bp at the 3′ end of the first *βftz-f1* exon. From there on, *αftz-f1* and *βftz-f1* share the same seven exons (Exons 3–9) of 218, 371, 57, 199, 259, 101, and 982 bp, respectively (Consortium, 2003). Analysis of *βftz-f1* cDNAs has identified two classes, suggesting that the longer form of the *βftz-f1* mRNA results from the use of an alternative polyadenylation signal about 800 bp downstream from that used for the short *βftz-f1* mRNA and for the *αftz-f1* mRNA. The long and short *βftz-f1* mRNAs encode the same protein (Fig. 1). The DBD and the LBD, as well as the intervening hinge region, are identical in αFtz-F1 and βFtz-F1. The DBD is encoded by the final 216 bp of Exon 2 and the first 81 bp of Exon 3. The LDB is encoded by the last 83 bp of Exon 6 through the rest of the open reading frame. The hinge region is coded for by the sequences in between those that encode the DBD and the LBD (Fig. 1).

The DNA-binding properties of *Drosophila* Ftz-F1, and of family members from vertebrates, have been characterized in some detail. Ftz-F1 proteins from *Bombyx*, *Drosophila*, *Xenopus*, and rodents were all shown to bind DNA as monomers with a PyCAAGGPyCPu consensus (Ueda et al., 1992; Lynch et al., 1993; Ellinger-Ziegelbauer et al., 1994). Ftz-F1 family proteins have been shown to be strong activators of transcription in a variety of systems, including mammalian cells, yeast cells, *Drosophila* cells, and in the *Drosophila* embryo (see Section 3).

The Ftz-F1 DBD contains two Cys_2–Cys_2-type zinc fingers, followed by a "Ftz-F1-box" that distinguishes this family of NRs. The Ftz-F1 box was originally identified as a conserved stretch of 23 amino acids followed by an additional 30 amino acids enriched in basic residues (Ueda et al., 1992). The N-terminal half of this Ftz-F1 box is equivalent to the C-terminal extension (CTE), seen in a number of NRs such as ER, and ERR (Melvin et al., 2004). This CTE is thought to extend the DNA recognition sequence of this family of

---

[1]Note that NCBI AAA28542 contains a sequencing error that incorrectly predicts a Ftz-F1 protein lacking the AF-2 domain.

receptors by an additional three bases, compared to the standard NR DBD recognition sequence of six nucleotides, thereby increasing DNA-binding specificity, consistent with the Ftz-F1 consensus site. The Ftz-F1 box was originally shown to be necessary for high-affinity DNA binding of Ftz-F1 *in vitro*, with deletions or mutations in this region causing up to 1000-fold decrease in binding affinity (Ueda et al., 1992). DNA-binding assays using mutations in different regions of the Ftz-F1–binding site further suggested that the zinc fingers and the Ftz-F1 box contact different parts of the binding site (Ueda et al., 1992). The Ftz-F1 box was also found to harbor a bipartite nuclear localization signal (Li et al., 1999). Structural analysis of the human LRH-1 (see later) DBD revealed that the Ftz-F1–specific portion of the Ftz-F1 box, referred to as the Ftz-F1 domain, folds into a helix that packs against the DBD but does not contact the DNA directly. Rather, this interaction allows the CTE to make specific contacts in the minor groove (Solomon et al., 2005). These studies also confirmed the importance of the Ftz-F1 box for transcriptional activity and provided evidence that the Ftz-F1 domain also plays a role in recruitment of coactivators by the LBD (Solomon et al., 2005).

2.3. DHR39 is related to Ftz-F1

A second *Drosophila* NR, termed Ftz-F1β or DHR39, was isolated on the basis of homology to Ftz-F1 (Ohno and Petkovich, 1992) and, independently, by screening an expression library with a Ftz-F1–binding site (Ayer et al., 1993). We refer to this protein here as DHR39 to avoid confusion with βFtz-F1 (Fig. 1). DHR39 and Ftz-F1 are 63% identical in the DBD and 22% identical in the LBD (Ohno and Petkovich, 1992). The two proteins bind the same DNA sequences *in vitro*. Ayer et al. (1993) found that DHR39 represses transcription of an Adh-promoter element in *Drosophila* cell lines (Ayer et al., 1993), while Ohno et al. (1994) found that both αFtz-F1 and DHR39 can activate transcription via a Ftz-F1 recognition sequence from the *ftz* zebra element, in cotransfection studies using a mammalian cell line (Ohno et al., 1994). While it remains unclear whether DHR39 has different activities on different promoters or in different cell lines, an understanding of its function awaits characterization of a *DHR39* mutant in *Drosophila* embryos. A genetic screen using a *piggyBac*-based element for insertional mutagenesis identified an insertion in this gene that is associated with embryonic lethality (Horn et al., 2003), so information regarding DHR39 function *in vivo* should be forthcoming.

3. Ftz-F1 is a cofactor for the homeodomain protein Ftz

3.1. Ftz-F1 binds to the ftz upstream element enhancer

As discussed earlier, Ftz-F1 was initially isolated on the basis of its binding to the *ftz* zebra element. A separate study using *Drosophila* embryo nuclear extracts to identify *trans*-regulatory factors controlling *ftz* stripe expression through the *ftz* upstream element identified additional binding sites for Ftz-F1 (Han et al., 1993, 1998). The upstream element is a positive autoregulatory enhancer element, located between 3.6 and 6 kb upstream of the transcription start site (Hiromi and Gehring, 1987). Ftz protein directly binds to multiple sites in the upstream element, via its homeodomain, and raises levels of *ftz* expression in stripes (Pick et al., 1990; Schier and Gehring, 1992). The upstream element is composed of two separable enhancers, a distal enhancer and a proximal enhancer, each of which is bound by Ftz directly (Pick et al., 1990). Studies by Han et al. (1993) identified a core 323-bp region of the proximal enhancer (*323-bp* ftz *proximal enhancer*; *323 fPE*) that contains 5 Ftz-binding sites and also binding sites for multiple DNA-binding proteins present in *Drosophila* embryo nuclear extracts. This short region of the upstream element was sufficient to direct reporter gene expression in a *ftz*-like seven-stripe pattern in transgenic embryos. One of the proteins binding to the enhancer turned out to be Ftz-F1, which bound three sites in the *323 fPE*. These sites were functionally redundant and necessary for expression of *ftz-lacZ* reporter fusion genes. Wild-type proximal enhancer-*lacZ* fusion genes directed expression in seven *ftz*-like stripes, and mutation of one Ftz-F1–binding site had no effect on gene expression. However, simultaneous mutation of two sites drastically reduced expression and mutation of all three sites abolished detectable expression in the embryo (Han et al., 1998). This demonstrated that Ftz-F1–binding sites are necessary *in vivo* for expression directed by this *ftz* enhancer element.

These Ftz-F1–binding sites share the consensus sequence defined in other studies. In addition to binding Ftz-F1, these DNA sites were bound by other nuclear proteins, two of which were purified and shown to be Tramtrack (Ttk) and Adf-1 (Han et al., 1998). The Ttk core-binding site is a subset of the Ftz-F1 site (AAGG) such that all Ftz-F1 sites represent potential Ttk-binding sites. As Ttk is thought to be a repressor (Read, 1992; Brown and Wu, 1993; Xiong and Montell, 1993), while Ftz-F1 is an activator, an intriguing but untested model is that Ttk represses transcription of the *ftz* gene via the same sites used by Ftz-F1 to activate it, as a mechanism of turning *ftz* stripes off in mid-embryogenesis. The other protein binding to the Ftz-F1 sites, Adf-1, is also a transcription activator that is expressed ubiquitously in *Drosophila* embryos (England et al., 1990, 1992). Adf-1 and Ftz-F1 were proposed to have redundant roles in activating *ftz* gene expression as *adf-1* mutant embryos did not show any detectable change in the expression of *ftz* or *ftz-lacZ* fusion genes (Landrigan, 1998).

3.2. Ftz-F1 and Ftz coordinately regulate gene expression

The mechanism underlying Ftz-F1 function in the embryo was elucidated from a combination of biochemical and genetic studies in different laboratories. *Drosophila ftz* is a member of a group of *Hox* genes that reside in an evolutionarily conserved cluster of homeotic genes, known as HOM-C in *Drosophila*. The genes in HOM-C encode transcription factors that share a highly conserved homeobox, encoding their DNA-binding homeodomains. In keeping with this sequence conservation, different Hox proteins bind to very similar DNA sequences. In addition, the binding specificity is rather loose such that Hox proteins have the potential to bind millions of sites in a genome. This loose and overlapping binding specificity contrasts with the exquisite biological specificity of Hox function *in vivo*. This contrast, known as the "Hox paradox," suggested that the specificity of action of Hox proteins might be achieved by their differential interaction with DNA-binding cofactors that would limit the number of genomic-binding sites utilized by each Hox protein, allowing different Hox proteins to select different target genes for regulation. To explore the possibility that Ftz interacts with such a DNA-binding cofactor, Yan Yu in the Pick laboratory developed a modified yeast interaction screen to isolate DNA-binding partners of Ftz that coordinately activated transcription via the *323 fPE* from the *ftz* autoregulatory upstream element (Yu et al., 1999; Pick et al., 2000). This screen identified Ftz-F1 as a partner for Ftz (Yu et al, 1997). While earlier studies (Section 3.3.1) had identified Ftz-F1–binding sites within the *323 fPE* regulatory element used for this yeast screen, the isolation of Ftz-F1 in the screen suggested that the two proteins interact in gene regulation.

As mentioned earlier, the *323 fPE* contains three Ftz-F1–binding sites and five Ftz-binding sites. One Ftz site is adjacent to an Ftz-F1 site, named F1F, with seven nucleotides between the core Ftz and Ftz-F1–binding sites (Yu et al., 1997). Gel retardation assays with bacterially expressed Ftz and Ftz-F1 proteins demonstrated that the two proteins bind F1F cooperatively. Ftz-F1 dramatically enhanced the ability of Ftz to bind this heterodimeric Ftz/Ftz-F1–binding site, as formation of a ternary complex (Ftz/Ftz-F1/DNA) was observed at a 20- to 50-fold lower concentration of Ftz than is required for Ftz binding alone. Conversely, Ftz enhanced Ftz-F1 binding to a lesser extent, roughly fivefold. Formation of the ternary complex required intact binding sites for each protein, further supporting a cooperative mechanism. Next, Ftz and Ftz-F1 were shown to form a stable, immunoprecipitable complex in wild-type *Drosophila* embryos and to synergistically activate transcription in yeast cells. Independently, studies in Henry Krause's laboratory demonstrated *in vitro* interactions of Ftz and Ftz-F1 using affinity chromatography and Far Western assays (Guichet et al., 1997). Also independently in the same year, studies in Allen Laughon's laboratory showed that Ftz regulates the expression of a downstream target gene, *engrailed* (*en*), through cooperative binding with Ftz-F1 to heterodimeric sites in an Ftz-dependent enhancer (Florence et al., 1997). This was the first demonstration of direct regulation of a downstream target gene by Ftz and Ftz-F1. The finding that Ftz-F1 raises the affinity of Ftz for heterodimeric sites provided an explanation of how Ftz selects specific target sites in the genome for regulation: the cooperative interaction with

Ftz-F1 selectively raises the affinity of Ftz for specific sites in the genome, thereby influencing binding site selection and specificity.

3.3. Ftz is required for Ftz-F1 activity in the Drosophila embryo

While the finding that Ftz-F1 raises the affinity of Ftz binding to hetero-dimeric binding sites was important for understanding Ftz-binding site selection *in vivo*, it also provided unexpected insights into Ftz-F1 function. Ftz-F1 is present in the nuclei of all somatic cells in the embryo (Fig. 2, green). Ftz-F1 is maternally deposited and thus is localized to the somatic nuclei before *ftz* expression begins. This broad expression of Ftz-F1 led to the expectation that mutations in the *ftz-f1* gene would have global effects on embryonic development. In two independent genetic screens, in the laboratories of Norbert Perrimon and Anne Ephrussi (Guichet et al., 1997; Yu et al., 1997), it was found that *ftz-f1* mutations[2] that affect maternally deposited αftz-f1 cause *ftz*-like pair-rule segmentation defects (Figs. 1 and 3). One of these was a maternal-specific allele, *ftz-f1*[209] (Guichet et al., 1997). The other—*ftz-f1*[P1598]—was isolated in the Perrimon laboratory in a screen for maternal effects of zygotic lethal alleles (Perrimon et al., 1996). The pair-rule for phenotype associated with loss-of-maternal *ftz-f1* function has since been confirmed by the isolation of additional *ftz-f1* alleles in other screens. Four maternal/zygotic lethal alleles were isolated by Luschnig et al. (2004) and two other alleles have also been characterized: *ftz-f1*[ex9] (Yamada et al., 2000; Suzuki et al., 2002) was derived from *ftz-f1*[209] and removes the region encoding the DBD. Another excision allele, *ftz-f1*[19], also generated a deletion of the DBD, and germ line clone embryos display a *ftz*-like pair-rule phenotype (Fig. 1C). These findings demonstrate that loss-of-*ftz-f1* function in the embryo affects only those domains in the embryo where Ftz and Ftz-F1 are coexpressed (Fig. 2, yellow). Thus, the earliest function of Ftz-F1 during embryonic development is entirely Ftz dependent: those regions of the embryo that express Ftz-F1 but not Ftz appear morphologically unaffected in the mutant embryos. This suggests that Ftz-F1 is unable to regulate gene expression in the absence of Ftz, at least at this stage of development. This coordinate regulation is visualized clearly by examination of the Ftz/Ftz-F1 downstream target gene *en* (Fig. 3A–D). *en* is expressed in 14 stripes in the embryo, in the primordia of the posterior portion of every segment. In *ftz* mutant embryos, alternate *en* stripes are missing; these seven stripes are those that overlap precisely with the anterior border of each *ftz* stripe (Lawrence et al., 1987). In *ftz-f1* mutant embryos, the same set of "Ftz-dependent" *en* stripes is missing, while the

---

[2]Embryos derived from homozygous *ftz-f1* mutant germ line clones using the FRT-OvoD system (Chou and Perrimon, 1996) are referred to in this text as "*ftz-f1* mutants" for the sake of simplicity.

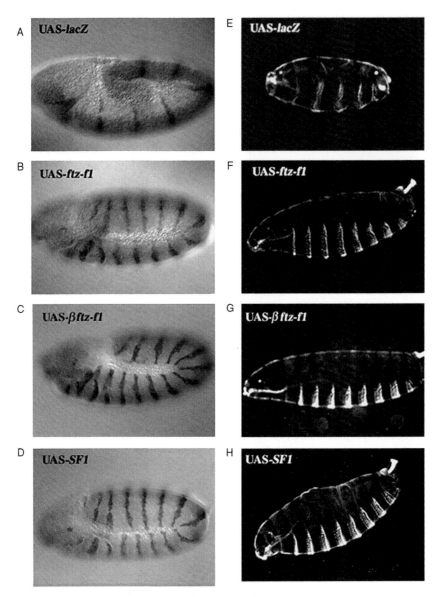

Fig. 3. Evolutionarily conserved protein domains are necessary for Ftz-F1 function *in vivo*. (A–D) Expression of En and (E–H) pair-rule defects in maternal-specific *ftz-f1²⁰⁹* mutants were unaffected by expression of *lacZ* but were rescued by expression of *ftz-f1*, βftz-f1, or the murine orthologue SF-1, using the GAL4-UAS system with an NGT40 driver (Tracey et al., 2000). (See Color Insert.)

"Ftz-independent" *en* stripes are present and correctly positioned (Fig. 3A). Studies using bioinformatics to identify new targets of Ftz/Ftz-F1 have identified additional genes coregulated by Ftz and Ftz-F1 in a similar fashion (Bowler et al., in press). Finally, *323 fPE-lacZ* fusion genes described earlier that contain binding sites for both Ftz and Ftz-F1 are expressed in seven *ftz*-like stripes in wild-type embryos, but stripes are missing in either *ftz* or *ftz-f1* mutants (Yussa et al., 2001). Other transcription factors that bind to the Ftz-F1 sites in the *323 fPE* (Ttk and Adf-1; see Section 3.3.1) are not functionally redundant in this situation because they fail to interact with Ftz.

In sum, Ftz-F1 is present in all somatic nuclei before Ftz is expressed, but it does not activate target genes, such as *en* or the *323 fPE-lacZ* fusion gene, before Ftz is expressed. Later, when Ftz is expressed in stripes, Ftz-F1 remains ubiquitous (Fig. 2, green), but it activates target genes only in those cells also expressing Ftz (Fig. 2, yellow). Thus, it appeared that Ftz serves as a temporally and spatially limiting cofactor required for target gene activation. How does Ftz modulate Ftz-F1 function? One possibility is that Ftz-F1 requires Ftz for DNA-binding site selection in the embryo, just as Ftz-F1 is required for Ftz-binding site selection. Ftz raises the affinity of Ftz-F1 for heterodimeric sites approximately fivefold in gel retardation assays. However, Ftz-F1 alone binds DNA with high affinity and specificity as a monomer and Ftz-F1 proteins from different species activate gene expression strongly, without the presence of added protein partner, in a number of cellular systems (Lala et al., 1992; Tsukiyama et al., 1992; Ueda et al, 1992; Ayer et al., 1993; Honda et al., 1993; Ohno et al., 1994; Ellinger-Ziegelbauer et al., 1994; Halvorson et al., 1996; Shapiro et al., 1996; Han et al., 1998; Nachtigal et al., 1998). This family is thus often considered a constitutively active group of NRs. These findings do not strongly support the model that Ftz-F1 requires a cofactor for DNA target site selection *in vivo*, although they do not rule out this mechanism.

An alternate, although not mutually exclusive, model was raised by the observation that Ftz has an evolutionarily conserved LRALL sequence at its N-terminus (Brown et al., 1994), as this sequence conforms to the LXXLL sequence utilized by NR coactivators to contact the AF-2 domains of NRs (Darimont et al., 1998; McInerney et al., 1998; Nolte et al., 1998). Biochemical experiments confirmed that the LRALL sequence of Ftz binds directly to the AF-2 motif at the C-terminus of Ftz-F1. In yeast two-hybrid assays, an N-terminal fragment of Ftz, including this LRALL sequence, bound full length Ftz-F1 via its C-terminal AF-2 domain. Mutation of the LRALL sequence to LRA<u>AA</u> dramatically reduced interaction with Ftz-F1 (>2000 vs 170 units) and Ftz LRA<u>AA</u> failed to interact with a Ftz-F1 C-terminal peptide (Yussa et al., 2001). Similarly, Schwartz et al. (2001) reported that the Ftz LRALL and the Ftz-F1 AF-2 domain interact in Far Western assays. Finally, Ftz protein carrying the LRA<u>AA</u> mutation is severely inhibited in its ability to carry out segmentation functions in *Drosophila* embryos (Suzuki et al., 2002; Lohr and Pick, 2005). These results indicate that there is a direct and functional interaction between the AF2 domain of Ftz-F1 and the LXXLL motif of Ftz. Thus, Ftz may function as an NR coactivator, relieving inhibition imposed on Ftz-F1 by corepressors (Fig. 2B).

According to this model, Ftz-F1 is nuclear localized and possibly bound to DNA prior to Ftz expression in stripes and in all cells, including those that never express Ftz. DNA-bound Ftz-F1 recruits corepressors that silence its transcriptional potential. As Ftz becomes expressed in stripes, it binds the AF-2 domain of Ftz-F1, mimicking a classical NR coactivator and displacing corepressors. Together Ftz and Ftz-F1 activate transcription of downstream target genes, such as *en*, to promote the development of alternate body segments in the embryo. Evidence has been presented that both Ftz (Colgan et al., 1993) and Ftz-F1 (Li et al., 1999) can directly contact components of the basal transcription machinery; however, it is equally likely that Ftz/Ftz-F1 engage additional coactivators or bridging proteins (such as MBF-1, see later) to activate target gene expression. This model relies on the well-established corepressor–coactivator exchange that regulates NR function in most meta-zoans (reviewed in Glass and Rosenfeld, 2000). Although Ftz-F1 corepressors have not yet been identified in *Drosophila*, several candidates have been identified, including Alien, which was shown to bind Ftz-F1 in GST pull-down assays (Dressel et al., 1999), and SMRTER, which binds EcR (Tsai et al., 1999). The interaction of Ftz-F1 with Ftz protein replaces the need for ligand-induced release of corepressors, as Ftz binding functions as an ON versus OFF switch for Ftz-F1 function. Since Ftz can contact Ftz-F1 through both its homeodomain and its LXXLL motif, it is possible that cooperative binding to DNA utilizing the homeodomain induces a conforma-tional change in Ftz-F1 that exposes its AF2 domain for interaction with the Ftz LXXLL motif.

3.4. *Ftz-F1 promoted the evolution of arthropod Ftz from an ancestral homeotic gene to a segmentation gene in* Drosophila

Drosophila ftz is a *Hox* gene located within a cluster of *Hox* genes (HOM-C) that are thought to have evolved by duplication and diversification (reviewed in Akam, 1995). Most of the genes in HOM-C have homeotic function, defined as the ability to transform one body part into an alternate body part. A small number of HOM-C genes have nonhomeotic function, such as *ftz*, which is required for segmentation (see earlier). Lohr et al. (2001) found that *ftz* genes isolated from other insects—a grasshopper (Sg-Ftz) and a beetle (Tc-Ftz)—have homeotic function when misexpressed in *Drosophila*, whereas *Drosophila ftz* is capable only of segmentation function (Lohr et al., 2001). Tc-Ftz also possessed segmentation activity when expressed in *Drosophila* embryos, while Sg-Ftz, the more distantly related *ftz* gene, had marginal segmentation potential. These studies suggested that *ftz* genes from divergent insects retain an ancestral homeotic function, while in more derived insects, *ftz* has taken on a role in segmentation. The switch in Ftz function within the insects correlates with the acquisition of the LRALL sequence that mediates strong interaction with Ftz-F1. *In vitro, Drosophila* Ftz and Tc-Ftz both interacted with Ftz-F1 much more strongly than did Sg-Ftz (Lohr et al., 2001). In addition, Sg-Ftz and Tc-Ftz retain a YPWM motif present in other homeotic Hox proteins that mediates interaction with a different cofactor, Extradenticle (Exd) (reviewed in Mann and Chan, 1996). This observation led to the hypothesis that a switch in cofactor interaction, from one with Exd to one with Ftz-F1, drove the evolution of Ftz from a homeotic to a segmentation protein. Gain- and loss-of-function mutations were then generated in both Dm-Ftz and Tc-Ftz to test this hypothesis (Lohr and Pick, 2005). These experiments confirmed that the presence of the LXXLL and strong interaction with Ftz-F1 are necessary and sufficient to explain the switch in Ftz function from an exclusive role in homeosis to one in segmentation per se (Lohr and Pick, 2005). Loss of the YPWM motif from *Drosophila ftz* resulted in loss of homeotic potential, generating a protein that functions exclusively in segmentation. Thus, the interaction of the LXXLL motif was a gain-of-function mutation that occurred during evolution and allowed a new interaction between a Hox protein and an NR, thereby qualitatively switching the function of this Hox protein.

4. βFtz-F1 regulates metamorphosis

The late form of Ftz-F1, βFtz-F1, is expressed during late embryogenesis and then again in larvae and prepupae. Both expression studies and analysis of *βftz-f1* mutant phenotypes suggest that βFtz-F1 is required for metamorphosis, functioning as a competence factor for ecdysone responses (Woodard et al., 1994; Broadus et al., 1999). The major transitions toward development of the *Drosophila* adult are the development and molting of three larval instars followed by pupal formation (reviewed in Riddiford, 1993). Each of these transitions is hormonally triggered via pulses of the steroid hormone ecdysone. Ecdysone binds to a heterodimeric NR composed of EcR and USP, which induces a cascade of gene expression, including the *BR-C, E74A, E75A,* and *E93* genes (reviewed in King-Jones and Thummel, 2005). βFtz-F1 is expressed toward the end of embryogenesis, with protein detectable by Western analysis between 16- and 19-hours AEL. It is then expressed transiently during the remainder of fly development with pulses of expression occurring before each larval ecdysis and during the mid-prepupal period, at a time when ecdysone levels are low (Woodard et al., 1994; Yamada et al., 2000). This timing of expression provided a clue to βFtz-F1 function; studies of Ashburner and Richards (Ashburner and Richards, 1976; Richards, 1976a,b) had predicted the existence of a competence factor for the prepupal response to ecdysone that is necessary for the induction of specific polytene chromosomal puffs by the hormone. Specifically, while the process of pupariation is initiated by high titers of ecdysone at late larval stages, competence to respond to this ecdysone pulse at the end of the prepupal stage was shown to require both a time period when ecdysone levels are low and protein synthesis. βFtz-F1 expression is induced precisely during this time window when ecdysone levels fall. Further, ectopic βFtz-F1 expression induced higher levels of early ecdysone-response gene expression in late third instar larval salivary glands than are seen in the absence of βFtz-F1, and βFtz-F1 misexpressed at this stage also prematurely induced E93, an ecdysone-responsive gene normally only expressed following the prepupal rise in ecdysone 12 hours later (Woodard et al., 1994). Finally, βFtz-F1 repressed its own transcription, and is also repressed by ecdysone, possibly explaining its own transient expression throughout development. The hypothesis that Ftz-F1 is required for morphogenesis, functioning as the competence factor during the lull in ecdysone levels, was strongly supported by analysis of βftz-f1 mutant phenotypes (Broadus et al., 1999; Yamada et al., 2000). Broadus et al. (1999) examined a ftz-f1 hypomorph able to survive to pupal formation and found that the majority of these pupae failed to develop to adults; rather, they showed numerous defects in the prepupal to pupal transition, including incomplete leg development, failure of histolysis of salivary glands, and submaximal induction of the ecdysone-induced cascade of genes BR-C, E74A, E75A, and E93, which normally peaks in late prepupae. In addition, there is failure to induce the cell death gene diap2 (Jiang et al., 2000). Further analysis indicated that βFtz-F1 functions in the salivary gland to initiate a process of autophagic cell death, which is integral to the process of metamorphosis (Lee and Beahrecke, 2001). βFtz-F1 was shown to be an indirect activator of cell death here, via its induction of E93, as E93 was required for Ftz-F1–induced cell death and for the transcription of other genes directly involved in programmed cell death (Lee et al., 2002). The presence of Ftz-F1–binding sites surrounding E93 suggests that βFtz-F1 may directly regulate E93 in this scenario.

In a separate study, Ueda's group used a genetic trick to rescue ftz-f1–associated lethality by expressing βftz-f1 as a heat inducible transgene and was thus able to assess its later functions in larval and pupal development (Yamada et al., 2000). These experiments demonstrated that βFtz-F1 is required at each larval molt as well as for the prepupal to pupal transition. Arrested development was observed in most prepupal tissues, corresponding to their finding that βFtz-F1 is expressed in all somatic tissues examined at the time frames it is expressed during the larval and pupal development. They further showed that premature expression of βFtz-F1 is lethal, in keeping with the earlier finding that premature βFtz-F1 expression ectopically induces E93 (Woodard et al., 1994). βFtz-F1 was also shown to directly regulate the pupal cuticle gene *EDG84A* (Murata et al., 1996) Thus, βFtz-F1 regulates other regulatory genes involved in ecdysis, such as *E74*, *E75*, and *E93*, but also structural genes that are required for pupal formation. It has been shown that βFtz-F1 regulates the expression of genes involved in ecdysone biosynthesis in the ring gland (Parvy et al., 2005). This finding is particularly interesting in light of the fact that mammalian Ftz-F1, steroido-genic factor-1 (SF-1), was isolated as a regulatory of cytochrome P450 steroid hydroxylases (Parker et al., 2002; see later).

A role for βFtz-F1 in ecdysis appears to be conserved in other Diptera, as the mosquito, *Aedes aegypti* βFtz-F1 also appears to function as a competence factor (Li et al., 2000). βFtz-F1 is expressed with a time-course consistent with this role. RNAi knockdown experiments demonstrated a requirement for βFtz-F1 in activation of target genes such as *E75* and the *vitellogenin* gene (*Vg*). Further evidence from mosquito suggested that competence to respond to ecdysone is a result of Ftz-F1 fat body expression, which is regulated posttranscriptionally by juvenile hormone (Li et al., 2000; Zhu et al., 2003).

Taken together, these findings suggest that βFtz-F1 expression is tightly controlled at the transcriptional level to mediate appropriate responses to ecdysone by regulating the expression of genes directly involved in metamor-phosis. Once expressed, however, βFtz-F1 is not sufficient to activate its putative direct target genes (e.g., *BR-C*, *E74A*, *E75A*, *E93*, and *EDG84A*) (Murata et al., 1996; Kawasaki et al., 2002). Rather, high titers of ecdy-sone are required, suggesting that transcriptional activation of target genes requires the coordinate action of βFtz-F1 and the EcR/USP, or another factor induced or present in the same time window. There is no evidence that βFtz-F1 itself responds directly to ecdysone and no other hormonal pulse is known to correlate with the activity of βFtz-F1. Therefore, it is likely that, as for αFtz-F1 in the embryo, βFtz-F1 activity is released by a protein–protein interaction. Candidate partners in this scenario are EcR/USP itself, Bonus, an NR cofactor required for molting (Beckstead et al., 2001), and Rigor mortis, which binds both isoforms of Ftz-F1 in GST pull-down assays and which is required for metamorphosis, displaying phenotypes similar to EcR and *βftz-f1* mutants (Gates et al., 2004).

In sum, *Drosophila* Ftz-F1 functions as a transcription factor for seemingly unrelated physiological functions in *Drosophila*: segmentation during embry-onic development and ecdysis and regulation of programmed cell death at pupal stages. What these events share is that Ftz-F1 is necessary but not sufficient for activation of target genes. In both cases, the expression of Ftz-F1 precedes—with a distinct time lag—the activation of target genes, thus impart-ing competence for transcription activation without actually carrying it out. While this lag—the time between expression of nuclear Ftz-F1 and activation of target genes—could in theory be explained by a ligand-induced switch, a protein partner that participates in transcription activation seems to better explain both the early and late functions of *Drosophila* Ftz-F1.

5. Ftz-F1 is highly conserved throughout the animal kingdom

5.1. Evolution of the Ftz-F1 family

The evolution of NRs and the *ftz-f1* group in particular have been studied extensively by Laudet's group (Laudet, 1997; Escriva et al., 2000, 2004; Bertrand et al., 2004; Laudet and Bonneton, 2005). *ftz-f1* genes have been found in a broad range of metazoans in both protostomes and deuterostomes (Fig. 4). All major phyla harbor *ftz-f1* genes. A candidate *ftz-f1* gene was found in the fluke *Schistosoma mansoni*, a lophotrochozoan (De Mendonca et al., 2002; Bertin et al., 2004). This gene appears to be more closely related to *Drosophila DHR39* than *ftz-f1* (Fig. 4; De Mendonca et al., 2002; Bertin et al., 2004). *ftz-f1* genes are well represented throughout the ecdysozoa. This includes *C. elegans* (Sluder et al., 1999; Asahina et al., 2000; Gissendanner and Sluder, 2000) and representative arthropods, including insects, with *Drosophila ftz-f1* being the founding member of the family. Other insects harboring *ftz-f1* genes include other Diptera, such as mosquito (Li et al., 2000; Zhu et al., 2003), and divergent insect groups such as Lepidoptera [*Bombyx mori* (Ueda and Hirose, 1991), *Manduca sexta* (Weller et al., 2001)] and Hymenoptera (Hepperle and Hartfelder, 2001). A *ftz-f1* gene was also found in a crustacean *Metapenaeus ensis* (shrimp), where it is expressed in mature oocytes and larvae (Chan and Chan, 1999). The *ftz-f1* family is found broadly in deuterostomes, where many members have been identified. This includes mammals, which harbor two related NRs, SF-1 and LRH-1 (see later); frogs (Ellinger-Ziegelbauer et al., 1994, 1995; Takase et al., 2000); and various fish (Liu et al., 1997; reviewed in von Hofsten and Olsson, 2005). In zebrafish, four *ftz-f1* family genes have been found; *ff1a* belongs to the *NR5A2* (*LRH-1*) group, while *ff1b* and *ff1d* appear to have arisen as duplications within a teleost lineage from an *NR5A1* (*SF-1*) family ancestor (Kuo et al., 2005; Fig. 4). *ff1c* does not align well with any mammalian family member and presumably arose after the divergence of mammals and fish (Kuo et al., 2005).

In many cases, these orthologues were first found as DNA-binding activities in cellular extracts, as for *Drosophila*, although others have been identified by homology. Among the earliest identified Ftz-F1 orthologues were those from the silkworm, *Bombyx mori* (BmFtz-F1, Ueda and Hirose, 1990); *Xenopus* (Ellinger-Ziegelbauer et al., 1994), and rodents. The latter include the independent identification of embryonal long terminal repeat-binding protein (ELP) (Tsukiyama et al., 1992), SF-1 (Lala et al., 1992), and Ad4BP (Honda et al., 1993). The broad representation of *ftz-f1* genes across phyla suggests that Ftz-F1 is an old NR, perhaps lying at the base of metazoan NRs (Bertrand et al., 2004; Escriva et al., 2004; Laudet and Bonneton, 2005). Although its origin likely predates the split between protostomes and deuterostomes, it has been proposed that the primordial function of *ftz-f1*

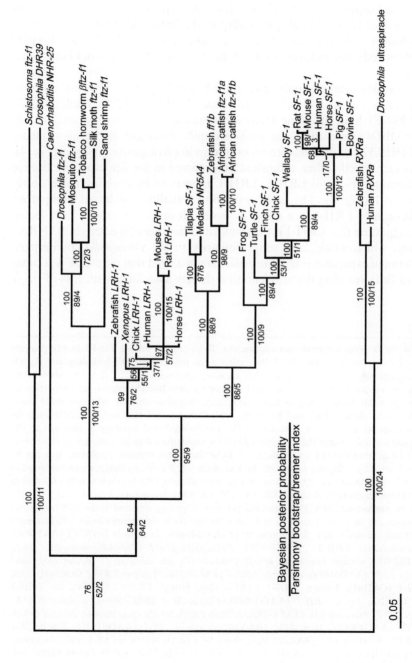

Fig. 4. (*continued*)

in ecdysozoa was in molting, as it plays roles in these processes in *Drosophila* as well as *C. elegans* (Asahina et al., 2000; Gissendanner and Sluder, 2000). A second dipteran, the mosquito *Aedes aegypti*, also utilizes *ftz-f1* in this process (see later; Li et al., 2000; Zhu et al., 2003) and the honey bee *ftz-f1* was shown to be responsive to ecdysteroids, suggesting that it may also play a role in ecdysone cascades in this hymenopteran (Hepperle and Hartfelder, 2001).

Drosophila harbors only one Ftz-F1 family gene, with two isoforms (Fig. 1) as well as a closely related gene, *DHR39* (*NR5B*). While DHR39 and Ftz-F1 bind the same DNA sequences and share strong similarity in their DBDs (see earlier), the genes are quite divergent in overall sequence and DHR39 clearly falls into an outgroup in phylogenetic analysis (Fig. 4).

As mentioned earlier, a duplication occurred in the deuterostome lineage to generate SF-1 and LRH-1. LRH-1 appears to be evolving more slowly than SF-1 (Fig. 4). The early embryonic lethality and broad, ubiquitous expression of LRH-1 in a range of tissues during development have led to the suggestion that LRH-1 is the primordial member of the vertebrate NRA5 family, which would bear most similarity to the *Drosophila* counterpart. Functional experiments in *Drosophila* demonstrated that mouse SF-1 is capable of fully rescuing the segmentation defects associated with loss-of-*ftz-f1*

Fig. 4. *ftz-f1* genes are evolutionarily ancient. Relationships among Ftz-F1 family proteins derived from Bayesian analysis of 293 amino acids representing DNA-binding and ligand-binding regions. Numbers above branches are Bayesian posterior probabilities; numbers below branches are parsimony bootstraps and Bremer indices. Branch lengths are proportion to Bayesian mean branch lengths (note scale bar). Bayesian analysis was conducted using MrBayes, 3.1 (Huelsenbeck and Ronquist, 2001; Ronquist and Huelsenbeck, 2003) using the Poisson model and default parameters. The parsimony-based topology is identical to the Bayesian topology except that *C. elegans* Ftz-F1 is reconstructed as the sister group to all SF-1, LRH-1 + arthropod Ftz-F1 sequences; otherwise the strict-consensus parsimony tree can be reconstructed by collapsing nodes with Bremer index = 0. Parsimony analysis was conducted in PAUP*, ver. 4 (Swofford, 2002) using 1000 random-addition TBR heuristic searches resulted in 40 minimal-length trees (length, 1326; CI, 0.76, RI, 0.81). Bootstraps derived from 1000 reps with 10 random-addition TBR searches per rep. Topology depicted is an MP tree; strict consensus can be reconstructed by collapsing the one node with Bremer index = 0. Terminals (GenBank Accession no): *Schistosoma ftz-f1* (AAG49449), *Drosophila DHR39* (AAA28464), *Caenorhabditis NHR-25* (AAF67038), *Drosophila ftz-f1* (P33244), mosquito *ftz-f1* (AAF82307), tobacco hornworm *βftz-f1* (AAL50351), silk moth *ftz-f1* (BAA01745), sand shrimp *ftz-f1* (AAD41899), zebrafish *LRH-1* (AAC60274), *Xenopus LRH-1* (A56543), chick *LRH-1* (O42101), human *LRH-1* (AAD03155), mouse *LRH-1* (S27874), rat *LRH-1* (NP_068510), horse *LRH-1* (AAG35649), tilapia *SF-1* (BAC75890), medaka *NR5A4* (BAA32394), zebrafish *ff1b* (AAF43283), African catfish *ftz-f1*a (AAG49004), African catfish *ftz-f1*b (AAG49005), frog *SF-1* (BAA36789), turtle *SF-1* (AAD01975), Finch *SF-1* (AAK97659), chick *SF-1* (BAA22839), wallaby *SF-1* (AAK94918), rat *SF-1* (P50569), mouse *SF-1* (A40716), human *SF-1* (AAH32501), horse *SF-1* (AAG35648), pig *SF-1* (AAC64209), and bovine *SF-1* (Q04752).

function in the early embryo (Fig. 3). Thus, despite the evolutionary divergence, key regulatory properties of the two proteins—Ftz-F1 and SF-1—have been preserved. Although LRH-1 has been shown to bind Ftz *in vitro* (Schwartz et al., 2001), rescue experiments will be necessary to determine if it also functions in a Ftz-F1-like fashion when expressed in *Drosophila*.

5.2. ftz-f1 *genes play diverse roles in metazoan development*

5.2.1. C. elegans ftz-f1 *is required for embryonic development and molting*

The *C. elegans* orthologue of *ftz-f1*, *nrh-25*, is the only other invertebrate *ftz-f1* gene for which genetic analysis has been carried out. The gene is closely related to *Drosophila ftz-f1*, with 80% identity in the DBD, but only 31% identity in the LBD. Mutations and/or RNAi experiments by three independent groups demonstrate conclusively that *nhr-25* mutations cause embryonic lethality, with defects in various epidermal/hypodermal cell types (Asahina et al., 2000; Gissendanner and Sluder, 2000; Chen et al., 2004b). These defects appear to reflect failure of cell–cell fusion, suggesting a role for *nhr-25* in controlling cell junctions or fusion (Chen et al., 2004b). RNAi, hypomorphic escapers, and/or partial rescue of a deletion allele were used to assess later functions of *nrh-25* by these groups. Adult escapers in these experiments were often sterile, displaying multiple defects in somatic gonad and male tail development, including excess proliferation of gonad cells. Additional defects were noted during larval development, which are reminiscent of the role of *βftz-f1* in *Drosophila*; specifically, these animals failed to molt, dying as L1–L2 larvae. This defect appeared to be due to incomplete cuticle synthesis and failure to shed old cuticle (Asahina et al., 2000; Gissendanner and Sluder, 2000). Thus, it was proposed that regulation of genes involved in molting is the primordial function of Ftz-F1 proteins in the ecdysozoa (Asahina et al., 2000; Gissendanner and Sluder, 2000). However, it is also of interest that *C. elegans* Ftz-F1 plays a role in somatic gonad development, as does the mammalian orthologue (see later).

5.2.2. Zebrafish Ftz-F1 *plays a role in sex determination*

Zebrafish have four *ftz-f1* genes, *ff1a–d* (Liu et al., 1997; Chai and Chan, 2000; Lin et al., 2000; Kuo et al., 2005). The developmental expression patterns of each gene are distinct, suggesting unique roles for each family member. *ff1a* includes four variants resulting from alternative promoter utilization and alternative splicing, and its expression pattern is suggestive of roles in steroidogenesis and gonadogenesis (von Hofsten et al., 2001, 2005). A *ff1* gene from Arctic char showed a cyclical expression pattern during sexual maturation, possibly supporting this link (von Hofsten et al., 2003). *ff1b* and *ff1d* are closely related genes. Antisense knockdown experiments indicated a role for *ff1b* in the development and maturation of a steroidogenic tissue, the interrenal organ, paralleling *SF-1* function in mammals (Chai et al., 2003; Liu et al., 2003b). Further, in cooperation with the homeodomain protein Prox1 (see later), Ff1b regulates the maturation of the interrenal organ (Kuo et al., 2005). ff1d expression in embryos overlaps with that of ff1b; in adults, ff1d is expressed in the brain, gonads, and liver, and it has been suggested that ff1d could regulate AMH, as does mammalian SF-1 (von Hofsten et al., 2005). ff1c is expressed only in adults but functional data is not yet available. Further, a model proposes that *ftz-f1* genes in fish play an important role in sex determination, by regulating both development of urogenital tissues and steroidogenesis (von Hofsten and Olsson, 2005).

5.2.3. Mammalian Ftz-F1 orthologues regulate steroidogenesis, cholesterol metabolism, and sexual differentiation

The best characterized member of the *ftz-f1* family is steroidogenic factor-1 (SF-1, NR5A1) which was originally identified and purified as a DNA-binding protein that regulates the expression of cytochrome P450 steroid hydroxylases (Rice et al., 1990, 1991; reviewed in Parker et al., 2002). Cloning of the cDNA encoding this activity revealed homology to *Drosophila* Ftz-F1, and thus, the gene was named *Ftz-F1* (Lala et al., 1992). Independently, a DNA-binding protein found in undifferentiated mouse embryonal carcinoma cells was found and named ELP (Tsukiyama et al., 1992), and Ad4BP, a bovine regulator of steroidogenic genes, was also identified (Honda et al., 1993). All of these turned out to be isoforms encoded by the *FTZ-F1* gene (Ikeda et al., 1993). ELP is represented by three isoforms (ELP1–3), and Ad4BP is identical to SF-1 and has its initiator methionine downstream of the ELP ATG (Ikeda et al., 1993; Ninomiya et al., 1995). SF-1 has been shown to regulate a large number of genes including those encoding steroidogenic cytochrome P450 enzyme genes, such as *CYP17* (Bakke and Lund, 1995), *CYP11A* and *-B* (Morohashi et al., 1993), the cholesterol side-chain cleavage cytochrome P450 (CYP11A; P450scc) (Clemens et al., 1994), cytochrome P450c17 (Zhang and Mellon, 1996), and other genes involved in steroidogenesis such as the steroidogenic acute regulatory protein (StAR) (Sugawara et al., 1996), mullerian inhibitory substance (MIS) (Shen et al., 1994), the β-subunit of luteinizing hormone (Halvorson et al., 1998), aromatase (Lynch et al., 1993; Michael et al., 1995), and the NR Dax-1 (Yu et al., 1998). Evidence for direct regulation by SF-1 was initially based on colocalization and *in vitro* or cell culture assays for most of these genes, and many of these interactions have been verified in whole animals (see later).

Comprehensive analysis in Keith Parker's laboratory and others has elucidated the function of SF-1 during development. *SF-1* is expressed in the developing adrenal gland and gonads of both males and females and in the primordia of the endocrine hypothalamus and anterior pituitary gland (Ikeda et al., 1994). Mouse knockout experiments revealed a requirement for SF-1 in the development of the adrenal gland and gonads. *SF-1* mutants died after birth lacking adrenal glands, and both internal and external urogenital tracks of female and male mice displayed female character (Luo et al., 1994; Sadovsky et al., 1995). Treatment of newborn mice with corticosteroids prolonged the survival of these mice, suggesting that hormone deficiency is the cause of early postnatal lethality (Luo et al., 1995). In addition, these mice failed to express gonadotrope-specific genes (Ingraham et al., 1994) and lacked the ventromedial hypothalamic nucleus (VMH) (Ikeda et al., 1995; Shinoda et al., 1995). SF-1–deficient mice that were partially rescued by corticosteroid injection to adulthood displayed late-onset obesity, resulting from decreased activity levels, a result consistent with older studies that had suggested a role for the VMH in weight control (Majdic et al., 2002). In a conditional knockout experiment, it was shown that pituitary expression of SF-1 is required for sex steroid production, explaining the failure of secondary sexual development and sterility in these mutants (Zhao et al., 2004). In humans, three cases of XY sex reversal have been reported in the literature due to different mutations in SF-1 (Achermann et al., 1999, 2001b, 2002; Correa et al., 2004). One of these appears to be a dominant negative protein truncated before the AF-2 domain (Correa et al., 2004).

In both mice and human patients, dose-sensitive effects of SF-1 have been observed. Mice that are heterozygous for *FTZ-F1* (SF-1) loss-of-function mutations display defects in adrenal development, with small adrenal glands (hypoplasia) and adrenal insufficiency caused by loss of adrenal precursor cells. These mice also display impaired responses to a variety of stresses (Bland et al., 2000, 2004). There appears to be a compensation mechanism that raises expression of SF-1 target genes and steroidogenic capacity and results in cellular hypertrophy in the gland such that, although development is abnormal, function is close to normal in adult animals (Bland et al., 2000; Babu et al., 2002; Bland et al., 2004). Finally, increased dosage of SF-1 was observed in a cadre of patients displaying childhood adrenocortical tumors (Figueiredo et al., 2005), suggesting a role for SF-1 in cell proliferation, and consistent with the hypoplasia seen in mice heterozygous for the *FTZ-F1* loss-of-function mutation.

LRH-1/FTF (NR5A2) was first identified in mouse on the basis of its ability to activate transcription of the α1-fetoprotein gene (Galarneau et al., 1996). As for the SF-1 family, LRH-1/FTF orthologues are found in a range of vertebrates, including chicken, fish, frogs, and humans (reviewed in Fayard et al., 2004; Fig. 4). Also like SF-1, LRH-1/FTF was independently isolated in different laboratories on the basis of its ability to bind to DNA and/or activate transcription of a number of genes, including α-fetoprotein (Galarneau et al., 1996); HNF3β, -4α, and -1α (Pare et al., 2001); aromatase; cholesterol 7α-hydroxylase (Nitta et al., 1999; del Castillo-Olivares et al., 2004); and others, and was thus independently named pancreas homologue receptor 1, α-fetoprotein transcription factor (FTF), human B1-binding factor, or CYP7A1 promoter-binding factor (reviewed in Fayard et al., 2004).

LRH-1/FTF is expressed not only in tissues of endodermal origin, such as liver and intestine, but also in the ovary. Targeted inactivation of LRH-1/FTF resulted in embryonic lethality at embryonic days 6.5–7.5 (E6.5–E7.5); these embryos display defects suggestive of visceral endoderm dysfunction (Pare et al., 2004). This suggests a critical role for LRH-1/FTF in embryonic development, the details of which remain to be worked out.

The role of LRH-1/FTF has been more extensively studied in the adult, where its role in regulating cholesterol homeostasis and bile acid metabolism has been characterized in some detail (Lu et al., 2000; reviewed in Fayard et al., 2004). LRH-1/FTF also appears to play a role in directly regulating cell cycle and thus potentially controlling cell proliferation (Botrugno et al., 2004; Schoonjans et al., 2005). In this capacity, LRH-1/FTF controls renewal of intestinal crypt cells by functioning as a tissue-specific coactivating partner of β-catenin (Botrugno et al., 2004). An intriguing role for LRH-1/FTF in breast cancer development came from studies of the transcriptional regulation of the *CYP19* gene, which encodes aromatase cytochrome P450, an enzyme responsible for estrogen biosynthesis (reviewed in Simpson et al., 2002). It was found that aromatase is upregulated in adipose tissue surrounding breast tumors, and it was proposed that this increases local estrogen levels, which stimulates tumor growth and progression (reviewed in Clyne et al., 2004). LRH-1/FTF was shown to be an activator of aromatase expression in the human breast pre-adipocytes, whereas SF-1, which was also capable of activating this promoter, was not coexpressed with aromatase in these cells (Clyne et al., 2002).

5.3. Posttranslational modifications influence SF-1 activity

A number of reports have suggested that the activity of mammalian SF-1 is regulated by phosphorylation (Hammer et al., 1999; Aesoy et al., 2002), acetylation (Jacob et al., 2001), and sumoylation (Chen et al., 2004a, Lee et al., 2005). The function of these modifications were elucidated in cell culture assays that measured transcription activation potential of SF-1. Phosphorylation of a Ser in the hinge region, between the DBD and LBD, increased transcriptional activity of SF-1, apparently through a MAPK signaling pathway (Hammer et al., 1999). In addition, multiple SF-1 protein species were detected in mouse adrenal extracts (Babu et al., 2000). A different group suggested that phosphorylation influences SF-1 function by increasing its stability (Aesoy et al., 2002). Although this Ser residue is conserved in *Drosophila* Ftz-F1, it is not embedded within a MAPK consensus site. However, Ftz-F1 does contain two other potential MAPK consensus site Ser's in its hinge region that may serve similar roles; their function *in vivo* remains to be determined.

The hinge region of SF-1 includes a repression domain, mapped in transcription assays in cell culture (Ou et al., 2001), which has been shown to be a target of sumoylation. Sumoylation repressed transcription, by altering protein-partner interactions (Komatsu et al., 2004; Lee et al., 2005) and/or by localizing SF-1 to nuclear speckles (Chen et al., 2004a). The two sumoylation sites in SF-1 are not conserved in *Drosophila* Ftz-F1, but two consensus sumoylation sites (Melchior, 2000) are present: one in the hinge region and one at the C-terminus of the DBD. The function of these sites in whole animals, either *Drosophila* or mammals, remains to be determined.

6. Ftz-F1 activity is modulated by novel protein–protein interactions

Like other NRs, the Ftz-F1 family proteins have been found to interact with NR coactivators such as CBP/p300 (Monte et al., 1998), SRC-1 (Ito et al., 1998), RIP140 (Sugawara et al., 2001), TIF2 (Borud et al., 2002), and components of the general transcription machinery such as TFIIB (Li et al., 1999). In addition, members of the Ftz-F1 family are also regulated by interaction with DNA-binding transcription factors. For instance, *Drosophila* Ftz-F1 activity is regulated by interaction with the Hox protein Ftz. SF-1 was shown to physically interact with c-Jun, via the Ftz-F1 box, and SF-1 and c-Jun coactivated transcription in cell culture (Li et al., 1999). SF-1 also interacted with Egr-1, a zinc finger transcription factor to synergistically activate LHβ (Halvorson et al., 1998), and SOX9 to regulate expression of MIS (AMH) (De Santa Barbara et al., 1998) and the testis-specific *vanin-1* gene (Wilson et al., 2005). Other partners are discussed in more detail later.

6.1. Ftz-F1 protein interactions with homeodomain proteins

Drosophila Ftz-F1 interacts directly with the homeodomain-containing protein Ftz to regulate embryonic development. Since that observation, Ftz-F1 family proteins from each major clade have been found to interact with homeodomain proteins.

6.1.1. C. elegans *nhr-25 interacts with LIN-39 and*
NOB-1 in vivo
Min Han's group has shown direct physical and genetic interactions between NHR-25 and two Hox proteins, LIN-39 and NOB-1, which display phenotypes similar to *nhr-25* mutants (Chen et al., 2004b). These interactions were specific both *in vitro* and *in vivo*, as other Hox proteins tested failed to interact. Neither of these Hox proteins possess an LXXLL motif, thought to be important for transcription activation by the Ftz/Ftz-F1 heterodimer in the *Drosophila* embryo (see earlier). The interaction of NHR-25 with Hox proteins is critical for function during development; interaction with LIN-39

is necessary for vulval cell differentiation, while interaction with NOB-1 is thought to be critical during embryogenesis (Chen et al., 2004b).

6.1.2. Zebrafish Ff1 interacts with Prox1

The zebrafish Ff1 proteins were found to interact with Prox1, a prospero-type homeodomain protein, both *in vitro* and *in vivo* (Liu et al., 2003b). Like *Drosophila* Ftz, Prox1 has NR coactivator potential in that it contains LXXLL motifs. Unlike the situation for *Drosophila* Ftz/Ftz-F1, Prox1 appears to inhibit transcriptional activation by Ff1 proteins, and also by SF-1, when proteins are coexpressed in zebrafish embryos (Liu et al., 2003b). In the future, it will be of interest to compare the detailed mechanisms of regulation underlying repression versus activation by NR/Hox protein partner pairs.

6.1.3. Prox1 interacts with LRH-1

As its zebrafish counterpart, mammalian LRH-1 was also found to inter-act with the homeodomain-containing protein, Prox1 (Steffensen et al., 2004). Also as for the *Drosophila* case, this interaction was first detected in a yeast two-hybrid screen (Johansson et al., 1999). LRH-1 and Prox1 were then shown to colocalize in the liver, pancreas, and other tissues in mouse and human. In cellular transfection assays and using GAL4 fusion proteins, it was shown that Prox1 functions as a repressor, inhibiting the transcription activity of LRH-1, most likely via recruitment of HDAC3 (Steffensen et al., 2004). Prox1 contains a single LXXLL motif (Oliver et al., 1993), as does *Drosophila* Ftz. However, in the case of the Ftz/Ftz-F1 interaction, this motif appears to contribute to transcriptional activation, whereas in the case of LRH-1/Prox1, a repressor domain in the N-terminus of the protein results in repression by the partner pair (Steffensen et al., 2004). This is consistent with similar findings in the zebrafish model (see earlier). Despite the difference in outcome (i.e., repression vs activation), the LRH-1 LBD interacted with *Drosophila* Ftz in Far Western assays (Schwartz et al., 2001).

6.1.4. SF-1 interacts with Ptx-1 to activate
POMC expression

One of the early observations of a regulatory protein–protein interaction for the Ftz-F1 family came from the work of Jacque Tremblay, Jacques Drouin, and others studying Ptx-1. Ptx-1 (Pituitary homeobox 1) is a homeodomain-containing transcription factor expressed in all pituitary cell types that was first identified on the basis of its ability to bind to and activate the POMC gene (Lamonerie et al., 1996). Ptx-1 is expressed in all pituitary cell types but activates transcription of different target genes in a cell-type restricted fashion, by interacting with specific partner proteins (Tremblay et al., 1998). Of particu-lar interest here was the finding of a physical and synergistic interaction between Ptx-1 and SF-1 in regulation of the pituitary *LHβ*, *MIS*, and *Egr-1* genes (Tremblay and Drouin, 1999; Tremblay et al., 1999). These authors proposed that Ptx-1 functions as a cofactor for SF-1 that unmasks its transcriptional activation domain, thereby obviating a requirement for ligand. This model closely parallels that proposed for the *Drosophila* Ftz-F1/Ftz interaction (Fig. 2).

6.2. Other novel protein–protein interactions of Ftz-F1 family members

6.2.1. MBF-1: A novel Ftz-F1 partner

Multiprotein bridging factors 1 and 2 (MBF-1 and MBF-2) were isolated as small polypeptide partners of BmFtz-F1 by Susumu Hirose's group in the mid-1990s (Li et al., 1994). MBF-1 functions as a bridging protein that binds directly to both Ftz-F1 and TATA-binding protein (TBP), and also recruits a second bridging protein, MBF-2 to this complex, resulting in selective activation of Ftz-F1–dependent transcription (Takemaru et al., 1997). Further mechanistic studies demonstrated that MBF-1 and MBF-2 are nuclear localized only when both are coexpressed with Ftz-F1, suggesting regulation of their transcription activation function at the level of nuclear localization (Liu et al., 2000). MBF-1 is an evolutionarily conserved protein found in yeast and humans (Takemaru et al., 1998; Kabe et al., 1999). In mammalian cells, MBF-1 interacts with SF-1 (Kabe et al., 1999) and LRH-1, via the Ftz-F1 box, as well as with other NRs (Brendel et al., 2002). Studies from Johan Auwerx' laboratory confirmed that MBF-1 lacks enzymatic activities typical of NR coactivators and that it forms a direct bridge to the basal transcriptional machinery by binding TFIID (Brendel et al., 2002). *Drosophila mbf-1* mutants did not show a *ftz-f1*-like mutant phenotype and in fact were viable and fertile (Liu et al., 2003a), suggesting that other factors play redundant roles in bridging Ftz-F1 protein complexes to the basal transcription machinery in *Drosophila*.

6.2.2. WT-1 and SF-1 coordinately activate transcription

An interesting cofactor for SF-1 was isolated in the laboratory of Holly Ingraham in an attempt to explain the tissue-specific function of SF-1 in the gonads. This group had shown that SF-1 directly regulates the sex-specific activation and maintenance of the MIS gene during gonad development (Ingraham et al., 1994; Shen et al., 1994) but noted that expression of SF-1 is broader than the expression domain of MIS, implicating a spatially localized partner for SF-1 in this process. WT1 was known to be important for gonadal and kidney development, and mutations in WT1 were associated with Denys-Drash syndrome, a syndrome manifesting urogenital defects that include ambiguous 46,XY genitalia and male pseudohermaphroditism (reviewed in Reddy and Licht, 1996). Nachitigal et al. (1998) provided strong evidence that WT1, a zinc finger transcription factor, functions as a partner for SF-1 in regulating MIS expression: the proteins colocalize *in vivo*, interact directly, and activate transcription synergistically. WT1 protein carrying the DDS mutation failed to synergize with SF-1, providing support for the cooperative function of these proteins *in vivo* (Nachtigal et al., 1998). Another group has provided evidence that WT1 can function as an inhibitor of SF-1 activation of P450 aromatase gene expression in human endometrial cells, suggesting different modes of regulation by the WT1/SF-1 partner pairs in different cellular contexts (Gurates et al., 2002).

6.2.3. GATA-4 interacts with SF-1

SF-1 was found to interact with GATA-4 in positively regulating the expression of MIS (Tremblay and Viger, 1999). This novel interaction between SF-1 and another zinc finger-containing transcription factor was shown to be direct and, in cell culture experiments, synergistic activation required DNA binding by SF-1 but could be achieved in the absence of a GATA-4–binding site (Tremblay and Viger, 1999). A mutant form of SF-1 associated with XY sex reversal that carries a single amino acid change (G to E) in the DBD (Achermann et al., 1999), which did not affect DNA binding to the MIS promoter, failed to cooperate with GATA-4 to synergistically activate MIS transcription, possibly explaining the sex reversal occurring in this patient (Tremblay and Viger, 2003).

6.2.4. Dax-1 and SHP repress the activity of SF-1 and LRH-1

One of the first identified partners of SF-1 was *Dax-1* (NROB1), an X-linked gene implicated in dosage-sensitive sex reversal and adrenal hypoplasia congenita (AHC) (Muscatelli et al., 1994; Zanaria et al., 1994). *Dax-1* encodes an NR that lacks a canonical DBD and that functions as a transcriptional repressor in conjunction with other NRs, including SF-1 and LRH-1 (Ito et al., 1997; Crawford et al., 1998; Nachtigal et al., 1998; Lopez et al., 2001). Dax-1 inhibits the ability of SF-1 to activate transcription in cell culture experiments but does not appear to inhibit its DNA-binding activity. A repression domain was identified in the C-terminus of Dax-1; this domain is deleted in many cases of ACH (Ito et al., 1997) and missense mutations found in other ACH patients also abrogate repression potential (Achermann et al., 2001a), implicating interaction with SF-1 and repression of its transcriptional activity as a major physiological function of Dax-1 (reviewed in Achermann et al., 2001b). Interaction with SF-1 and repression of transcription potential require three LXXLL motifs located in the N-terminus of Dax-1 (Suzuki et al., 2003), a scenario similar to that seen for Ftz/Ftz-1, except that the interaction with Dax-1 results in repression, while the interaction with Ftz results in synergistic activation. Crawford et al. (1998) presented evidence suggesting that Dax-1 mediates repression by recruitment of N-CoR to a complex with SF-1 and that ACH mutations which decrease repression activity do so by decreasing N-CoR recruitment. Dax-1 was also shown to antagonize the synergistic transcription activation of MIS by WT-1 and SF-1 (see earlier). The Ingraham group proposed that a competition between WT1 and Dax-1 for interaction with SF-1 regulates male sexual development: in XY males, the relative gene dosage of WT1:Dax-1 is 2:1 and male-specific genes are activated by SF-1; in XY dosage sensitive sex reversal, the ratio is 2:2 and Dax-1 inhibits activation of male-specific genes; in DDS, the ratio is 1:1 again allowing Dax-1 to repress activation by SF-1/WT1 (Nachtigal et al., 1998). Analysis of the genetic interaction between *Dax-1* and *SF-1* in mice demonstrated that loss of *Dax-1* partially reversed adrenal defects seen in *SF-1*$^{+/-}$ mice, providing strong support for the model that *Dax-1* counters *SF-1* function *in vivo* (Babu et al., 2002). However, despite the fact that this genetic interaction supports the *in vitro* studies demonstrating a direct and antagonistic interaction between SF-1 and Dax-1, the result is counter to what would be expected on the basis of ACH patients in which loss of Dax-1 repression potential results in adrenal hypoplasia. These apparently opposite findings in mouse and human may represent species differences or may be due to a novel splice variant of Dax-1 found in human tissue, Dax-1A (reviewed in Else and Hammer, 2005), which does not repress SF-1 and may antagonize Dax-1 activity (Hossain et al., 2004).

Small heterodimer partner (SHP, NROB2) is an NR closely related to Dax-1 that also possesses an LBD but lacks a standard DBD and contains two LXXLL motifs (reviewed in Bavner et al., 2005). SHP functions as a transcriptional repressor in conjunction with many NRs, but its interaction with LRH-1 in regulating cholesterol homeostasis is particularly well established (Goodwin et al., 2000; Lu et al., 2000; Denson et al., 2001; Lee and Moore, 2002; Kovacic et al., 2004; reviewed in Fayard et al., 2004). SHP interacts with LRH-1 by virtue of contacts between its LXXLL motif(s) and the LRH-1 AF-2 domain (Lee and Moore, 2002; Ortlund et al., 2005; Li et al., 2005b). LRH-1 and SHP participate in a negative regulatory cascade that maintains hepatic cholesterol catabolism. Further, SHP is present in human adipose tissue and may serve to repress LRH-1 activation of aromatase in breast preadipocytes (Clyne et al., 2004).

7. Ligand for Ftz-F1 family proteins

As for other orphan NRs, different approaches have been taken to identify a ligand for the Ftz-F1 family. Early reports of an oxysterol ligand remain controversial as these steroids have not been found to modulate SF-1 activity in various cellular contexts (Lala et al., 1997; Mellon and Bair, 1998). Further, the broad cellular contexts in which Ftz-F1 proteins bind DNA and activate transcription suggested that this family of proteins is constitutively active and may not require a classical ligand for its activity. This notion was supported by the finding that the activity of Ftz-F1 proteins is regulated by protein–protein interactions that could obviate the need for a ligand-dependent switch in Ftz-F1 activity (see earlier). Strong support for ligand independence for this family of proteins came from crystal structures of the mouse LRH-1 LBD, which revealed a large but empty ligand-binding pocket that nonetheless adopted an active conformation. This would allow for stable interaction of coactivators with the AF-2 region of the receptor in the absence of bound ligand (Sablin et al., 2003; reviewed in Li et al., 2003). Consistent with this, mutations in the LRH-1 ligand-binding pocket did not affect transcriptional activity (Sablin et al., 2003; Krylova et al., 2005). However, a surprise came in 2005 when four groups independently reported the presence of a phospholipid ligand in the ligand-binding pockets of human LRH-1 and both mouse and human SF-1 (Krylova et al., 2005; Ortlund et al., 2005; Wang et al., 2005; Li et al., 2005a). That the LBD of SF-1 was occupied had also been suggested by a 2004 study that reported the presence of a fortuitous but unidentified ligand bound to SF-1 (Madauss et al., 2004). Bound phospholipids, and binding pocket residues in SF-1 and human LRH-1, were required for maximal transcription activity in cell systems (Krylova et al., 2005; Wang et al., 2005; Li et al., 2005a). SF-1 bound eukaryotic phospholipids with highest affinity for PI(3,5)P3 and PI(3,4,5)P3 (Krylova et al., 2005). These findings link Ftz-F1 family NR function with PI3 kinase signaling, leading to the intriguing suggestion that PI3K signaling might play a role in SF-1 and LRH-1 control of cell proliferation.

The presence of phospholipid molecules in a ligand-binding pocket of Ftz-F1 family proteins is intriguing and may suggest novel modes of regulation for this family of orphan receptors. However, fortuitous ligands have been reported in crystal structures for other NRs: in these cases, a ligand is incorporated into the LBD of the NR during the course of expression and purification of proteins expressed in bacteria (Bourguet et al., 2000; Billas et al., 2001; Stehlin et al., 2001). These fortuitous lipids may bind because they bear resemblance to endogenous ligands and/or they may stabilize either active or inactive structures of these NRs. In the case of human LRH-1 and mouse and human SF-1, phospholipid binding appears to be required for receptor activity. This may indicate that phospholipids function as classical ligands that activate these NRs in specific cellular contexts. Alternatively, these phospholipids could be constitutively bound to these NRs, stabilizing their structures in all cells, with protein–protein interactions functioning as the ON/OFF switches for NR activity. It will be of particular interest to distinguish between these scenarios in *Drosophila*, where Ftz-F1 is active only in cells that express Ftz. A ligand-dependent activation of Ftz-F1 in this scenario would require that either (1) Ftz activates genes involved in ligand synthesis in an Ftz-F1–independent fashion or (2) Ftz bound to Ftz-F1 recruits an enzyme to the complex that generates a local ligand which binds to Ftz-F1. There is as yet no evidence to suggest that either of these scenarios occur in the blastoderm embryo. The first scenario is extremely unlikely as Ftz requires Ftz-F1 for target gene activation (see earlier).

Krylova et al. (2005) have argued that the common ancestor of LRH-1 and SF-1 was ligand dependent and that murine LRH-1 has lost ligand-binding potential.

Laudet has argued that the primordial member of the Ftz-F1 family was a ligand-independent receptor (Laudet, 1997). Further analysis of potential involvement of a ligand in the regulation of *Drosophila*, or other arthropod, Ftz-F1 proteins would go a long way toward resolving this issue.

Acknowledgments

We thank Dana Cruz for working to determine deletion break points and Antonina Kruppa for help with making figures. We thank Eric Baehrecke for comments on the manuscript.

References

Achermann, J.C., Ito, M., Ito, M., Hindmarsh, P.C., Jameson, J.L. 1999. A mutation in the gene encoding steroidogenic factor-1 causes XY sex reversal and adrenal failure in humans. Nat. Genet. 22, 125–126.

Achermann, J.C., Ito, M., Silverman, B.L., Habiby, R.L., Pang, S., Rosler, A., Jameson, J.L. 2001a. Missense mutations cluster within the carboxyl-terminal region of DAX-1 and impair transcriptional repression. J. Clin. Endocrinol. Metab. 86, 3171–3175.

Achermann, J.C., Meeks, J.J., Jameson, J.L. 2001b. Phenotypic spectrum of mutations in DAX-1 and SF-1. Mol. Cell Endocrinol. 185, 17–25.

Achermann, J.C., Ozisik, G., Ito, M., Orun, U.A., Harmanci, K., Gurakan, B., Jameson, J.L. 2002. Gonadal determination and adrenal development are regulated by the orphan nuclear receptor steroidogenic factor-1, in a dose-dependent manner. J. Clin. Endocrinol. Metab. 87, 1829–1833.

Aesoy, R., Mellgren, G., Morohashi, K.-I., Lund, J. 2002. Activation of cAMP-dependent protein kinase increases the protein level of steroidogenic factor-1. Endocrinology 143, 295–303.

Akam, M. 1995. Hox genes and the evolution of diverse body plans. Philos. Trans. R. Soc. Lond. B. Biol. Sci. 349, 313–319.

Altschul, S.F., Gish, W., Miller, W., Myers, E.W., Lipman, D.J. 1990. Basic local alignment search tool. J. Mol. Biol. 215, 403–410.

Asahina, M., Ishihara, T., Jindra, M., Kohara, Y., Katsura, I., Hirose, S. 2000. The conserved nuclear receptor Ftz-F1 is required for embryogenesis, moulting and reproduction in *Caenorhabditis elegans*. Genes Cells 5, 711–723.

Ashburner, M., Richards, G. 1976. Sequential gene activation by ecdysone in polytene chromosomes of *Drosophila melanogaster*. III. Consequences of ecdysone withdrawal. Dev. Biol. 54, 241–255.

Ayer, S., Walker, N., Mosammaparast, M., Nelson, J.P., Shilo, B.Z., Benyajati, C. 1993. Activation and repression of *Drosophila* alcohol dehydrogenase distal transcription by two steroid hormone receptor superfamily members binding to a common response element. Nucleic Acids Res. 21, 1619–1627.

Babu, P.S., Bavers, D.L., Shah, S., Hammer, G.D. 2000. Role of phosphorylation, gene dosage and Dax-1 in SF-1 mediated steroidogenesis. Endocr. Res. 26, 985–994.

Babu, P.S., Bavers, D.L., Beuschlein, F., Shah, S., Jeffs, B., Jameson, J.L., Hammer, G.D. 2002. Interaction between Dax-1 and steroidogenic factor-1 *in vivo*: Increased adrenal responsiveness to ACTH in the absence of Dax-1. Endocrinology 143, 665–673.

Bakke, M., Lund, J. 1995. Mutually exclusive interactions of two nuclear orphan receptors determine activity of a cyclic adenosine $3',5'$-monophosphate-responsive sequence in the bovine CYP17 gene. Mol. Endocrinol. 9, 327–339.

Bavner, A., Sanyal, S., Gustafsson, J.A., Treuter, E. 2005. Transcriptional corepression by SHP: Molecular mechanisms and physiological consequences. Trends Endocrinol. Metab. 16, 478–488.

Beckstead, R., Ortiz, J.A., Sanchez, C., Prokopenko, S.N., Chambon, P., Losson, R., Bellen, H.J. 2001. Bonus, a Drosophila homolog of TIF1 proteins, interacts with nuclear receptors and can inhibit betaFTZ-F1-dependent transcription. Mol. Cell 7, 753–765.

Bertin, B., Sasorith, S., Caby, S., Oger, F., Cornette, J., Wurtz, J.M., Pierce, R.J. 2004. Unique functional properties of a member of the Fushi Tarazu-Factor 1 family from Schistosoma mansoni. Biochem. J. 382, 337–351.

Bertrand, S., Brunet, F.G., Escriva, H., Parmentier, G., Laudet, V., Robinson-Rechavi, M. 2004. Evolutionary genomics of nuclear receptors: From twenty-five ancestral genes to derived endocrine systems. Mol. Biol. Evol. 21, 1923–1937.

Billas, I.M.L., Moulinier, L., Rochel, N., Moras, D. 2001. Crystal structure of the ligand-binding domain of the Ultraspiracle protein USP, the ortholog of retinoid X receptors in insects. J. Biol. Chem. 276, 7465–7474.

Bland, M.L., Jamieson, C.A., Akana, S.F., Bornstein, S.R., Eisenhofer, G., Dallman, M.F., Ingraham, H.A. 2000. Haploinsufficiency of steroidogenic factor-1 in mice disrupts adrenal development leading to an impaired stress response. Proc. Natl. Acad Sci. USA 97, 14488–14493.

Bland, M.L., Fowkes, R.C., Ingraham, H.A. 2004. Differential requirement for steroidogenic factor-1 gene dosage in adrenal development versus endocrine function. Mol. Endocrinol. 18, 941–952.

Borud, B., Hoang, T., Bakke, M., Jacob, A.L., Lund, J., Mellgren, G. 2002. The nuclear receptor coactivators p300/CBP/cointegrator-associated protein (p/CIP) and transcription intermediary factor 2 (TIF2) differentially regulate PKA-stimulated transcriptional activity of steroidogenic factor 1. Mol. Endocrinol. 16, 757–773.

Botrugno, O.A., Fayard, E., Annicotte, J.S., Haby, C., Brennan, T., Wendling, O., Tanaka, T., Kodama, T., Thomas, W., Auwerx, J., Schoonjans, K. 2004. Synergy between LRH-1 and beta-catenin induces G1 cyclin-mediated cell proliferation. Mol. Cell 15, 499–509.

Bourguet, W., Andry, V., Iltis, C., Klaholz, B., Potier, N., Van Dorsselaer, A., Chambon, P., Gronemeyer, H., Moras, D. 2000. Heterodimeric complex of RAR and RXR nuclear receptor ligand-binding domains: Purification, crystallization, and preliminary X-ray diffraction analysis. Protein Expr. Purif. 19, 284–288.

Bowler, T., Kosman, D., Licht, J.D., Pick, L. 2006. Computational identification of Ftz/Ftz-F1 downstream target genes. Dev. Biol. (in press.)

Brendel, C., Gelman, L., Auwerx, J. 2002. Multiprotein bridging factor-1 (MBF-1) is a cofactor for nuclear receptors that regulate lipid metabolism. Mol. Endocrinol. 16, 1367–1377.

Broadus, J., McCabe, J.R., Endrizzi, B., Thummel, C.S., Woodard, C.T. 1999. The Drosophila βFTZ-F1 orphan nuclear receptor provides competence for stage-specific responses to the steroid hormone ecdysone. Mol. Cell 3, 143–149.

Brown, J.L., Wu, C. 1993. Repression of Drosophila pair-rule segmentation genes by ectopic expression of tramtrack. Development 117, 45–58.

Brown, S.J., Hilgenfeld, R.B., Denell, R.E. 1994. The beetle Tribolium castaneum has a fushi tarazu homolog expressed in stripes during segmentation. Proc. Natl. Acad. Sci. USA 91, 12922–12926.

Chai, C., Chan, W.K. 2000. Developmental expression of a novel Ftz-F1 homologue, ff1b (NR5A4) in the zebrafish Danio rerio. Mech. Dev. 91, 421–426.

Chai, C., Liu, Y.W., Chan, W.K. 2003. Ff1b is required for the development of steroidogenic component of the zebrafish interrenal organ. Dev. Biol. 260, 226–244.

Chan, S.M., Chan, K.M. 1999. Characterization of the shrimp eyestalk cDNA encoding a novel fushi tarazu-factor 1 (FTZ-F1). FEBS Lett. 454, 109–114.

Chen, W.-Y., Lee, W.-C., Hsu, N.-C., Huang, F., Chung, B.-c. 2004a. SUMO modification of repression domains modulates function of nuclear receptor 5A1 (steroidogenic factor-1). J. Biol. Chem. 279, 38730–38735.

Chen, Z., Eastburn, D.J., Han, M. 2004b. The *Caenorhabditis elegans* nuclear receptor gene nhr-25 regulates epidermal cell development. Mol. Cell. Biol. 24, 7345–7358.

Chou, T.-B., Perrimon, N. 1996. The autosomal FLP-DFS technique for generating germline mosaics in *Drosophila melanogaster*. Genetics 144, 1673–1679.

Clemens, J.W., Lala, D.S., Parker, K.L., Richards, J.S. 1994. Steroidogenic factor-1 binding and transcriptional activity of the cholesterol side-chain cleavage promoter in rat granulosa cells. Endocrinology 134, 1499–1508.

Clyne, C.D., Speed, C.J., Zhou, J., Simpson, E.R. 2002. Liver receptor homologue-1 (LRH-1) regulates expression of aromatase in preadipocytes. J. Biol. Chem. 277, 20591–20597.

Clyne, C.D., Kovacic, A., Speed, C.J., Zhou, J., Pezzi, V., Simpson, E.R. 2004. Regulation of aromatase expression by the nuclear receptor LRH-1 in adipose tissue. Mol. Cell. Endocrinol. 215, 39–44.

Colgan, J., Wampler, S., Manley, J.L. 1993. Interaction between a transcriptional acitvator and transcription factor IIB *in vivo*. Nature 362, 549–553.

Consortium, T.F. 2003. The FlyBase database of the *Drosophila* genome projects and community literature (http://flybase.org/). Nucleic Acids Res. 31, 172–175.

Correa, R.V., Domenice, S., Bingham, N.C., Billerbeck, A.E., Rainey, W.E., Parker, K.L., Mendonca, B.B. 2004. A microdeletion in the ligand binding domain of human steroidogenic factor 1 causes XY sex reversal without adrenal insufficiency. J. Clin. Endocrinol. Metab. 89, 1767–1772.

Crawford, P.A., Dorn, C., Sadovsky, Y., Milbrandt, J. 1998. Nuclear receptor DAX-1 recruits nuclear receptor corepressor N-CoR to steroidogenic factor 1. Mol. Cell. Biol. 18, 2949–2956.

Cruz, D.L. 2000. Using the *Drosophila* Genome Project to map a mutation at the molecular level. Undergraduate Thesis. South Hadley, MA: Mount Holyoke College.

Darimont, B.D., Wagner, R.L., Aprilette, J.W., Stallcup, M.R., Kushnre, P.J., Baxter, J.D., Fletterick, R.J., Yamamoto, K.R. 1998. Structure and specificity of nuclear receptor-coactivator interactions. Genes Dev. 12, 3343–3356.

De Mendonca, R.L., Bouton, D., Bertin, B., Escriva, H., Noel, C., Vanacker, J.M., Cornette, J., Laudet, V., Pierce, R.J. 2002. A functionally conserved member of the FTZ-F1 nuclear receptor family from Schistosoma mansoni. Eur. J. Biochem. 269, 5700–5711.

De Santa Barbara, P., Bonneaud, N., Boizet, B., Desclozeaux, M., Moniot, B., Sudbeck, P., Scherer, G., Poulat, F., Berta, P. 1998. Direct interaction of SRY-related protein SOX9 and steroidogenic factor 1 regulates transcription of the human anti-Mullerian hormone gene. Mol. Cell. Biol. 18, 6653–6665.

del Castillo-Olivares, A., Campos, J.A., Pandak, W.M., Gil, G. 2004. The role of alpha1-fetoprotein transcription factor/LRH-1 in bile acid biosynthesis: A known nuclear receptor activator that can act as a suppressor of bile acid biosynthesis. J. Biol. Chem. 279, 16813–16821.

Denson, L.A., Sturm, E., Echevarria, W., Zimmerman, T.L., Makishima, M., Mangelsdorf, D.J., Karpen, S.J. 2001. The orphan nuclear receptor, shp, mediates bile acid-induced inhibition of the rat bile acid transporter, ntcp. Gastroenterology 121, 140–147.

Dressel, U., Thormeyer, D., Altncicek, B., Paululat, A., Eggert, M., Schneider, S., Tenbaum, S.P, Renkawitz, R., Baniahmad, A. 1999. Alien, a highly conserved protein with characteristics of a corepressor for members of the nuclear hormone receptor superfamily. Mol. Cell. Biol. 19, 3383–3394.

Ellinger-Ziegelbauer, H., Hihi, A.K., Laudet, V., Keller, H., Wahli, W., Dreyer, C. 1994. FTZ-F1-related orphan recpetors in *Xenopus laevis*: Transcriptional regulators differentially expressed during early embryogenesis. Mol. Cell. Biol. 14, 2786–2797.

Ellinger-Ziegelbauer, H., Glaser, B., Dreyer, C. 1995. A naturally occuring short variant of the FTZ-F1-related nuclear orphan receptor xFF1rA and interactions between domains of xFF1rA. Mol. Endocrinol. 9, 872–886.

Else, T., Hammer, G.D. 2005. Genetic analysis of adrenal absence: Agenesis and aplasia. Trends Endocrinol. Metab. 16, 458–468.

England, B.P., Heberlein, U., Tjian, R. 1990. Purified *Drosophila* transcription factor, *Adh* distal factor-1 (Adf-1), binds to sites in several *Drosophila* promoters and activates transcription. J. Biol. Chem. 265, 5086–5094.

England, B.P., Admon, A., Tjian, R. 1992. Cloning of *Drosophila* transcription factor Adf-1 reveals homology to Myb oncoprotein. Proc. Natl. Acad. Sci. USA 89, 683–687.

Escriva, H., Delaunay, F., Laudet, V. 2000. Ligand binding and nuclear receptor evolution. Bioessays 22, 717–727.

Escriva, H., Bertrand, S., Laudet, V. 2004. The evolution of the nuclear receptor superfamily. Essays Biochem. 40, 11–26.

Fayard, E., Auwerx, J., Schoonjans, K. 2004. LRH-1: An orphan nuclear receptor involved in development, metabolism and steroidogenesis. Trends Cell Biol. 14, 250–260.

Figueiredo, B.C., Cavalli, L.R., Pianovski, M.A., Lalli, E., Sandrini, R., Ribeiro, R.C., Zambetti, G., DeLacerda, L., Rodrigues, G.A., Haddad, B.R. 2005. Amplification of the steroidogenic factor 1 gene in childhood adrenocortical tumors. J. Clin. Endocrinol. Metab. 90, 615–619.

Florence, B., Guichet, A., Ephrussi, A., Laughon, A. 1997. Ftz-F1 is a cofactor in Ftz activation of the *Drosophila engrailed* gene. Development 124, 839–847.

Galarneau, L., Pare, J.F., Allard, D., Hamel, D., Levesque, L., Tugwood, J.D., Green, S., Belanger, L. 1996. The alpha1-fetoprotein locus is activated by a nuclear receptor of the Drosophila FTZ-F1 family. Mol. Cell. Biol. 16, 3853–3865.

Gates, J., Lam, G., Ortiz, J.A., Losson, R., Thummel, C.S. 2004. *Rigor mortis* encodes a novel nuclear receptor interacting protein required for ecdysone signaling during *Drosophila* larval development. Development 131, 25–36.

Gissendanner, C.R., Sluder, A.E. 2000. *nhr-25*, the *Caenorhabditis elegans* ortholog of *ftz-f1*, is required for epidermal and somatic gonad development. Dev. Biol. 221, 259–272.

Glass, C.K., Rosenfeld, M.G. 2000. The coregulator exchange in transcriptional functions of nuclear receptors. Genes Dev. 14, 121–141.

Goodwin, B., Jones, S.A., Price, R.R., Watson, M.A, McKee, D.D., Moore, L.B., Galardi, C., Wilson, J.G., Lewis, M.C., Roth, M.E., Maloney, P.R., Willson, T.M., et al. 2000. A regulatory cascade of the nuclear receptors FXR, SHP-1, and LRH-1 represses bile acid biosynthesis. Mol. Cell 6, 517–526.

Guichet, A., Copeland, J.W.R., Erdelyi, M., Hlousek, D., Zavorszky, P., Ho, J., Brown, S., Percival-Smith, A., Krause, H.M., Ephrussi, A. 1997. The nuclear receptor homologue Ftz-F1 and the homeodomain protein Ftz are mutually dependent cofactors. Nature 385, 548–552.

Gurates, B., Sebastian, S., Yang, S., Zhou, J., Tamura, M, Fang, Z., Suzuki, T., Sasano, H., Bulun, S.E. 2002. WT1 and DAX-1 inhibit aromatase P450 expression in human endometrial and endometriotic stromal cells. J. Clin. Endocrinol. Metab. 87, 4369–4377.

Halvorson, L.M., Kaiser, U.B., Chin, W.W. 1996. Stimulation of luteinizing hormone β gene promoter activity by the orphan nuclear receptor, steroidogenic factor-1. J. Biol. Chem. 271, 6645–6650.

Halvorson, L.M., Ito, M., Jameson, J.L., Chin, W.W. 1998. Steroidogenic factor-1 and early growth response protein 1 act through two composite DNA binding sites to regulate luteinizing hormone beta-subunit gene expression. J. Biol. Chem. 273, 14712–14720.

Hammer, G.D., Krylova, I., Zhang, Y., Darimont, B.D., Simpson, K., Weigel, N.L., Ingraham, H.A. 1999. Phosphorylation of the nuclear receptor SF-1 modulates cofactor recruitment: Integration of hormone signaling in reproduction and stress. Mol. Cell 3, 521–526.

Han, W., Yu, Y., Altan, N., Pick, L. 1993. Multiple proteins interact with the *fushi tarazu* proximal enhancer. Mol. Cell. Biol. 13, 5549–5559.

Han, W., Yu, Y., Su, K., Kohanski, R.A., Pick, L. 1998. A binding site for multiple transcriptional activators in the *fushi tarazu* proximal enhancer is essential for gene expression *in vivo*. Mol. Cell. Biol. 18, 3384–3394.

Hepperle, C., Hartfelder, K. 2001. Differentially expressed regulatory genes in honey bee caste development. Naturwissenschaften 88, 113–116.

Hiromi, Y., Gehring, W.J. 1987. Regulation and function of the *Drosophila* segmentation gene *fushi tarazu*. Cell 50, 963–974.

Hiromi, Y., Kuroiwa, A., Gehring, W.J. 1985. Control elements of the *Drosophila* segmentation gene *fushi tarazu*. Cell 43, 603–613.

Honda, S.-I., Morohashi, K.-I., Nomura, M., Takeya, H., Kitajima, M., Omura, T. 1993. Ad4BP regulating steroidogenic P-450 gene is a member of steroid hormone receptor superfamily. J. Biol. Chem. 268, 7494–7502.

Horn, C., Offen, N., Nystedt, S., Hacker, U., Wimmer, E.A. 2003. piggyBac-based insertional mutagenesis and enhancer detection as a tool for functional insect genomics. Genetics 163, 647–661.

Hossain, A., Li, C., Saunders, G.F. 2004. Generation of two distinct functional isoforms of dosage-sensitive sex reversal-adrenal hypoplasia congenita-critical region on the X chromosome gene 1 (DAX-1) by alternative splicing. Mol. Endocrinol. 18, 1428–1437.

Huelsenbeck, J.P., Ronquist, F. 2001. MRBAYES: Bayesian inference of phylogenetic trees. Bioinformatics 17, 754–755.

Ikeda, Y., Lala, D.S., Luo, X., Kim, E., Moisan, M.-P., Parker, K.L. 1993. Characterization of the mouse FTZ-F1 gene, which encodes a key regulator of steroid hydroxylase gene expression. Mol. Endocrinol. 7, 852–860.

Ikeda, Y., Shen, W.-H., Ingraham, H.A., Parker, K.L. 1994. Developmental expression of mouse steroidogenic factor-1, as essential regulator of the steroid hydroxylases. Mol. Endocrinol. 8, 654–662.

Ikeda, Y., Luo, X., Abbud, R., Nilson, J.H., Parker, K.L. 1995. The nuclear receptor steroidogenic factor 1 is essential for the formation of the ventromedial hypothalamic nucleus. Mol. Endocrinol. 9, 478–486.

Ingraham, H.A., Lala, D.S., Ikeda, Y., Luo, X., Shen, W.-H., Nachtigal, M.W., Abbud, R., Nilson, J.H., Parker, K.L. 1994. The nuclear receptor steroidogenic factor 1 acts at multiple levels of the reproductive axis. Genes Dev. 8, 2302–2312.

Ito, M., Yu, R., Jameson, J.L. 1997. DAX-1 inhibits SF-1-mediated transactivation via a carboxy-terminal domain that is deleted in adrenal hypoplasia congenita. Mol. Cell. Biol. 17, 1476–1483.

Ito, M., Yu, R.N., Jameson, J.L. 1998. Steroidogenic factor-1 contains a carboxy-terminal transcriptional activation domain that interacts with steroid receptor coactivator-1. Mol. Endocrinol. 12, 290–301.

Jacob, A.L., Lund, J., Martinez, P., Hedin, L. 2001. Acetylation of steroidogenic factor 1 protein regulates its transcriptional activity and recruits the coactivator GCN5. J. Biol. Chem. 276, 37659–37664.

Jiang, C., Lamblin, A.F., Steller, H., Thummel, C.S. 2000. A steroid-triggered transcriptional hierarchy controls salivary gland cell death during *Drosophila* metamorphosis. Mol. Cell 5, 445–455.

Johansson, L., Thomsen, J.S., Damdimopoulos, A.E., Spyrou, G., Gustafsson, J.A., Treuter, E. 1999. The orphan nuclear receptor SHP inhibits agonist-dependent transcriptional activity of estrogen receptors ERalpha and ERbeta. J. Biol. Chem. 274, 345–353.

Kabe, Y., Goto, M., Shima, D., Imai, T., Wada, T., Morohashi, K., Shirakawa, M., Hirose, S., Handa, H. 1999. The role of human MBF1 as a transcriptional coactivator. J. Biol. Chem. 274, 34196–34202.

Kawasaki, H., Hirose, S., Ueda, H. 2002. BetaFTZ-F1 dependent and independent activation of Edg78E, a pupal cuticle gene, during the early metamorphic period in *Drosophila melanogaster*. Dev. Growth Differ. 44, 419–425.

King-Jones, K., Thummel, C.S. 2005. Nuclear receptors—a perspective from *Drosophila*. Nat. Rev. Genet. 6, 311–323.

Komatsu, T., Mizusaki, H., Mukai, T., Ogawa, H., Baba, D., Shirakawa, M., Hatakeyama, S., Nakayama, K.I., Yamamoto, H., Kikuchi, A., Morohashi, K.-I. 2004. Small ubiquitin-like modifier 1 (SUMO-1) modification of the synergy control motif of Ad4 binding protein/ steroidogenic factor 1 (Ad4BP/SF-1) regulates synergistic transcription between Ad4BP/ SF-1 and Sox9. Mol. Endocrinol. 18, 2451–2462.

Kostrouchova, M., Krause, M., Kostrouch, Z., Rall, J.E. 1998. CHR3: A *Caenorhabditis elegans* orphan nuclear hormone receptor required for proper epidermal development and molting. Development 125, 1617–1626.

Kovacic, A., Speed, C.J., Simpson, E.R., Clyne, C.D. 2004. Inhibition of aromatase transcription via promoter II by short heterodimer partner in human preadipocytes. Mol. Endocrinol. 18, 252–259.

Krylova, I.N., Sablin, E.P., Moore, J., Xu, R.X., Waitt, G.M., MacKay, J.A., Juzumiene, D., Bynum, J.M., Madauss, K., Montana, V., Lebedeva, L.A., Suzawa, M., et al. 2005. Structural analyses reveal phosphatidyl inositols as ligands for the NR5 orphan receptors SF-1 and LRH-1. Cell 120, 343–355.

Kuo, M.W., Postlethwait, J., Lee, W.C., Lou, S.W., Chan, W.K., Chung, B.C. 2005. Gene duplication, gene loss and evolution of expression domains in the vertebrate nuclear receptor NR5A (Ftz-F1) family. Biochem. J. 389, 19–26.

Lala, D.S., Rice, D.A., Parker, K.L. 1992. Steroidogenic Factor I, a key regulator of steroidogenic enzyme expression, is the mouse homolog of *fushi tarazu*-Factor I. Mol. Endocrinol. 6, 1249–1258.

Lala, D.S., Syka, P.M., Lazarchik, S., Magelsdorf, D.J., Parker, K.L., Heyman, R.A. 1997. Activation of the orphan nuclear receptor steroidogenic factor 1 by oxysterols. Proc. Natl. Acad. Sci. USA 94, 4895–4900.

Lam, G.T., Jiang, C., Thummel, C.S. 1997. Coordination of larval and prepupal gene expression by the DHR3 orphan receptor during *Drosophila* metamorphosis. Development 124, 1757–1769.

Lamonerie, T., Tremblay, J.J., Lanctot, C., Therrien, M., Gauthier, Y., Drouin, J. 1996. Ptx1, a bicoid-related homeo box transcription factor involved in transcription of the pro-opiomelanocortin gene. Genes Dev. 10, 1284–1295.

Landrigan, M.F. 1998. Potential Drosophila melanogaster *fushi tarazu* (*ftz*) regulators interacting with the *ftz* proximal enhancer. Ph.D. Thesis.

Laudet, V. 1997. Evolution of the nuclear receptor superfamily: Early diversification from an ancestral orphan receptor. J. Mol. Endocrinol. 19, 207–226.

Laudet, V., Bonneton, F. 2005. Evolution of nuclear hormone receptors in insects. In: *Comprehensive Molecular Insect Science* (L.I. Gilbert, K. Iatrou, S.S. Gill, Eds.), Oxford: Elsevier, vol. 3, pp. 287–318.

Lavorgna, G., Leda, H., Clos, J., Wu, C. 1991. FTZ-F1, a steroid hormone receptor-like protein implicated in the activation of *fushi tarazu*. Science 252, 848–851.

Lavorgna, G., Karim, F.D., Thummel, C.S., Wu, C. 1993. Potential role for a FTZ-F1 steroid receptor superfamily member in the control of *Drosophila* metamorphosis. Proc. Natl. Acad. Sci. USA 90, 3004–3008.

Lawrence, P.A., Johnston, P., Macdonald, P., Struhl, G. 1987. Borders of parasegments are delimited by the *fushi tarazu* and *even-skipped* genes. Nature 328, 440–445.

Lee, C.-Y., Beahrecke, E.H. 2001. Steroid regulation of autophagic programmed cell death during development. Development 128, 1443–1455.

Lee, C.-Y., Simon, C.R., Woodard, C.T., Baehrecke, E.H. 2002. Genetic mechanism for the stage- and tissue-specific regulation of steroid triggered programmed cell death in *Drosophila*. Dev. Biol. 252, 138–148.

Lee, M.B., Lebedeva, L.A., Suzawa, M., Wadekar, S.A., Desclozeaux, M., Ingraham, H.A. 2005. The DEAD-Box protein DP103 (Ddx20 or Gemin-3) represses orphan nuclear receptor activity via SUMO modification. Mol. Cell. Biol. 25, 1879–1890.

Lee, Y.K., Moore, D.D. 2002. Dual mechanisms for repression of the monomeric orphan receptor liver receptor homologous protein-1 by the orphan small heterodimer partner. J. Biol. Chem. 277, 2463–2467.

Levis, R., Hazelrigg, T., Rubin, G.M. 1985. Separable cis-acting control elements for expression of the white gene of *Drosophila*. EMBO J. 4, 3489–3499.

Li, C., Kapitskaya, M.Z., Zhu, J., Miura, K., Segraves, W., Raikhel, A.S. 2000. Conserved molecular mechanism for the stage specificity of the mosquito vitellogenic response to ecdysone. Dev. Biol. 224, 96–110.

Li, F.Q., Ueda, H., Hirose, S. 1994. Mediators of activation of fushi tarazu gene transcription by BmFTZ-F1. Mol. Cell. Biol. 14, 3013–3021.

Li, L.-A., Chiang, E.F.-L., Chen, J.-C., Hsu, N.-C., Chen, Y.-J., Chung, B.-C. 1999. Function of steroidogenic factor 1 domains in nuclear localization, transactivation, and interaction with transcription factor TFIIB and c-Jun. Mol. Endocrinol. 13, 1588–1598.

Li, Y., Lambert, M.H., Xu, H.E. 2003. Activation of nuclear receptors: A perspective from structural genomics. Structure 11, 741–746.

Li, Y., Choi, M., Cavey, G., Daugherty, J., Suino, K., Kovach, A., Bingham, N.C., Kliewer, S.A., Xu, H.E. 2005a. Crystallographic identification and functional characterization of phospholipids as ligands for the orphan nuclear receptor Steriodogenic Factor-1. Mol. Cell 17, 4911–4502.

Li, Y., Choi, M., Suino, K., Kovach, A., Daugherty, J., Kliewer, S.A., Xu, H.E. 2005b. Structural and biochemical basis for selective repression of the orphan nuclear receptor liver receptor homolog 1 by small heterodimer partner. Proc. Natl. Acad. Sci. USA 102, 9505–9510.

Lin, W., Wang, H.W., Sum, C., Liu, D., Hew, C.L., Chung, B. 2000. Zebrafish ftz-f1 gene has two promoters, is alternatively spliced, and is expressed in digestive organs. Biochem. J. 348 (Pt. 2), 439–446.

Liu, D., Le Drean, Y., Ekker, M., Xiong, F., Hew, C.L. 1997. Teleost FTZ-F1 homolog and its splicing variant determine the expression of the salmon gonadotropin IIbeta subunit gene. Mol. Endocrinol. 11, 877–890.

Liu, Q.X., Ueda, H., Hirose, S. 2000. MBF2 is a tissue- and stage-specific coactivator that is regulated at the step of nuclear transport in the silkworm Bombyx mori. Dev. Biol. 225, 437–446.

Liu, Q.X., Jindra, M., Ueda, H., Hiromi, Y., Hirose, S. 2003a. *Drosophila* MBF1 is a co-activator for tracheae defective and contributes to the formation of tracheal and nervous systems. Development 130, 719–728.

Liu, Y.W., Gao, W., Teh, H.L., Tan, J.H., Chan, W.K. 2003b. Prox1 is a novel coregulator of Ff1b and is involved in the embryonic development of the zebrafish interrenal primordium. Mol. Cell. Biol. 23, 7243–7255.

Lohr, U., Pick, L. 2005. Cofactor-interaction motifs and the cooption of a homeotic Hox protein into the segmentation pathway of *Drosophila melanogaster*. Curr. Biol. 15, 643–649.

Lohr, U., Yussa, M., Pick, L. 2001. *Drosophila fushi tarazu*: A gene on the border of homeotic function. Curr. Biol. 11, 1403–1412.

Lopez, D., Shea-Eaton, W., Sanchez, M.D., McLean, M.P. 2001. DAX-1 represses the high-density lipoprotein receptor through interaction with positive regulators sterol regulatory element-binding protein-1a and steroidogenic factor-1. Endocrinology 142, 5097–5106.

Lu, T.T., Makishima, M., Repa, J.J., Schoonjans, K., Kerr, T.A., Auwerx, J., Mangelsdorf, D.J. 2000. Molecular basis for feedback regulation of bile acid synthesis by nuclear receptors. Mol. Cell 6, 507–515.

Luo, X., Ikeda, Y., Parker, K.L. 1994. A cell-specific nuclear receptor is essential for adrenal and gonadal development and sexual differentiation. Cell 77, 481–490.

Luo, X., Ikeda, Y., Schlosser, D.A., Parker, K.L. 1995. Steroidogenic factor 1 is the essential transcript of the mouse Ftz-F1 gene. Mol. Endocrinol. 9, 1233–1239.

Luschnig, S., Moussian, B., Krauss, J., Desjeux, I., Perkovic, J., Nusslein-Volhard, C. 2004. An F1 genetic screen for maternal-effect mutations affecting embryonic pattern formation in *Drosophila melanogaster*. Genetics 167, 325–342.

Lynch, J.P., Lala, D.S., Peluso, J.J., Luo, W., Parker, K.L., White, B.A. 1993. Steroidogenic factor 1, an orphan nuclear receptor, regulates the expression of the rat aromatase gene in gonadal tissues. Mol. Endocrinol. 7, 776–786.

Madauss, K., Juzumiene, D., Waitt, G., Williams, J., Williams, S. 2004. Generation and characterization of human steroidogenic factor 1 LBD crystals with and without bound cofactor peptide. Endocr. Res. 30, 775–785.

Majdic, G., Young, M., Gomez-Sanchez, E., Anderson, P., Szczepaniak, L.S., Dobbins, R.L., McGarry, J.D., Parker, K.L. 2002. Knockout mice lacking steroidogenic factor 1 are a novel genetic model of hypothalamic obesity. Endocrinology 143, 607–614.

Mann, R.S., Chan, S.-K. 1996. Extra specificity from *extradenticle*: The partnership between HOX and PBX/EXD homeodomain proteins. TIGS 12, 258–262.

McInerney, E.M., Rose, D.W., Flynn, S.E., Westin, S., Mullen, T.-M., Krones, A., Inostroza, J., Torchia, J., Nolte, R.T., Assa-Munt, N., Milburn, M.V., Glass, C.K., et al. 1998. Determinants of coactivator LXXLL motif specificity in nuclear receptor transcriptional acitvation. Genes Dev. 12, 3357–3368.

Melchior, F. 2000. SUMO—nonclassical ubiquitin. Annu. Rev. Cell Dev. Biol. 16, 591–626.

Mellon, S.H., Bair, S.R. 1998. 25-hydroxycholesterol is not a ligand for the orphan nuclear receptor Steroidogenic Factor-1 (SF-1). Endocrinology 139, 3026–3029.

Melvin, V.S., Harrell, C., Adelman, J.S., Kraus, W.L., Churchill, M., Edwards, D.P. 2004. The role of the C-terminal extension (CTE) of the estrogen receptor alpha and beta DNA binding domain in DNA binding and interaction with HMGB. J. Biol. Chem. 279, 14763–14771.

Michael, M.D., Kilgore, M.W., Morohashi, K., Simpson, E.R. 1995. Ad4BP/SF-1 regulates cyclic AMP-induced transcription from the proximal promoter (PII) of the human aromatase P450 (CYP19) gene in the ovary. J. Biol. Chem. 270, 13561–13566.

Monte, D., DeWitte, F., Hum, D.W. 1998. Regulation of the human P450scc gene by steroidogenic factor a is mediated by CBP/p300. J. Biol. Chem. 273, 4585–4591.

Morohashi, K., Zanger, U.M., Honda, S., Hara, M., Waterman, M.R., Omura, T. 1993. Activation of CYP11A and CYP11B gene promoters by the steroidogenic cell-specific transcription factor, Ad4BP. Mol. Endocrinol. 7, 1196–1204.

Murata, T., Kageyama, Y., Hirose, S., Ueda, H. 1996. Regulation of the EDG84A gene by FTZ-F1 during metamorphosis in *Drosophila melanogaster*. Mol. Cell. Biol. 16, 6509–6515.

Muscatelli, F., Strom, T.M., Walker, A.P., Zanaria, E., Recan, D., Meindl, A., Bardoni, B., Guioli, S., Zehetner, G., Rabl, W., et al. 1994. Mutations in the DAX-1 gene give rise to both X-linked adrenal hypoplasia congenita and hypogonadotropic hypogonadism. Nature 372, 672–676.

Nachtigal, M.W., Hirokawa, Y., Enyeart-vanHouten, D.L., Flanagan, J.N., Hammer, G.D., Ingraham, H.A. 1998. Wilms' tumor 1 and Dax-1 modulate the orphan nuclear receptor SF-1 in sex-specific gene expression. Cell 93, 445–454.

Ninomiya, Y., Okada, M., Kotomura, N., Suzuki, K., Tsukiyama, T., Niwa, O. 1995. Genomic organization and isoforms of the mouse ELP gene. J. Biochem. (Tokyo) 118, 380–389.

Nitta, M., Ku, S., Brown, C., Okamoto, A.Y., Shan, B. 1999. CPF: An orphan nuclear receptor that regulates liver-specific expression of the human cholesterol 7alpha-hydroxylase gene. Proc. Natl. Acad. Sci. USA 96, 6660–6665.

Nolte, R.T., Wisley, G.B., Westin, S., Cobb, J.E., Lambert, M.H., Kurokawa, R., Rosenfeld, M.G., Willson, T.M., Glass, C.K., Milburn, M.V. 1998. Ligand binding and co-activator assembly of the peroxisome proliferator-activated receptor-γ. Nature 395, 137–143.

Ohno, C., Ueda, H., Petkovich, M. 1994. The *Drosophila* nuclear receptors *FTZ-F1α* and *FTZ-F1β* compete as monomers for binding to a site in the *fushi tarazu* gene. Mol. Cell. Biol. 14, 3166–3175.

Ohno, C.K., Petkovich, M. 1992. *FTZ-F1β* a novel member of the *Drosophila* nuclear receptor family. Mech. Dev. 40, 13–24.

Oliver, G., Sosa-Pineda, B., Geisendorf, S., Spana, E.P., Doe, C.Q., Gruss, P. 1993. Prox 1, a prospero-related homeobox gene expressed during mouse development. Mech. Dev. 44, 3–16.

Ortlund, E.A., Lee, Y., Solomon, I.H., Hager, J.M., Safi, R., Choi, Y., Guan, Z., Tripathy, A., Raetz, C.R., McDonnell, D.P., Moore, D.D., Redinbo, M.R. 2005. Modulation of human nuclear receptor LRH-1 activity by phospholipids and SHP. Nat. Struct. Mol. Biol. 12, 357–363.

Ou, Q., Mouillet, J.-F., Yan, X., Dorn, C., Crawford, P.A., Sadovsky, Y. 2001. The DEAD box protein DP103 is a regulatory of steroidogenic factor-1. *Mol. Endocrin.* Mol. Endocrinol. 15, 69–79.

Pare, J.F., Roy, S., Galarneau, L., Belanger, L. 2001. The mouse fetoprotein transcription factor (FTF) gene promoter is regulated by three GATA elements with tandem E box and Nkx motifs, and FTF in turn activates the Hnf3beta, Hnf4alpha, and Hnf1alpha gene promoters. J. Biol. Chem. 276, 13136–13144.

Pare, J.F., Malenfant, D., Courtemanche, C., Jacob-Wagner, M., Roy, S., Allard, D., Belanger, L. 2004. The fetoprotein transcription factor (FTF) gene is essential to embryogenesis and cholesterol homeostasis and is regulated by a DR4 element. J. Biol. Chem. 279, 21206–21216.

Parker, K.L., Rice, D.A., Lala, D.S., Ikeda, Y., Luo, X., Wong, M., Bakke, M., Zhao, L., Frigeri, C., Hanley, N.A., Stallings, N., Schimmer, B.P. 2002. Steroidogenic factor 1: An essential mediator of endocrine development. Recent Prog. Horm. Res. 57, 19–36.

Parvy, J.P., Blais, C., Bernard, F., Warren, J.T., Petryk, A., Gilbert, L.I., O'Connor, M.B., Dauphin-Villemant, C. 2005. A role for betaFTZ-F1 in regulating ecdysteroid titers during post-embryonic development in *Drosophila melanogaster*. Dev. Biol. 282, 84–94.

Perrimon, N., Lanjuin, A., Arnold, C., Noll, E. 1996. Zygotic lethal mutations with maternal effect phenotypes in *Drosophila melanogaster*. II. Loci on the second and third chromosomes identified by P-element induced mutations. Genetics 144, 1681–1692.

Pick, L., Schier, A., Affolter, M., Schmidt-Glenewinkel, T., Gehring, W.J. 1990. Analysis of the *ftz* upstream element: Germ layer-specific enhancers are independently autoregulated. Genes Dev. 4, 1224–1239.

Pick, L., Lohr, U., Yu, Y. 2000. A double interaction screen to isolate DNA binding and protein-tethered transcription factors. In: *Yeast Hybrid Technologies* (L. Zhu, G.J. Hannon, Eds.), Natick, MA: Eaton Publishing.

Read, D.B. 1992. *Expression and Function of the Tramtrack Gene of Drosophila.* New York: Columbia University.

Reddy, J.C., Licht, J.D. 1996. The WT1 Wilms' tumor suppressor gene: How much do we really know? Biochim. Biophys. Acta 1287, 1–28.

Rice, D.A., Kirkman, M.S., Aitken, L.D., Mouw, A.R., Schimmer, B.P., Parker, K.L. 1990. Analysis of the promoter region of the gene encoding mouse cholesterol side-chain cleavage enzyme. J. Biol. Chem. 265, 11713–11720.

Rice, D.A., Mouw, A.R., Bogerd, A.M., Parker, K.L. 1991. A shared promoter element regulates the expression of three steroidogenic enzymes. Mol. Endocrinol. 5, 1552–1561.

Richards, G. 1976a. Sequential gene activation by ecdysone in polytene chromosomes of *Drosophila melanogaster*. IV. The mid prepupal period. Dev. Biol. 54, 256–263.

Richards, G. 1976b. Sequential gene activation by ecdysone in polytene chromosomes of *Drosophila melanogaster*. V. The late prepupal puffs. Dev. Biol. 54, 264–275.

Riddiford, L.M. 1993. Hormone receptors and the regulation of insect metamorphosis. Receptor 3, 203–209.

Ronquist, F., Huelsenbeck, J.P. 2003. MrBayes 3: Bayesian phylogenetic inference under mixed models. Bioinformatics 19, 1572–1574.

Sablin, E.P., Krylova, I.N., Fletterick, R.J., Ingraham, H.A. 2003. Structural basis for ligand-dependent activation of the orphan nuclear receptor LRH-1. Mol. Cell 11, 1575–1585.

Sadovsky, Y., Crawford, P.A., Woodson, K.G., Polish, J.A., Clements, M.A., Tourtellotte, L.M., Simburger, K., Milbrandt, J. 1995. Mice deficient in the orphan receptor steroidogenic factor 1 lack adrenal glands and gonads but express P450 side-chain-cleavage enzyme in the placenta and have normal embryonic serum levels of corticosteroids. Proc. Natl. Acad. Sci. USA 92, 10939–10943.

Schier, A.F., Gehring, W.J. 1992. Direct homeodomain-DNA interaction in the autoregulation of the *fushi tarazu* gene. Nature 356, 804–807.

Schoonjans, K., Dubuquoy, L., Mebis, J., Fayard, E., Wendling, O., Haby, C., Geboes, K., Auwerx, J. 2005. Liver receptor homolog 1 contributes to intestinal tumor formation through effects on cell cycle and inflammation. Proc. Natl. Acad. Sci. USA 102, 2058–2062.

Schwartz, C.J.E., Sampson, H.M., Hlousek, D., Percival-Smith, A., Copeland, J.W.R., Simmonds, A.J., Krause, H.M. 2001. FTZ-Factor 1 and Fushi Tarazu interact via conserved nuclear receptor and coactivator motifs. EMBO J. 20, 510–519.

Shapiro, D.B., Pappalardo, A., White, B.A., Peluso, J.J. 1996. Steroidogenic factor-1 as a positive regulator of rat granulosa cell differentiation and a negative regulator of mitosis. Endocrinology 137, 1187–1195.

Shen, W.-H., Moore, C.C.D., Ikeda, Y., Parker, K.L., Ingraham, H.A. 1994. Nuclear receptor Steroidogenic Factor 1 regulates the Mullerian Inhibiting Substance gene: A link to the sex determination cascade. Cell 77, 651–661.

Shinoda, K., Lei, H., Yoshii, H., Nomura, M., Nagano, M., Shiba, H., Sasaki, H., Osawa, Y., Ninomiya, Y., Niwa, O., et al. 1995. Developmental defects of the ventromedial hypothalamic nucleus and pituitary gonadotroph in the Ftz-F1 disrupted mice. Dev. Dyn. 204, 22–29.

Simpson, E.R., Clyne, C., Rubin, G., Boon, W.C., Robertson, K., Britt, K., Speed, C., Jones, M. 2002. Aromatase—a brief overview. Annu. Rev. Physiol. 64, 93–127.

Sluder, A.E., Mathews, S.W., Hough, D., Yin, V.P., Maina, C.V. 1999. The nuclear receptor superfamily has undergone extensive proliferation and diversification in nematodes. Gen. Res. 9, 103–120.

Solomon, I.H., Hager, J.M., Safi, R., McDonnell, D.P., Redinbo, M.R., Ortlund, E.A. 2005. Crystal structure of the human LRH-1 DBD-DNA complex reveals Ftz-F1 domain positioning is required for receptor activity. J. Mol. Biol. 354, 1091–1102.

Steffensen, K.R., Holter, E., Bavner, A., Nilsson, M., Pelto-Huikko, M., Tomarev, S., Treuter, E. 2004. Functional conservation of interactions between a homeodomain cofactor and a mammalian FTZ-F1 homologue. EMBO Rep. 5, 613–619.

Stehlin, C., Wurtz, J.-M., Steinmetz, A., Greiner, E., Schule, R., Moras, D., Renaud, J.-P. 2001. X-ray structure of the orphan nuclear receptor RORβ ligand-binding domain in the active conformation. EMBO J. 20, 5822–5831.

Sugawara, T., Holt, J.A., Kiriakidou, M., Strauss, J.F., III 1996. Steroidogenic factor 1-dependent promoter activity of the human steroidogenic acute regulatory protein (StAR) gene. Biochemistry 35, 9052–9059.

Sugawara, T., Abe, S., Sakuragi, N., Fujimoto, Y., Nomura, E., Fujieda, K., Saito, M., Fujimoto, S. 2001. RIP 140 modulates transcription of the steroidogenic acute regulatory protein gene through interactions with both SF-1 and DAX-1. Endocrinology 142, 3570–3577.

Suzuki, T., Kawasaki, H., Yu, R.T., Ueda, Umesono, K. 2002. Segmentation gene product Fushi tarazu is an LXXLL motif-dependent coactivator for orphan receptor FTZ-F1. Proc. Natl. Acad. Sci. USA 98, 12403–12408.

Suzuki, T., Kasahara, M., Yoshioka, H., Morohashi, K., Umesono, K. 2003. LXXLL-related motifs in Dax-1 have target specificity for the orphan nuclear receptors Ad4BP/SF-1 and LRH-1. Mol. Cell Biol. 23, 238–249.

Swofford, D.L. 2002. *PAUP*: Phylogenetic Analysis Using Parsimony (And Other Methods)*. Sunderland, MA: Sinauer Associates.

Takase, M., Nakajima, T., Nakamura, M. 2000. FTZ-F1alpha is expressed in the developing gonad of frogs. Biochim. Biophys. Acta 1494, 195–200.

Takemaru, K., Li, F.Q., Ueda, H., Hirose, S. 1997. Multiprotein bridging factor 1 (MBF1) is an evolutionarily conserved transcriptional coactivator that connects a regulatory factor and TATA element-binding protein. Proc. Natl. Acad. Sci. USA 94, 7251–7256.

Takemaru, K., Harashima, S., Ueda, H., Hirose, S. 1998. Yeast coactivator MBF1 mediates GCN4-dependent transcriptional activation. Mol. Cell. Biol. 18, 4971–4976.

Tracey, W.D., Ning, X., Klingler, M., Kramer, S.G., Gergen, J.P. 2000. Quantitative analysis of gene function in the *Drosophila* embryo. Genetics 154, 273–284.

Tremblay, J.J., Drouin, J. 1999. Egr-1 is a downstream effector of GnRH and synergizes by direct interaction with Ptx1 and SF-1 to enhance luteinizing hormone beta gene transcription. Mol. Cell. Biol. 19, 2567–2576.

Tremblay, J.J., Viger, R.S. 1999. Transcription factor GATA-4 enhances Mullerian inhibiting substance gene transcription through a direct interaction with the nuclear receptor SF-1. Mol. Endocrinol. 13, 1388–1401.

Tremblay, J.J., Viger, R.S. 2003. A mutated form of steroidogenic factor 1 (SF-1 G35E) that causes sex reversal in humans fails to synergize with transcription factor GATA-4. J. Biol. Chem. 278, 42637–42642.

Tremblay, J.J., Lanctot, C., Drouin, J. 1998. The pan-pituitary activator of transcription, Ptx1 (Pituitary Homeobox 1), acts in synergy with SF-1 and Pit1 and is an upstream regulator of the Lim-homeodomain gene Lim3/Lhx3. Mol. Endocrinol. 12, 428–441.

Tremblay, J.J., Marcil, A., Gauthier, Y., Drouin, J. 1999. Ptx1 regulates SF-1 activity by an interaction that mimics the role of the ligand-binding domain. EMBO J. 18, 3431–3441.

Tsai, C.-C., Kao, H.-Y., Yao, T.-P., Mckeown, M., Evans, R.M. 1999. SMRTER, a *Drosophila* nuclear receptor coregulator, reveals that EcR-mediated repression is critical for development. Mol. Cell 4, 175–186.

Tsukiyama, T., Ueda, H., Hirose, S., Niwa, O. 1992. Embryonic long terminal repeat-binding protein is a murine homolog of FTZ-F1, a member of the steroid receptor superfamily. Mol. Cell. Biol. 12, 1286–1291.

Ueda, H., Hirose, S. 1990. Identification and purification of a *Bombyx mori* homologue of FTZ-F1. Nucleic Acids Res. 18, 7229–7234.

Ueda, H., Hirose, S. 1991. Defining the sequence recognized with BmFTZ-F1, a sequence specific DNA binding factor in the silkworm, *Bombyx mori*, as revealed by direct sequencing of bound oligonucleaotides and gel mobility shift competition analysis. Nucleic Acids Res. 19, 3689–3693.

Ueda, H., Sonoda, S., Brown, J.L., Scott, M.P., Wu, C. 1990. A sequence-specific DNA-binding protein that activates *fushi tarazu* segmentation gene expression. Genes Dev. 4, 624–635.

Ueda, H., Sun, G.-C., Murata, T., Hirose, S. 1992. A novel DNA-binding motif abuts the zinc finger domain of insect nuclear hormone receptor FTZ-F1 and mouse embryonal long erminal repeat-binding protein. Mol. Cell. Biol. 12, 5667–5672.

von Hofsten, J., Olsson, P.E. 2005. Zebrafish sex determination and differentiation: Involvement of FTZ-F1 genes. Reprod. Biol. Endocrinol. 3, 63.

von Hofsten, J., Jones, I., Karlsson, J., Olsson, P.E. 2001. Developmental expression patterns of FTZ-F1 homologues in zebrafish (Danio rerio). Gen. Comp. Endocrinol. 121, 146–155.

von Hofsten, J., Karlsson, J., Olsson, P.E. 2003. Fushi tarazu factor-1 mRNA and protein is expressed in steroidogenic and cholesterol metabolising tissues during different life stages in Arctic char (Salvelinus alpinus). Gen. Comp. Endocrinol. 132, 96–102.

von Hofsten, J., Larsson, A., Olsson, P.E. 2005. Novel steroidogenic factor-1 homolog (ff1d) is coexpressed with anti-Mullerian hormone (AMH) in zebrafish. Dev. Dyn. 233, 595–604.

Wakimoto, B.T., Turner, F.R., Kaufman, T.C. 1984. Defects in embryogenesis in mutants associated with the antennapedia gene complex of Drosophila melanogaster. Dev. Biol. 102, 147–172.

Wang, W., Zhang, C., Marimuthu, A., Krupka, H.I., Tabrizizad, M., Shelloe, R., Mehra, U., Eng, K., Nguyen, H., Settachatgul, C., Powell, B., Milburn, M.V., et al. 2005. The crystal structures of human steroidogenic factor-1 and lever receptor homologue-1. Proc. Natl. Acad. Sci. USA 102, 7505–7510.

Weller, J., Sun, G.C., Zhou, B., Lan, Q., Hiruma, K., Riddiford, L.M. 2001. Isolation and developmental expression of two nuclear receptors, MHR4 and betaFTZ-F1, in the tobacco hornworm, Manduca sexta. Insect Biochem. Mol. Biol. 31, 827–837.

White, K.P., Hurban, P., Watanabe, T., Hogness, D.S. 1997. Coordination of Drosophila metamorphosis by two ecdysone-induced nuclear receptors. Science 276, 114–117.

Wilson, M.J., Jeyasuria, P., Parker, K.L., Koopman, P. 2005. The transcription factors steroidogenic factor-1 and SOX9 regulate expression of Vanin-1 during mouse testis development. J. Biol. Chem. 280, 5917–5923.

Woodard, C.T., Baehrecke, E.H., Thummel, C.S. 1994. A molecular mechanism for the stage specificity of the Drosophila prepupal genetic response to ecdysone. Cell 79, 607–615.

Xiong, W.C., Montell, C. 1993. tramtrack is a transcriptional repressor required for cell fate determination in the Drosophila eye. Genes Dev. 7, 1085–1096.

Yamada, M., Murata, T., Hirose, S., Lavorgna, G., Suzuki, E., Ueda, H. 2000. Temporally restricted expression of transcription factor betaFTZ-F1: Significance for embryogenesis, molting and metamorphosis in Drosophila melanogaster. Development 127, 5083–5092.

Yu, R.N., Ito, M., Jameson, J.L. 1998. The murine Dax-1 promoter is stimulated by SF-1 (steroidogenic factor-1) and inhibited by COUP-TF (chicken ovalbumin upstream promoter-transcription factor) via a composite nuclear receptor-regulatory element. Mol. Endocrinol. 12, 1010–1022.

Yu, Y., Li, W., Su, K., Han, W., Yussa, M, Perrimon, N., Pick, L. 1997. The nuclear hormone receptor FTZ-F1 is a cofactor for the Drosophila homeodomain protein Ftz. Nature 385, 552–555.

Yu, Y., Yussa, M., Song, J., Hirsch, J., Pick, L. 1999. A double interaction screen identifies positive and negative ftz gene regulators and Ftz-interacting proteins. Mech. Dev. 83, 95–105.

Yussa, M., Lohr, U., Su, K., Pick, L. 2001. The nuclear receptor Ftz-F1 and homeodomain protein Ftz interact through evolutionarily conserved protein domains. Mech. Dev. 107, 39–53.

Zanaria, E., Muscatelli, F., Bardoni, B., Strom, T.M., Guioli, S., Guo, W., Lalli, E., Moser, C., Walker, A.P., McCabe, E.R. 1994. An unusual member of the nuclear hormone receptor superfamily responsible for X-linked adrenal hypoplasia congenita. Nature 372, 635–641.

Zhang, P., Mellon, S.H. 1996. The orphan nuclear receptor steroidogenic factor-1 regulates the cyclic adenosine 3′,5′-monophosphate-mediated transcriptional activation of rat cytochrome P450c17 (17 alpha-hydroxylase/c17–20 lyase). Mol. Endocrinol. 10, 147–158.

Zhao, L., Bakke, M., Hanley, N.A., Majdic, G., Stallings, N.R., Jeyasuria, P., Parker, K.L. 2004. Tissue-specific knockouts of steroidogenic factor 1. Mol. Cell. Endocrinol. 215, 89–94.

Zhu, J., Chen, L., Raikhel, A.S. 2003. Posttranscriptional control of the competence factor betaFTZ-F1 by juvenile hormone in the mosquito Aedes aegypti. Proc. Natl. Acad. Sci. USA 100, 13338–13343.

Role of chicken ovalbumin upstream promoter-transcription factor I in the development of nervous system

Ke Tang,[1] Fu-Jung Lin,[1]
Sophia Y. Tsai[1,2] and Ming-Jer Tsai[1,2]

[1]Department of Molecular and Cellular Biology,
Baylor College of Medicine, Houston, Texas
[2]Program of Development, Baylor College of Medicine, Houston, Texas

Contents

1. Introduction

Chicken *o*valbumin *u*pstream *p*romoter-*t*ranscription *f*actors (COUP-TFs) belong to the steroid/thyroid hormone receptor superfamily (NR2F subgroup according to the nuclear receptor nomenclature, 1999). Since their

Advances in Developmental Biology
Volume 16 ISSN 1574-3349
DOI: 10.1016/S1574-3349(06)16009-3

ligands are still unknown, they are designated as orphan receptors. Like other typical members in this family, COUP-TFs contain a highly conserved DNA-binding domain (DBD) and a putative ligand-binding domain (LBD) (Tsai and O'Malley, 1994; Tsai and Tsai, 1997).

COUP-TFs were originally identified during the investigation on the expression of the chicken ovalbumin gene. The chicken ovalbumin gene possesses a duplicate G(A)T(G)GTCA box (*GGTGTCAAAGGTCAAACT*), on its promoter, named the *c*hicken *o*valbumin *u*pstream *p*romoter (COUP) element, which is essential for efficient transcription both *in vivo* and *in vitro* (Pastorcic et al., 1986; Sagami et al., 1986). COUP-TFs bind to the COUP sequence and stimulate the initiation of transcription, *in vitro*, in conjunction with S300-II, which was later identified as TFIIB (Tsai et al., 1987). In 1987, the first COUP transcription factor, human COUP-TFI (hCOUP-TFI), was purified and characterized from HeLa cell nuclear extract (Wang et al., 1987). The amino acid analysis of COUP-TF revealed that it is a member of the steroid/thyroid hormone receptor superfamily (Tsai et al., 1987). Subsequently, the first human complementary DNA clone encoding *COUP-TFI* was isolated from a HeLa cell cDNA library (Wang et al., 1989). It was also independently cloned from a human-embryo fibroblast cDNA library via homology to human *erbA* and named *EAR3* (Miyajima et al., 1988). Genomic assay indicated that there were two related *COUP-TF* genes in the human genome (Ritchie et al., 1990). In 1991, the second human family member, *hCOUP-TFII*, was identified based on its high sequence homology to *hCOUP-TFI* (Wang et al., 1991). *hCOUP-TFII* was also isolated as an apolipoprotein AI-regulatory protein-1 (ARP-1) from a placental library (Ladias and Karathanasis, 1991). *EAR2*, a distant cousin of COUP-TFs, was identified by cDNA cloning (Miyajima et al., 1988). Based on the sequence homology and its ability to inhibit retinoic acid–induced activation of target genes, EAR2 was proposed to be the third COUP-TF member (Ladias et al., 1992; Jonk et al., 1994). Following the cloning of *hCOUP-TFI*, *hCOUP-TFII*, and human *EAR2*, COUP-TFs homologues were subsequently identified in a wide range of organisms, including the *seven-up* gene (*svp*) from *Drosophila* (Mlodzik et al., 1990), *SpCOUP-TF* from sea urchin (Chan et al., 1992), *xCOUP-TFI, xCOUP-TFII*, and *xCOUP-TFIII* from *Xenopus* (Matharu and Sweeney, 1992; van der Wees et al., 1996) (Laudet, V., personal communication), *bCOUP-TFI* and *bCOUP-TFII* from bovine (Wehrenberg et al., 1992), the *zCOUP-TFI/svp [44]*, *zCOUP-TFII/svp[40]*, and *zCOUP-TFIV/svp[46]* from zebrafish (Fjose et al., 1993; Fjose et al., 1995), *mCOUP-TFI* and *mCOUP-TFII* from mouse (Jonk et al., 1994; Qiu et al., 1994), *cCOUP-TFII* from chick (Lutz et al., 1994), *rCOUP-TFI* from rat (Connor et al., 1995), *hamCOUP-TFI* from hamster (Tsai, M.-J., unpublished result), *AmphiCOUP-TF* from amphioxus (Langlois et al., 2000), and *AaSvp* from mosquito (Miura et al., 2002). Thus, COUP-TF (NR2F) subgroup of orphan nuclear receptors has been cloned from more species than any other member within the nuclear receptor superfamily

(Laudet, 1997; Tsai and Tsai, 1997; Giguere, 1999; Pereira et al., 2000). In the past decades, accumulating evidence indicates that COUP-TFs are transcriptional regulators required in development, differentiation, and homeostasis (Tsai and Tsai, 1997).

In mouse, there are three COUP transcription factors, COUP-TFI/EAR3, COUP-TFII/ARP-1, and EAR2. It has been shown that *COUP-TFII* plays a crucial role in angiogenesis, heart development, venous and arterial specification, female reproduction, limb myoblast migration, stomach patterning, and diaphragm development (Pereira et al., 1999; Lee et al., 2004; Takamoto et al., 2005a,b; You et al., 2005a,b), and *EAR2*-null mutant mouse exhibits abnormal development of the locus coeruleus and impaired regulation of the forebrain clock (Warnecke et al., 2005). In this chapter, we focus largely on the first identified member in COUP-TF subfamily, COUP-TFI, with respect to its structural homology to members of the NR2F subgroup, its expression profiles during mouse development, and, most importantly, its potential physiological functions during the development of nervous system.

2. *COUP-TFI* gene

The first COUP-TF member, hCOUP-TFI, was purified from HeLa cell nuclear extract in a screen for high-affinity COUP element-binding proteins using a combination of conventional column chromatography and sequence-specific DNA binding (Wang et al., 1987). The human complementary DNA clone encoding COUP-TF, *hCOUP-TFI*, was isolated from a HeLa cell cDNA library (Wang, et al., 1989). *COUP-TFI* gene maps to the distal region of chromosome 13 in the mouse and chromosome 5 in human (Qiu et al., 1995). Comparing with other steroid/thyroid hormone receptor genes (Tsai and O'Malley, 1994), the gene structure of *COUP-TFI* is relatively simple which encompasses approximately 10 kb and contains only 3 exons (Ritchie et al., 1990). *COUP-TFI* and *COUP-TFII* share 80% identity at DNA level (Qiu et al., 1994).

COUP-TFI encodes a protein of a predicted size of approximately 46 kD (Wang et al., 1989). As a typical member of the steroid/thyroid hormone receptor superfamily, COUP-TFI contains a highly conserved DBD with two type II zinc fingers, which is responsible for DNA recognition and dimerization, and a putative LBD, which is crucial for dimerization, nuclear localization, transactivation, and intermolecular silencing (Tsai and O'Malley, 1994; Tsai and Tsai, 1997). Within COUP-TF subgroup, both the DBD and the putative LBD exhibit a high degree of homology (Fig. 1). mCOUP-TFI and mCOUP-TFII are strikingly homologous at the amino acid level with 98% identity in the DBD and 96% identity in the putative LBD (Qiu et al., 1994; Tsai and Tsai, 1997). COUP-TFI and EAR2 also show high homology at the amino acid level with 86% identity in the DBD

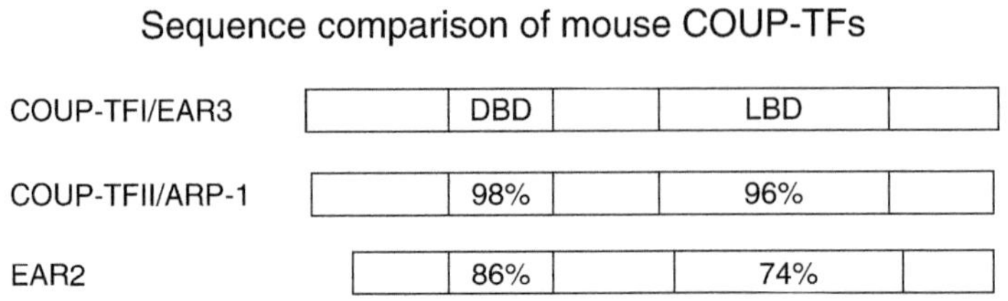

Sequence comparison of mouse COUP-TFs

COUP-TFI/EAR3   [| DBD | | LBD |]

COUP-TFII/ARP-1   [| 98% | | 96% |]

EAR2   [| 86% | | 74% |]

Fig. 1. Homology of the mouse COUP-TFs, COUP-TFI/EAR3, COUP-TFII/ARP-1, and EAR2. There are two highly conserved regions of homology, referred as DBD region and putative LBD region. The DBD region is responsible for DNA recognition and dimerization, and the putative LBD is important for heat-shock protein association, dimerization, nuclear localization, transactivation, and intermolecular silencing (Tsai and O'Malley, 1994; Tsai and Tsai, 1997).

and 74% identity in the putative LBD (Qiu et al., 1994; Tsai and Tsai, 1997). In contrast, the N-terminal domain of COUP-TFI and COUP-TFII are significantly different, and only exhibit a 40% identity, which may indicate distinct functions for the two different members (Qiu et al., 1994; Tsai and Tsai, 1997).

3. Expression of *COUP-TFI* in the developing mouse brain

The expression of *COUP-TFI* in mouse embryo is first detected at 1–2 somite stage (embryonic day 7.5, E7.5), upregulated sharply at E9.5, reaches maximal levels around E13.5, and is downregulated gradually thereafter (Qiu et al., 1994). *COUP-TFI* transcripts were distributed throughout the developing neural tissues in mouse (Table 1).

3.1. Forebrain

COUP-TFI is expressed anteriorly in the telencephalon and diencephalon. At E9.0, *COUP-TFI* expression is seen in the forebrain (Qiu et al., 1997). At E10.5, *COUP-TFI* is detected in the optic stalk, the dorsocaudal part of the telencephalon and diencephalon (Qiu et al., 1994). At E11.5, the expression of *COUP-TFI* shows similar pattern as that of E10.5. In neocortex, *COUP-TFI* exhibits high caudoventral expression, and the graded pattern is maintained even after birth (Liu et al., 2000; Zhou et al., 2001). Later, the expression of *COUP-TFI* expanded anteriorly and dorsally in the pallium and, by E13.5, the transcripts are localized throughout the pallium (Qiu et al., 1994). At this stage, the expression of *COUP-TFI* is evident in all layers of the neocortex, the ventricular and subventricular zones of the lateral ganglionic eminence, the ventricular zone of the medial ganglionic eminence, the preoptic area,

Table 1
The expression profiles of *COUP-TFI* in developing mouse brain

Stage	Expression profile	References
E8.0	Presumptive rhombomeres (r) 1–3 in the hindbrain, premigratory neural crest cell	Qiu et al., 1997
E8.5	High in dorsal edges of the presumptive r1–r4, low in the dorsal edges of the r5–r6	Qiu et al., 1997
E9.0	Forebrain, midbrain, entire hindbrain neuroepithelium, migrating NCC from r2, r4, and r6	Qiu et al., 1997
E10.5	Optic stalk, dorsocaudal part of telencephalon, diencephalon (future D1/D2 regions), tectum, spinal motor neurons, spinal posterior somatic motor components	Qiu et al., 1994
E11.5	Similar as that of E10.5, high caudoventral gradient in neocortex (maintained even after birth), D1/D2 regions of diencephalon, high in the neuroepithelium of the cranial motor nuclei, low in other hindbrain neuroepithelium, neural tube, dorsal root ganglion, trigeminal ganglion	Jonk et al., 1994; Qiu et al., 1994; Liu et al., 2000; Zhou et al., 2001
E13.5	Pallium, all layers of neocortex, ventricular and subventricular zones of the lateral ganglion eminence, ventricular zone of the medial ganglionic eminence, preoptic area, scattered cells in basal telencephalon, high in D1 and D2 regions, low in D3 and D4 areas, ventral retina, central part of dorsal retina	Qiu et al., 1994; Tripodi et al., 2004; Tang, Tsai, and Tsai, unpublished data
E14.5	Pallium, lateral and medial ganglionic eminences, high in D1, D2 regions, basal level in D3 and D4 regions, posterior commissure, cerebellum, most regions of spinal cord, retina	Jonk et al., 1994; Qiu et al., 1994; Tang, Tsai, and Tsai, unpublished data
E18.5	High caudoventral gradient in neocortex, D1, D2, D3, and D4 in diencephalon, low level throughout the spinal cord	Qiu et al., 1994; Zhou et al., 2001

and in the basal telencephalon (Tripodi et al., 2004). At E14.5, *COUP-TFI* expression is maintained in the pallium, and in both the lateral and medial ganglionic eminences. By E18.5, the expression of *COUP-TFI* in the telencephalon decreases to the basal level (Qiu et al., 1994).

In the developing diencephalic neuromeres, the expression of *COUP-TFI* shows a segmental manner. Based on the definition by Figdor and Stern (1993) for chick embryos, the neuromeres are designated as D1 (hypothalamus and ventral thalamus), D2 (dorsal thalamus), and D3 and D4 (pretectal region). At E10.5, *COUP-TFI* is expressed highly in the future D1 and D2 areas (Qiu et al., 1994). While D1 and D2 become evident at E11.5, the expression of *COUP-TFI* is maintained highly in both regions. At E14.5, the D3/D4 region is visible, and the zona limitans intrathalamica appears between D1 and D2. The expression of *COUP-TFI* remains high in D1 and D2 regions with the basal level in D3/D4 region; however, the expression of *COUP-TFI* is undetectable in zona limitans intrathalamica. Such expression profiles are consistent before birth (Qiu et al., 1994).

3.2. Midbrain

The expression of *COUP-TFI* is detected in midbrain at E9.0 (Qiu et al., 1997). By E10.5, *COUP-TFI* expression shows high anterior and low posterior gradient in the tectum, with the highest intensity at the rostral end (Qiu et al., 1994). From E16.5, the expression of *COUP-TFI* is restricted rostrally and, by E18.5, it is limited to an anterior strip of the tectum (Qiu et al., 1994).

3.3. Hindbrain and spinal cord

At 4–6 somite stage (E8.0), the expression of *COUP-TFI* is seen at the presumptive rhombomeres (r) 1–3 in the hindbrain, and *COUP-TFI* is expressed in premigratory neural crest cells (NCC). At 10–20 somite stage (E8.5), the *COUP-TFI* transcripts are readily detected in the dorsal edges of the presumptive r1–r4, and lowly in the dorsal edges of the presumptive r5–r6. At E9.0, *COUP-TFI* is detected in the entire hindbrain neuroepithelium and in the migrating NCC from r2, r4, and r6 (Qiu et al., 1997). At E11.5, *COUP-TFI* is highly expressed in the neuroepithelium of the cranial motor nuclei and at a lower level in the rest of the hindbrain neuroepithelium. In the spinal cord, *COUP-TFI* is expressed in spinal motor neurons and posterior somatic motor components from E10.5. At E14.5, *COUP-TFI* expression is seen in most regions of the spinal cord. *COUP-TFI* expression is undetectable in the somatic motor neurons at E16, but continues to be expressed in the lateral horn, where sympathetic neurons reside. By E18.5, *COUP-TFI* is expressed evenly throughout the spinal cord (Qiu et al., 1994).

In situ hybridization and immunohistochemical analysis reveals that *COUP-TFI* also exhibits dynamic expression profiles in somites, olfactory epithelium, foregut, brachial arch, heart, as well as the mesenchymal compartment of the nasal septum, the follicles of vibrissae, and the cochlea (Jonk et al., 1994; Qiu et al., 1994; Pereira et al., 1995; Qiu et al., 1997; Tang et al., 2005). Clearly, the expression pattern of *COUP-TFI* in developing nervous system strongly suggests that it may be required for early neural development and differentiation.

4. Multiple roles of the *COUP-TFI* in brain development

4.1. COUP-TFI *in PNS development*

COUP-TFI is highly expressed in the peripheral nervous system (PNS) during embryonic stages, and *COUP-TFI* gene plays a crucial role for the PNS development. *COUP-TFI$^{-/-}$* mutant mice die perinatally, apparently due to starvation and dehydration (Qiu et al., 1997). Whole-mount immunohistochemical analysis with 165-kD neurofilament protein reveals the severe defects at the IX cranial ganglion (the glossopharyngeal ganglion), which provides both sensory and motor innervations to the pharynx and the root of the tongue. In control embryos, ganglion IX and X are localized next to each other with few nerve fiber connections in between, and the axons from glossopharyngeal ganglion project to the hindbrain. However, in *COUP-TFI$^{-/-}$* mutants, ganglion IX is isolated as independent neuron cluster or is completely fused with vagus (X) cranial ganglion. In previous cases, axon connections between ganglion IX and hindbrain fail to form. The majority of mutant embryos show abnormal glossopharyngeal ganglion on one side of the embryos, while a few severely affected mutants exhibit defects on both sides. *COUP-TFI* expression is readily detected in the premigratory and migratory NCC (Qiu et al., 1997). In *COUP-TFI$^{-/-}$* mutant embryos, the NCC migrate properly as evidenced by the expression of CRABPI, the marker for the migrating NCC (Maden et al., 1992); however, TUNEL assay shows that mutant embryos exhibit excessive cell death in the region just dorsal to the developing ninth ganglion. Most likely, the superior component of the glossopharyngeal ganglion is compromised during the migratory phase of the NCC. The pronounced cell death results in a lower number of neurons specifically in the ganglia IX but not in the ganglia X of the *COUP-TFI$^{-/-}$* mutants (Qiu et al., 1997). The unusual development of glossopharyngeal ganglion impairs pharynx and tongue functions, which result in failure to swallow, leading to the death of *COUP-TFI$^{-/-}$* mutants (Qiu et al., 1997; Tsai and Tsai, 1997).

In addition to the malformation of the superior ganglion in ganglion IX, abnormal axon guidance and arborization are observed in several PNS regions of $COUP\text{-}TFI^{-/-}$ mutant embryos (Qiu et al., 1997). In E10.5 control embryos, ganglion IX sends projections to the hindbrain; by contrast, some $COUP\text{-}TFI^{-/-}$ mutant embryos lack nerve projections between the ninth ganglion and the hindbrain, and in some others, axons from glossopharyngeal ganglion fasciculate with vagus cranial ganglion (ganglion X). The mutants also exhibit shortened and broadened oculomotor nerve (ganglion III) in one side of the brain. At later developmental stages (E11.5–E13.5), about half of the $COUP\text{-}TFI^{-/-}$ mutant embryos display an unusual extent of arborization or branching of the axonal trees at the facial and cervical plexus regions, as well as at the ophthalmic branch of the trigeminal nerve. The nerve fibers from these regions have thicker primary axon but with less secondary branches and much less tertiary or higher order branches (Qiu et al., 1997). All the observations above demonstrate clearly that $COUP\text{-}TFI$ is required for axonal guidance and arborization.

4.2. COUP-TFI *in regionalization of the cerebral cortex*

The intrinsic mechanisms, involved in the early patterning of the cerebral cortex, are well studied. Patterning centers, which produce SHH, TGF-β, WNT, and FGF secreted proteins, partly direct the regionalization and morphogenesis in the cortex. These signaling centers regulate the graded expression of the transcription factors that conduct histogenetic programs for neurogenesis (Grove and Fukuchi-Shimogori, 2003; Sur and Rubenstein, 2005). At E11.5, the onset of cerebral corticogenesis, $COUP\text{-}TFI$ expression exhibits a graded pattern in the neocortex with high lateral to low medial expression and high caudal to low rostral expression. The high caudolateral expression gradient of $COUP\text{-}TFI$ is maintained in the cortical plate after birth (Liu et al., 2000; Zhou et al., 2001). FGF8 signaling is important for the regionalization of the rostral telencephalon (Grove and Fukuchi-Shimogori, 2003). It has been demonstrated that FGF signaling can modulate the expression of $COUP\text{-}TFI$ in the developing cortical primordium. Ectopic expression of FGF8 repressed COUP-TFI expression; in contrast, ectopic expression of sFGFR3c activated COUP-TFI expression (Fukuchi-Shimogori and Grove, 2003; Shimogori and Grove, 2005). The spatial–temporal expression pattern of $COUP\text{-}TFI$ indicates that it is a likely candidate for regulating regionalization in the developing cortex.

Indeed, analysis on $COUP\text{-}TFI^{-/-}$ null mutant mice revealed that the expression profiles of region-specific marker genes, such as *Id2*, *ROR-β*, and *Cadherin 8*, are largely disturbed on the developing cortex (Zhou et al., 2001). For example, expression of *Id2*, a helix-loop-helix transcription factor, displays a caudal–rostral decreasing gradient in the control cortex at E17.5; in contrast, it was expressed evenly along the whole cortex in mutant embryos. In postnatal day 3/4 (P3/P4) mice, *Id2* transcripts were detected in the subplate and layers 6, 5, and 2/3. In layer 5, the expression of *Id2* is restricted to the region caudal to the boundary separating motor cortex and somatosensory cortex. In layers 2/3, *Id2* is expressed rostrally with the boundary in the somatosensory area. However, in the *COUP-TFI* mutants, *Id2* transcripts are distributed uniformly throughout the layer 2/3 and layer 5. Similar to *Id2*, the region- and lamina-specific expression patterns of *ROR-β*, an orphan nuclear receptor, and *Cadherin 8*, a member of type II class cadherin, were altered in *COUP-TFI*$^{-/-}$ mutants (Zhou et al., 2001).

Furthermore, dye-tracing experiments show that in *COUP-TFI*$^{-/-}$ mutant, there are aberrant connections between visual cortex and ventrobasal thalamus (VB), which normally generates connections with somatosensory cortex. It suggests that the cortical area should develop into visual cortex having the characteristic of somatosensory cortex (Zhou et al., 2001). Taken together, the altered expression profiles of regional marker genes and miswired corticothalamic connections indicate that the normal function of *COUP-TFI* gene is required for the correct patterning of the cerebral cortex. Since the graded expression of *Emx2* and *Pax6*, two important intrinsic factors, does not change in *COUP-TFI*$^{-/-}$ mutants (Zhou et al., 2001), this result indicates that *COUP-TFI*, *Emx2*, and *Pax6* might function independently or in a cooperative manner to regulate the early regionalization of cortex, and COUP-TFI is one of the regulators for the formation of the caudoventral pallium and its boundaries (Sur and Rubenstein, 2005).

4.3. COUP-TFI *in neurogenesis of the central nervous system*

COUP-TFI gene plays a crucial role in determining the caudoventral identity of the cortex, and it is also important for the neurogenesis. In *COUP-TFI*$^{-/-}$ mutants, the differentiation of both layer IV neurons and subplate neurons are compromised (Zhou et al., 1999). *COUP-TFI* is highly expressed in the developing central nervous system, and the expression remains highest in the layer IV of the cortex at later stages (Qiu et al., 1994; Qiu et al., 1997; Liu et al., 2000). Analysis of the cortical laminar structure reveals a thinner cortical plate at P21 of *COUP-TFI*$^{-/-}$ mutants. The expression of *RORβ*, a layer IV marker gene, is significantly reduced at P21, which further confirms that *COUP-TFI*$^{-/-}$ mutants fail to form or maintain the small granular neuron of layer IV. BrdU-labeling experiments demonstrate that in *COUP-TFI*$^{-/-}$ mutants, layer IV neurons migrate and proliferate appropriately before E15.5; however, excessive cell death in layer IV neurons between E15.5 and P0 was detected in *COUP-TFI* mutants. The loss of the layer IV neurons might be caused by the lack of innervations from the thalamus (Zhou et al., 1999) or through a cell autonomous process (Liu et al., 2000).

COUP-TFI is highly expressed in the preplate, which separates into upper marginal zone and lower subplate layer later during corticogenesis. Subplate defect was clearly displayed in *COUP-TFI*$^{-/-}$ mutants (Zhou et al., 1999). At P0, subplate neurons are packed into a thin layer between the cortical plate and the intermediate zone in control animals; by contrast, this subplate layer was hardly detected in the mutants (Zhou et al., 1999). Calretinin, a calcium-binding protein, is expressed highly in both subplate neurons and Cajal-Retzius (CR) cells in the marginal zone of the developing cortex (Fonseca et al., 1995). In *COUP-TFI*$^{-/-}$ mutant embryos (E14.5, E15.5, E16.5), the expression of calretinin in the marginal zone is normal; however, it is undetectable in the cortical plate of the *COUP-TFI* mutants at all stages examined (Zhou et al., 1999). Moreover, gain-of-function approach was used to investigate the role of *COUP-TFI* in neuronal specification of the cortex. *COUP-TFI* expression vector was electroporated into the outermost layers of the cerebral cortex from E12.5 brain slices, and the expression of CR marker genes, such as *reelin, calretinin*, and *Tbr1*, was reduced. Therefore, the differentiation of CR cells was inhibited by the ectopic expression of *COUP-TFI* (Studer et al., 2005). *COUP-TFI* is also necessary for the proper development of oligodendrocyte in the optic nerve. *COUP-TFI* is expressed in optic nerves, which contain nerve fibers of retina ganglion cells (RGC), oligodendrocytes, and astrocytes. *COUP-TF*$^{-/-}$ mutants exhibit delayed differentiation of oligodendrocytes and delayed axon myelination in the CNS (Yamaguchi et al., 2004).

4.4. COUP-TFI *in axon guidance, innervations, and arborization*

Proper connection of thalamocortical, intracortical, and corticothalamic are important during CNS development. Axons from subplate and layer VI project to the thalamus; axons of layer V neurons target to basal ganglia, thalamus, midbrain, and brain stem; layer IV receives projections from thalamus; and layers II and III send projections to other cortical regions. The reciprocal connections between thalamus and neocortex are generated from E13 to E18 in mice (Lopez-Bendito and Molnar, 2003; Sur and Rubenstein, 2005). The absence of layer IV neurons and subplate neurons in *COUP-TFI* mutants strongly suggests that there are severe defects on axon connections between cortical areas and thalamus (Zhou et al., 1999). At E17.5 and P0, thalamocortical axon projections were anterogradely traced by placing DiI crystals in the VB, which sends the projections to the primary somatosensory cortex. In control mice, the connections were readily observed all the way from the VB to the primary somatosensory cortex, and thalamocortical axons generated local branches in the target regions. In the *COUP-TFI*$^{-/-}$ mutants, only very few thalamocortical axons from VB ever reached their target areas. The projections from the VB to the internal capsule were normal; however, very few thalamocortical axons were able to go beyond the internal capsule and extended to the intermediate zone. Thalamocortical axons displayed similar axon guidance defects as that reported previously in the PNS in the mutants (Qiu et al., 1997). After reaching the internal capsule, a majority of axons do not go further and some even turn back. Moreover, the abnormality on arborization of thalamocortical axon is also readily detected in the $COUP-TFI^{-/-}$ mutants. Very few branches develop from the axon, and no innervation was observed at P0 mutant mice (Zhou et al., 1999). The observations from $COUP-TFI^{-/-}$ mouse further support the hypothesis that thalamic axons need the guidance from preplate axons to reach the final destination in order to innervate layer IV neurons (Molnar and Blakemore, 1995; Lopez-Bendito and Molnar, 2003).

The *in vivo* data clearly suggest that $COUP-TFI$ is essential for the proper axon guidance in both CNS and PNS. Furthermore, evidence from cell culture experiments also supports the importance of $COUP-TFI$ in axon innervation. The expression of $COUP-TFI$ gene is induced sharply by the retinoic acid treatment in P19 embryonic carcinoma (EC) cells (Jonk et al., 1994). Inhibition of endogenous $COUP-TFI$ in P19 EC cell with low levels of dominant-negative protein damages neurite extension (Adam et al., 2000). In this aspect, surprisingly, $COUP-TFI$ is isolated as a neurite outgrowth inhibitor by an expression-screening method. Overexpression of $COUP-TFI$ in NIH-3T3 fibroblast cells results in a reduction of stable contact formation between neurites from NG108-15 cells and transfected cells (Connor et al., 1995). Thus, both over- and underexpression of $COUP-TFI$ disrupt the proper axon innervation of target cells. Therefore, accumulating evidence, *in vivo* and *in vitro*, indicates that $COUP-TFI$ is a potential regulatory factor for axon guidance, innervations, and arborization. It would be interesting to know how $COUP-TFI$ is involved in the regulation of axon projections.

4.5. COUP-TFI *in cell migration*

Analysis of cell migration in brain slice culture system has shed new light on the role of $COUP-TFI$ during the neurogenesis (Tripodi et al., 2004). $COUP-TFI$ is expressed in migrating cells in the basal telencephalon at early embryonic stage. Brain slice culture and graft experiments reveal that these $COUP-TFI$ positive cells not only migrate dorsally to the cortical plate but also migrate ventrally to preoptic and hypothalamic areas in a tangential migration fashion. Ectopic expression of $COUP-TFI$ in the basal forebrain promotes cell migration in both directions (Tripodi et al., 2004). *In vitro*, overexpression of $COUP-TFI$ in P19 EC cells also promotes neuronal migration after RA treatment (Adam et al., 2000). The expression of vitronectin, an extracellular matrix (ECM) protein, which regulates the SHH activities in both neural tube and cerebellum (Pons and Marti, 2000; Pons et al., 2001), is dramatically stimulated in *COUP-TFI* overexpressing cells. It seems that *COUP-TFI* may modulate cell migration through regulating synthesis of ECM proteins.

5. Summary and perspective

In the past years, growing evidence supports that *COUP-TFI* gene is required for the early neural development, especially for cortex patterning, neuronal fate specification, neural cell migration, as well as axon guidance and arborization (Fig. 2). We also observed malformation of the hippocampus, abnormality of the corpus callosum, and hippocampal commissures, as well as premature fusion of exoccipital bones with basioccipital bone (Tsai and Tsai, 1997; Pereira et al., 2000; Zhou et al., 2000; Park et al., 2003). Furthermore, *COUP-TFI* expression is evident in the developing eye, heart, stomach, kidney, nasal septum, tongue, follicles of vibrissae, and cochlea (Jonk et al., 1994; Pereira et al., 1995; Tsai and Tsai, 1997; Tsai, unpublished observation), which indicates that *COUP-TFI* may also play an important role in organogenesis. Nevertheless, the perinatal lethality of $COUP\text{-}TFI^{-/-}$ mutants hampers further exploration of its function in late development stages. In addition, *COUP-TFI* and *COUP-TFII* genes are highly homologous. Even though they display unique expression profiles in many areas of

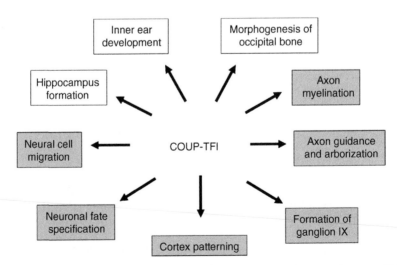

Fig. 2. Potential physiological functions of *COUP-TFI* during mouse development. Using *COUP-TFI* null mutants, our laboratories have shown that *COUP-TFI* plays a role in the development of nervous system.

the developing brain, their transcripts are colocalized in other areas of the CNS. Therefore, it is possible that *COUP-TFI* and *COUP-TFII* compensate each other in those regions. Since *COUP-TFII*$^{-/-}$ mutants also exhibit embryonic lethality, the generation of *COUP-TFI* and *COUP-TFII* flox mouse models and inducible deletion of COUP-TFs will enable us to fully understand their developmental and physiological functions.

Acknowledgments

This work was supported by NIH grants HL076448 to S.Y.T. and DK 45641 and HD17379 to M.J.T. We are very grateful to C. Zapien for helping in the preparation of this chapter.

References

Adam, F., Sourisseau, T., Metivier, R., Le Page, Y., Desbois, C., Michel, D., Salbert, G. 2000. *COUP-TFI* (chicken ovalbumin upstream promoter-transcription factor I) regulates cell migration and axogenesis in differentiating P19 embryonic carcinoma cells. Mol. Endocrinol. 14, 1918–1933.

Chan, S.M., Xu, N., Niemeyer, C.C., Bone, J.R., Flytzanis, C.N. 1992. SpCOUP-TF: A sea urchin member of the steroid/thyroid hormone receptor family. Proc. Natl. Acad. Sci. USA 89, 10568–10572.

Connor, H., Nornes, H., Neuman, T. 1995. Expression screening reveals an orphan receptor chick ovalbumin upstream promoter transcription factor I as a regulator of neurite/substrate-cell contacts and cell aggregation. J. Biol. Chem. 270, 15066–15070.

Figdor, M.C., Stern, C.D. 1993. Segmental organization of embryonic diencephalon. Nature 363, 630–634.

Fjose, A., Nornes, S., Weber, U., Mlodzik, M. 1993. Functional conservation of vertebrate *seven-up* related genes in neurogenesis and eye development. EMBO J. 12, 1403–1414.

Fjose, A., Weber, U., Mlodzik, M. 1995. A novel vertebrate svp-related nuclear receptor is expressed as a step gradient in developing rhombomeres and is affected by retinoic acid. Mech. Dev. 52, 233–246.

Fonseca, M., del Rio, J.A., Martinez, A., Gomez, S., Soriano, E. 1995. Development of calretinin immunoreactivity in the neocortex of the rat. J. Comp. Neurol. 361, 177–192.

Fukuchi-Shimogori, T., Grove, E.A. 2003. *Emx2* patterns the neocortex by regulating FGF positional signaling. Nat. Neurosci. 6, 825–831.

Giguere, V. 1999. Orphan nuclear receptors: From gene to function. Endocr. Rev. 20, 689–725.

Grove, E.A., Fukuchi-Shimogori, T. 2003. Generating the cerebral cortical area map. Annu. Rev. Neurosci. 26, 355–380.

Jonk, L.J., de Jonge, M.E., Pals, C.E., Wissink, S., Vervaart, J.M., Schoorlemmer, J., Kruijer, W. 1994. Cloning and expression during development of three murine members of the COUP family of nuclear orphan receptors. Mech. Dev. 47, 81–97.

Ladias, J.A., Karathanasis, S.K. 1991. Regulation of the apolipoprotein AI gene by ARP-1, a novel member of the steroid receptor superfamily. Science 251, 561–565.

Ladias, J.A., Hadzopoulou-Cladaras, M., Kardassis, D., Cardot, P., Cheng, J., Zannis, V., Cladaras, C. 1992. Transcriptional regulation of human apolipoprotein genes *ApoB*,

ApoCIII, and *ApoAII* by members of the steroid hormone receptor superfamily HNF-4, ARP-1, EAR-2, and EAR-3. J. Biol. Chem. 267, 15849–15860.

Langlois, M.C., Vanacker, J.M., Holland, N.D., Escriva, H., Queva, C., Laudet, V., Holland, L.Z. 2000. Amphicoup-TF, a nuclear orphan receptor of the lancelet Branchiostoma floridae, is implicated in retinoic acid signalling pathways. Dev. Genes Evol. 210, 471–482.

Laudet, V. 1997. Evolution of the nuclear receptor superfamily: Early diversification from an ancestral orphan receptor. J. Mol. Endocrinol. 19, 207–226.

Lee, C.T., Li, L., Takamoto, N., Martin, J.F., Demayo, F.J., Tsai, M.J., Tsai, S.Y. 2004. The nuclear orphan receptor *COUP-TFII* is required for limb and skeletal muscle development. Mol. Cell. Biol. 24, 10835–10843.

Liu, Q., Dwyer, N.D., O'Leary, D.D. 2000. Differential expression of *COUP-TFI, CHL1*, and two novel genes in developing neocortex identified by differential display PCR. J. Neurosci. 20, 7682–7690.

Lopez-Bendito, G., Molnar, Z. 2003. Thalamocortical development: How are we going to get there? Nat. Rev. Neurosci. 4, 276–289.

Lutz, B., Kuratani, S., Cooney, A.J., Wawersik, S., Tsai, S.Y., Eichele, G., Tsai, M.J. 1994. Developmental regulation of the orphan receptor *COUP-TF II* gene in spinal motor neurons. Development 120, 25–36.

Maden, M., Horton, C., Graham, A., Leonard, L., Pizzey, J., Siegenthaler, G., Lumsden, A., Eriksson, U. 1992. Domains of cellular retinoic acid-binding protein I (CRABP I) expression in the hindbrain and neural crest of the mouse embryo. Mech. Dev. 37, 13–23.

Matharu, P.J., Sweeney, G.E. 1992. Cloning and sequencing of a COUP transcription factor gene expressed in *Xenopus* embryos. Biochim. Biophys. Acta 1129, 331–334.

Miura, K., Zhu, J., Dittmer, N.T., Chen, L., Raikhel, A.S. 2002. A *COUP-TF/Svp* homolog is highly expressed during vitellogenesis in the mosquito Aedes aegypti. J. Mol. Endocrinol. 29, 223–238.

Miyajima, N., Kadowaki, Y., Fukushige, S., Shimizu, S., Semba, K., Yamanashi, Y., Matsubara, K., Toyoshima, K., Yamamoto, T. 1988. Identification of two novel members of *erbA* superfamily by molecular cloning: The gene products of the two are highly related to each other. Nucleic Acids Res. 16, 11057–11074.

Mlodzik, M., Hiromi, Y., Weber, U., Goodman, C.S., Rubin, G.M. 1990. The *Drosophila seven-up* gene, a member of the steroid receptor gene superfamily, controls photoreceptor cell fates. Cell 60, 211–224.

Molnar, Z., Blakemore, C. 1995. How do thalamic axons find their way to the cortex? Trends Neurosci. 18, 389–397.

Park, J.I., Tsai, S.Y., Tsai, M.J. 2003. Molecular mechanism of chicken ovalbumin upstream promoter-transcription factor (COUP-TF) actions. Keio J. Med. 52, 174–181.

Pastorcic, M., Wang, H., Elbrecht, A., Tsai, S.Y., Tsai, M.J., O'Malley, B.W. 1986. Control of transcription initiation *in vitro* requires binding of a transcription factor to the distal promoter of the ovalbumin gene. Mol. Cell. Biol. 6, 2784–2791.

Pereira, F.A., Qiu, Y., Tsai, M.J., Tsai, S.Y. 1995. Chicken ovalbumin upstream promoter transcription factor (COUP-TF): Expression during mouse embryogenesis. J. Steroid Biochem. Mol. Biol. 53, 503–508.

Pereira, F.A., Qiu, Y., Zhou, G., Tsai, M.J., Tsai, S.Y. 1999. The orphan nuclear receptor *COUP-TFII* is required for angiogenesis and heart development. Genes Dev. 13, 1037–1049.

Pereira, F.A., Tsai, M.J., Tsai, S.Y. 2000. COUP-TF orphan nuclear receptors in development and differentiation. Cell. Mol. Life Sci. 57, 1388–1398.

Pons, S., Marti, E. 2000. Sonic hedgehog synergizes with the extracellular matrix protein vitronectin to induce spinal motor neuron differentiation. Development 127, 333–342.

Pons, S., Trejo, J.L., Martinez-Morales, J.R., Marti, E. 2001. Vitronectin regulates Sonic hedgehog activity during cerebellum development through CREB phosphorylation. Development 128, 1481–1492.

Qiu, Y., Cooney, A.J., Kuratani, S., DeMayo, F.J., Tsai, S.Y., Tsai, M.J. 1994. Spatiotemporal expression patterns of chicken ovalbumin upstream promoter-transcription factors in the developing mouse central nervous system: Evidence for a role in segmental patterning of the diencephalon. Proc. Natl. Acad. Sci. USA 91, 4451–4455.

Qiu, Y., Krishnan, V., Zeng, Z., Gilbert, D.J., Copeland, N.G., Gibson, L., Yang-Feng, T., Jenkins, N.A., Tsai, M.J., Tsai, S.Y. 1995. Isolation, characterization, and chromosomal localization of mouse and human *COUP-TF I* and *II* genes. Genomics 29, 240–246.

Qiu, Y., Pereira, F.A., DeMayo, F.J., Lydon, J.P., Tsai, S.Y., Tsai, M.J. 1997. Null mutation of *mCOUP-TFI* results in defects in morphogenesis of the glossopharyngeal ganglion, axonal projection, and arborization. Genes Dev. 11, 1925–1937.

Ritchie, H.H., Wang, L.H., Tsai, S., O'Malley, B.W., Tsai, M.J. 1990. *COUP-TF* gene: A structure unique for the steroid/thyroid receptor superfamily. Nucleic Acids Res. 18, 6857–6862.

Sagami, I., Tsai, S.Y., Wang, H., Tsai, M.J., O'Malley, B.W. 1986. Identification of two factors required for transcription of the ovalbumin gene. Mol. Cell. Biol. 6, 4259–4267.

Shimogori, T., Grove, E.A. 2005. Fibroblast growth factor 8 regulates neocortical guidance of area-specific thalamic innervation. J. Neurosci. 25, 6550–6560.

Studer, M., Filosa, A., Rubenstein, J.L. 2005. The nuclear receptor *COUP-TFI* represses differentiation of Cajal-Retzius cells. Brain Res. Bull. 66, 394–401.

Sur, M., Rubenstein, J.L.R. 2005. Patterning and plasticity of the cerebral cortex. Science 310, 805–810.

Takamoto, N., Kurihara, I., Lee, K., Demayo, F.J., Tsai, M.J., Tsai, S.Y. 2005a. Haploinsufficiency of chicken ovalbumin upstream promoter transcription factor II in female reproduction. Mol. Endocrinol. 19, 2299–2308.

Takamoto, N., You, L.R., Moses, K., Chiang, C., Zimmer, W.E., Schwartz, R.J., DeMayo, F.J., Tsai, M.J., Tsai, S.Y. 2005b. *COUP-TFII* is essential for radial and anteroposterior patterning of the stomach. Development 132, 2179–2189.

Tang, L.S., Alger, H.M., Lin, F., Pereira, F.A. 2005. Dynamic expression of *COUP-TFI* and *COUP-TFII* during development and functional maturation of the mouse inner ear. Gene Expr. Patterns 5, 587–592.

Tripodi, M., Filosa, A., Armentano, M., Studer, M. 2004. The *COUP-TF* nuclear receptors regulate cell migration in the mammalian basal forebrain. Development 131, 6119–6129.

Tsai, S.Y., Sagami, I., Wang, H., Tsai, M.J., O'Malley, B.W. 1987. Interactions between a DNA-binding transcription factor (COUP) and a non-DNA binding factor (S300-II). Cell 50, 701–709.

Tsai, M.J., O'Malley, B.W. 1994. Molecular mechanisms of action of steroid/thyroid receptor superfamily members. Annu. Rev. Biochem. 63, 451–486.

Tsai, S.Y., Tsai, M.J. 1997. Chick ovalbumin upstream promoter-transcription factors (COUP-TFs): Coming of age. Endocr. Rev. 18, 229–240.

van der Wees, J., Matharu, P.J., de Roos, K., Destree, O.H., Godsave, S.F., Durston, A.J., Sweeney, G.E. 1996. Developmental expression and differential regulation by retinoic acid of *Xenopus COUP-TF-A* and *COUP-TF-B*. Mech. Dev. 54, 173–184.

Wang, L.H., Tsai, S.Y., Sagami, I., Tsai, M.J., O'Malley, B.W. 1987. Purification and characterization of chicken ovalbumin upstream promoter transcription factor from HeLa cells. J. Biol. Chem. 262, 16080–16086.

Wang, L.H., Tsai, S.Y., Cook, R.G., Beattie, W.G., Tsai, M.J., O'Malley, B.W. 1989. COUP transcription factor is a member of the steroid receptor superfamily. Nature 340, 163–166.

Wang, L.H., Ing, N.H., Tsai, S.Y., O'Malley, B.W., Tsai, M.J. 1991. The COUP-TFs compose a family of functionally related transcription factors. Gene Expr. 1, 207–216.

Warnecke, M., Oster, H., Revelli, J.P., Alvarez-Bolado, G., Eichele, G. 2005. Abnormal development of the locus coeruleus in *Ear2*(Nr2f6)-deficient mice impairs the functionality of the forebrain clock and affects nociception. Genes Dev. 19, 614–625.

Wehrenberg, U., Ivell, R., Walther, N. 1992. The COUP transcription factor (COUP-TF) is directly involved in the regulation of oxytocin gene expression in luteinizing bovine granulosa cells. Biochem. Biophys. Res. Commun. 189, 496–503.

Yamaguchi, H., Zhou, C., Lin, S.C., Durand, B., Tsai, S.Y., Tsai, M.J. 2004. The nuclear orphan receptor *COUP-TFI* is important for differentiation of oligodendrocytes. Dev. Biol. 266, 238–251.

You, L.R., Lin, F.J., Lee, C.T., DeMayo, F.J., Tsai, M.J., Tsai, S.Y. 2005a. Suppression of Notch signalling by the *COUP-TFII* transcription factor regulates vein identity. Nature 435, 98–104.

You, L.R., Takamoto, N., Yu, C.T., Tanaka, T., Kodama, T., Demayo, F.J., Tsai, S.Y., Tsai, M.J. 2005b. Mouse lacking *COUP-TFII* as an animal model of Bochdalek-type congenital diaphragmatic hernia. Proc. Natl. Acad. Sci. USA 102, 16351–16356.

Zhou, C., Qiu, Y., Pereira, F.A., Crair, M.C., Tsai, S.Y., Tsai, M.J. 1999. The nuclear orphan receptor *COUP-TFI* is required for differentiation of subplate neurons and guidance of thalamocortical axons. Neuron 24, 847–859.

Zhou, C., Tsai, S.Y., Tsai, M. 2000. From apoptosis to angiogenesis: New insights into the roles of nuclear orphan receptors, chicken ovalbumin upstream promoter-transcription factors, during development. Biochim. Biophys. Acta 1470, M63–M68.

Zhou, C., Tsai, S.Y., Tsai, M.J. 2001. *COUP-TFI*: An intrinsic factor for early regionalization of the neocortex. Genes Dev. 15, 2054–2059.

Retinoid-related orphan receptors (RORs): Roles in cellular differentiation and development

Anton M. Jetten and Joung Hyuck Joo

Cell Biology Section, Division of Intramural Research,
National Institute of Environmental Health Sciences,
National Institutes of Health, Research Triangle Park, North Carolina

Contents

Advances in Developmental Biology
Volume 16 ISSN 1574-3349
DOI: 10.1016/S1574-3349(06)16010-X

Retinoid-related orphan receptors RORα, RORβ, and RORγ are transcription factors belonging to the steroid hormone receptor superfamily. During embryonic development, RORs are expressed in a spatial and temporal manner and are critical in the regulation of cellular differentiation and the development of several tissues. RORα plays a key role in the development of the cerebellum particularly in the regulation of the maturation and survival of Purkinje cells. In RORα-deficient mice, the reduced production of Sonic Hedgehog by these cells appears to be the major cause of the decreased proliferation of granule cell precursors and the observed cerebellar atrophy. RORα has been implicated in the regulation of a number of other physiological processes, including bone formation. RORβ expression is largely restricted to several regions of the brain, retina, and pineal gland. Mice deficient in RORβ develop retinal degeneration that results in blindness. RORγ is essential for lymph node organogenesis. In the intestine, RORγ is required for the formation of several other lymphoid tissues: Peyer's patches, cryptopatches, and isolated lymphoid follicles. RORγ plays a key role in the generation of lymphoid tissue inducer (LTi) cells that are essential for the development of these lymphoid tissues. In addition, RORγ is a critical regulator of thymopoiesis. It controls the differentiation of immature single-positive thymocytes into double-positive (DP) thymocytes and promotes the survival of DP thymocytes by inducing the expression of the antiapoptotic gene Bcl-X_L. All three RORs appear to play a role in the control of circadian rhythms. RORα positively regulates the expression of Bmal1, a transcription factor that is critical in the control of the circadian clock. This chapter intends to provide an overview of the status of the functions RORs have in these biological processes.

1. Introduction

Retinoid-related orphan receptors (RORs) form a subgroup of the nuclear receptor superfamily (Evans, 1988; Willy and Mangelsdorf, 1998; Desvergne and Wahli, 1999; Giguere, 1999; Kumar and Thompson, 1999; Aranda and Pascual, 2001; Jetten et al., 2001; Jetten, 2004; Novac and Heinzel, 2004). This subfamily consists of three members: RORα (Becker-Andre et al., 1993; Giguere et al., 1994, 1995), RORβ (Carlberg et al., 1994; Schaeren-Wiemers et al., 1997; Andre et al., 1998b), and RORγ (Hirose et al., 1994; Ortiz et al., 1995; Medvedev et al., 1996; He et al., 1998; Jetten et al., 2001; Jetten and Ueda, 2002). These isotypes are also referred to as NR1F1–3 (Nuclear Receptor Nomenclature Committee) and RORA-C (Human Gene Nomenclature Committee) and have been cloned from many mammalian species, including mouse, rat, and human (Becker-Andre et al., 1993; Carlberg et al., 1994; Giguere et al., 1994; Hirose et al., 1994; Ortiz et al., 1995; Medvedev et al., 1996; He et al., 1998; Koibuchi and Chin, 1998; Jetten et al., 2001; Jetten, 2004). Homologues of RORs have been identified in several lower species. *Drosophila melanogaster* (Koelle et al., 1992; Horner et al., 1995; Thummel, 1995; Carney et al., 1997; Sullivan and Thummel, 2003; Gates et al., 2004), the nematode *Caenorhabditis elegans*, the spruce budworm *Choristoneura fumiferana* (Kostrouch et al., 1995; Palli et al., 1996, 1997), and the tobacco hawkmoth *Manduca sexta* (Palli et al., 1992; Lan et al., 1997, 1999; Langelan et al., 2000; Riddiford et al., 2003; Hiruma and Riddiford, 2004) express an ROR homologue, referred to as DHR3, CHR3, and MHR3, respectively.

The RORα, RORβ, and RORγ genes have been mapped to human chromosomes 15q22.2, 9q21.13, and 1q21.3, respectively. The RORα gene, which comprises 12 exons encoding a coding region of about 1.7 kb, spans a relatively large 730 kb region of the genome. The RORβ gene covers 188 kb, while the RORγ spans only 24 kb. The RORα gene is located in the middle of the common fragile site FRA15A (Smith et al., 2006; Zhu et al., 2006). Genomic instability, including DNA breakage and rearrangements, at common fragile sites may result in changes in the expression and function of genes encoded within these regions. Altered expression of genes associated with common fragile sites, such as the fragile histidine triad (FHIT) and WW domain-containing oxidoreductase 1 (WOX1) gene, has been implicated in human disease and particularly in the development of different types of cancer (Smith et al., 2006). The association of the RORα gene with FRA15A opens the possibility that genomic instability in this region may lead to changes in the expression of RORα and be a factor in the development of certain cancers.

Each ROR gene produces several variants or isoforms that are generated by a combination of alternative promoter usage and exon splicing (Giguere et al., 1994; Hamilton et al., 1996; Matysiak-Scholze and Nehls, 1997; He et al., 1998; Andre et al., 1998b; Villey et al., 1999). These isoforms differ only in their N-terminal A/B domain. In humans, four different RORα isoforms, referred to as RORα1–ROR4, have been identified, while only two isoforms, RORα1 and RORα4, have been reported for mouse. Each of RORβ and RORγ genes generates two different isoforms, 1 and 2 (He et al., 1998; Andre et al., 1998b; Villey et al., 1999). The variants differ in their pattern of tissue-specific expression and can regulate distinct physiological processes and target genes. For example, RORα1 and RORα4 are coexpressed in mouse cerebellum, while other mouse tissues express predominantly RORα4 (Matysiak-Scholze and Nehls, 1997; Chauvet et al., 2002). The expression of RORγ2, also referred to as RORγt, is highly restricted to $CD4^+CD8^+$ thymocytes in the thymus and to lymphoid tissue inducer (LTi) cells, while other tissues express RORγ1 (He et al., 1998; Eberl et al., 2004). Expression of RORβ2 appears to be restricted to the pineal gland and the retina (Andre et al., 1998b).

2. Regulation of gene transcription by RORs

RORs exhibit a structural architecture that is typical of nuclear receptors. RORs contain four major functional domains: an N-terminal (A/B) domain, a DNA-binding domain (DBD), a hinge domain, and a ligand-binding domain

(LBD) (Evans, 1988; Moras and Gronemeyer, 1998; Willy and Mangelsdorf, 1998; Giguere, 1999; Pike et al., 2000; Jetten et al., 2001; Steinmetz et al., 2001). The DBD consists of two highly conserved zinc finger motifs involved in the recognition of ROR response elements (ROREs) that consist of the consensus motif AGGTCA preceded by an AT-rich sequence (Carlberg et al., 1994; Giguere et al., 1994; Ortiz et al., 1995; Greiner et al., 1996; Medvedev et al., 1996; Schrader et al., 1996; Andre et al., 1998a; Moraitis and Giguere, 1999; Jetten et al., 2001). The RORs bind ROREs as a monomer (Carlberg et al., 1994; Giguere et al., 1994; Ortiz et al., 1995; Greiner et al., 1996; Medvedev et al., 1996; Schrader et al., 1996; Andre et al., 1998b; Moraitis and Giguere, 1999). This conclusion was supported by crystal structure analyses that indicated the presence of a kink in H10 of the RORβ(LBD) at A^{411}-K^{412} and of the RORα(LBD) at C^{505}-K^{506} (Stehlin et al., 2001; Kallen et al., 2002; Stehlin-Gaon et al., 2003). Because H10 plays a critical role in the homo- and heterodimerization of nuclear receptors, the presence of a kink would greatly affect the dimerization capability of RORs. Therefore, it is unlikely that RORs are able to form receptor homo- or heterodimers. The P-box, the loop between the last two cysteins within the first zinc finger, recognizes the core motif in the major groove (Giguere et al., 1995; McBroom et al., 1995; Jetten et al., 2001). Residues just downstream from the two zinc fingers, referred to as C-terminal extension (CTE), play a critical role in determining the DNA binding specificity of RORs (Giguere et al., 1994, 1995; Vu-Dac et al., 1997; Andre et al., 1998b; Sundvold and Lien, 2001). The CTE makes contact with the 5′-AT-rich segment of the RORE in the adjacent minor groove. The N-terminus (A/B domain) also influences the binding affinity of RORs. This modulation likely accounts for the distinct binding specificities exhibited by the different ROR variants (Giguere et al., 1994, 1995; Vu-Dac et al., 1997; Andre et al., 1998b; Sundvold and Lien, 2001).

The nuclear receptors Rev-ErbAα and Rev-Erbβ (NR1D1 and D2, respectively) have very similar binding specificities for ROREs as the RORs (Giguere et al., 1995). Since Rev-Erb receptors act as constitutive transcriptional repressors, they are able to inhibit ROR-mediated transcriptional activation by competing with RORs for the same DNA response element (Forman et al., 1994; Retnakaran et al., 1994; Downes et al., 1996; Medvedev et al., 1997; Austin et al., 1998; Bois-Joyeux et al., 2000). Studies have indicated the physiological significance of such interplay in the control of circadian rhythm. Rev-Erb and RORα, respectively, repress and activate transcription of the Bmal1 gene, which encodes a transcription factor that is critical in the control of the circadian clock, by competing for ROREs in the promoter region of the Bmal1 gene (Albrecht, 2002; Preitner et al., 2002; Gachon et al., 2004; Nakajima et al., 2004; Triqueneaux et al., 2004; Akashi and Takumi, 2005; Guillaumond et al., 2005). Such cross talk may be implicated in the regulation of other physiological processes as well.

The LBDs of nuclear receptors play a role in ligand binding, nuclear localization, receptor dimerization, contain a transactivation function, and provide an interface for the interaction with coactivators and corepressors. The LBDs of RORs are moderately conserved; the LBD of RORα exhibits a 63% and 58% identity with that of RORβ and RORγ, respectively (Jetten et al., 2001; Jetten, 2004). Although the LBDs of nuclear receptors do not exhibit a high degree of homology, their secondary structure is very similar and usually contains 12 α-helices (H1–H12). H12, which contains the activation function 2 (AF-2) consensus motif ΦΦXE/DΦΦ (where Φ is a hydrophobic amino acid and X is any amino acid), is 100% conserved among RORs. X-ray structural analysis has demonstrated that RORs have two additional helices, H2′ and H11′ (Stehlin et al., 2001; Kallen et al., 2002; Stehlin-Gaon et al., 2003). In addition, these studies provided evidence indicating that the activity of RORs is controlled by ligands. Cholesterol, 7-dehydrocholesterol, and cholesterol sulfate have been shown to bind RORα in a reversible manner and to enhance RORE-dependent transcriptional activation by RORα (Kallen et al., 2002, 2004). Whether cholesterol, cholesterol sulfate, or other (sulfated) lipid metabolites function as physiological ligands for RORα or whether they are merely structural cofactors that stabilize the conformation of RORα has yet to be established.

Initial studies of the X-ray structure of the RORβ(LBD) identified stearic acid as a fortuitously captured ligand that appeared to act as a filler and stabilizer rather than as a functional ligand (Stehlin et al., 2001). Follow-up studies demonstrated that several retinoids, including all-*trans* retinoic acid (ATRA) and the synthetic retinoid ALRT 1550 (ALRT), were able to act as functional ligands for RORβ (Stehlin-Gaon et al., 2003). These retinoids were able to bind RORβ(LBD) with high affinity and in a reversible manner while the RAR-selective antagonist RO 41–5253 was unable to bind. ATRA and ALRT 1550 were also able to bind RORγ but not RORα (Stehlin-Gaon et al., 2003). Both ATRA and ALRT inhibited RORβ-mediated transcriptional activation suggesting that these retinoids act as partial antagonists. Future studies are needed to determine whether cholesterol(sulfate) and retinoic acid are genuine physiological ligands for RORs and, if so, what physiological functions and target genes they regulate. Regardless of whether these agents are physiological ligands, these studies have demonstrated that the activity of RORs can be modulated by (synthetic) ligands and, therefore, support the concept that RORs might be potential targets for pharmacological intervention of pathological processes.

For many receptors, binding of a ligand functions as a switch that induces a conformational change in the receptor that involves a repositioning of H12 (AF-2) (Harding et al., 1997; Heery et al., 1997; Darimont et al., 1998; McInerney et al., 1998; Nolte et al., 1998; Nagy et al., 1999; Xu et al., 1999; Glass and Rosenfeld, 2000; Heery et al., 2001). Binding of an agonist induces a transcriptionally active conformation of the receptor. The activated state allows recruitment of coactivator complexes that, through their histone acetylase activity, induce chromatin remodeling and subsequently an increase in the transcription of target genes (Glass and Rosenfeld, 2000; McKenna and O'Malley, 2002; Xu, 2005). Recruitment of a corepressor complex induces, through its histone deacetylase activity, compactation of chromatin and repression of gene expression (Horlein et al., 1995; Nagy et al., 1999; Hu and Lazar, 2000). RORs have been reported to be able to interact with corepressors as well as coactivators suggesting that RORs can function as repressors and activators of gene transcription. Whether the recruitment of corepressors and coactivators by RORs is dependent on, respectively, the absence and presence of physiological ligands has yet to be determined. However, study of RORα-mediated transcriptional regulation shows (Gold et al., 2003) that RORs recruit different coactivator complexes when bound to ROREs in the promoter region of different genes. This suggests that the promoter context plays an important role in determining which coactivators are recruited by RORs (Jetten et al., 2001; Jetten, 2004). Table 1 shows a summary of proteins that have been reported to interact with ROR receptors. NCOA1, NCOA2, PGC-1α, p300, and CBP are among the coactivators reported to mediate transcriptional activation by RORs (Harding et al., 1997; Atkins et al., 1999; Lau et al., 1999; Littman et al., 1999; Harris et al., 2002; Gold et al., 2003; Jetten, 2004; Kurebayashi et al., 2004; Xie et al., 2005; Jetten et al., unpublished obervations). RORs have also been demonstrated to interact with a number of corepressors, including NCOR1, NCOR2, RIP140, and the neuronal interacting factor X (NIX1) (Harding et al., 1997; Littman et al., 1999; Greiner et al., 2000; Jetten, 2004; Johnson et al., 2004; Jetten et al., unpublished observations) (Table 1).

RORs also interact with several proteins that are involved in ubiquitination or are part of the proteasome complex (Atkins et al., 1999; Jetten et al., unpublished obervations) (Table 1). The ubiquitin (Ub)-proteasome system is intimately involved in regulating chromatin structure remodeling and in the transcriptional control by a number of nuclear receptors (Poukka et al., 1999; Dace et al., 2000; Dennis et al., 2001; Wallace and Cidlowski, 2001; Ismail and Nawaz, 2005; Kinyamu et al., 2005). Ubiquitination is a multistep process that involves three types of enzymes: ubiquitin-activating enzymes (E1), ubiquitin-conjugating enzymes (E2), and ubiquitin ligases (E3) (Weissman, 2001). E3 ligases link ubiquitin covalently to Lys residues in target proteins and then successively attach additional ubiquitins forming a polyubiquitin chain. Polyubiquitin serves as a recognition signal and targets the proteins to the proteasome for rapid degradation (Adams, 2003; Muratani and Tansey, 2003). Several studies have indicated a link between ROR signaling and the ubiquitin-proteasome system. The proteasome subunit β type 6 (PSMB6) and the proteasome 26S ATPase subunit PSMC5 have been shown to interact with ROR receptors (Atkins et al., 1999; Jetten et al., unpublished obervations). In addition, inhibition of the 26S proteasome complex by the proteasome

Table 1
Summary of proteins recruited by ROR receptors

Protein symbols	References
Coactivators	
CREBBP (CBP)	Gold et al., 2003; Jetten et al., unpublished obervations
P300	Lau et al., 1999; Harris et al., 2002; Gold et al., 2003
NCOA1 (SRC1)	Littman et al., 1999; Moraitis et al., 2002; Gold et al., 2003; Kurebayashi et al., 2004; Xie et al., 2005
NCOA2 (GRIP1, TIF2)	Atkins et al., 1999; Harris et al., 2002; Moraitis et al., 2002; Gold et al., 2003; Xie et al., 2005; Jetten et al., unpublished observations
NCOA3 (SRC3)	Moraitis et al., 2002; Gold et al., 2003
NCOA6 (AIB3, PRIB)	Jetten et al., unpublished obervations
PPARGC1A (PGC-1α)	Jetten et al., unpublished obervations
CTNNB1 (β-catenin)	Gold et al., 2003
HTATIP (TIP60)	Gold et al., 2003
TRIP11 (TRIP230)	Atkins et al., 1999
PPARBP (TRIP2)	Atkins et al., 1999; Harris et al., 2002
Corepressors	
NCOR1 (N-CoR)	Harding et al., 1997; Jetten et al., unpublished obervations
NCOR2 (SMRT)	Harding et al., 1997
NRIP1 (RIP140)	Littman et al., 1999; Jetten et al., unpublished obervations
CDH4 (Mi-2b)	Johnson et al., 2004
NIX1	Greiner et al., 1996
HR (Hairless)	Moraitis et al., 2002; Moraitis and Giguere, 2003
Ubiquitin-proteasome system	
UBE21 (UBC9)	Jetten et al., unpublished obervations
PSMB6 (macropain)	Jetten et al., unpublished obervations
PSMC5 (TRIP1, SUG1)	Atkins et al., 1999; Jetten et al., unpublished obervations
Others	
TRIM24 (TIF1)	Atkins et al., 1999
PNRC-1/2	Jetten et al., unpublished obervations
MyoD	Lau et al., 1999

inhibitor MG-132 has been reported to increase the level of ubiquitinated RORα protein and to inhibit RORα-mediated transcriptional activation (Moraitis and Giguere, 2003). These studies suggest that the ubiquitin-proteasome system is an integral part of the mechanism by which RORs control transcription. The latter is supported by studies showing that Hairless (*Hr*) (Cachon-Gonzalez et al., 1994) functions as an effective repressor of ROR-induced transcriptional activation (Moraitis et al., 2002; Hsieh et al., 2003; Moraitis and Giguere, 2003). This repressor activity appears to involve protection of RORs from degradation by the ubiquitin (Ub)-proteasome system. The ubiquitin-conjugating enzyme I (UBE2I or UBC9), which catalyzes sumoylation, also interacts with RORs (Jetten et al., unpublished obervations). Whether UBE2I plays a role in the sumoylation of RORs has yet to be established.

RORs have been shown to induce RORE-dependent transcriptional activation of a reporter gene in a cell type-dependent manner (Giguere et al., 1994; Ortiz et al., 1995; Greiner et al., 1996; Medvedev et al., 1996; Schrader et al., 1996; Austin et al., 1998; Atkins et al., 1999; Delerive et al., 2001; Gawlas and Stunnenberg, 2001) and functionally active ROREs have been identified in many putative target genes, including apolipoproteins A-I, C-III, and A-V, prosaposin, and IκBα (Carlberg and Wiesenberg, 1995; Steinhilber et al., 1995; Tini et al., 1995; Matsui, 1996; Paravicini et al., 1996; Schrader et al., 1996; Dussault and Giguere, 1997; Matsui, 1997; Vu-Dac et al., 1997; He et al., 1998; Jin et al., 1998, 2001; Chu and Zingg, 1999; Littman et al., 1999; Villey et al., 1999; Bois-Joyeux et al., 2000; Delerive et al., 2001, 2002; Raspe et al., 2001; Sundvold and Lien, 2001; Sun et al., 2003; Genoux et al., 2005; Lind et al., 2005). However, further studies are needed to determine whether these genes are genuine physiological targets of RORs. Microarray analysis identified a number of authentic RORα target genes in cerebellar Purkinje cells. These include Sonic Hedgehog (Shh) and Purkinje cell protein 2 (Pcp2) (Gold et al., 2003). This will be discussed later in more detail.

3. Functions of RORα in brain development

RORα is expressed in a number of adult tissues. In testis, RORα is expressed in the peritubular cells but not in the seminiferous tubules (Steinmayr et al., 1998). In skin, RORα is expressed in the suprabasal cells of the epidermis, sebaceous glands, and anagen hair follicles. Spermatogenesis is unaffected in mice deficient in RORα expression (RORα$^{-/-}$ mice) and mice are fertile (Steinmayr et al., 1998). In addition, RORα$^{-/-}$ mice do not display changes in epidermal differentiation; however, these mice have been reported to develop a less dense fur that grows back slowly after shaving, suggesting a regulatory role in hair development.

Several studies have shown that RORα is most highly expressed in several regions of the brain, particularly the cerebellum and thalamus (Becker-Andre et al., 1993; Carlberg et al., 1994; Hamilton et al., 1996; Matysiak-Scholze and Nehls, 1997; Nakagawa et al., 1997, 1998; Dussault et al., 1998; Vogel et al., 2000). In the thalamus, RORα mRNA is highly expressed in neurons and in the cerebellum it is detected in Purkinje cells but not in the granule cell layer (Sotelo and Wassef, 1991; Matsui et al., 1995; Sashihara et al., 1996; Nakagawa et al., 1997; Ino, 2004). RORα is expressed at moderate levels in the olfactory bulb and at low levels in layer IV of the neocortex (Monnier et al., 1999; Michel et al., 2000; Nakagawa and O'Leary, 2003). Low levels of expression of RORα can also be observed in the pituitary, the superficial region of the dorsal cochlear nucleus, the suprachiasmatic nucleus, the superior colliculus, and the spinal trigeminal nucleus. In the retina, the ganglion cells and cells of the inner nuclear layer contain low levels of RORα protein and mRNA (Steinmayr et al., 1998; Ino, 2004).

During forebrain development, RORα mRNA is expressed in a spatiotemporal manner (Sashihara et al., 1996; Nakagawa and O'Leary, 2003). At embryonic day 11.5 (E11.5), RORα mRNA is not detectable in the dorsal thalamus but by E12.5 is expressed in a narrow ventrolateral region in the mantle zone where the LIM homeobox 9 gene (*Lhx9*) is not expressed. At E17.5 of mouse development, RORα is expressed strongly throughout the ventroposterior nucleus, the dorsal lateral geniculate nucleus, and the ventral part of the medial geniculate nucleus (Nakagawa and O'Leary, 2003; Ino, 2004). These regions relay somatosensory, visual, and auditory information to the neocortex. It was concluded that expression of RORα in the principle sensory neurons begins shortly after they become postmitotic. In the neocortex, RORα is first detected at E18.5 in the middle layers of the middle part of the neocortex and at postnatal day (P) 0 (day of birth) weakly in the somatosensory area (Nakagawa and O'Leary, 2003). By P5, a moderate level of expression is seen in cortical layer IV.

RORα-null mice, generated by targeted disruption of the RORα gene, display ataxia that is correlated with severe cerebellar atrophy (Dussault et al., 1998; Steinmayr et al., 1998). The ataxic and cerebellar phenotypes are identical to that observed in homozygous *staggerer* mutant mice (RORα$^{sg/sg}$) which carry a deletion within the RORα gene that prevents translation of the LBD (Sidman et al., 1962; Landis and Sidman, 1978; Herrup and Mullen, 1981; Hamilton et al., 1996). Moreover, the electrophysiological characteristics are indistinguishable between the two mouse strains. The ataxic phenotype observed in RORα-deficient mice is at least partially due to the development of cerebellar atrophy (Gold et al., in press). The cerebellum of RORα-deficient mice contains significantly fewer Purkinje cells and exhibits a loss of cerebellar granule cells (Sidman et al., 1962; Landis and Reese, 1977; Landis and Sidman, 1978; Herrup and Mullen, 1979, 1981; Hamilton et al., 1996; Doulazmi et al., 1999). It has been reported that the cerebellum of adult RORα$^{sg/sg}$ mice contains 80% fewer calbindin-positive Purkinje cells compared to wild-type mice (Doulazmi et al., 2001). Although the morphology of the cerebellar cortex in young heterozygous

($ROR\alpha^{sg/+}$) mice appears normal, on aging an accelerated loss of Purkinje cells is noticed. This loss occurs earlier in males than in females, suggesting that gender is a factor (Hadj-Sahraoui et al., 1997, 2001; Doulazmi et al., 1999).

The reduction in the number of Purkinje cells could be due to either a defect in the control of proliferation or differentiation of Purkinje cell precursors or to a defect in the maturation or survival of Purkinje cells (Landis and Sidman, 1978; Herrup and Mullen, 1979, 1981; Bouvet et al., 1987; Messer et al., 1990; Sotelo and Wassef, 1991; Boukhtouche et al., 2006a,b; Gold et al., in press). All cells in the cerebellum are derived from the germinal matrix region of the metencephalon (Goldowitz and Hamre, 1998). The nuclear and Golgi neurons, and Purkinje cells are generated from the ventricular neuroepithelium. The granule precursors arise from the rhombic lip and form the external granular layer (EGL). During early postnatal development, granule cell precursors in the outer zone of the EGL proliferate, then exit the cell cycle and differentiate. They subsequently migrate through the molecular layer (ML) past the Purkinje cells to their destination, the internal granule cell layer (IGL). During E11–E13 of mouse development, Purkinje cell precursors exit the mitotic cycle, leave the ventricular zone, and migrate along radial glia to form a temporary cerebellar plate-like structure (Goldowitz and Hamre, 1998; Gold et al., 2003). At E13–E14, they begin to express $ROR\alpha$. In $ROR\alpha^{sg/sg}$ mice, at E17.5 the normal number of Purkinje cells has been generated (Vogel et al., 2000; Doulazmi et al., 2001). However, by P5 the number of Purkinje cells is dramatically reduced, the external granular layer is significantly thinner, and a dramatic cerebellar hypoplasia is apparent. The surviving Purkinje cells have stunted dendritic arbors lacking distal spiny branchlets and are deficient in the assembly of mature synapses with granule cells suggesting that their dendritic differentiation is impaired (Sotelo and Changeux, 1974; Landis and Sidman, 1978; Sotelo and Wassef, 1991; Shirley and Messer, 2004). In addition, the remaining $ROR\alpha^{sg/sg}$ Purkinje cells do not express a number of genes normally expressed in mature postnatal Purkinje cells, including Purkinje cell-specific protein (Pcp2), calmodulin, Zebrin I, and the N-methyl-D-aspartate (NMDA) receptor (Messer et al., 1990; Sotelo and Wassef, 1991; Hamilton et al., 1996; Nakagawa et al., 1996a,b; Gold et al., 2003). These observations indicate that the surviving $ROR\alpha^{sg/sg}$ Purkinje cells do not mature. Because a reduction in $ROR\alpha^{sg/sg}$ Purkinje cells is observed only after E17.5, the lack of $ROR\alpha$ expression does not affect the genesis of Purkinje cells but rather their survival. It seems that the absence of $ROR\alpha$ is not directly responsible for the death of $ROR\alpha^{sg/sg}$ Purkinje cells but a consequence of the block or inhibition of Purkinje cell maturation (Fig. 1). This concept is supported by the reduced expression of genes normally expressed in mature Purkinje cells and the impaired dendritogenesis observed in $ROR\alpha^{sg/sg}$ Purkinje cells. This conclusion was further strengthened by a study examining the effects of lentivirus-mediated overexpression of $ROR\alpha$ on dendritic differentiation in $ROR\alpha^{sg/sg}$ Purkinje cells using organotypic cultures (Boukhtouche et al., 2006b).

A

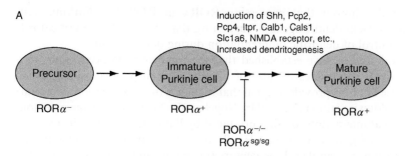

Induction of Shh, Pcp2,
Pcp4, Itpr, Calb1, Cals1,
Slc1a6, NMDA receptor, etc.,
Increased dendritogenesis

B

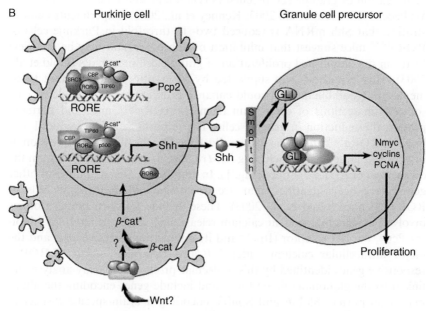

Fig. 1. Functions of RORα in the development of the cerebellum. (A) RORα plays a critical role in the maturation of Purkinje cells. Immature Purkinje cells arise from RORα⁻ precursor cells. RORα becomes highly expressed in postmitotic Purkinje cells at E12.5 of mouse embryonic development. RORα is required for dendritogenesis to proceed and for the induction of a number of genes normally expressed in mature Purkinje cells. Dendritogenesis and the expression of several genes normally expressed in mature Purkinje cells are inhibited in RORα-deficient mice. (B) RORα regulates expression of several genes, including *Pcp2* and *Shh*, directly by binding to ROREs in their promoter regions. The composition of the coactivator complexes assembled by RORα appears to be distinct for each gene, suggesting that the promoter context plays a critical role. These coactivators include β-catenin (β-cat), SRC1, p300, CBP, and Tip60. Activation of β-catenin by Wnt signaling might influence the transcriptional activity of RORα. RORα may act downstream of Wnt and because *Shh* is an RORα target gene, RORα may function as a mediator between the Wnt and Shh signaling pathways. Shh released by Purkinje cells interacts with Patched (Ptch) receptors present on granule cell precursors and lead to the activation of GLI transcription factors. Subsequently, this results in the transcriptional activation of several growth regulatory genes, including PCNA and cyclins, and an induction of the proliferation of these cells. The reduced Shh expression in Purkinje cells from RORα-deficient mice is a major cause of the cerebellar atrophy observed in these mice (Landis and Sidman, 1978; Herrup and Mullen, 1979; Hamilton et al., 1996; Goldowitz and Hamre, 1998; Steinmayr et al., 1998; Wallace, 1999; Gold et al., 2003, 2006; Kenney et al., 2003; Boukhtouche et al., 2006b). (See Color Insert.)

This study showed that expressing RORα in RORα$^{sg/sg}$ Purkinje cells restores normal dendritogenesis suggesting that RORα plays a critical role in the control of dendritic differentiation during development.

Several studies have established that the number of Purkinje cells, to a large extent, determines the size of the granule cell population suggesting that these neurons provide a signal that regulates the proliferation of granule cells (Goldowitz and Hamre, 1998; Vogel et al., 2000). Studies have demonstrated that interactions of Shh, produced by Purkinje cells, with its receptor Patched (Ptch), present on granule cells, provides an important signal for the proliferation of granule cell precursors (Dahmane and Ruiz-i-Altaba, 1999; Wallace, 1999; Gold et al., 2003; Kenney et al., 2003). Experiments demonstrating that Shh mRNA is reduced two- to threefold in Purkinje cells of RORα$^{sg/sg}$ mice suggest that inhibition of Shh expression may be a critical factor in the diminished proliferation of granule cell precursors (Gold et al., 2003). This hypothesis was supported by observations showing that treatment with recombinant Shh could enhance proliferation of granule precursor cells in sections of cerebellum from P4 RORα$^{sg/sg}$ mice and partially prevent the reduction in granule cells. The importance of the reduced Shh levels in RORα$^{sg/sg}$ mice was further illustrated by the observed repression of several Shh target genes, including *N-Myc*, *Baf53*, and various cyclins, in the cerebellum of RORα$^{sg/sg}$ mice (Fig. 1). In addition to *Shh*, a number of other genes have been identified that are downregulated in Purkinje cells of RORα$^{sg/sg}$ mice (Gold et al., 2003). These include genes-encoding proteins involved in signal-dependent calcium release such as the calmodulin inhibitor Pcp4, the IP3 receptor (Itpr1) and its interacting partner Cals1, and the major intracellular calcium buffer 1 (Calb1). Another group of RORα-responsive genes identified by this molecular profile microarray analysis are linked to the glutamatergic pathway and include genes encoding the glutamate transporters Slc1a6 and Spnb3, encoding a brain-specific β-spectrin (Gold et al., 2003). The observation that several of these genes are downregulated in RORα$^{sg/sg}$ Purkinje cells is in agreement with the hypothesis that lack of RORα inhibits the progression of the maturation of these cells.

The repression of these genes in RORα$^{sg/sg}$ Purkinje cells raised the question of whether they are regulated by RORα by a direct or indirect mechanism. Analysis of the promoter sequence of a number of these genes indicated the presence of several putative ROREs (Gold et al., 2003). Chromatin immunoprecipitation (CHIP) assays subsequently demonstrated that RORα antibodies pulled down the promoters of Pcp2, Pcp4, Itpr1, Shh, and Slc1a6 from chromatin isolated from Purkinje cells of wild type but not from that of mutant mice (Gold et al., in press). These results support the conclusion that RORα binds to these ROREs in these promoter regions *in vivo* and suggest that these genes are directly regulated by RORα.

Further analysis of the protein complexes recruited by RORα to these different ROREs demonstrated that RORα recruited distinct coactivators and cofactors to different promoter regions. RORα recruited β-catenin and p300 as coactivators to the Shh promoter, β-catenin, Tip60, and SRC-1 to the Pcp2 promoter, CBP, Tip60, SRC1, and GRIP-1 to the Slc1a6 promoter, and β-catenin, SRC-1, p300, and Tip60 to the Pcp4 promoter (Gold et al., 2003) (Fig. 1). Microinjection of blocking antibodies against SRC1, TIP60, and β-catenin greatly inhibited RORα-mediated activation of the Pcp2 promoter while antibodies against CBP and p/CIP did not have any effect. These results confirmed the functional role of these coactivators in RORα-mediated activation of this promoter. The recruitment of different coactivator complexes by RORα is an important and intriguing finding and suggests that the sequence and/or the promoter context of the RORE plays a critical role in determining what coactivator complex is assembled by RORα.

The association of β-catenin with RORα/coactivator complexes may suggest a possible link between Wnt and ROR signaling. Activation of β-catenin by Wnt signaling might influence the transcriptional activity of RORα. RORα may act downstream of Wnt and because *Shh* is an RORα target gene, RORα may function as a mediator between the Wnt and Shh signaling pathways (Fig. 1). This would be consistent with other studies indicating an interaction between Wnt and Shh signaling pathways (Borycki et al., 2000) and a role for Wnt signaling in cerebellar development (Salinas et al., 1994; Wang et al., 2001).

The function of RORα in brain development has not been studied much beyond the cerebellum. However, several studies have indicated a role for RORα in the development of the olfactory bulb (Monnier et al., 1999; Michel et al., 2000). RORα mRNA and protein have been detected at moderate levels in the mitral cell layer and in tufted and periglomerular cells of the olfactory bulb (Matsui et al., 1995; Sashihara et al., 1996; Ino, 2004). Moreover, cytological alterations have been observed in the olfactory bulb of ROR$\alpha^{sg/sg}$ mice (Monnier et al., 1999; Michel et al., 2000). The olfactory bulb of ROR$\alpha^{sg/sg}$ mice is structurally disorganized and the number of neurons and periglomerular neurons in particular mitral cells are significantly decreased (Monnier et al., 1999). Moreover, changes in the granular zone are observed typified by an abundance of glial cell process extension and a reduction of the astrocyte network. Such changes might lead to a failure in normal neuronal input processing. The latter is supported by studies showing a significant reduction in interneural interactions in the olfactory bulb of ROR$\alpha^{sg/sg}$ mice after odorant stimulation (Michel et al., 2000). This reduction may be related to a decrease in postsynaptic elements and due to the decline in mitral cells. Moreover, these changes might be responsible for some of the observed behavioral abnormalities such as impaired interindividual recognition.

4. Physiological functions of RORβ

RORβ exhibits a rather restricted pattern of expression and is particularly highly expressed in certain regions of the brain and in the retina (Carlberg et al., 1994; Schaeren-Wiemers et al., 1997; Andre et al., 1998a,b; Azadi et al., 2002). In the retina, RORβ expression is highly dynamic during embryonic development. At E15, all cells in the retina appear to express RORβ while at E17.5 and P5, RORβ is detected only in the inner and outer nuclear layers (Schaeren-Wiemers et al., 1997). At P9, RORβ is preferentially expressed in the inner nuclear layer. This expression is greatly reduced at P16 correlating with the eye-opening. With maturation, the adult retina shows RORβ expression to be highest in the outermost part of the outer nuclear layer where the photoreceptor cones reside, with weaker expression in the inner nuclear layer and ganglion cells (Schaeren-Wiemers et al., 1997). In adult brain, RORβ mRNA is most highly expressed in the cortex and the thalamus. Expression is also found in the hypothalamus, suprachiasmatic nucleus (SCN), and pineal gland, while little expression was detected in the cerebellum or hippocampus.

Nakagawa and O'Leary (2003) extensively examined the expression of RORβ in the developing mouse neocortex and dorsal thalamus from E12.5 to P5. These studies indicated that during brain development RORβ expression is highly dynamic and that RORβ mRNA is expressed in a spatial and temporal manner. RORβ expression was undetectable in the neocortical ventricular zone at any of the embryonic stages examined. At E12.5, RORβ le in the dorsal telencephalon but by E14.5 high levels of expression are evident in the lateral and rostral parts of the neocortex. By E16.5, levels of RORβ mRNA are increased and expressed in a graded fashion: from strong high lateral to low medial and from high rostral to low caudal expression. At birth, high levels of expression were observed in the somatosensory area and putative auditory and visual areas in the caudal neocortex. At P2, RORβ mRNA is highly expressed in the primary visual, auditory, and somatosensory regions in layer IV and sporadically in layer V (Nakagawa and O'Leary, 2003). In mice at P5, RORβ is highly expressed in layers IV and V of the primary sensory areas.

During the development of the dorsal thalamus, RORβ mRNA was undetectable at E11.5 and E12.5. At E17.5, RORβ mRNA was expressed at low to moderate levels in the ventroposterior nucleus and the ventral part of the medial geniculate nucleus, respectively, but was undetectable in the dorsal lateral geniculate nucleus (Nakagawa and O'Leary, 2003). A high level of expression was also observed in the centromedial nucleus. By P2, RORβ is highly expressed in the ventroposterior nucleus and the ventral part of the medial geniculate nucleus and minimally expressed in the dorsal lateral geniculate nucleus.

Disruption of RORβ gene expression in mice causes a number of abnormalities both structural and functional in nature (Andre et al., 1998a). RORβ-null mice exhibit a "duck-like" gait but maintain normal reflexes and balancing responses. No gross anatomical changes were observed in the brain and spinal cord of adult ROR$\beta^{-/-}$ mice. Although at birth, the retina of RORβ-null mice appears morphologically very similar to that of wild-type mice, the retina in adult $ROR\beta^{-/-}$ mice is disorganized and lacks the normal layer structure. Retinal degeneration occurs during the first weeks after birth and finally results in blindness. The precise molecular mechanism underlying this retinal degeneration is not yet understood.

5. Role of RORs in circadian rhythm

Circadian rhythms are daily cycles of behavioral and physiological changes that are driven by an endogenous oscillator within a 1-day period (Schibler and Sassone-Corsi, 2002; Isojima et al., 2003; Schibler and Naef, 2005). Circadian rhythm impacts behavior, a variety of physiological functions, drug metabolism, and has been implicated in disease, including sleep disorders and cancer. In mammals, light is the principal signal received by the non-cone, non-rod system in the retina. This signal is transmitted through the retinohypothalamic tract to the SCN. Mammalian circadian rhythms are governed by a central circadian clock that resides in the SCN of the anterior hypothalamus (Ripperger and Schibler, 2001; Isojima et al., 2003). This master oscillator can maintain circadian rhythms in the absence of light input. Many physiological processes in most, if not all, tissues are controlled by circadian oscillators, the phase of which is regulated by signals originating from the SCN.

Although their precise role is still far from clear, several studies have implicated RORs in the regulation of circadian behavior (Schaeren-Wiemers et al., 1997; Preitner et al., 2002; Ueda et al., 2002b; Akashi and Takumi, 2005; Guillaumond et al., 2005; Jetten et al., unpublished obervations). This is indicated by an aberrant circadian behavior observed in mice deficient in $ROR\alpha$ or $ROR\beta$ (Schaeren-Wiemers et al., 1997; Sato et al., 2004; Akashi and Takumi, 2005). Although examination of circadian-related behavior in $ROR\alpha$-deficient ($ROR\alpha^{sg/sg}$) mice has been difficult given the locomotor difficulties related to their *staggerer* phenotype, reports indicate that the time period of circadian-dependent locomotor activity is decreased both in the $ROR\alpha^{sg/+}$ and in $ROR\alpha^{sg/sg}$ mice that may be related to altered circadian rhythm or just altered motor function in $ROR\alpha$-deficient mice (Akashi and Takumi, 2005). $ROR\alpha^{sg/sg}$ mice display anomalous free-running locomotor activity rhythms and an altered feeding pattern during the circadian cycle (Guastavino et al., 1991; Sato et al., 2004; Akashi and Takumi, 2005).

In several tissues, expression of RORs shows an oscillating pattern during the circadian cycle. $ROR\beta$ is particularly highly expressed in tissues that play a key role in the generation and maintenance of circadian rhythms, including the SCN, the pineal gland, and the retina (Schaeren-Wiemers et al., 1997; Andre et al., 1998a,b; Sumi et al., 2002). The expression of $ROR\beta2$ mRNA in the pineal gland and retina reaches a maximum during the night period at circadian time CT18 (Andre et al., 1998a; Ueda et al., 2002b). In the SCN, the expression of $ROR\alpha$ and $ROR\beta$ display a similar circadian profile with peaks during the day and troughs during the night, while RORγ is not expressed in the SCN. RORγ mRNA shows a circadian expression pattern in both liver and kidney but not in thymus (Ueda et al., 2002b, 2005; Jetten et al., unpublished observations). RORγ is expressed at low levels during the day and at optimum levels at CT18 trailing the peak expression of Rev-ErbAα (at CT8) and Per2 (at CT13). RORα mRNA expression shows little circadian oscillation in liver, kidney, and lung (Ueda et al., 2002b; Akashi and Takumi, 2005).

Although many aspects of the regulation of the circadian oscillator still have to be elucidated, studies have made great advances in understanding the control mechanism of the circadian clock (Ripperger and Schibler, 2001; Piggins and Loudon, 2005; Roenneberg and Merrow, 2005; Schibler and Naef, 2005). The main feature of this mechanism is that the circadian oscillators consist of several interconnected molecular feedback loops. The positive loop consists of the basic helix-loop-helix/PAS-type transcription factors Bmal1 and CLOCK, while two cryptochrome (Cry) and three period proteins (Per) are involved in the negative control of the oscillator (Albrecht, 2002; Reppert and Weaver, 2002; Schibler and Sassone-Corsi, 2002; Isojima et al., 2003; Schibler and Naef, 2005). Heterodimers of Bmal1 and Clock enhance the transcription of Per and Cry genes through their interaction with E-box enhancers (5′-CACGTG) in the 5′-upstream regulatory regions of these genes. After being synthesized in the cytoplasm, Per1 and Per2 become phosphorylated by casein kinase Iε/δ. This promotes their dimerization with Cry1 and allows transfer of the complex into the nucleus where they repress the transcription of Bmal1. It has been proposed that Cry1 represses the activation of their own expression by binding to the Bmal1:CLOCK complex. The turnover of these proteins then leads to a new cycle of activation of Bmal1 and CLOCK, while in turn increasing levels of Per2 and Cry1 suppress the expression of Bmal1 and CLOCK. The repression of Bmal1 appears to involve the nuclear orphan receptors Rev-ErbAα and Rev-Erbβ, which function as transcriptional repressors. The periodic expression of Rev-ErbAα is regulated negatively by Per:Cry complexes through E-box elements in its promoter while CLOCK:Bmal1 stimulates Rev-ErbAα promoter activity. In turn, an increase in Rev-ErbAα results in the repression of Bmal1 transcription. This repression is mediated through the binding of Rev-Erb to several RORE response elements (AAAGTAGGTCA) in the promoter of the Bmal1 gene (Preitner et al., 2002; Ueda et al., 2002b; Nakajima et al., 2004; Sato et al., 2004; Guillaumond et al., 2005). Mutation of theseROREs abolishes the circadian rhythmicity of the Bmal1 promoter indicating the importance of this response element in Bmal1 transcriptional regulation.

The ROREs in the Bmal1 promoter also bind ROR receptors. Overexpression of RORα1 and RORα4 induces Bmal1 promoter activity by interacting with these ROREs, this induction can be inhibited by coexpression of Rev-ErbAα (Ueda et al., 2002b; Nakajima et al., 2004; Sato et al., 2004;

Akashi and Takumi, 2005; Guillaumond et al., 2005; Ueda et al., 2005). RORβ and RORγ are also able to induce Bmal1 activity; however, RORα4 appears to be the most effective in inducing this activity. Studies examining the activation of a luciferase reporter, under the control of the Bmal1 (-3465 to $+57$) promoter region, in NIH3T3 cells after serum shock revealed an oscillating profile of transcriptional activation. Coexpression of a dominant-negative RORα lacking its LBD caused a severe weakening of this oscillation. In addition, a knockdown of RORα expression using siRNA also attenuated the amplitude of the oscillation. The transcriptional oscillation of Bmal1 promoter activity was muted in mouse embryo fibroblasts isolated from ROR$\alpha^{sg/sg}$ mice compared to wild-type fibroblasts. Mutation of one of the RORE sites in the Bmal1 promoter abolished the transcriptional oscillation suggesting the importance of these sites in the regulation of Bmal1 by RORα (Ueda et al., 2002b; Akashi and Takumi, 2005). This is further supported by findings showing that at CT18 the expression of Bmal1 in the SCN of ROR$\alpha^{sg/sg}$ mice was significantly lower compared to that of wild-type mice (Sato et al., 2004). These observations demonstrate that RORα functions as a positive regulator of Bmal1 and that cross talk between RORα and Rev-ErbAα nuclear receptors plays a critical role in the control of Bmal1 expression. At high levels, Rev-ErbAα competes with RORα for binding to these ROREs and represses Bmal1 expression while at low levels it allows RORα to bind and activate Bmal1 transcription (Preitner et al., 2002; Isojima et al., 2003; Sato et al., 2004; Triqueneaux et al., 2004; Akashi and Takumi, 2005; Guillaumond et al., 2005). In addition, the circadian regulation of RORα itself may be a factor in the control of Bmal1 expression as well. Studies have shown that a loss of RORα reduces the expression of several other genes in the SCN, including tubulin alpha 8 (Tuba8) and Rasd1 (RAS, dexamethasone-induced 1) (Ueda et al., 2002b; Sato et al., 2004). These genes contain ROREs in their promoter regulatory region suggesting that they might be regulated by RORα directly. Observations showing that in the absence of RORα these genes still retain a certain degree of an oscillating pattern of expression suggest that other factors play a role in the control of their circadian expression.

6. Role of RORγ in the development of secondary lymphoid tissues

Study of RORγ-null mice has revealed several important clues about the function of RORγ in the development of secondary lymphoid tissues (He et al., 2000; Jetten et al., 2001; He, 2002; Jetten and Ueda, 2002; Eberl and Littman, 2003; Lipp and Muller, 2004). These studies have demonstrated that mice deficient in RORγ lack lymph nodes and Peyer's patches, suggesting a critical role for RORγt (RORγ2) in lymph node organogenesis (Littman et al., 1999; Kurebayashi et al., 2000; Sun et al., 2000; Jetten et al., 2001; He, 2002; Jetten and Ueda, 2002; Eberl and Littman, 2003; Eberl et al., 2004; Lipp and Muller, 2004).

In recent years, our understanding of lymph node development has been greatly enhanced (Mebius et al., 1997, 2001, Rennert et al., 1998; Fu and Chaplin, 1999; Cyster, 2003; Eberl and Littman, 2003; Muller et al., 2003; Nishikawa et al., 2003; Eberl et al., 2004; Cupedo and Mebius, 2005; Eberl, 2005). In mice, lymph node organogenesis is initiated around E10.5 with the invagination of endothelial cells resulting in the development of lymphoid sacs. This process involves the conversion of endothelial cells into cells with a lymphatic phenotype. The homeobox protein Prox-1 has been demonstrated to play a critical role in this conversion (Wigle et al., 2002). From these early lymph sacs, lymphatic vessels grow out and eventually form the complete lymphatic network by E15.5. The formation of the lymphatic network is not affected in $ROR\gamma^{-/-}$ mice, suggesting that lack of lymph nodes appears not to be due to defects in the formation of lymph sacs and lymphatic vessels (Sun et al., 2000). The earliest anlagen of lymph nodes are formed after infiltration of organizer cells. It is believed that these specialized cells are derived from differentiating mesenchymal cells; however, much about this differentiation process is still poorly understood (Cupedo and Mebius, 2005; Eberl, 2005). During ontogeny, LTi cells, characterized as $CD45^+CD4^+CD3^-$ $IL-7R\alpha^+$ cells, are among the earliest hematopoietic cells to colonize sites destined to develop into secondary lymphoid organs. Although the precise origin of LTi cells has not yet been established, it has been suggested that LTi cells may originate from $IL-7R\alpha^+Sca-1^{low}c-Kit^{low}$ hematopoietic precursor cells in the fetal liver (Cupedo and Mebius, 2005; Eberl, 2005). By E12.5, LTi cells are detected in spleen, fetal blood, and lymph node anlagen and by E16 in Peyer's patch anlagen. LTi cells are present in very low numbers after birth. The migration of LTi cells into the anlagen and the cross talk between LTi and mesenchymal organizer cells play a key role in the development of lymph nodes (Cupedo et al., 2002; Eberl and Littman, 2003; Lipp and Muller, 2004; Cupedo and Mebius, 2005; Eberl, 2005) (Fig. 2). The interaction between these two cell types is mediated by a variety of NF family ligands, chemokines, adhesion proteins, and their corresponding receptors and involves several positive feedback loops and amplifying mechanisms. Binding of membrane-bound lymphotoxin (LT) $\alpha1\beta2$ heterotrimers on the LTi cells to lymphotoxin β receptors (LTβR) on mesenchymal organizer cells is critical in the formation of lymph nodes and Peyer's patches (De Togni et al., 1994; Koni et al., 1997; Futterer et al., 1998; Rennert et al., 1998; Fu and Chaplin, 1999; Ruddle, 1999; Cupedo and Mebius, 2005). This interaction results in the activation of the NF-κB1 pathway through the phosphorylation and destruction of IκBα and activation of the NF-κB–inducing kinase (NIK) in the NF-κB2 pathway. Mice deficient in the expression of lymphotoxin LTα, the lymphotoxin receptor LTRβ, or proteins acting downstream of LTR signaling, such as NIK, are deficient in lymph nodes and Peyer's patches due to the inability of the LTi and

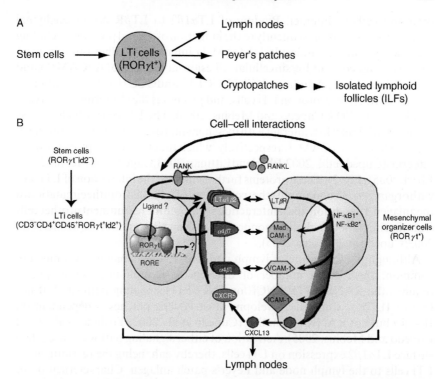

Fig. 2. RORγt is essential for the development of secondary lymphoid tisues. (A) Lymphoid tissue inducer (LTi) cells are derived from hematopoietic stem cells and are essential in the development of lymph nodes, Peyer's patches, and cryptopatches. The isolated lymphoid follicles (ILFs) are thought to be derived from cryptopatches after the colonization of the intestine by bacteria. RORγt is required for the generation and/or survival of LTi cells. The absence of these lymphoid tissues in RORγ-deficient mice is due to a deficiency of LTi cells in these mice. (B) Role of RORγt and LTi cells in lymph node development. The transcription factors Id2 and RORγt are both essential for the generation of LTi cells (CD4⁺CD3⁻ CD45⁺IL-7Rα⁺RORγt⁺Id2⁺). LTi cells are recruited to lymph node anlagen through their interaction with mesenchymal organizer cells. Interactions between these two cell types are key in the recruitment of LTi cell from the circulation to the lymph node anlagen. Activation of the RANK signaling pathway enhances the expression of lymphotoxins (LT) in LTi cells and promotes lymph node development. Binding of LTα1β2 on LTi cells to the LTβR on mesenchymal organizer cells plays a key role in lymph node development. The latter results in the activation of NF-κB pathways and induction of VCAM-1, ICAM-1, and MAdCAM-1 in mesenchymal organizer cells and their subsequent binding to integrins on LTi cells further promote their interaction and the recruitment of additional LTi cells. In addition, induction of various chemokines in mesenchymal organizer cells, including CXCL13 and CCL19, interact with their corresponding receptors on LTi cells thereby promoting their interaction and allow the recruitment of monocytes, T and B lymphocytes. (Yokota et al., 1999; Kurebayashi et al., 2000; Sun et al., 2000; Jetten et al., 2001; Cupedo et al., 2002; Eberl and Littman, 2003; Eberl et al., 2004; Jetten, 2004; Lipp and Muller, 2004; Eberl, 2005). (See Color Insert.)

organizer cells to interact. Binding of LTα1β2 to LTβR on mesenchymal organizer cells leads subsequently to the induction of several proteins, including vascular cell adhesion molecule-1 (VCAM-1), intercellular adhesion molecule-1 (ICAM-1), mucosal addressin cellular adhesion molecule-1 (MAdCAM-1), and several chemokines, including CXCL13, CCL19, and CCL21. CXCL13 interacts with CXCR5, its receptor on LTi cells, and promotes the clustering of LTi cells (Muller et al., 2003; Cupedo and Mebius, 2005). The LTi cells, which express integrin α4β7 and α4β1, interact with mesenchymal organizer cells by binding to MadCAM-1 and VCAM-1, respectively, which function as receptors for these integrins (Cupedo et al., 2002; Eberl and Littman, 2003; Cupedo and Mebius, 2005; Eberl, 2005) (Fig. 2). These proteins further enhance the interaction of LTi cells with organizer cells and amplify the recruitment of LTi cells from the circulation to the anlagen. In addition, these interactions promote the recruitment of other cells from the circulation, including monocytes, and T and B lymphocytes (Muller et al., 2003; Cupedo and Mebius, 2005).

Although the formation of lymph nodes and Peyer's patches has much in common, there are a number of differences. Lymph node development requires the RANKL (TRANCE or TNFSF11) signaling pathway but not that of IL-7Rα, while the development of Peyer's patches is dependent on IL-7Rα but not RANKL signaling (Cupedo et al., 2002; Yoshida et al., 2002; Cupedo and Mebius, 2005). Activation of either signaling pathway upregulate surface LTα1β2 expression on LTi cells, thereby enhancing the recruitment of LTi cells to the lymph node and Peyer's patch anlagen. Characterization of RORγ$^{-/-}$ mice showed that LTi cells are absent from spleen, mesentery, and intestine of RORγ$^{-/-}$ E18.5 embryos (Sun et al., 2000). These observations indicated that RORγt (RORγ2) plays a critical role in the generation or in the survival of LTi cells. Creation of RORγt$^{gfp/+}$ and RORγt$^{gfp/gfp}$ knock-in mice, expressing the enhanced green fluorescent protein (EGFP) under the control of the RORγt regulatory elements, allowed tracing of cells expressing RORγt (Eberl and Littman, 2003; Eberl et al., 2004; Eberl, 2005). These studies have provided valuable new insights into the roles of RORγt and LTi cells in lymph node development (Fig. 2). Visualization of EGFP in RORγt$^{gfp/+}$ mice showed that at E16.5 of fetal development RORγt is exclusively expressed in LTi cells present in the lymph node anlagen, the submucosal region of the intestine, and around large vessels in spleen. No EGFP-positive cells were detectable at E16.5 RORγt-deficient RORγt$^{gfp/gfp}$ fetuses indicating the absence of LTi cells (Eberl and Littman, 2003; Eberl et al., 2004). In addition, adult RORγt-deficient RORγt$^{gfp/gfp}$ mice lacked lymph nodes and Peyer's patches confirming that RORγt is essential for the development of lymph nodes and Peyer's patches (Kurebayashi et al., 2000; Sun et al., 2000). From these studies several conclusions were drawn: (1) LTi cells are defined by the expression of RORγt, (2) LTi cells are essential for the development of lymph nodes and Peyer's patches, and (3) the lack of lymph nodes and Peyer's patches in RORγt- or RORγ-deficient mice is due to the absence of LTi cells (Fig. 2).

The basic helix-loop-helix transcription factor Id2 also plays a critical role in lymph node organogenesis (Yokota et al., 1999). As demonstrated for ROR$\gamma^{-/-}$ mice, Id2-null mice do not develop lymph nodes and Peyer's patches (Yokota et al., 1999; Sun et al., 2000; Eberl and Littman, 2003). In addition, Id2 has been reported to be expressed in LTi cells and these cells were shown to be absent in Id2-null mice. These findings suggest a critical role for Id2 in the generation and/or survival of LTi cells (Yokota et al., 1999, 2001; Eberl, 2005). In contrast to ROR$\gamma^{-/-}$ mice, Id2-null mice do not contain nasal-associated lymphoid tissue (NALT) and have a greatly reduced number of natural killer cells and splenic CD8α^+ dendritic cells (Fukuyama et al., 2002; Harmsen et al., 2002). These observations indicate that Id2 is involved in the regulation of several other differentiation programs. Moreover, since NALT is formed in ROR$\gamma^{-/-}$ mice in the absence of LTi cells, the organogenesis of this lymphoid tissue seems to involve a different mechanism.

Besides Peyer's patches and mesenteric lymph nodes, the immune system in the gut contains several other lymphoid structures: cryptopatches and isolated lymphoid follicles (ILFs) (Eberl, 2005; Taylor and Williams, 2005). In addition, the gut contains intraepithelial T lymphocytes (IELs), a rather unique group of T lymphocytes that contains a high percentage of CD8α^+TCR$\alpha\beta$ and TCR$\gamma\delta$ T cells. Cryptopatches consist of clusters of hematopoietic cells (Lineage (Lin)$^-$ Kit$^+$IL-7Rα^+CD44$^+$) that appear about 1 week after birth along the gut epithelium between the crypts in the lamina propria (Eberl, 2005). IL-7Rα plays an important role in the development of cryptopatches since these structures are absent in mice deficient in this receptor (Kanamori et al., 1996). Studies have demonstrated that cells in cryptopatches and in ILFs express RORγt in conjunction with the helix-loop-helix transcription factor Id2, c-Kit, CD44, and IL-7Rα (Eberl and Littman, 2004; Eberl, 2005). Moreover, it was shown that cryptopatches and ILFs were absent in mice deficient in either RORγt or Id2. Targeted expression of Bcl-X$_L$ under the control of the RORγt promoter in RORγt-deficient mice has been reported to rescue RORγt$^{-/-}$ double-positive (DP) thymocytes from undergoing apoptosis but failed to restore the development of Lin$^-$Kit$^+$IL-7Rα^+, cryptopatches, and ILFs. These studies demonstrated that adult intestinal Lin$^-$Kit$^+$IL-7Rα^+CD44$^+$ cells share many similarities with LTi cells, both express RORγt and Id2 and require these transcription factors for their development. Based on these observations, it was concluded that the Lin$^-$Kit$^+$IL-7Rα^+CD44$^+$ cells constitute the adult counterpart of LTi cells (Eberl and Littman, 2004; Eberl, 2005) (Fig. 2). It was further proposed that ILFs develop from cryptopatches in response to inflammatory innate immune signals generated by the colonization of the intestine by bacteria.

7. Critical functions of RORγ in thymopoiesis

Initial reports showed that RORγ is highly expressed in the thymus suggesting that RORγ might have a role in the regulation of thymopoiesis (Hirose et al., 1994; Ortiz et al., 1995; Medvedev et al., 1996; He et al., 1998; He, 2000; Guo et al., 2002). This was supported by studies demonstrating that the size of the thymus and number of thymocytes in RORγ-null mice was greatly reduced compared to those of wild-type mice (Kurebayashi et al., 2000; Sun et al., 2000; Jetten et al., 2001; Jetten and Ueda, 2002). This reduction was due to a dramatic decrease in the number of DP CD4$^+$CD8$^+$ and mature single-positive (SP) CD4$^-$CD8$^+$ and CD4$^+$CD8$^-$ thymocytes. Mice deficient in the expression of the RORγt (RORγ2) isoform exhibited the same thymic phenotype as RORγ-null mice that are deficient in both RORγ1 and RORγt (Eberl and Littman, 2003), suggesting that the observed thymic phenotype is due to the loss of RORγt. These observations indicated that RORγt plays a critical role in thymocyte homeostasis by regulating differentiation, proliferation, and/or apoptosis of thymocytes (Jetten et al., 2001; He, 2002; Winoto and Littman, 2002; Jetten, 2004).

Pluripotent lymphocyte progenitors, after their migration from fetal liver or adult bone marrow to the thymus, become committed to the T cell lineage, predominantly T cell receptor (TCR) $\alpha\beta^+$ T cells (SP cells, CD4$^+$CD25$^+$T$_{reg}$, and NK1.1$^+$ natural T cells). This program of differentiation proceeds via a highly coordinated series of steps and involves differentiation, expansion of subpopulations through cell proliferation, and gene recombination (Jameson et al., 1995; Kisielow and von Boehmer, 1995; Anderson et al., 1996; Fehling and von Boehmer, 1997; Baird et al., 1999; Ellmeier et al., 1999; Sebzda et al., 1999). In addition, it involves a number of checkpoints to either select thymocytes or eliminate them by apoptosis (Starr et al., 2003; Zhang et al., 2005). Thymopoiesis can be divided into several major stages based on the presence of cell surface antigens, including CD44, CD25, CD3, CD8, and CD4. Components of the TCR complex and several cytokine signaling pathways play a critical role in controlling the transition between the various stages. A number of transcription factors, including several nuclear receptors, have been identified as key regulators of different steps of thymopoiesis (He, 2000, 2002; Jetten et al., 2001; Yokota et al., 2001; Jetten and Ueda, 2002; Winoto and Littman, 2002; Eberl and Littman, 2003). Double-negative (DN) CD4$^-$CD8$^-$ thymocytes constitute 3–5% of the total thymocyte population that can be divided into four subsets, DN1–DN4, based on the presence of CD25 and CD44. When committed to the T cell lineage, the T cell precursor CD25$^-$ CD44$^+$ cells (DN1) differentiate successively via two intermediate stages, CD25$^+$CD44$^+$ (DN2) and CD25$^+$CD44$^-$ (DN3), into CD44$^-$CD25$^-$ (DN4) thymocytes (Fig. 3). During these stages TCRγ, TCRδ, and TCRβ gene rearrangements, and β-selection occurs (Jameson et al., 1995; Kisielow and von Boehmer, 1995; Anderson et al., 1996; Fehling and von Boehmer, 1997;

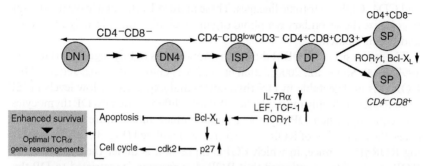

Fig. 3. RORγt exhibits multiple functions in thymopoiesis. CD4⁻CD8⁻CD25⁻CD44⁺ (DN1) cells, hematopoietic precursor cells, differentiate via DN2 and DN3 into CD4⁻CD8⁻CD25⁻CD44⁻ (DN4) cells. The DN4 cells then give rise to immature single-positive (ISP) cells (CD3⁻CD4⁻CD8^{low}). The ISP cells subsequently differentiate into CD3⁺CD4⁺CD8⁺, DP thymocytes. Expression of RORγt, as well as Bcl-X$_L$, LEF, and TCF-1, are induced during the ISP–DP transition and again downregulated during the differentiation of DP into SP cells. RORγt promotes the differentiation of ISP into DP cells and is a positive regulator of Bcl-X$_L$ expression. The latter stabilizes the cdk inhibitor p27, which subsequently inhibits cdk2 activity and increases the life span of DP thymocytes. Lack of RORγt expression inhibits the ISP to DP transition. In addition, expression of Bcl-X$_L$ in DP thymocytes is reduced resulting in increased apoptosis, reduced the life span of DP thymocytes, and consequently impaired TCRα rearrangements (He et al., 1998, 2000; Kurebayashi et al., 2000; Sun et al., 2000; Jetten et al., 2001; Guo et al., 2002; He, 2002; Jetten and Ueda, 2002; Eberl and Littman, 2003).

Baird et al., 1999; Ellmeier et al., 1999). A pre-TCR signal induces the DN thymocytes to undergo proliferation and differentiation into a distinct, intermediate-stage cell type, referred to as immature single-positive (ISP), CD3⁻CD4⁻CD8^{low} cells (Miyazaki, 1997). Without a proper pre-TCR signal DN cells undergo apoptosis. The Bcl-2 family member Bfl-1 plays an important role in protecting DN cells from undergoing apoptosis (Chen et al., 2000; Zhang et al., 2005). The ISP cells subsequently differentiate into CD4⁺CD8⁺ (DP) thymocytes. DP thymocytes constitute the majority of thymocytes (75%) in the thymus. After successful TCRα gene rearrangement, DP cells expressing TCRαβ receptor undergo a careful selection process to eliminate thymocytes expressing nonfunctional or autoreactive TCR (Starr et al., 2003; Zhang et al., 2005). DP thymocytes with low affinity for self-peptide–major histocompatibility complex (MHC) undergo apoptosis by a process referred to as death by neglect, while DP thymocytes that express potentially self-reactive T cell antigen receptors are eliminated by apoptotic process referred to as negative selection. It is estimated that the majority (90%) of thymocytes are eliminated by death by neglect and only a small fraction of thymocytes (about 5%), exhibiting intermediate affinities for self-peptide–MHCs, undergo positive selection (Sebzda et al., 1999). The molecular mechanisms underlying these apoptotic processes are not yet fully understood. The positive selected DP thymocytes mature into single-positive (SP) CD4⁺CD8⁻ helper and CD4⁻CD8⁺ cytotoxic lineages. These mature T cells then leave the thymus to colonize the secondary lymphoid organs, including the spleen, lymph nodes, and Peyers's patches.

The expression of RORγt during thymopoiesis is tightly regulated (He et al., 1998, 2000; Sun et al., 2000; Jetten et al., 2001; Guo et al., 2002; Jetten, 2004). RORγt is undetectable in DN thymocytes and expressed at low levels in ISP cells. RORγt is highly induced when ISP cells differentiate into DP thymocytes. Subsequently, when DP thymocytes differentiate into mature SP T lymphocytes, the expression of RORγt is again downregulated (Fig. 3). Studies analyzing RORγt$^{gfp/+}$ mice, in which EGFP expression is under the control of the RORγt promoter, confirmed that RORγt expression is restricted to DP thymocytes (Eberl et al., 2004). Analysis of different thymocyte subpopulations in RORγ$^{-/-}$ mice showed that the percentage of ISP thymocytes is greatly increased. The accumulation of ISP cells in RORγ$^{-/-}$ mice appears to be due to a delay in the differentiation of ISP into DP cells suggesting a role for RORγt in the regulation of the ISP–DP transition (Guo et al., 2002). In addition to RORγt, the HMG-box transcription factors TCF-1 and LEF-1, and the IL-7Rα signaling pathway have been implicated in the control of the transition of ISP to DP thymocytes (Verbeek et al., 1995; He et al., 2000; Yu et al., 2004). IL-7Rα is highly expressed in DN and SP thymocytes, and absent in DP thymocytes. Activation of the IL-7Rα signaling pathway in DN thymocytes positively regulates the levels of the antiapoptotic protein Bcl2 that is needed for the survival and proliferation of these cells. In contrast, IL-7Rα activation suppresses the expression of TCF-1, LEF-1, and RORγt. Turning off the activation of IL-7Rα signaling appears to be a prerequisite for the induction of TCF-1, LEF-1, and RORγt, and for the differentiation of ISP cells into DP thymocytes to proceed (Yu et al., 2004) (Fig. 3).

In addition to the increase in ISP thymocytes, the number of DP and SP thymocytes in RORγ$^{-/-}$ mice was dramatically reduced while the percentages of DN cells were significantly enhanced (He, 2000; Kurebayashi et al., 2000; Sun et al., 2000; Jetten et al., 2001; Guo et al., 2002; Jetten and Ueda, 2002; Winoto and Littman, 2002; Jetten, 2004). Cell cycle analysis indicated that the number of thymocytes in S phase was dramatically increased in the thymus of RORγ$^{-/-}$ mice (Kurebayashi et al., 2000; Sun et al., 2000). The latter may partially reflect the observed increase in the percentage of DN thymocytes; however, it might also involve an aberrant regulation of the proliferation of RORγ$^{-/-}$ thymocytes.

TUNEL staining provided the first indication that apoptosis was significantly increased in sections from thymus of RORγ$^{-/-}$ mice (Kurebayashi et al., 2000; Sun et al., 2000; Jetten et al., 2001; Jetten, 2004). Flow cytometric analysis demonstrated that the cells undergoing apoptosis were DP thymocytes. These observations indicated that the reduction in DP thymocytes in RORγ$^{-/-}$ mice was in part related to increased apoptosis. In addition to enhanced apoptosis *in vivo*, RORγ$^{-/-}$ thymocytes placed in culture undergo accelerated apoptosis as indicated by the rapid loss of mitochondrial membrane potential ($\Delta\Psi_m$), an increase in annexin IV binding, and the activation of several caspases (Kurebayashi et al., 2000; Sun et al., 2000; Jetten and Ueda, 2002; Ueda et al., 2002a). The decreased life span of DP thymocytes was shown to be related to a dramatic reduction in the expression of the antiapoptotic gene Bcl-X_L. Little change was observed in the expression of Bax and Bak mRNAs. The expression pattern of Bcl-X_L during thymopoiesis is very similar to that of RORγt; expression of both genes is induced during the ISP to DP transition and again downregulated during the DP to SP transition. Based on their coexpression and the repression of Bcl-X_L expression in ROR$\gamma^{-/-}$ DP thymocytes, it was concluded that RORγ acts as a positive regulator of Bcl-X_L expression and consequently promotes survival of these cells (Ma et al., 1995; He et al., 1998; Kurebayashi et al., 2000; Jetten et al., 2001; Jetten and Ueda, 2002; Eberl et al., 2004; Jetten, 2004) (Fig. 3). This is supported by findings demonstrating that a transgene encoding RORγt restored the expression of Bcl-X_L in DP thymocytes of ROR$\gamma^{-/-}$ mice as well as the survival of these cells (Xie et al., 2005). In addition, studies showing that overexpression of RORγt in T cell hybridomas protects cells from activation-induced apoptosis are in agreement with this concept (He et al., 1998). Whether RORγ regulates Bcl-X_L directly, by binding to a specific RORE in the regulatory region of the Bcl-X_L gene, or indirectly has yet to be established.

Bcl-X_L positively regulates the stability of the cyclin-dependent kinase 2 (cdk2) inhibitor p27 by inhibiting its degradation via the proteasome pathway (Gil-Gomez et al., 1998). The reduction in Bcl-X_L expression in ROR$\gamma^{-/-}$ DP thymocytes results in a decrease in the level of p27 that consequently causes an increase in cdk2 activity (Sun et al., 2000; Jetten and Ueda, 2002). These results indicate that by regulating the expression of Bcl-X_L, RORγt promotes cell survival and inhibits cell division (Fig. 3).

Rearrangements of genes encoding the TCRα chain is a distinctive feature of DP cells that are controlled by the T early α promoter and the TCRα enhancer. Rearrangement of TCR Vα and Jα gene segments begins with the proximal segments recombining before the distal ones. The shortened life span of ROR$\gamma^{-/-}$ DP thymocytes has been causally linked to the impaired rearrangements of 3' Jα segments and the bias for Jα rearrangements to the proximal 5' end of the Jα locus observed in these cells (Guo et al., 2002). Constitutive expression of the early growth response gene 3 (Egr3), a zinc finger transcription factor that promotes proliferation when DN thymocytes differentiate into ISP cells, in thymocytes increases apoptosis among DP cells and shortens their life span *in vitro*. This is accompanied by reduced expression of both RORγt and Bcl-X_L (Xi and Kersh, 2004). This correlation is in agreement with the concept that RORγt and Bcl-X_L act as promoters of cell survival. Future studies will determine whether there is a direct link between Egr3 and the repression of RORγt transcription.

Targeted expression of Bcl-X$_L$ to immature thymocytes rescues ROR$\gamma^{-/-}$ DP thymocytes from undergoing accelerated apoptosis and restores their normal life span and levels of p27 protein (Sun et al., 2000). These results indicate that the reduction in Bcl-X$_L$ expression is an early and key event in causing accelerated apoptosis in ROR$\gamma^{-/-}$ thymocytes. The expression of Bcl-X$_L$ also restores Jα rearrangements of more distal 3' Jα segments (Guo et al., 2002). However, expression of Bcl-X$_L$ does not change the delay in the ISP–DP transition, suggesting that this impairment occurs independently of Bcl-X$_L$ (Guo et al., 2002).

Although ROR$\gamma^{-/-}$ mice appear initially healthy, by the age of 4 months about 50% of the mice have died of thymic lymphoma (Jetten and Ueda, 2002; Ueda et al., 2002a). The lymphoma cells metastasized to the kidney, liver, and spleen. The thymic lymphomas contained a greatly increased number of mitotic as well as apoptotic cells. Analysis of the CD4/CD8 phenotype showed that the thymic lymphomas are heterogeneous and often contained greatly increased numbers of DN and SP CD8$^+$ cells. The molecular mechanism underlying this increased susceptibility to T cell lymphoma development is not yet understood but may relate to defects in the control of cellular proliferation and differentiation of thymocytes observed in ROR$\gamma^{-/-}$ thymocytes. Increased proliferation may result in an enhanced probability to acquire genetic alterations that lead to additional changes in the expression of tumor suppressor and proto-oncogenes (Jetten and Ueda, 2002; Ueda et al., 2002a).

8. Role of RORα in lymphocyte development

Both the spleen and the thymus are significantly smaller in RORα-deficient mice suggesting that RORα may have a role in the regulation of thymopoiesis and lymphocyte development (Trenkner and Hoffmann, 1986; Dzhagalov et al., 2004). RORα mRNA is expressed at low levels in DP thymocytes and at high levels in SP thymocytes. RORα expression is decreased in mature CD8$^+$ and increased in mature CD4$^+$ T lymphocytes. B220$^+$ B lymphocytes express low levels of RORα. The total number of thymocytes and splenocytes is greatly diminished in ROR$\alpha^{-/-}$ mice. The DP thymocyte population in the thymus was dramatically reduced whereas the percentages of DN, CD4$^+$, and CD8$^+$ SP cells were greatly enhanced. Although the total number of CD4$^+$ and CD8$^+$ T lymphocytes in the spleen was decreased, their percentage was higher than in wild-type mice, while the percentage of B cells was reduced. These findings suggest that RORα plays an important role in lymphocyte development. Reconstitution of lymphocyte development in Rag-2$^{-/-}$ mice with bone marrow cells of ROR$\alpha^{-/-}$ showed that these cells could completely restore normal T and B cell development (Dzhagalov et al., 2004). The results indicate that the impaired lymphocyte development in

ROR$\alpha^{-/-}$ mice is not due to intrinsic defects in these cells. It has been suggested that RORα indirectly regulates T and B lymphocyte development by providing an appropriate microenvironment (Dzhagalov et al., 2004). In addition, it is well known that defects in innervation can affect the size of tissues. Therefore, reduced innervation of thymus and spleen in ROR$\alpha^{-/-}$ mice due to cerebellar defects may be partially responsible for the reduced size of these tissues.

No differences were observed in the induction of proliferation of CD4$^+$ and CD8$^+$ T lymphocytes in response to lymphopenic environment or the addition of anti-CD3 or lipopolysaccharide (LPS) between ROR$\alpha^{-/-}$ and wild-type mice. Also, there was little difference in the induction of the interleukins IL-2 and IL-6, and of TNFα. This is in contrast to ROR$\alpha^{-/-}$ mast cells and macrophages, in which the production of IL-6 and TNFα was greatly increased after LPS treatment (Kopmels et al., 1992; Dzhagalov et al., 2004). Similarly, CD8$^+$ T cells from ROR$\alpha^{-/-}$ mice but not CD4$^+$ T cells produce higher levels of interferon γ after LPS stimulation. These results indicate that at least in certain immune cells RORα functions as a negative regulator of cytokine expression. Although the mechanism of this repression is not fully understood, RORα has been reported to positively regulate the expression of IκBα, a negative regulator of the NF-κB signaling pathway (Delerive et al., 2001). These observations suggest that RORα may function as a negative regulator of inflammation. The latter is supported by studies showing that ROR$\alpha^{sg/sg}$ mice exhibit an enhanced susceptibility to LPS-induced lung inflammation (Stapleton et al., 2005).

9. Role of RORs in mesenchymal differentiation

Although several studies have provided evidence for regulatory roles of RORs in adipocytes, myoblast differentiation, and bone formation, knowledge about their precise molecular and physiological functions is still limited. Both RORα and RORγ have been reported to be induced during differentiation of 3T3L1 preadipocytes (Adachi et al., 1996; Austin et al., 1998). This induction is promoted by peroxisome proliferator-activated receptor γ (PPARγ) and PPARγ agonists, and inhibited by retinoic acid. The induction of RORα and RORγ expression appears to be a late event and unrelated to the commitment of preadipocytes to undergo adipocyte differentiation (Adachi et al., 1996; Austin et al., 1998). In adult mice, RORγ is expressed in brown adipose tissue and not in white adipose tissue, suggesting that it may regulate a function that is typical for brown adipose tissue. Brown adipose tissue is a major site for adaptive thermogenesis (Lowell and Spiegelman, 2000). However, RORγ does not appear to play a major role in regulating adaptive thermogenesis since no significant difference was observed between wild-type and RORγ-deficient mice in the induction of uncoupling protein 1 during cold-induced thermogenesis (Jetten et al., unpublished observations).

Moreover, RORγ expression in brown adipose tissue did not change in wild-type mice during cold-induced thermogenesis.

A role for RORα in bone metabolism was suggested by studies showing that RORα$^{sg/sg}$ mice are osteopenic (Meyer et al., 2000). The total mineral content and bone density are significantly reduced in bones of RORα-deficient mice compared to those of wild-type mice. These observations indicate an imbalance between bone formation and bone resorption and suggest a positive regulatory function for RORα in bone development. This is supported by studies showing that expression of RORα mRNA is significantly enhanced during differentiation of mesenchymal stem cells into osteoblasts and that RORα is able to activate the promoter of the bone sialoprotein gene (Meyer et al., 2000). Although these observations are in agreement with the hypothesis that RORα regulates osteoblast function and activity, further studies are needed to determine the precise role of RORα in bone metabolism.

RORα and RORγ mRNA have been reported to be expressed in skeletal muscle and in mouse myoblast C2C12 cells suggesting a possible regulatory role for these receptors in myogenesis or muscle function (Becker-Andre et al., 1993; Lau et al., 1999). This was supported by studies showing that ectopic expression of a dominant-negative form of RORα repressed myogenesis as indicated by an inhibition of the induction of the muscle-specific helix-loop-helix transcription factors MyoD and MYF4, and the cdk inhibitor p21 (Lau et al., 1999). This inhibition appears to involve a direct interaction of RORα1 with the coactivator p300 and MyoD. The interaction of RORα and MyoD is mediated through their DBD and N-terminal region, respectively. These findings are consistent with the hypothesis that RORα functions as a positive regulator of myogenesis. Since Rev-ErbA represses myogenesis and is able to compete with RORs for RORE binding, regulation of myogenesis may involve cross talk between these two receptor pathways (Burke et al., 1996; Downes et al., 1996; Lau et al., 1999).

Studies have demonstrated that expression of a dominant-negative form of RORα in myoblast C2C12 cells attenuates the expression of several genes involved in the regulation of lipid metabolism suggesting a role of RORα in the control of lipid homeostasis (Lau et al., 2004). Evidence was provided indicating that muscle carnitine palmitoyltransferase-1 and caveolin-3 promoters are regulated directly by RORα and involve recruitment of the coactivators p300 and PGC-1. Reports showing that RORα functions as a positive regulator of the expression of several apolipoproteins further support a role of RORα in lipid metabolism (Vu-Dac et al., 1997; Mamontova et al., 1998; Raspe et al., 2001; Rader, 2002; Genoux et al., 2005; Lind et al., 2005; Jakel et al., 2006).

10. Role of the ROR homologues DHR3 and CHR3 in development

DHR3, the ROR homologue in *D. melanogaster*, has been reported to play an important role in *Drosophila* development (Koelle et al., 1992; Horner et al., 1995; Thummel, 1995; Carney et al., 1997; Sullivan and

Thummel, 2003; Gates et al., 2004; King-Jones and Thummel, 2005). DHR3 plays a critical role during embryogenesis prior to molting and at the onset of metamorphosis. During embryogenesis, DHR3 mRNA is detectable between 6 and 12 hours after egg lay (Sullivan and Thummel, 2003). Through this stage, DHR3 functions as transcriptional repressor and cooperates with other repressors. During larval and prepupal development, DHR3 is expressed in the second half of the second instar and in the early prepupa stage (Lam et al., 1997, 1999). During development, expression of DHR3 mRNA is in synchrony with the pulses of steroid hormone 20-hydroxyecdysone (20-E) suggesting that its expression is regulated by 20-E. 20-E acts by binding to ecdysteroid receptors (EcRs), nuclear hormone receptors that occur in many isoforms. The ecdysone receptors form a heterodimer with ultraspiracle (USP), the insect homologue of the retinoid X receptor (RXR). Activation of this heterodimer by 20-E and binding to ecdysone response elements in the promoter region of target genes results in enhanced gene transcription. However, optimal induction of DHR3 is dependent on protein synthesis suggesting that its transcriptional regulation by 20-E occurs at least partially by an indirect mechanism. DHR3 upregulates the expression of another nuclear receptor βFTZ-F1, the homologue of the mammalian nuclear receptor SF-1 (NR5A3) (Kageyama et al., 1997). This induction is mediated through the binding of DHR3 to DHR3-binding sites in the βFTZ-F1 promoter. The nuclear receptor E75B, a homologue of the mammalian nuclear receptor Rev-Erb (NR1D) that acts as a transcriptional repressor, can repress βFTZ-F1 expression by competing with DHR3 for the same DNA binding sites. It is interesting to note that a similar antagonism, as observed between DHR3 and E75B, has been reported for their mammalian homologues, the ROR and Rev-Erb receptors (Forman et al., 1994; Downes et al., 1996; Medvedev et al., 1997; Austin et al., 1998; Bois-Joyeux et al., 2000; Akashi and Takumi, 2005; Guillaumond et al., 2005).

Ecdysone is important in orchestrating the reorganization of neural circuits during metamorphosis (Weeks, 2003). Analysis of *Drosophila* DHR3 mutants showed that they form a normal larval cuticle but that few survive to hatching (Carney et al., 1997). DHR3 mutants display various defects in the peripheral nervous system that include a general disorganization of the neuronal clusters and an absence or mislocalization of neurons. Since RORα has been shown to regulate dendritic arborization of Purkinje cells in mice, a similar role has been suggested for the ROR homologue DHR3 in the regulation of remodeling of neurons in *Drosophila* (Boukhtouche et al., 2006b).

As has been demonstrated for DHR3, the expression of CHR3 and MHR3, the ROR homologues identified in *C. elegans*, *C. fumiferana* (Palli et al., 1996, 1997), and the tobacco hawkmoth *M. sexta* (Kostrouch et al., 1995), respectively, is also regulated by 20-E (Langelan et al., 2000; Hiruma and Riddiford, 2004). CHR3 regulates molting at all four larval stages of

C. elegans. Disruption of CHR3 in *C. elegans* results in several developmental changes, including incomplete molting and generation of a short, fat pheno-type (Kostrouchova et al., 1998, 2001). The CHR3 receptor is also required for proper epidermal development. In *M. sexta*, MHR3 is induced by 20-E in the epidermis. Evidence has been provided indicating that this induction is regulated by ecdysone receptor-USP-1 heterodimers (Riddiford et al., 2003). These results demonstrate that as their mammalian homologues DHR3, CHR3, and MHR3 are essential for the development of several tissues.

11. Summary

As illustrated in this chapter, ROR receptors are critical regulators of cellular differentiation and the development of several tissues. This includes a role for RORα in the development of the cerebellum and bone formation, a role for RORβ in the brain and retina, and a role for RORγ in the development of several secondary lymphoid tissues and in the control of thymopoi-esis. In addition, RORs play an important role in the regulation of circadian rhythms. Whether ROR activity *in vivo* and consequently the physiological processes controlled by RORs are regulated by endogenous ligands has yet to be determined. However, evidence has been provided indicating that in cultured cells ROR transcriptional activity can be modulated by ligands. These observations leave open the possibility that synthetic (ant)agonists might be useful in the development of new therapeutic strategies for several human diseases in which RORs are implicated. This overview shows that great insights have been obtained into the physiological function of RORs in several tissues; however, RORs are expressed in many other tissues in which the physiological function of RORs still needs to be uncovered.

Acknowledgments

The author would like to thank Drs. D. Germolec, J. Harry, and E. Allen for their comments on the chapter. This research was supported by the Intramural Research Program of the NIEHS, NIH.

References

Adachi, H., Dawson, M.I., Jetten, A.M. 1996. Suppression by retinoids of the induction of the CCAAT/enhancer-binding protein α and the nuclear receptors PPARγ and RORγ during adipocyte differentiation of 3T3-L1 cells. Mol. Cell Differ. 4, 365–381.

Adams, J. 2003. The proteasome: Structure, function, and role in the cell. Cancer Treat. Rev. 29 (Suppl. 1), 3–9.

Akashi, M., Takumi, T. 2005. The orphan nuclear receptor RORalpha regulates circadian transcription of the mammalian core-clock Bmal1. Nat. Struct. Mol. Biol. 12, 441–448.

Albrecht, U. 2002. Invited review: Regulation of mammalian circadian clock genes. J. Appl. Physiol. 92, 1348–1355.

Anderson, G., Moore, N.C., Owen, J.J., Jenkinson, E.J. 1996. Cellular interactions in thymocyte development. Annu. Rev. Immunol. 14, 73–99.

Andre, E., Conquet, F., Steinmayr, M., Stratton, S.C., Porciatti, V., Becker-Andre, M. 1998a. Disruption of retinoid-related orphan receptor beta changes circadian behavior, causes retinal degeneration and leads to vacillans phenotype in mice. EMBO J. 17, 3867–3877.

Andre, E., Gawlas, K., Becker-Andre, M. 1998b. A novel isoform of the orphan nuclear receptor RORbeta is specifically expressed in pineal gland and retina. Gene 216, 277–283.

Aranda, A., Pascual, A. 2001. Nuclear hormone receptors and gene expression. Physiol. Rev. 81, 1269–1304.

Atkins, G.B., Hu, X., Guenther, M.G., Rachez, C., Freedman, L.P., Lazar, M.A. 1999. Coactivators for the orphan nuclear receptor RORalpha. Mol. Endocrinol. 13, 1550–1557.

Austin, S., Medvedev, A., Yan, Z.H., Adachi, H., Hirose, T., Jetten, A.M. 1998. Induction of the nuclear orphan receptor RORgamma during adipocyte differentiation of D1 and 3T3-L1 cells. Cell Growth Differ. 9, 267–276.

Azadi, S., Zhang, Y., Caffe, A.R., Holmqvist, B., van Veen, T. 2002. Thyroid-beta2 and the retinoid RAR-alpha, RXR-gamma and ROR-beta2 receptor mRNAs; expression profiles in mouse retina, retinal explants and neocortex. Neuroreport 13, 745–750.

Baird, A.M., Gerstein, R.M., Berg, L.J. 1999. The role of cytokine receptor signaling in lymphocyte development. Curr. Opin. Immunol. 11, 157–166.

Becker-Andre, M., Andre, E., DeLamarter, J.F. 1993. Identification of nuclear receptor mRNAs by RT-PCR amplification of conserved zinc-finger motif sequences. Biochem. Biophys. Res. Commun. 194, 1371–1379.

Bois-Joyeux, B., Chauvet, C., Nacer-Cherif, H., Bergeret, W., Mazure, N., Giguere, V., Laudet, V., Danan, J.L. 2000. Modulation of the far-upstream enhancer of the rat alpha-fetoprotein gene by members of the ROR alpha, Rev-erb alpha, and Rev-erb beta groups of monomeric orphan nuclear receptors. DNA Cell Biol. 19, 589–599.

Borycki, A., Brown, A.M., Emerson, C.P., Jr. 2000. Shh and Wnt signaling pathways converge to control Gli gene activation in avian somites. Development 127, 2075–2087.

Boukhtouche, F., Doulazmi, M., Frederic, F., Dusart, I., Brugg, B., Mariani, J. 2006a. RORalpha, a pivotal nuclear receptor for Purkinje neuron survival and differentiation: From development to ageing. Cerebellum 5, 97–104.

Boukhtouche, F., Janmaat, S., Vodjdani, G., Gautheron, V., Mallet, J., Dusart, I., Mariani, J. 2006b. Retinoid-related orphan receptor alpha controls the early steps of Purkinje cell dendritic differentiation. J. Neurosci. 26, 1531–1538.

Bouvet, J., Usson, Y., Legrand, J. 1987. Morphometric analysis of the cerebellar Purkinje cell in the developing normal and hypothyroid chick. Int. J. Dev. Neurosci. 5, 345–355.

Burke, L., Downes, M., Carozzi, A., Giguere, V., Muscat, G.E. 1996. Transcriptional repression by the orphan steroid receptor RVR/Rev-erb beta is dependent on the signature motif and helix 5 in the E region: Functional evidence for a biological role of RVR in myogenesis. Nucleic Acids Res. 24, 3481–3489.

Cachon-Gonzalez, M.B., Fenner, S., Coffin, J.M., Moran, C., Best, S., Stoye, J.P. 1994. Structure and expression of the hairless gene of mice. Proc. Natl. Acad. Sci. USA 91, 7717–7721.

Carlberg, C., Hooft van Huijsduijnen, R., Staple, J.K., DeLamarter, J.F., Becker-Andre, M. 1994. RZRs, a new family of retinoid-related orphan receptors that function as both monomers and homodimers. Mol. Endocrinol. 8, 757–770.

Carlberg, C., Wiesenberg, I. 1995. The orphan receptor family RZR/ROR, melatonin and 5-lipoxygenase: An unexpected relationship. J. Pineal Res. 18, 171–178.

Carney, G.E., Wade, A.A., Sapra, R., Goldstein, E.S., Bender, M. 1997. DHR3, an ecdysone-inducible early-late gene encoding a *Drosophila* nuclear receptor, is required for embryogenesis. Proc. Natl. Acad. Sci. USA 94, 12024–12029.

Chauvet, C., Bois-Joyeux, B., Danan, J.L. 2002. Retinoic acid receptor-related orphan receptor (ROR) alpha4 is the predominant isoform of the nuclear receptor RORalpha in the liver and is up-regulated by hypoxia in HepG2 human hepatoma cells. Biochem. J. 364, 449–456.

Chen, C., Edelstein, L.C., Gelinas, C. 2000. The Rel/NF-kappaB family directly activates expression of the apoptosis inhibitor Bcl-x(L). Mol. Cell. Biol. 20, 2687–2695.

Chu, K., Zingg, H.H. 1999. Activation of the mouse oxytocin promoter by the orphan receptor RORalpha. J. Mol. Endocrinol. 23, 337–346.

Cupedo, T., Kraal, G., Mebius, R.E. 2002. The role of CD45+CD4+CD3− cells in lymphoid organ development. Immunol. Rev. 189, 41–50.

Cupedo, T., Mebius, R.E. 2005. Cellular interactions in lymph node development. J. Immunol. 174, 21–25.

Cyster, J.G. 2003. Lymphoid organ development and cell migration. Immunol. Rev. 195, 5–14.

Dace, A., Zhao, L., Park, K.S., Furuno, T., Takamura, N., Nakanishi, M., West, B.L., Hanover, J.A., Cheng, S. 2000. Hormone binding induces rapid proteasome-mediated degradation of thyroid hormone receptors. Proc. Natl. Acad. Sci. USA 97, 8985–8990.

Dahmane, N., Ruiz-i-Altaba, A. 1999. Sonic hedgehog regulates the growth and patterning of the cerebellum. Development 126, 3089–3100.

Darimont, B.D., Wagner, R.L., Apriletti, J.W., Stallcup, M.R., Kushner, P.J., Baxter, J.D., Fletterick, R.J., Yamamoto, K.R. 1998. Structure and specificity of nuclear receptor-coactivator interactions. Genes Dev. 12, 3343–3356.

De Togni, P., Goellner, J., Ruddle, N.H., Streeter, P.R., Fick, A., Mariathasan, S., Smith, S.C., Carlson, R., Shornick, L.P., Strauss-Schoenberger, J., Russell, J.H., Karr, R., et al. 1994. Abnormal development of peripheral lymphoid organs in mice deficient in lymphotoxin. Science 264, 703–707.

Delerive, P., Monte, D., Dubois, G., Trottein, F., Fruchart-Najib, J., Mariani, J., Fruchart, J.C., Staels, B. 2001. The orphan nuclear receptor ROR alpha is a negative regulator of the inflammatory response. EMBO Rep. 2, 42–48.

Delerive, P., Chin, W.W., Suen, C.S. 2002. Identification of Reverb(alpha) as a novel ROR (alpha) target gene. J. Biol. Chem. 277, 35013–35018.

Dennis, A.P., Haq, R.U., Nawaz, Z. 2001. Importance of the regulation of nuclear receptor degradation. Front. Biosci. 6, D954–D959.

Desvergne, B., Wahli, W. 1999. Peroxisome proliferator-activated receptors: Nuclear control of metabolism. Endocr. Rev. 20, 649–688.

Doulazmi, M., Frederic, F., Lemaigre-Dubreuil, Y., Hadj-Sahraoui, N., Delhaye-Bouchaud, N., Mariani, J. 1999. Cerebellar Purkinje cell loss during life span of the heterozygous staggerer mouse (Rora(+)/Rora(sg)) is gender-related. J. Comp. Neurol. 411, 267–273.

Doulazmi, M., Frederic, F., Capone, F., Becker-Andre, M., Delhaye-Bouchaud, N., Mariani, J. 2001. A comparative study of Purkinje cells in two RORalpha gene mutant mice: Staggerer and RORalpha(−/−). Brain Res. Dev. Brain Res. 127, 165–174.

Downes, M., Burke, L.J., Muscat, G.E. 1996. Transcriptional repression by Rev-erbA alpha is dependent on the signature motif and helix 5 in the ligand binding domain: Silencing does not involve an interaction with N-CoR. Nucleic Acids Res. 24, 3490–3498.

Dussault, I., Giguere, V. 1997. Differential regulation of the N-myc proto-oncogene by ROR alpha and RVR, two orphan members of the superfamily of nuclear hormone receptors. Mol. Cell. Biol. 17, 1860–1867.

Dussault, I., Fawcett, D., Matthyssen, A., Bader, J.A., Giguere, V. 1998. Orphan nuclear receptor ROR alpha-deficient mice display the cerebellar defects of staggerer. Mech. Dev. 70, 147–153.

Dzhagalov, I., Giguere, V., He, Y.W. 2004. Lymphocyte development and function in the absence of retinoic acid-related orphan receptor alpha. J. Immunol. 173, 2952–2959.

Eberl, G. 2005. Inducible lymphoid tissues in the adult gut: Recapitulation of a fetal developmental pathway? Nat. Rev. Immunol. 5, 413–420.

Eberl, G., Littman, D.R. 2003. The role of the nuclear hormone receptor RORgammat in the development of lymph nodes and Peyer's patches. Immunol. Rev. 195, 81–90.

Eberl, G., Littman, D.R. 2004. Thymic origin of intestinal alphabeta T cells revealed by fate mapping of RORgammat+ cells. Science 305, 248–251.

Eberl, G., Marmon, S., Sunshine, M.J., Rennert, P.D., Choi, Y., Littman, D.R. 2004. An essential function for the nuclear receptor RORgammat in the generation of fetal lymphoid tissue inducer cells. Nat. Immunol. 5, 64–73.

Ellmeier, W., Sawada, S., Littman, D.R. 1999. The regulation of CD4 and CD8 coreceptor gene expression during T cell development. Annu. Rev. Immunol. 17, 523–554.

Evans, R.M. 1988. The steroid and thyroid hormone receptor superfamily. Science 240, 889–895.

Fehling, H.J., von Boehmer, H. 1997. Early alpha beta T cell development in the thymus of normal and genetically altered mice. Curr. Opin. Immunol. 9, 263–275.

Forman, B.M., Chen, J., Blumberg, B., Kliewer, S.A., Henshaw, R., Ong, E.S., Evans, R.M. 1994. Cross-talk among ROR alpha 1 and the Rev-erb family of orphan nuclear receptors. Mol. Endocrinol. 8, 1253–1261.

Fu, Y.X., Chaplin, D.D. 1999. Development and maturation of secondary lymphoid tissues. Annu. Rev. Immunol. 17, 399–433.

Fukuyama, S., Hiroi, T., Yokota, Y., Rennert, P.D., Yanagita, M., Kinoshita, N., Terawaki, S., Shikina, T., Yamamoto, M., Kurono, Y., Kiyono, H. 2002. Initiation of NALT organogenesis is independent of the IL-7R, LTbetaR, and NIK signaling pathways but requires the Id2 gene and CD3(−)CD4(+)CD45(+) cells. Immunity 17, 31–40.

Futterer, A., Mink, K., Luz, A., Kosco-Vilbois, M.H., Pfeffer, K. 1998. The lymphotoxin beta receptor controls organogenesis and affinity maturation in peripheral lymphoid tissues. Immunity 9, 59–70.

Gachon, F., Nagoshi, E., Brown, S.A., Ripperger, J., Schibler, U. 2004. The mammalian circadian timing system: From gene expression to physiology. Chromosoma 113, 103–112.

Gates, J., Lam, G., Ortiz, J.A., Losson, R., Thummel, C.S. 2004. Rigor mortis encodes a novel nuclear receptor interacting protein required for ecdysone signaling during *Drosophila* larval development. Development 131, 25–36.

Gawlas, K., Stunnenberg, H.G. 2001. Differential transcription of the orphan receptor RORbeta in nuclear extracts derived from Neuro2A and HeLa cells. Nucleic Acids Res. 29, 3424–3432.

Genoux, A., Dehondt, H., Helleboid-Chapman, A., Duhem, C., Hum, D.W., Martin, G., Pennacchio, L.A., Staels, B., Fruchart-Najib, J., Fruchart, J.C. 2005. Transcriptional regulation of apolipoprotein A5 gene expression by the nuclear receptor RORalpha. Arterioscler. Thromb. Vasc. Biol. 25, 1186–1192.

Giguere, V. 1999. Orphan nuclear receptors: From gene to function. Endocr. Rev. 20, 689–725.

Giguere, V., Tini, M., Flock, G., Ong, E., Evans, R.M., Otulakowski, G. 1994. Isoform-specific amino-terminal domains dictate DNA-binding properties of ROR alpha, a novel family of orphan hormone nuclear receptors. Genes Dev. 8, 538–553.

Giguere, V., Beatty, B., Squire, J., Copeland, N.G., Jenkins, N.A. 1995. The orphan nuclear receptor ROR alpha (RORA) maps to a conserved region of homology on human chromosome 15q21-q22 and mouse chromosome 9. Genomics 28, 596–598.

Gil-Gomez, G., Berns, A., Brady, H.J. 1998. A link between cell cycle and cell death: Bax and Bcl-2 modulate Cdk2 activation during thymocyte apoptosis. EMBO J. 17, 7209–7218.

Glass, C.K., Rosenfeld, M.G. 2000. The coregulator exchange in transcriptional functions of nuclear receptors. Genes Dev. 14, 121–141.

Gold, D.A., Baek, S.H., Schork, N.J., Rose, D.W., Larsen, D.D., Sachs, B.D., Rosenfeld, M.G., Hamilton, B.A. 2003. RORalpha coordinates reciprocal signaling in cerebellar development through sonic hedgehog and calcium-dependent pathways. Neuron 40, 1119–1131.

Gold, D.A., Gent, P.M., Hamilton, B.A. 2006. RORalpha in genetic control of cerebellum development: 50 staggering years. Brain Res. (in press).

Goldowitz, D., Hamre, K. 1998. The cells and molecules that make a cerebellum. Trends Neurosci. 21, 375–382.

Greiner, E.F., Kirfel, J., Greschik, H., Dorflinger, U., Becker, P., Mercep, A., Schule, R. 1996. Functional analysis of retinoid Z receptor beta, a brain-specific nuclear orphan receptor. Proc. Natl. Acad. Sci. USA 93, 10105–10110.

Greiner, E.F., Kirfel, J., Greschik, H., Huang, D., Becker, P., Kapfhammer, J.P., Schule, R. 2000. Differential ligand-dependent protein-protein interactions between nuclear receptors and a neuronal-specific cofactor. Proc. Natl. Acad. Sci. USA 97, 7160–7165.

Guastavino, J.M., Bertin, R., Portet, R. 1991. Effects of the rearing temperature on the temporal feeding pattern of the staggerer mutant mouse. Physiol. Behav. 49, 405–409.

Guillaumond, F., Dardente, H., Giguere, V., Cermakian, N. 2005. Differential control of Bmal1 circadian transcription by REV-ERB and ROR nuclear receptors. J. Biol. Rhythms 20, 391–403.

Guo, J., Hawwari, A., Li, H., Sun, Z., Mahanta, S.K., Littman, D.R., Krangel, M.S., He, Y.W. 2002. Regulation of the TCRalpha repertoire by the survival window of CD4(+)CD8(+) thymocytes. Nat. Immunol. 3, 469–476.

Hadj-Sahraoui, N., Frederic, F., Zanjani, H., Herrup, K., Delhaye-Bouchaud, N., Mariani, J. 1997. Purkinje cell loss in heterozygous staggerer mutant mice during aging. Brain Res. Dev. Brain Res. 98, 1–8.

Hadj-Sahraoui, N., Frederic, F., Zanjani, H., Delhaye-Bouchaud, N., Herrup, K., Mariani, J. 2001. Progressive atrophy of cerebellar Purkinje cell dendrites during aging of the heterozygous staggerer mouse (Rora(+/sg)). Brain Res. Dev. Brain Res. 126, 201–209.

Hamilton, B.A., Frankel, W.N., Kerrebrock, A.W., Hawkins, T.L., FitzHugh, W., Kusumi, K., Russell, L.B., Mueller, K.L., van Berkel, V., Birren, B.W., Kruglyak, L., Lander, E.S. 1996. Disruption of the nuclear hormone receptor RORalpha in staggerer mice. Nature 379, 736–739.

Harding, H.P., Atkins, G.B., Jaffe, A.B., Seo, W.J., Lazar, M.A. 1997. Transcriptional activation and repression by RORalpha, an orphan nuclear receptor required for cerebellar development. Mol. Endocrinol. 11, 1737–1746.

Harmsen, A., Kusser, K., Hartson, L., Tighe, M., Sunshine, M.J., Sedgwick, J.D., Choi, Y., Littman, D.R., Randall, T.D. 2002. Cutting edge: Organogenesis of nasal-associated lymphoid tissue (NALT) occurs independently of lymphotoxin-alpha (LT alpha) and retinoic acid receptor-related orphan receptor-gamma, but the organization of NALT is LT alpha dependent. J. Immunol. 168, 986–990.

Harris, J.M., Lau, P., Chen, S.L., Muscat, G.E. 2002. Characterization of the retinoid orphan-related receptor-alpha coactivator binding interface: A structural basis for ligand-independent transcription. Mol. Endocrinol. 16, 998–1012.

He, Y.W. 2000. The role of orphan nuclear receptor in thymocyte differentiation and lymphoid organ development. Immunol. Res. 22, 71–82.

He, Y.W. 2002. Orphan nuclear receptors in T lymphocyte development. J. Leukoc. Biol. 72, 440–446.

He, Y.W., Deftos, M.L., Ojala, E.W., Bevan, M.J. 1998. RORgamma t, a novel isoform of an orphan receptor, negatively regulates Fas ligand expression and IL-2 production in T cells. Immunity 9, 797–806.

He, Y.W., Beers, C., Deftos, M.L., Ojala, E.W., Forbush, K.A., Bevan, M.J. 2000. Down-regulation of the orphan nuclear receptor ROR gamma t is essential for T lymphocyte maturation. J. Immunol. 164, 5668–5674.

Heery, D.M., Kalkhoven, E., Hoare, S., Parker, M.G. 1997. A signature motif in transcriptional co-activators mediates binding to nuclear receptors. Nature 387, 733–736.

Heery, D.M., Hoare, S., Hussain, S., Parker, M.G., Sheppard, H.M. 2001. Core LXXLL motif sequences in CBP, SRC1 and RIP140 define affinity and selectivity for steroid and retinoid receptors. J. Biol. Chem. 276, 6695–6702.

Herrup, K., Mullen, R.J. 1979. Staggerer chimeras: Intrinsic nature of Purkinje cell defects and implications for normal cerebellar development. Brain Res. 178, 443–457.

Herrup, K., Mullen, R.J. 1981. Role of the Staggerer gene in determining Purkinje cell number in the cerebellar cortex of mouse chimeras. Brain Res. 227, 475–485.

Hirose, T., Smith, R.J., Jetten, A.M. 1994. ROR gamma: The third member of ROR/RZR orphan receptor subfamily that is highly expressed in skeletal muscle. Biochem. Biophys. Res. Commun. 205, 1976–1983.

Hiruma, K., Riddiford, L.M. 2004. Differential control of MHR3 promoter activity by isoforms of the ecdysone receptor and inhibitory effects of E75A and MHR3. Dev. Biol. 272, 510–521.

Horlein, A.J., Naar, A.M., Heinzel, T., Torchia, J., Gloss, B., Kurokawa, R., Ryan, A., Kamei, Y., Soderstrom, M., Glass, C.K., Rosenfeld, M.G. 1995. Ligand-independent repression by the thyroid hormone receptor mediated by a nuclear receptor co-repressor. Nature 377, 397–404.

Horner, M.A., Chen, T., Thummel, C.S. 1995. Ecdysteroid regulation and DNA binding properties of *Drosophila* nuclear hormone receptor superfamily members. Dev. Biol. 168, 490–502.

Hsieh, J.C., Sisk, J.M., Jurutka, P.W., Haussler, C.A., Slater, S.A., Haussler, M.R., Thompson, C.C. 2003. Physical and functional interaction between the vitamin D receptor and hairless corepressor, two proteins required for hair cycling. J. Biol. Chem. 278, 38665–38674.

Hu, I., Lazar, M.A. 2000. Transcriptional Repression by Nuclear Hormone Receptors. Trends Endocrinol. Metab. 11, 6–10.

Ino, H. 2004. Immunohistochemical characterization of the orphan nuclear receptor ROR alpha in the mouse nervous system. J. Histochem. Cytochem. 52, 311–323.

Ismail, A., Nawaz, Z. 2005. Nuclear hormone receptor degradation and gene transcription: An update. IUBMB Life 57, 483–490.

Isojima, Y., Okumura, N., Nagai, K. 2003. Molecular mechanism of mammalian circadian clock. J. Biochem. (Tokyo) 134, 777–784.

Jakel, H., Fruchart-Najib, J., Fruchart, J.C. 2006. Retinoic acid receptor-related orphan receptor alpha as a therapeutic target in the treatment of dyslipidemia and atherosclerosis. Drug News Perspect. 19, 91–97.

Jameson, S.C., Hogquist, K.A., Bevan, M.J. 1995. Positive selection of thymocytes. Annu. Rev. Immunol. 13, 93–126.

Jetten, A.M. 2004. Recent advances in the mechanisms of action and physiological functions of the retinoid-related orphan receptors (RORs). Curr. Drug Targets Inflamm. Allergy 3, 395–412.

Jetten, A.M., Ueda, E. 2002. Retinoid-related orphan receptors (RORs): Roles in cell survival, differentiation and disease. Cell Death Differ. 9, 1167–1171.

Jetten, A.M., Kurebayashi, S., Ueda, E. 2001. The ROR nuclear orphan receptor subfamily: Critical regulators of multiple biological processes. Prog. Nucleic Acid Res. 69, 205–247.

Jetten, A.M., Kim, Y.S., Nakajima, T., Ueda, E., Kim, S.-C., Kang, H.S., Koegl, M., unpublished observations.

Jin, P., Sun, Y., Grabowski, G.A. 1998. Role of Sp proteins and RORalpha in transcription regulation of murine prosaposin. J. Biol. Chem. 273, 13208–13216.

Jin, P., Sun, Y., Grabowski, G.A. 2001. *In vivo* roles of RORalpha and Sp4 in the regulation of murine prosaposin gene. DNA Cell Biol. 20, 781–789.

Johnson, D.R., Lovett, J.M., Hirsch, M., Xia, F., Chen, J.D. 2004. NuRD complex component Mi-2beta binds to and represses RORgamma-mediated transcriptional activation. Biochem. Biophys. Res. Commun. 318, 714–718.

Kageyama, Y., Masuda, S., Hirose, S., Ueda, H. 1997. Temporal regulation of the mid-prepupal gene FTZ-F1: DHR3 early late gene product is one of the plural positive regulators. Genes Cells 2, 559–569.

Kallen, J.A., Schlaeppi, J., Bitsch, F., Geisse, S., Geiser, M., Delhon, I., Fournier, B. 2002. X-ray structure of the RORα LBD at 1.63A: Structural and functional data that cholesterol or a cholesterol derivative is the natural ligand of RORα. Structure 10, 1697–1707.

Kallen, J., Schlaeppi, J.M., Bitsch, F., Delhon, I., Fournier, B. 2004. Crystal structure of the human RORalpha Ligand binding domain in complex with cholesterol sulfate at 2.2 A. J. Biol. Chem. 279, 14033–14038.

Kanamori, Y., Ishimaru, K., Nanno, M., Maki, K., Ikuta, K., Nariuchi, H., Ishikawa, H. 1996. Identification of novel lymphoid tissues in murine intestinal mucosa where clusters of c-kit+IL-7R+Thy1+ lympho-hemopoietic progenitors develop. J. Exp. Med. 184, 1449–1459.

Kenney, A.M., Cole, M.D., Rowitch, D.H. 2003. Nmyc upregulation by sonic hedgehog signaling promotes proliferation in developing cerebellar granule neuron precursors. Development 130, 15–28.

King-Jones, K., Thummel, C.S. 2005. Nuclear receptors—a perspective from *Drosophila*. Nat. Rev. Genet. 6, 311–323.

Kinyamu, H.K., Chen, J., Archer, T.K. 2005. Linking the ubiquitin-proteasome pathway to chromatin remodeling/modification by nuclear receptors. J. Mol. Endocrinol. 34, 281–297.

Kisielow, P., von Boehmer, H. 1995. Development and selection of T cells: Facts and puzzles. Adv. Immunol. 58, 87–209.

Koelle, M.R., Segraves, W.A., Hogness, D.S. 1992. DHR3: A *Drosophila* steroid receptor homolog. Proc. Natl. Acad. Sci. USA 89, 6167–6171.

Koibuchi, N., Chin, W.W. 1998. ROR alpha gene expression in the perinatal rat cerebellum: Ontogeny and thyroid hormone regulation. Endocrinology 139, 2335–2341.

Koni, P.A., Sacca, R., Lawton, P., Browning, J.L., Ruddle, N.H., Flavell, R.A. 1997. Distinct roles in lymphoid organogenesis for lymphotoxins alpha and beta revealed in lymphotoxin beta-deficient mice. Immunity 6, 491–500.

Kopmels, B., Mariani, J., Delhaye-Bouchaud, N., Audibert, F., Fradelizi, D., Wollman, E.E. 1992. Evidence for a hyperexcitability state of staggerer mutant mice macrophages. J. Neurochem. 58, 192–199.

Kostrouch, Z., Kostrouchova, M., Rall, J.E. 1995. Steroid/thyroid hormone receptor genes in *Caenorhabditis elegans*. Proc. Natl. Acad. Sci. USA 92, 156–159.

Kostrouchova, M., Krause, M., Kostrouch, Z., Rall, J.E. 1998. CHR3: A *Caenorhabditis elegans* orphan nuclear hormone receptor required for proper epidermal development and molting. Development 125, 1617–1626.

Kostrouchova, M., Krause, M., Kostrouch, Z., Rall, J.E. 2001. Nuclear hormone receptor CHR3 is a critical regulator of all four larval molts of the nematode *Caenorhabditis elegans*. Proc. Natl. Acad. Sci. USA 98, 7360–7365.

Kumar, R., Thompson, E.B. 1999. The structure of the nuclear hormone receptors. Steroids 64, 310–319.

Kurebayashi, S., Ueda, E., Sakaue, M., Patel, D.D., Medvedev, A., Zhang, F., Jetten, A.M. 2000. Retinoid-related orphan receptor gamma (RORgamma) is essential for lymphoid organogenesis and controls apoptosis during thymopoiesis. Proc. Natl. Acad. Sci. USA 97, 10132–10137.

Kurebayashi, S., Nakajima, T., Kim, S.C., Chang, C.Y., McDonnell, D.P., Renaud, J.P., Jetten, A.M. 2004. Selective LXXLL peptides antagonize transcriptional activation by the retinoid-related orphan receptor RORgamma. Biochem. Biophys. Res. Commun. 315, 919–927.

Lam, G.T., Jiang, C., Thummel, C.S. 1997. Coordination of larval and prepupal gene expression by the DHR3 orphan receptor during *Drosophila* metamorphosis. Development 124, 1757–1769.

Lam, G., Hall, B.L., Bender, M., Thummel, C.S. 1999. DHR3 is required for the prepupal-pupal transition and differentiation of adult structures during *Drosophila* metamorphosis. Dev. Biol. 212, 204–216.

Lan, Q., Wu, Z., Riddiford, L.M. 1997. Regulation of the ecdysone receptor, USP, E75 and MHR3 mRNAs by 20-hydroxyecdysone in the GV1 cell line of the tobacco hornworm, *Manduca sexta*. Insect Mol. Biol. 6, 3–10.

Lan, Q., Hiruma, K., Hu, X., Jindra, M., Riddiford, L.M. 1999. Activation of a delayed-early gene encoding MHR3 by the ecdysone receptor heterodimer EcR-B1-USP-1 but not by EcR-B1-USP-2. Mol. Cell. Biol. 19, 4897–4906.

Landis, D.M., Reese, T.S. 1977. Structure of the Purkinje cell membrane in staggerer and weaver mutant mice. J. Comp. Neurol. 171, 247–260.

Landis, D.M., Sidman, R.L. 1978. Electron microscopic analysis of postnatal histogenesis in the cerebellar cortex of staggerer mutant mice. J. Comp. Neurol. 179, 831–863.

Langelan, R.E., Fisher, J.E., Hiruma, K., Palli, S.R., Riddiford, L.M. 2000. Patterns of MHR3 expression in the epidermis during a larval molt of the tobacco hornworm *Manduca sexta*. Dev. Biol. 227, 481–494.

Lau, P., Bailey, P., Dowhan, D.H., Muscat, G.E. 1999. Exogenous expression of a dominant negative RORalpha1 vector in muscle cells impairs differentiation: RORalpha1 directly interacts with p300 and myoD. Nucleic Acids Res. 27, 411–420.

Lau, P., Nixon, S.J., Parton, R.G., Muscat, G.E. 2004. RORalpha regulates the expression of genes involved in lipid homeostasis in skeletal muscle cells: Caveolin-3 and CPT-1 are direct targets of ROR. J. Biol. Chem. 279, 36828–36840.

Lind, U., Nilsson, T., McPheat, J., Stromstedt, P.E., Bamberg, K., Balendran, C., Kang, D. 2005. Identification of the human ApoAV gene as a novel RORalpha target gene. Biochem. Biophys. Res. Commun. 330, 233–241.

Lipp, M., Muller, G. 2004. Lymphoid organogenesis: Getting the green light from RORgammat. Nat. Immunol. 5, 12–14.

Littman, D.R., Sun, Z., Unutmaz, D., Sunshine, M.J., Petrie, H.T., Zou, Y.R. 1999. Role of the nuclear hormone receptor ROR gamma in transcriptional regulation, thymocyte survival, and lymphoid organogenesis. Cold Spring Harb. Symp. Quant. Biol. 64, 373–381.

Lowell, B.B., Spiegelman, B.M. 2000. Towards a molecular understanding of adaptive thermogenesis. Nature 404, 652–660.

Ma, A., Pena, J.C., Chang, B., Margosian, E., Davidson, L., Alt, F.W., Thompson, C.B. 1995. Bclx regulates the survival of double-positive thymocytes. Proc. Natl. Acad. Sci. USA 92, 4763–4767.

Mamontova, A., Seguret-Mace, S., Esposito, B., Chaniale, C., Bouly, M., Delhaye-Bouchaud, N., Luc, G., Staels, B., Duverger, N., Mariani, J., Tedgui, A. 1998. Severe atherosclerosis and hypoalphalipoproteinemia in the staggerer mouse, a mutant of the nuclear receptor RORalpha. Circulation 98, 2738–2743.

Matsui, T. 1996. Differential activation of the murine laminin B1 gene promoter by RAR alpha, ROR alpha, and AP-1. Biochem. Biophys. Res. Commun. 220, 405–410.

Matsui, T. 1997. Transcriptional regulation of a Purkinje cell-specific gene through a functional interaction between ROR alpha and RAR. Genes Cells 2, 263–272.

Matsui, T., Sashihara, S., Oh, Y., Waxman, S.G. 1995. An orphan nuclear receptor, mROR alpha, and its spatial expression in adult mouse brain. Brain Res. Mol. Brain Res. 33, 217–226.

Matysiak-Scholze, U., Nehls, M. 1997. The structural integrity of ROR alpha isoforms is mutated in staggerer mice: Cerebellar coexpression of ROR alpha1 and ROR alpha4. Genomics 43, 78–84.

McBroom, L.D., Flock, G., Giguere, V. 1995. The nonconserved hinge region and distinct amino-terminal domains of the ROR alpha orphan nuclear receptor isoforms are required for proper DNA bending and ROR alpha-DNA interactions. Mol. Cell. Biol. 15, 796–808.

McInerney, E.M., Rose, D.W., Flynn, S.E., Westin, S., Mullen, T.M., Krones, A., Inostroza, J., Torchia, J., Nolte, R.T., Assa-Munt, N., Milburn, M.V., Glass, C.K., et al. 1998. Determinants of coactivator LXXLL motif specificity in nuclear receptor transcriptional activation. Genes Dev. 12, 3357–3368.

McKenna, N.J., O'Malley, B.W. 2002. Minireview: Nuclear receptor coactivators—an update. Endocrinology 143, 2461–2465.

Mebius, R.E., Rennert, P., Weissman, I.L. 1997. Developing lymph nodes collect CD4+CD3− LTbeta+ cells that can differentiate to APC, NK cells, and follicular cells but not T or B cells. Immunity 7, 493–504.

Mebius, R.E., Miyamoto, T., Christensen, J., Domen, J., Cupedo, T., Weissman, I.L., Akashi, K. 2001. The fetal liver counterpart of adult common lymphoid progenitors gives rise to all lymphoid lineages, CD45+CD4+CD3− cells, as well as macrophages. J. Immunol. 166, 6593–6601.

Medvedev, A., Yan, Z.H., Hirose, T., Giguere, V., Jetten, A.M. 1996. Cloning of a cDNA encoding the murine orphan receptor RZR/ROR gamma and characterization of its response element. Gene 181, 199–206.

Medvedev, A., Chistokhina, A., Hirose, T., Jetten, A.M. 1997. Genomic structure and chromosomal mapping of the nuclear orphan receptor ROR gamma (RORC) gene. Genomics 46, 93–102.

Messer, A., Plummer-Siegard, J., Eisenberg, B. 1990. Staggerer mutant mouse Purkinje cells do not contain detectable calmodulin mRNA. J. Neurochem. 55, 293–302.

Meyer, T., Kneissel, M., Mariani, J., Fournier, B. 2000. *In vitro* and *in vivo* evidence for orphan nuclear receptor RORalpha function in bone metabolism. Proc. Natl. Acad. Sci. USA 97, 9197–9202.

Michel, V., Monnier, Z., Guastavino, J.M., Propper, A., Math, F. 2000. Functional alterations in the olfactory bulb of the staggerer mutant mouse. Neurosci. Lett. 280, 1–4.

Miyazaki, T. 1997. Two distinct steps during thymocyte maturation from CD4−CD8− to CD4+CD8+ distinguished in the early growth response (Egr)-1 transgenic mice with a recombinase-activating gene-deficient background. J. Exp. Med. 186, 877–885.

Monnier, Z., Bahjaoui-Bouhaddi, M., Bride, J., Bride, M., Math, F., Propper, A. 1999. Structural and immunohistological modifications in olfactory bulb of the staggerer mutant mouse. Biol. Cell 91, 29–44.

Moraitis, A.N., Giguere, V. 1999. Transition from monomeric to homodimeric DNA binding by nuclear receptors: Identification of RevErbAalpha determinants required for RORalpha homodimer complex formation. Mol. Endocrinol. 13, 431–439.

Moraitis, A.N., Giguere, V. 2003. The corepressor hairless protects RORa orphan nuclear receptor from proteasome-mediated degradation. J. Biol. Chem. 278, 52511–52518.

Moraitis, A.N., Giguere, V., Thompson, C.C. 2002. Novel mechanism of nuclear receptor corepressor interaction dictated by activation function 2 helix determinants. Mol. Cell. Biol. 22, 6831–6841.

Moras, D., Gronemeyer, H. 1998. The nuclear receptor ligand-binding domain: Structure and function. Curr. Opin. Cell Biol. 10, 384–391.

Muller, G., Hopken, U.E., Lipp, M. 2003. The impact of CCR7 and CXCR5 on lymphoid organ development and systemic immunity. Immunol. Rev. 195, 117–135.

Muratani, M., Tansey, W.P. 2003. How the ubiquitin-proteasome system controls transcription. Nat. Rev. Mol. Cell Biol. 4, 192–201.

Nagy, L., Kao, H.Y., Love, J.D., Li, C., Banayo, E., Gooch, J.T., Krishna, V., Chatterjee, K., Evans, R.M., Schwabe, J.W. 1999. Mechanism of corepressor binding and release from nuclear hormone receptors. Genes Dev. 13, 3209–3216.

Nakagawa, S., Watanabe, M., Inoue, Y. 1996a. Altered gene expression of the *N*-methyl-D-aspartate receptor channel subunits in Purkinje cells of the staggerer mutant mouse. Eur. J. Neurosci. 8, 2644–2651.

Nakagawa, S., Watanabe, M., Inoue, Y. 1996b. Regional variation in expression of calbindin and inositol 1,4,5-trisphosphate receptor type 1 mRNAs in the cerebellum of the staggerer mutant mouse. Eur. J. Neurosci. 8, 1401–1407.

Nakagawa, S., Watanabe, M., Inoue, Y. 1997. Prominent expression of nuclear hormone receptor ROR alpha in Purkinje cells from early development. Neurosci. Res. 28, 177–184.

Nakagawa, S., Watanabe, M., Isobe, T., Kondo, H., Inoue, Y. 1998. Cytological compartmentalization in the staggerer cerebellum, as revealed by calbindin immunohistochemistry for Purkinje cells. J. Comp. Neurol. 395, 112–120.

Nakagawa, Y., O'Leary, D.D. 2003. Dynamic patterned expression of orphan nuclear receptor genes RORalpha and RORbeta in developing mouse forebrain. Dev. Neurosci. 25, 234–244.

Nakajima, Y., Ikeda, M., Kimura, T., Honma, S., Ohmiya, Y., Honma, K. 2004. Bidirectional role of orphan nuclear receptor RORalpha in clock gene transcriptions demonstrated by a novel reporter assay system. FEBS Lett. 565, 122–126.

Nishikawa, S., Honda, K., Vieira, P., Yoshida, H. 2003. Organogenesis of peripheral lymphoid organs. Immunol. Rev. 195, 72–80.

Nolte, R.T., Wisely, G.B., Westin, S., Cobb, J.E., Lambert, M.H., Kurokawa, R., Rosenfeld, M.G., Willson, T.M., Glass, C.K., Milburn, M.V. 1998. Ligand binding and co-activator assembly of the peroxisome proliferator-activated receptor-gamma. Nature 395, 137–143.

Novac, N., Heinzel, T. 2004. Nuclear receptors: Overview and classification. Curr. Drug Targets 3, 335–346.

Ortiz, M.A., Piedrafita, F.J., Pfahl, M., Maki, R. 1995. TOR: A new orphan receptor expressed in the thymus that can modulate retinoid and thyroid hormone signals. Mol. Endocrinol. 9, 1679–1691.

Palli, S.R., Hiruma, K., Riddiford, L.M. 1992. An ecdysteroid-inducible Manduca gene similar to the *Drosophila* DHR3 gene, a member of the steroid hormone receptor superfamily. Dev. Biol. 150, 306–318.

Palli, S.R., Ladd, T.R., Sohi, S.S., Cook, B.J., Retnakaran, A. 1996. Cloning and developmental expression of Choristoneura hormone receptor 3, an ecdysone-inducible gene and a member of the steroid hormone receptor superfamily. Insect Biochem. Mol. Biol. 26, 485–499.

Palli, S.R., Ladd, T.R., Retnakaran, A. 1997. Cloning and characterization of a new isoform of Choristoneura hormone receptor 3 from the spruce budworm. Arch. Insect Biochem. Physiol. 35, 33–44.

Paravicini, G., Steinmayr, M., Andre, E., Becker-Andre, M. 1996. The metastasis suppressor candidate nucleotide diphosphate kinase NM23 specifically interacts with members of the ROR/RZR nuclear orphan receptor subfamily. Biochem. Biophys. Res. Commun. 227, 82–87.

Piggins, H.D., Loudon, A. 2005. Circadian biology: Clocks within clocks. Curr. Biol. 15, R455–R457.

Pike, A.C., Brzozowski, A.M., Hubbard, R.E. 2000. A structural biologist's view of the oestrogen receptor. J. Steroid Biochem. Mol. Biol. 74, 261–268.

Poukka, H., Aarnisalo, P., Karvonen, U., Palvimo, J.J., Janne, O.A. 1999. Ubc9 interacts with the androgen receptor and activates receptor-dependent transcription. J. Biol. Chem. 274, 19441–19446.

Preitner, N., Damiola, F., Lopez-Molina, L., Zakany, J., Duboule, D., Albrecht, U., Schibler, U. 2002. The orphan nuclear receptor REV-ERBalpha controls circadian transcription within the positive limb of the mammalian circadian oscillator. Cell 110, 251–260.

Rader, D.J. 2002. High-density lipoproteins and atherosclerosis. Am. J. Cardiol. 90, 62i–70i.

Raspe, E., Duez, H., Gervois, P., Fievet, C., Fruchart, J.C., Besnard, S., Mariani, J., Tedgui, A., Staels, B. 2001. Transcriptional regulation of apolipoprotein C-III gene expression by the orphan nuclear receptor RORalpha. J. Biol. Chem. 276, 2865–2871.

Rennert, P.D., James, D., Mackay, F., Browning, J.L., Hochman, P.S. 1998. Lymph node genesis is induced by signaling through the lymphotoxin beta receptor. Immunity 9, 71–79.

Reppert, S.M., Weaver, D.R. 2002. Coordination of circadian timing in mammals. Nature 418, 935–941.

Retnakaran, R., Flock, G., Giguere, V. 1994. Identification of RVR, a novel orphan nuclear receptor that acts as a negative transcriptional regulator. Mol. Endocrinol. 8, 1234–1244.

Riddiford, L.M., Hiruma, K., Zhou, X., Nelson, C.A. 2003. Insights into the molecular basis of the hormonal control of molting and metamorphosis from *Manduca sexta* and *Drosophila melanogaster*. Insect Biochem. Mol. Biol. 33, 1327–1338.

Ripperger, J.A., Schibler, U. 2001. Circadian regulation of gene expression in animals. Curr. Opin. Cell Biol. 13, 357–362.

Roenneberg, T., Merrow, M. 2005. Circadian clocks: Translation. Curr. Biol. 15, R470–R473.

Ruddle, N.H. 1999. Lymphoid neo-organogenesis: Lymphotoxin's role in inflammation and development. Immunol. Res. 19, 119–125.

Salinas, P.C., Fletcher, C., Copeland, N.G., Jenkins, N.A., Nusse, R. 1994. Maintenance of Wnt-3 expression in Purkinje cells of the mouse cerebellum depends on interactions with granule cells. Development 120, 1277–1286.

Sashihara, S., Felts, P.A., Waxman, S.G., Matsui, T. 1996. Orphan nuclear receptor ROR alpha gene: Isoform-specific spatiotemporal expression during postnatal development of brain. Brain Res. Mol. Brain Res. 42, 109–117.

Sato, T.K., Panda, S., Miraglia, L.J., Reyes, T.M., Rudic, R.D., McNamara, P., Naik, K.A., FitzGerald, G.A., Kay, S.A., Hogenesch, J.B. 2004. A functional genomics strategy reveals Rora as a component of the mammalian circadian clock. Neuron 43, 527–537.

Schaeren-Wiemers, N., Andre, E., Kapfhammer, J.P., Becker-Andre, M. 1997. The expression pattern of the orphan nuclear receptor RORbeta in the developing and adult rat nervous system suggests a role in the processing of sensory information and in circadian rhythm. Eur. J. Neurosci. 9, 2687–2701.

Schibler, U., Naef, F. 2005. Cellular oscillators: Rhythmic gene expression and metabolism. Curr. Opin. Cell Biol. 17, 223–229.

Schibler, U., Sassone-Corsi, P. 2002. A web of circadian pacemakers. Cell 111, 919–922.

Schrader, M., Danielsson, C., Wiesenberg, I., Carlberg, C. 1996. Identification of natural monomeric response elements of the nuclear receptor RZR/ROR. They also bind COUP-TF homodimers. J. Biol. Chem. 271, 19732–19736.

Sebzda, E., Mariathasan, S., Ohteki, T., Jones, R., Bachmann, M.F., Ohashi, P.S. 1999. Selection of the T cell repertoire. Annu. Rev. Immunol. 17, 829–874.

Shirley, L.T., Messer, A. 2004. Early postnatal Purkinje cells from staggerer mice undergo aberrant development *in vitro* with characteristic morphologic and gene expression abnormalities. Brain Res. Dev. Brain Res. 152, 153–157.

Sidman, R.L., Lane, P.W., Dickie, M.M. 1962. Staggerer, a new mutation in the mouse affecting the cerebellum. Science 137, 610–612.

Smith, D.I., Zhu, Y., McAvoy, S., Kuhn, R. 2006. Common fragile sites, extremely large genes, neural development and cancer. Cancer Lett. 232, 48–57.

Sotelo, C., Changeux, J.P. 1974. Transsynaptic degeneration 'en cascade' in the cerebellar cortex of staggerer mutant mice. Brain Res. 67, 519–526.

Sotelo, C., Wassef, M. 1991. Cerebellar development: Afferent organization and Purkinje cell heterogeneity. Philos. Trans. R. Soc. Lond. B Biol. Sci. 331, 307–313.

Stapleton, C.M., Jaradat, M., Dixon, D., Kang, H.S., Kim, S.C., Liao, G., Carey, M.A., Cristiano, J., Moorman, M.P., Jetten, A.M. 2005. Enhanced susceptibility of staggerer

(RORalphasg/sg) mice to lipopolysaccharide-induced lung inflammation. Am. J. Physiol. Lung Cell. Mol. Physiol. 289, L144–L152:

Starr, T.K., Jameson, S.C., Hogquist, K.A. 2003. Positive and negative selection of T cells. Annu. Rev. Immunol. 21, 139–176.

Stehlin, C., Wurtz, J.M., Steinmetz, A., Greiner, E., Schule, R., Moras, D., Renaud, J.P. 2001. X-ray structure of the orphan nuclear receptor RORbeta ligand-binding domain in the active conformation. EMBO J. 20, 5822–5831.

Stehlin-Gaon, C., Willmann, D., Zeyer, D., Sanglier, S., Van Dorsselaer, A., Renaud, J.P., Moras, D., Schule, R. 2003. All-*trans* retinoic acid is a ligand for the orphan nuclear receptor RORbeta. Nat. Struct. Biol. 10, 820–825.

Steinhilber, D., Brungs, M., Werz, O., Wiesenberg, I., Danielsson, C., Kahlen, J.P., Nayeri, S., Schrader, M., Carlberg, C. 1995. The nuclear receptor for melatonin represses 5-lipoxygenase gene expression in human B lymphocytes. J. Biol. Chem. 270, 7037–7040.

Steinmayr, M., Andre, E., Conquet, F., Rondi-Reig, L., Delhaye-Bouchaud, N., Auclair, N., Daniel, H., Crepel, F., Mariani, J., Sotelo, C., Becker-Andre, M. 1998. Staggerer phenotype in retinoid-related orphan receptor alpha-deficient mice. Proc. Natl. Acad. Sci. USA 95, 3960–3965.

Steinmetz, A.C., Renaud, J.P., Moras, D. 2001. Binding of ligands and activation of transcription by nuclear receptors. Annu. Rev. Biophys. Biomol. Struct. 30, 329–359.

Sullivan, A.A., Thummel, C.S. 2003. Temporal profiles of nuclear receptor gene expression reveal coordinate transcriptional responses during *Drosophila* development. Mol. Endocrinol. 17, 2125–2137.

Sumi, Y., Yagita, K., Yamaguchi, S., Ishida, Y., Kuroda, Y., Okamura, H. 2002. Rhythmic expression of ROR beta mRNA in the mice suprachiasmatic nucleus. Neurosci. Lett. 320, 13–16.

Sun, Y., Witte, D.P., Jin, P., Grabowski, G.A. 2003. Analyses of temporal regulatory elements of the prosaposin gene in transgenic mice. Biochem. J. 370, 557–566.

Sun, Z., Unutmaz, D., Zou, Y.R., Sunshine, M.J., Pierani, A., Brenner-Morton, S., Mebius, R.E., Littman, D.R. 2000. Requirement for RORgamma in thymocyte survival and lymphoid organ development. Science 288, 2369–2373.

Sundvold, H., Lien, S. 2001. Identification of a novel peroxisome proliferator-activated receptor (PPAR) gamma promoter in man and transactivation by the nuclear receptor RORalpha1. Biochem. Biophys. Res. Commun. 287, 383–390.

Taylor, R.T., Williams, I.R. 2005. Lymphoid organogenesis in the intestine. Immunol. Res. 33, 167–181.

Thummel, C.S. 1995. From embryogenesis to metamorphosis: The regulation and function of *Drosophila* nuclear receptor superfamily members. Cell 83, 871–877.

Tini, M., Fraser, R.A., Giguere, V. 1995. Functional interactions between retinoic acid receptor-related orphan nuclear receptor (ROR alpha) and the retinoic acid receptors in the regulation of the gamma F-crystallin promoter. J. Biol. Chem. 270, 20156–20161.

Trenkner, E., Hoffmann, M.K. 1986. Defective development of the thymus and immunological abnormalities in the neurological mouse mutation "staggerer." J. Neurosci. 6, 1733–1737.

Triqueneaux, G., Thenot, S., Kakizawa, T., Antoch, M.P., Safi, R., Takahashi, J.S., Delaunay, F., Laudet, V. 2004. The orphan receptor Rev-erbalpha gene is a target of the circadian clock pacemaker. J. Mol. Endocrinol. 33, 585–608.

Ueda, E., Kurebayashi, S., Sakaue, M., Backlund, M., Koller, B., Jetten, A.M. 2002a. High incidence of T-cell lymphomas in mice deficient in the retinoid-related orphan receptor RORgamma. Cancer Res. 62, 901–909.

Ueda, H.R., Chen, W., Adachi, A., Wakamatsu, H., Hayashi, S., Takasugi, T., Nagano, M., Nakahama, K., Suzuki, Y., Sugano, S., Iino, M., Shigeyoshi, Y., Hashimoto, S. 2002b. A transcription factor response element for gene expression during circadian night. Nature 418, 534–539.

Ueda, H.R., Hayashi, S., Chen, W., Sano, M., Machida, M., Shigeyoshi, Y., Iino, M., Hashimoto, S. 2005. System-level identification of transcriptional circuits underlying mammalian clocks. Nature Genet. 37, 187–192.

Verbeek, S., Izon, D., Hofhuis, F., Robanus-Maandag, E., te Riele, H., van de Wetering, M., Oosterwegel, M., Wilson, A., MacDonald, H.R., Clevers, H. 1995. An HMG-box-containing T-cell factor required for thymocyte differentiation. Nature 374, 70–74.

Villey, I., de Chasseval, R., de Villartay, J.P. 1999. RORgammaT, a thymus-specific isoform of the orphan nuclear receptor RORgamma/TOR, is up-regulated by signaling through the pre-T cell receptor and binds to the TEA promoter. Eur. J. Immunol. 29, 4072–4080.

Vogel, M.W., Sinclair, M., Qiu, D., Fan, H. 2000. Purkinje cell fate in staggerer mutants: Agenesis versus cell death. J. Neurobiol. 42, 323–337.

Vu-Dac, N., Gervois, P., Grotzinger, T., De Vos, P., Schoonjans, K., Fruchart, J.C., Auwerx, J., Mariani, J., Tedgui, A., Staels, B. 1997. Transcriptional regulation of apolipoprotein A-I gene expression by the nuclear receptor RORalpha. J. Biol. Chem. 272, 22401–22404.

Wallace, A.D., Cidlowski, J.A. 2001. Proteasome-mediated glucocorticoid receptor degradation restricts transcriptional signaling by glucocorticoids. J. Biol. Chem. 276, 42714–42721.

Wallace, V.A. 1999. Purkinje-cell-derived Sonic hedgehog regulates granule neuron precursor cell proliferation in the developing mouse cerebellum. Curr. Biol. 9, 445–448.

Wang, Y., Huso, D., Cahill, H., Ryugo, D., Nathans, J. 2001. Progressive cerebellar, auditory, and esophageal dysfunction caused by targeted disruption of the frizzled-4 gene. J. Neurosci. 21, 4761–4771.

Weeks, J.C. 2003. Thinking globally, acting locally: Steroid hormone regulation of the dendritic architecture, synaptic connectivity and death of an individual neuron. Prog. Neurobiol. 70, 421–442.

Weissman, A.M. 2001. Themes and variations on ubiquitylation. Nat. Rev. Mol. Cell Biol. 2, 169–178.

Wigle, J.T., Harvey, N., Detmar, M., Lagutina, I., Grosveld, G., Gunn, M.D., Jackson, D.G., Oliver, G. 2002. An essential role for Prox1 in the induction of the lymphatic endothelial cell phenotype. EMBO J. 21, 1505–1513.

Willy, P.J., Mangelsdorf, D.J. 1998. Nuclear orphan receptors: The search for novel ligands and signaling pathways. In: *Hormones and Signaling* (B.W. O'Malley, Ed.), vol. 1, San Diego: Academic Press, pp. 308–358.

Winoto, A., Littman, D.R. 2002. Nuclear hormone receptors in T lymphocytes. Cell 109 (Suppl.), S57–S66.

Xi, H., Kersh, G.J. 2004. Sustained early growth response gene 3 expression inhibits the survival of CD4/CD8 double-positive thymocytes. J. Immunol. 173, 340–348.

Xie, H., Sadim, M.S., Sun, Z. 2005. RORgammat recruits steroid receptor coactivators to ensure thymocyte survival. J. Immunol. 175, 3800–3809.

Xu, L., Glass, C.K., Rosenfeld, M.G. 1999. Coactivator and corepressor complexes in nuclear receptor function. Curr. Opin. Genet. Dev. 9, 140–147.

Xu, W. 2005. Nuclear receptor coactivators: The key to unlock chromatin. Biochem. Cell Biol. 83, 418–428.

Yokota, Y., Mansouri, A., Mori, S., Sugawara, S., Adachi, S., Nishikawa, S., Gruss, P. 1999. Development of peripheral lymphoid organs and natural killer cells depends on the helix-loop-helix inhibitor Id2. Nature 397, 702–706.

Yokota, Y., Mori, S., Narumi, O., Kitajima, K. 2001. *In vivo* function of a differentiation inhibitor, Id2. IUBMB Life 51, 207–214.

Yoshida, H., Naito, A., Inoue, J., Satoh, M., Santee-Cooper, S.M., Ware, C.F., Togawa, A., Nishikawa, S. 2002. Different cytokines induce surface lymphotoxin-alphabeta on IL-7 receptor-alpha cells that differentially engender lymph nodes and Peyer's patches. Immunity 17, 823–833.

Yu, Q., Erman, B., Park, J.H., Feigenbaum, L., Singer, A. 2004. IL-7 receptor signals inhibit expression of transcription factors TCF-1, LEF-1, and RORgammat: Impact on thymocyte development. J. Exp. Med. 200, 797–803.

Zhang, N., Hartig, H., Dzhagalov, I., He, Y.W. 2005. The role of apoptosis in the development and function of T lymphocytes. Cell Res. 15, 749–769.

Zhu, Y., McAvoy, S., Kuhn, R., Smith, D.I. 2006. RORA, a large common fragile site gene, is involved in cellular stress response. Oncogene 25, 2901–2908.

Hairless: A nuclear receptor corepressor essential for skin function

Catherine C. Thompson[1,2,3] and
Gerard M. J. Beaudoin, III[1,2]

[1] Kennedy Krieger Research Institute,
Baltimore, Maryland
[2] Department of Neuroscience,
Johns Hopkins University School of Medicine, Baltimore, Maryland
[3] Department of Molecular Biology and Genetics,
Johns Hopkins University School of Medicine, Baltimore, Maryland

Contents

Advances in Developmental Biology
Volume 16 ISSN 1574-3349
DOI: 10.1016/S1574-3349(06)16011-1

1. Introduction

Mammalian development is the result of carefully coordinated outputs of multiple signaling pathways and processes. Among the diverse tissues and organs that comprise the organism, skin provides an exceptional opportunity to study many of the processes at play during development. In skin, many developmental processes occur throughout postnatal life and can be monitored noninvasively. Multipotent epithelial stem cells maintain the skin through the production of progenitor cells, providing precursors for all of the cell types present in the skin (Alonso and Fuchs, 2003). Signals acting on both the stem cells and the progenitor cells regulate the homeostatic maintenance of multiple structures, including the epidermis, hair follicles, and sebaceous glands (Alonso and Fuchs, 2003). In addition, epithelial–mesenchymal interactions important for morphogenesis are also required during regeneration, which includes hair follicles and epidermal wound healing (Millar, 2002; Alonso and Fuchs, 2003). Insight into all of these developmental paradigms has been garnered from analyses of both targeted and spontaneously occurring mutant mice (Nakamura et al., 2001).

Over 75 years ago, *Hairless* (*Hr*) mutant mice were identified as adult mice with no hair (Sumner, 1924; Brooke, 1926). Because *Hr* mutant mice lose their hair subsequent to normal hair follicle development, the *Hr* gene was implicated as a critical regulator of the hair cycle (Montagna et al., 1952; Orwin et al., 1967; Mann, 1971; Panteleyev et al., 1998b; Zarach et al., 2004). The identification of the rodent *Hr* genes (Stoye et al., 1988; Cachon-Gonzalez et al., 1994; Thompson, 1996) led to the identification of the human *Hr* gene, which revealed that *Hr* is mutated in congenital hair loss disorders (alopecia universalis, papular atrichia) that phenotypically resemble the mouse mutants (Ahmad et al., 1998a; Cichon et al., 1998). Based on the *Hr* mutant phenotype, *Hr* likely has a role in regulating homeostasis of the epithelial components of skin as well as a role in hair cycling. Work from this laboratory has shown that the protein encoded by *Hr* (HR) functions through nuclear receptors, acting as a corepressor for multiple receptors (Potter et al., 2001; Moraitis et al., 2002; Hsieh et al., 2003).

Nuclear receptors regulate numerous developmental and physiological processes through their function as ligand-activated transcription factors, by regulating the expression of specific target genes. Analysis of the mechanisms of nuclear receptor function as transcription factors has revealed an enormous complexity of proteins and protein complexes that influence nuclear receptor transcriptional activity (Mangelsdorf and Evans, 1995; Jepsen and Rosenfeld, 2002; Privalsky, 2004; Smith and O'Malley, 2004). By far the most prevalent are coactivators, dozens of proteins that facilitate transcriptional activation by liganded nuclear receptors, via various enzymatic activities including histone acetylation and methylation (McKenna and O'Malley, 2002; Perissi and Rosenfeld, 2005). Corepressors serve an opposing function, as they facilitate transcriptional repression by a subset of nuclear receptors, including thyroid hormone (TH), retinoic acid, and vitamin D receptors. These receptors function as heterodimers with retinoid X receptor (RXR) and can repress transcription in the absence of their ligand via their association with corepressors (Mangelsdorf and Evans, 1995; Jepsen and Rosenfeld, 2002; Privalsky, 2004; Smith and O'Malley, 2004). Another class of corepressors inhibits the activity of ligand-bound receptors, and includes RIP140 and L-CoR (Cavailles et al., 1995; Fernandes et al., 2003; Steel et al., 2005).

Nuclear receptor corepressor (N-CoR) and silencing mediator for retinoic acid and thyroid hormone receptors (SMRT) were the first nuclear receptor corepressors to be identified (Chen and Evans, 1995; Horlein et al., 1995; Sande and Privalsky, 1996). N-CoR and SMRT are 235-kDa proteins that share sequence identity, and extensive analysis has shown that they are functional paralogues (Jepsen and Rosenfeld, 2002; Privalsky, 2004). Both N-CoR and SMRT bind to retinoic acid receptors (RARs) and thyroid hormone receptors (TRs) in the absence of ligand, and mutant receptors that lack interaction with N-CoR and SMRT no longer repress (Chen and Evans, 1995; Horlein et al., 1995). Similar to coactivators, N-CoR and SMRT have been shown to be part of large multiprotein complexes that function via associated enzymatic activities that modify chromatin structure, such as histone deacetylation. For N-CoR and SMRT, functional domains and interacting protein complexes have been characterized in great detail (reviewed in Privalsky, 2004; Perissi and Rosenfeld, 2005).

A number of studies indicate the importance of transcriptional repression by nuclear receptors in development and disease (Privalsky, 2004). N-CoR and SMRT are broadly expressed and likely play important roles in many tissues. N-CoR has an essential role *in vivo*, as targeted disruption of N-CoR results in embryonic lethality (Jepsen et al., 2000). *Hr* is essential for proper skin function, and based on its expression in brain and a limited number of other tissues, may play a more specialized role than N-CoR and SMRT. This chapter will review the properties of HR function as a nuclear receptor corepressor (Fig. 1), highlighting similarities and differences with other corepressors. We will also explore in detail the role of *Hr in vivo* in regulating epithelial stem cell differentiation in the skin, and its potential role in other tissues.

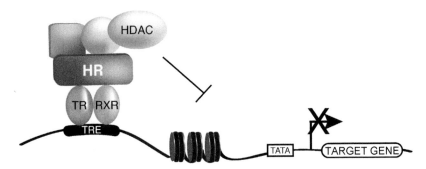

Fig. 1. Model for HR regulation of gene expression. Schematic representation of HR corepressor function. HR is found in a complex with TR and RXR bound to DNA via a TH response element (TRE). HR associates with HDACs; HDAC activity modifies chromatin structure and may also modify basal transcription factors, resulting in transcriptional repression. Unlabeled shapes represent other, as yet unidentified proteins, proposed to interact with the HR–TR complex.

2. Identifying the molecular function of HR

The murine *Hr* gene was isolated by cloning the retroviral insertion site in the original Hr^{hr} (formerly *hr*) allele (Stoye et al., 1988; Cachon-Gonzalez et al., 1994). The rat *Hr* gene was concurrently isolated as a gene that is upregulated by TH in neonatal rat brain, an organ that requires TH for proper development (Thompson, 1996). The *Hr* gene encodes a 130-kDa protein (HR) that does not show significant homology to known structural or functional motifs (Cachon-Gonzalez et al., 1994). Six cysteine residues that are conserved in HR and a rat testis protein of unknown function were noted and proposed to form a zinc finger domain, although zinc binding of this motif in HR has not been demonstrated (Cachon-Gonzalez et al., 1994). Thus, the primary sequence of HR did not provide insight into its function.

A breakthrough in determining the function of the HR protein was the discovery that it interacts with TR. HR interaction with TR was discovered in a yeast-two-hybrid screen for HR-interacting proteins (Thompson and Bottcher, 1997). Interaction with TR has been demonstrated *in vitro* using multiple biochemical assays and *in vivo* by coimmunoprecipitation with both overexpressed and endogenous proteins (Thompson and Bottcher, 1997; Potter et al., 2001, 2002). Interaction of HR with TR suggested that HR might influence TR function, and evidence that HR bound to TR more avidly in the absence of TH suggested that HR might function as a corepressor (Thompson and Bottcher, 1997). Consistent with the role of HR as a transcriptional regulator, HR expressed in cultured cells is localized to the nucleus, the site of action for transcriptional regulators (Thompson and Bottcher, 1997).

Although HR is not related to N-CoR and SMRT by primary amino acid sequence, HR has similar functional properties (Fig. 2). Critical properties common to nuclear receptor corepressors include: (1) interaction with nuclear receptors in the absence of hormone (ligand), (2) mediating repression by unliganded TR, (3) multiple independent domains capable of mediating repression, (4) multiple domains composed of conserved hydrophobic residues that mediate receptor interaction, and (5) interaction with histone deacetylases (HDACs) (Horlein et al., 1995; Li et al., 1997a; Nagy et al., 1997; Ordentlich et al., 1999; Perissi et al., 1999; Guenther et al., 2000; Huang et al., 2000; Kao et al., 2000; Wen et al., 2000; Xu et al., 2002). As detailed below, HR shares these properties, establishing it as a bona fide nuclear receptor corepressor.

2.1. HR mediates repression by TR in the absence of TH

In cotransfection assays, HR can mediate repression by unliganded TR, while expression of HR has little effect on transcriptional activation by TH (Fig. 3A). Repression by HR requires receptor binding, as a mutant HR that lacks TR binding can no longer repress. HR can mediate repression by unliganded TR on different TH response elements (TREs) (Potter et al., 2001, 2002). Notably, HR represses TR-mediated basal transcription most effectively in pituitary-derived cell lines, a cell type in which HR is endogenously expressed, and not in many other cell lines such as COS (kidney-derived). This may reflect the presence of tissue-specific factors, or may be due to relatively high levels of N-CoR and/or SMRT in COS and other cells (Misiti et al., 1998). Thus, HR mediates TR-dependent transcriptional repression in the absence of hormone, the functional definition of a nuclear receptor corepressor.

Evidence that HR expression is induced by TH in brain and other tissues suggests that HR would not act as a corepressor for TR *in vivo*. However, HR could function via TR in tissues in which HR expression is not TH

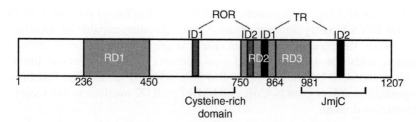

Fig. 2. Schematic representation of rat HR structural and functional domains. Repression domains (RD1, 236–450; RD2, 750–864; RD3, 864–981); TR-interacting domains (TR-ID1, 816–830; TR-ID2, 1026–1038); ROR-interacting domains, ROR-ID1, 586–590; ROR-ID2, 778–782); cysteine-rich domain, 587–712; JmjC domain, 964–1175. Note that rat HR is 1207 amino acids, mouse and human *Hr* initiate at an internal AUG (amino acid 27 in rat *Hr*) and are 1182 and 1189 amino acids, respectively.

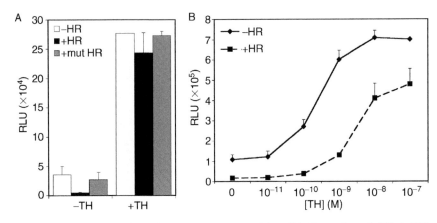

Fig. 3. HR reduces TH-dependent reporter gene activity. (A) Expression of HR (+HR) mediates basal repression by TR in the absence of TH (−TH) but does not influence TR activity in the presence of 200 nM L-T$_3$ (+TH). Interaction is required for repression as an HR derivative (mutHR) with mutated TR-IDs no longer represses. (B) HR inhibits TR activation at physiologic levels of TH. A TH-responsive reporter gene (MLV×2 tk-luc) was transfected into GH1 cells together with an expression vector for HR (+HR) or vector alone (−HR). Luciferase activity (RLU) was measured in the presence of the indicated concentrations of L-T$_3$ (TH) and normalized to an internal control. Note that HR inhibits TR activity approximately sixfold at nanomolar concentrations of TH.

dependent, such as skin and adult brain (Thompson, 1996). In addition, recent work indicates that HR can suppress TR-mediated transcriptional activation at physiological TH concentrations (Fig. 3B); HR inhibits TR activity approximately sixfold at 10^{-9} M L-T$_3$. Thus, HR may modulate TR/TH signaling in addition to repressing basal transcription.

2.2. HR has multiple repression domains

Analysis of deletion derivatives of HR expressed as fusion proteins with the GAL4 DNA-binding domain revealed three domains capable of mediating repression. Repression domains include a single domain in the N-terminus (amino acids 236-450; RD1) and two domains in the C-terminus (amino acids 750-864, RD2; amino acids 864-981, RD3) (Fig. 2). The functional organization differs from N-CoR and SMRT as RD2 overlaps with a receptor interaction domain.

2.3. HR has two TR-interacting domains

Deletion analysis showed that HR interacts with TR via two C-terminal regions (between amino acids 750-864 and 980-1084) (Fig. 2). The N-terminal

TR-interacting region appears to have higher affinity for TR (Potter et al., 2001). Specific amino acids required for interaction were identified using site-directed mutagenesis to change specific residues in these regions. Comparing essential residues in the two TR interaction domains (TR-ID1, 816-830; TR-ID2, 1026-1038) yielded a consensus motif for HR binding to TR (I/L-I-X-X-L/V-V) similar to those identified for N-CoR and SMRT binding to TR and RAR, including the L/I-X-X-I/V-I motif termed the coRNR box (Hu and Lazar, 1999; Nagy et al., 1999; Perissi et al., 1999; Webb et al., 2000; Cohen et al., 2001) and the extended amphipathic α helix predicted and subsequently demonstrated to mediate corepressor–receptor interaction (Perissi et al., 1999; Xu et al., 2002). The HR consensus defines a motif in which the spacing but not the identity of hydrophobic residues is conserved. The identity of specific residues may determine the specificity of corepressor–receptor interaction. For example, HR does not bind to RAR, and SMRT shows higher affinity for binding to RAR than TR (Potter et al., 2001; Privalsky, 2004). Thus, HR has two separable domains that mediate TR interaction, including overlapping interaction and repression domains (TR-ID1, RD2) (Fig. 2).

2.4. Interaction of HR with histone deacetylases

Transcriptional repression often results from the association of corepressors with HDACs (Pazin and Kadonaga, 1997; Burke and Baniahmad, 2000; Glass and Rosenfeld, 2000). Although HDACs 1 and 2 were first recognized to interact with corepressors, subsequent biochemical analyses have indicated that HDAC3 is found in an endogenous complex with corepressors and is a critical component of the corepressor complex (Li et al., 2000a; Wen et al., 2000; Guenther et al., 2001; Yoon et al., 2003). Among the HDACs tested, HR interacts indirectly with several HDACS, preferentially with HDACs 1, 3, and 5 (Potter et al., 2001, 2002). HR corepressor activity is reduced in the presence of an HDAC inhibitor, suggesting that HDAC activity is partially responsible for repression by HR (Potter et al., 2002). Biochemical studies have revealed the presence of N-CoR and SMRT in large multiprotein complexes that include HDACs, TBL1, TBLR1, as well as other proteins (reviewed in Privalsky, 2004). Such studies have not been performed with HR, but HR does colocalize with HDACs and other corepressors in matrix-associated deacetylase (MAD) bodies (Potter et al., 2001). MAD bodies were identified as subnuclear structures that contain multiple HDACs and core-pressors, and whose integrity is dependent on HDAC activity (Downes et al., 2000). Colocalization of HR in MAD bodies suggests that HR may be part of multiprotein complexes that may include other corepressors.

Analysis has revealed that the HR protein includes a C-terminal JmjC domain (Clissold and Ponting, 2001). The JmjC domain is a motif originally found in the protein encoded by *Jumonji* (JMJ), a gene whose loss causes developmental defects (Takeuchi et al., 1995, 1999; Clissold and Ponting, 2001). Studies have shown that JMJ acts as a transcriptional repressor (Kim et al., 2003; Toyoda et al., 2003), and the JmjC domain in JHDM1 [JmjC domain-containing histone demethylase 1; formerly F-box and leucine rich repeat protein 11 (FBXL11)] was shown to function in histone demethylation (Tsukada et al., 2006). By analogy, the JmjC domain in HR may have a similar enzymatic activity. Thus, part of the repression function of HR and other JmjC domain-containing proteins may lie in counterbalancing the methylation activity of coactivators such as CARM-1 and PRMT-1 (Chen et al., 1999; Wang et al., 2001), analogous to the opposing activity of coactivators and corepressors on histone acetylation.

3. HR is a corepressor for multiple nuclear receptors

3.1. HR interacts with an orphan nuclear receptor (RORα) important for cerebellar development

Although HR does not interact with RAR like other corepressors, HR does bind to retinoic acid receptor-related orphan receptor α (RORα). RORα is a constitutively active orphan nuclear receptor that plays a vital role in cerebellar development (Hamilton et al., 1996; Matysiak-Scholze and Nehls, 1997; Dussault et al., 1998). HR interacts with RORα and potently inhibits transcriptional activation by all ROR isoforms (α, β, γ) (Moraitis et al., 2002). Further analysis has shown that HR binding protects RORα from proteasome-mediated degradation (Moraitis and Giguere, 2003). The ability of HR to mediate ligand-independent repression by TR, and also influence the activity of a constitutively active orphan receptor, indicated that HR serves multiple roles in mediating transcriptional repression.

Mapping the domain of HR that interacts with ROR revealed two motifs with the LxxLL consensus sequence shown to mediate interaction with coactivators (Heery et al., 1997; McInerney et al., 1998) (Fig. 2). Mutations of both LxxLL motifs abolished the ability of HR to repress RORα activity (Moraitis et al., 2002). Furthermore, the specificity of HR corepressor action could be transferred to an RAR by exchanging the AF-2 domain (Moraitis et al., 2002). These results unexpectedly demonstrated that in the context of ROR, the repressive activity of HR is dependent on the integrity of the coactivator-type motifs rather than the TR-interacting motifs.

3.2. HR interacts with vitamin D receptor, a nuclear receptor important for skin function

Although interaction of HR with TR demonstrated its function as a corepressor, interaction with TR does not easily explain the prominent skin phenotype in *Hr* mutants. TH-deficient humans and mice do have some hair and skin defects, but none as striking as the complete hair loss and in some cases skin wrinkling observed in *Hr* mutants (Alonso and Rosenfield, 2003). Targeted deletion of RXRα and β in mice results in a hair loss phenotype (Li et al., 2000b), but HR does not directly interact with RXRs, and RXRs are heterodimeric partners for multiple nuclear receptors. Evidence is mounting that the receptor through which HR acts in the skin is most likely the vitamin D receptor (VDR). Like TR, VDR functions as a heterodimer with RXR, and VDR is structurally related to TR. In addition, mice and humans with mutations in VDR have a hair loss phenotype similar to that of mice and humans with mutations in *Hr* (Li et al., 1997b; Yoshizawa et al., 1997). VDR also has roles in mineral metabolism and bone formation (Li et al., 1997b; Yoshizawa et al., 1997).

HR interacts with VDR in multiple biochemical assays, including GST pull-down and coimmunoprecipitation (Hsieh et al., 2003). Interaction of HR with VDR is weaker than interaction with TR and may require other proteins. The region of VDR interaction with HR overlaps the sites responsible for interaction with TR and ROR, although the amino acids required for interaction of HR with VDR have not been determined. In contrast to its role with TR, HR represses VDR activity both in the absence and presence of vitamin D. The repressive effect of HR occurs on both synthetic and naturally occurring VDR-responsive promoters, and in cultured keratinocytes, a cell type in which HR and VDR are normally expressed (Hsieh et al., 2003; Beaudoin, G.M.J. and Thompson, C.C., unpublished observations). The biochemically defined interaction of HR and VDR can occur *in vivo*, as *Hr* and VDR mRNAs are coexpressed in skin, colocalizing in a subset of cells of the hair follicle (Hsieh et al., 2003). Although mutations in VDR cause hair loss, vitamin D deficiency does not result in alopecia either in mice or humans (Sakai et al., 2001). This suggests that transcriptional activation by vitamin D-bound VDR is not essential for hair cycle regulation and is supported by work showing that mutant VDRs that do not activate transcription in response to vitamin D can rescue the alopecia of VDR-null mice (Skorija et al., 2005). Notably, the mutant VDRs retain the ability to bind HR and repress basal transcription, suggesting that repression by unliganded VDR via HR is essential for VDR function in the skin (Skorija et al., 2005).

Thus, multiple studies have shown that HR functions as a corepressor for nuclear receptors. By analogy with other corepressors, HR may possess additional functional properties. For example, N-CoR and SMRT have been shown to bind to antagonist-bound steroid receptors and also act as corepressors for other transcription factors (Jepsen and Rosenfeld, 2002; Privalsky, 2004). Novel features of HR function include its role in inhibiting transcriptional activation by an orphan receptor (ROR) and ligand-bound VDR. In addition, the presence of a JmjC domain suggests that HR may contain an endogenous chromatin-modifying activity. The data summarized is from rat HR, but the human HR protein also functions as a corepressor (Hillmer, A. and Thompson, C.C., unpublished observations). Recent work shows that HR corepressor activity is important *in vivo*, as the proteins encoded by multiple mutant human *Hr* alleles are deficient for corepressor activity, indicating that disruption of corepressor function underlies congenital hair loss disorders (Thompson et al., 2006; Hsieh, J.-C., Haussler, M.H., Sisk, J.M., and Thompson, C.C., unpublished observations).

4. HR action in skin

Based on the phenotype of *Hr* mutant mice and humans, *Hr* has an important role in the skin. Skin provides an essential function, protecting the organism from environmental factors such as dehydration, mechanical damage, and infection. Mammalian skin is composed of three distinct structures (Fig. 4A). At the surface is the overlying epidermis, which provides a barrier to the elements (Alonso and Fuchs, 2003). Extending below the surface are the hair follicles, which produce hair, used for sensation and thermoregulation (Stenn and Paus, 2001). Associated with each hair follicle is a sebaceous gland, which produces oils that coat the hair (Stenn and Paus, 2001).

Production and maintenance of the epidermis, hair follicles, and sebaceous glands are achieved by undifferentiated, proliferating progenitors generating the cells that terminally differentiate into the mature cells comprising each structure (Alonso and Fuchs, 2003). Both the epidermis and the sebaceous glands are composed of epithelial cells or keratinocytes at various stages of differentiation. Proliferating progenitors that ultimately populate the epidermis and sebaceous gland are found in the basal layer of the epidermis and the follicle outer root sheath (ORS) (Fig. 4A). The hair follicle is composed of keratinocytes following one of several spatially segregated differentiation programs (Hardy, 1992). In actively growing hair, the undifferentiated cells of the lower hair bulb proliferate and then differentiate into multiple lineages which include the three layers of the hair and the three layers of the inner root sheath (IRS) (Hardy, 1992). Differentiation of the hair bulb keratinocytes is controlled in part by signals emanating from mesenchymally derived fibroblasts located at the center of the hair bulb, the dermal papilla (DP) (Fig. 4A) (Jahoda et al., 1984; Oliver and Jahoda, 1988; Hardy, 1992). The undifferentiated progenitors of all three components may originate as cells generated by stem cells in the stem cell niche, the bulge (Alonso and Fuchs, 2003; Fuchs et al., 2004).

Once established, the epidermis and sebaceous gland are always present in the skin, while the hair bulb and the associated hair shaft are generated

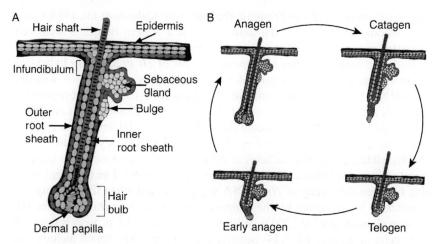

Fig. 4. Schematic diagram of the skin and hair cycle. (A) Schematic section through skin showing three major epithelial components, the epidermis, sebaceous gland, and hair follicle. The undifferentiated, progenitor keratinocytes (green) are found in the outer root sheath (ORS) and the basal layer of the epidermis. These progenitor cells are generated by the stem cells in the bulge (yellow) and migrate along the ORS to the hair bulb, to produce the hair shaft and inner root sheath, or to the sebaceous gland or through the infundibulum to the epidermis. The interaction of the mesenchymal cells (dermal papilla) and bulge stem cells is thought to regulate hair morphogenesis and cycling. (B) Schematic representation of the hair cycle. The hair follicle cycles from periods of hair growth (anagen) to an apoptosis-driven regression of the hair bulb and lower follicle (catagen) followed by a period of rest (telogen). The follicle exits telogen and reenters anagen to regenerate the hair bulb. Adapted from Beaudoin et al. (2005). (See Color Insert.)

cyclically (Fig. 4B) (Hardy, 1992; Stenn and Paus, 2001). Typically, the hair is periodically shed and regrows throughout the life of the organism (Stenn and Paus, 2001). The hair cycle can be influenced by environmental and endocrine factors but can continue in isolated hair follicles, indicating that control of the cycle ultimately resides within the follicle and attached DP (Hardy, 1992; Stenn and Paus, 2001). The hair cycle can be divided into distinct stages: *anagen, catagen,* and *telogen.* Active growth of hair occurs during *anagen* (Fig. 4B) (Hardy, 1992; Stenn and Paus, 2001). Once the hair has grown to an appropriate length, the hair follicle enters an apoptosis-driven period of regression called *catagen,* which results in the destruction of the hair bulb and retraction of the DP toward the bulge (Lindner et al., 1997; Stenn and Paus, 2001). On completion of catagen, the follicle remains in a period of rest, *telogen* (Stenn and Paus, 2001). Following telogen, hair growth is reinitiated by reentry into anagen and regeneration of the hair bulb (Hardy, 1992; Stenn and Paus, 2001). An exchange of signals between the DP and bulge appears to be critical for reentry into anagen, as removal of the DP blocks anagen reentry and insertion of a DP can reinitiate the hair cycle (Jahoda et al., 1984; Oliver and Jahoda, 1988). Thus, the hair follicle is a system in which stem cell-mediated organ regeneration relies on epithelial–mesenchymal interactions.

Initial molecular studies addressing the role of *Hr* used *in situ* hybridization to determine where *Hr* mRNA is expressed in the skin, and showed that *Hr* mRNA is expressed in epithelial cells (keratinocytes) and not mesenchymal cells (DP) (Cachon-Gonzalez et al., 1994, 1999; Panteleyev et al., 2000). Although *Hr* mRNA was detected throughout the initial development of the hair follicle (Cachon-Gonzalez et al., 1994, 1999; Panteleyev et al., 2000), the first defect in the hair follicles of *Hr* mutant mice does not arise until the onset of catagen (Mann, 1971; Panteleyev et al., 1999). The apparent paradox between the timing of *Hr* mRNA expression and when the phenotype is manifested was resolved by the temporal and spatial localization of the HR protein (Beaudoin et al., 2005). Within hair follicles, HR protein is found in the nuclei of keratin 14 (K14) positive cells in the ORS, which includes the bulge region. HR is also detected in K14-negative hair bulb cells but not in the DP. Follicles actively growing hair in anagen do not contain detectable HR protein (Beaudoin et al., 2005) despite expressing *Hr* mRNA (Panteleyev et al., 2000). HR protein is initially detected as follicles enter catagen, which coincides with the onset of phenotypic alterations in the hair follicles of *Hr* mutant mice (Panteleyev et al., 1999; Zarach et al., 2004). HR expression is maintained in the ORS through late catagen, into telogen, and during the transition to the next anagen. Once the hair bulb has reformed in mid anagen, HR protein is again undetectable. The discordance between *Hr* mRNA expression and protein expression during anagen points to posttranscriptional or posttranslational control of HR protein expression and/or stability. Thus, HR protein expression is regulated spatially and temporally during the hair cycle, consistent with HR functioning in both follicle regression and reinitiation of anagen.

5. Skin phenotype in *Hr* mutants

The skin phenotype in all known murine *Hr* alleles is initially similar. Animals appear normal for about the first 2 weeks after birth, after which they begin to lose their hair (Fig. 5A). All hair is lost and fails to regrow, caused by the failure of hair follicles to regenerate. In some alleles, the skin becomes wrinkled, which occurs progressively with age and can become quite severe. There are multiple spontaneous alleles of *Hr*, which differ in the severity of hair loss and skin wrinkling (Fig. 5B). Two different spontaneous *Hr* mutant mouse lines, *hairless* (Hr^{hr}) and *rhino* (Hr^{rh}), have been characterized in detail (Mann, 1971). In *hairless* mutant mice, after the initial hair loss

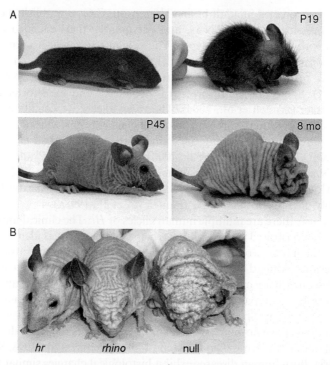

Fig. 5. Phenotype of *Hr* mutant mice. (A) $Hr^{-/-}$ mice show progressive hair loss. Initial hair growth is normal (P9), but after shedding the hair does not grow back (P21, P45). The skin becomes progressively more wrinkled with age (8 months). (B) Different *Hr* alleles show varying degrees of skin wrinkling. Adult mice homozygous for *hairless* (hr, Hr^{hr}), *rhino* (Hr^{rh}), and null (Hr^{-}) alleles; note that wrinkling is most severe in null. (See Color Insert.)

there is a slow regrowth of sparse hair and the skin remains relatively smooth. In *rhino* mutant mice, once hair is lost it never regrows, and the skin subsequently becomes wrinkled (Mann, 1971). These two mouse lines were shown to be alleles of the same gene by the inability of the mutations to complement (Mann, 1971). *Hairless* "knockout" mice ($Hr^{-/-}$) made by targeted deletion have a more severe wrinkling phenotype (Zarach et al., 2004) that is also observed in a spontaneous deletion allele (*rhino*-Yurlovo) (Panteleyev et al., 1998a).

Molecular analysis has shown that expression of *Hr* in mouse mutants correlates with the severity of the phenotype. There are two mRNA species transcribed from the *Hr* gene, a full-length message (exons 1–19) of approximately 6 kb, and a shorter message (3 kb) that includes exons 8–19 (Thompson, 1996; Potter et al., 2002; Zarach et al., 2004; Thompson, C.C., unpublished observations). The 3-kb message is present at low levels in skin and is most prominently expressed in adult brain (Thompson, C.C., unpublished observations). Expression of these RNAs, and presumably their encoded proteins, correlates with observed phenotypes. The original Hr^{hr} allele is caused by a retroviral insertion in intron 6 (Stoye et al., 1988) which results in reduction but not absence of full-length Hr mRNA (Cachon-Gonzalez et al., 1994; Zarach et al., 2004). Some full-length protein is produced, and the shorter message remains intact (Zarach et al., 2004). The *rhino* alleles are usually associated with nonsense mutations in the 5' end of the gene, which would not produce a full-length protein while leaving the 3' message intact (Ahmad et al., 1998b; Panteleyev et al., 1998b; Zarach et al., 2004). The Hr "knockout" allele is null as both mRNAs are disrupted and no protein is detected (Zarach et al., 2004). The *rhino*-Yurlovo allele is likely null as it is caused by an insertion in exon 16 which presumably disrupts both mRNAs (Panteleyev et al., 1998a).

As in mouse, there are multiple human alleles of Hr. The clinical importance of Hr was demonstrated when the human orthologue of Hr was identified as the gene mutated in two rare human diseases, alopecia universalis and atrichia with papular lesions (Ahmad et al., 1998a; Cichon et al., 1998; Sprecher et al., 1998). It was subsequently recognized that many cases of alopecia universalis may be misdiagnosed cases of papular atrichia, with the development of papular rash a phenotypic variation (Sprecher et al., 1999a,b). As in mice, humans with these diseases initially have normal hair, but within a few years the hair falls out and in general does not regrow (Ahmad et al., 1998a; Cichon et al., 1998; Sprecher et al., 1998). Both human diseases exhibit histological changes similar to those observed in Hr mutant mice (Ahmad et al., 1998a; Cichon et al., 1998; Sprecher et al., 1998). There are multiple mutant human Hr alleles, which include missense and nonsense mutations; unlike in mice, the human phenotype is not correlated with the severity of the mutation. Thus, the function of Hr appears to be conserved between mammalian species.

6. Role of HR in maintenance of the skin

Although hair loss and skin wrinkling are easily observable, the skin of Hr mutants also undergoes changes that can only be detected histologically. Comparing histology of Hr mutant mouse skin to wild type, the first difference identified is a widening of the top of the hair follicle, the infundibulum (Fig. 6A), to form what is termed a utricle (Mann, 1971; Zarach et al., 2004). Molecular analysis of $Hr^{-/-}$ skin has shown that the initial phenotypic change at the top of the hair follicle is caused by altered cell proliferation and differentiation. Follicles normally express keratin 17 (K17), a protein specific for the ORS (Fig. 6B). In $Hr^{-/-}$ skin, the infundibulum expresses an epidermal marker, keratin 10 (K10) (Fig. 6B). Based on BrdU incorporation and PCNA immunostaining, this is also a site of increased cell proliferation (Zarach et al., 2004). $Hr^{-/-}$ mice have normal barrier function, suggesting

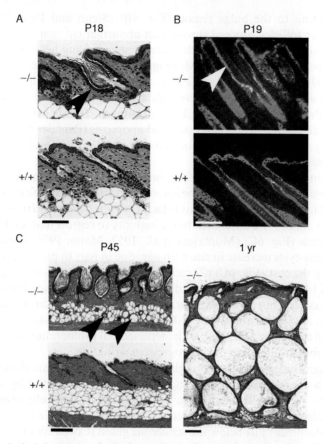

Fig. 6. Histological changes in skin of *Hr* mutant mice. (A) Hematoxylin–eosin stained sections from wild-type (+/+) and $Hr^{-/-}$ skin at P18. Arrowhead indicates widening of the top of the hair follicle adjacent to the epidermis (utricle) in $Hr^{-/-}(-/-)$ skin. Scale bar: 50 μm. (B) Utricles are composed of epidermal cells. Immunofluorescence for keratin expression demonstrates cells in the utricle wall express K10 (arrowhead, red), an epidermal marker, instead of K17 (green). Scale bar: 50 μm. (C) (Left) Dermal cysts develop in $Hr^{-/-}(-/-)$ skin. Hematoxylin–eosin stained sections of wild-type (+/+) and $Hr^{-/-}$ skin at P45. Arrowheads indicate cysts found in the dermis. (Right) Dermal cysts increase in size in $Hr^{-/-}$ skin. Hematoxylin–eosin stained section of skin from an $Hr^{-/-}(-/-)$ mouse at 1-year old. Scale bars: 20 μm. Adapted from Zarach et al. (2004). (See Color Insert.)

that epidermal function is not compromised. These results suggested that HR normally acts to suppress epidermal differentiation.

The first defect in the hair follicles in *Hr* mutant skin is the disintegration of the follicles between the bulge and the DP in catagen (Montagna et al., 1952; Orwin et al., 1967; Mann, 1971; Panteleyev et al., 1999). Normally, catagen is accompanied by the retraction of the DP from the hair bulb, loss of the keratinocytes in the hair bulb, and regulated loss of the ORS with the

DP retracting to the bulge region (Fig. 4B) (Stenn and Paus, 2001). *Hr* mutant hair follicles proceed through an abnormal catagen typified by the stranding of the ORS and DP below the dermis, leaving behind cell clusters in the subcutaneous fat layer (Montagna et al., 1952; Orwin et al., 1967; Mann, 1971; Panteleyev et al., 1999). *Hr^{hr}* hair follicles exhibit increased apoptosis at a time when hair bulb cells are still proliferating at the start of catagen (Panteleyev et al., 1999). Later in catagen, *Hr^{hr}* hair follicles exhibit increased cell proliferation and decreased apoptosis (Panteleyev et al., 1999), suggesting that *Hr* is necessary for coordinating the timing of proliferation and apoptosis during catagen.

Following catagen, hair follicles do not reform in *Hr* mutant mice. Molecular analysis has confirmed that the loss of hair follicles is complete as markers expressed in early regenerating hair follicles are not present in *Hr* mutant skin (Zarach et al., 2004). Subsequent to the inability to regrow hair, cysts develop in the dermis (Fig. 6C) (Montagna et al., 1952; Mann, 1971; Zarach et al., 2004). These cysts increase in size with age, due in part to proliferation of the cells lining the cyst (Fig. 6C) (Zarach et al., 2004). The cells in the cyst walls express K14, and have properties of sebaceous glands, as they contain lipids and express Scd1 but do not express adipocyte-specific markers (Mann, 1971; Bernerd et al., 1996; Zarach et al., 2004). The cysts in $Hr^{-/-}$ skin contribute to the overall thickness, and along with increased cell proliferation in the epidermis contribute to the severe and progressive wrinkling.

Although epidermis, sebaceous glands, and hair follicles are distinct structures with characteristic properties, they can be regenerated by the same population of epithelial stem cells. Most epithelial stem cells reside in the follicle bulge (Fig. 4), an outcropping of the ORS at the site of arrector pili muscle attachment just below the sebaceous gland (Cotsarelis et al., 1990). A number of studies have shown that cells in the bulge are bona fide stem cells, including that their progeny are found in both the hair bulb and epidermis (Taylor et al., 2000; Oshima et al., 2001), and that isolated bulge stem cells can regenerate skin that contains normal epidermis, sebaceous glands, and hair follicles (Blanpain et al., 2004; Morris et al., 2004). Thus, the current model for stem cell-mediated replenishment of the skin is that the epithelial stem cells, localized in the bulge, generate progenitor cells which can either migrate toward the surface to replenish the sebaceous gland and epidermis, or internally along the ORS to the hair bulb. Based on the increase in epidermal and sebaceous cells and decrease in hair cells in $Hr^{-/-}$ skin, we proposed a model in which the role of HR is to regulate the timing of epithelial stem cell differentiation, and that disruption of timing in *Hr* mutants leads to changes in cell fate favoring epidermis and sebaceous cells at the expense of the hair follicle (Zarach et al., 2004).

Historically, the role of *Hr* in skin has been assessed from the study of the aforementioned partial and complete loss of function mutations. To test the model derived from analysis of $Hr^{-/-}$ skin, we generated and characterized a gain-of-function mutant in which the rat *Hr* cDNA is expressed under the control of the K14 promoter (K14-rHr) to drive expression early in skin development in the undifferentiated cells of the epidermis and hair follicle. Analysis of $Hr^{-/-}$ mice had suggested that HR normally suppresses epidermal and sebaceous differentiation and promotes differentiation toward hair cell fate. If this is the case, increased expression of *Hr* should lead to a reduction and/or delay in epidermal and sebaceous cell differentiation, and a promotion and/or acceleration of hair cell differentiation.

Based on the analysis of K14-rHr skin, this model generally holds true (Beaudoin et al., 2005, unpublished observations). In K14-rHr epidermis, there is a greater number of undifferentiated (K14-positive) and a reduced number of terminally differentiated (filaggrin-positive) keratinocytes compared to wild type, indicating a delay in epidermal differentiation. The hair of the K14-rHr transgenic mice is shorter than normal due to reduced production of each lineage of the hair bulb. The reduced production of each lineage appears to be the result of reduced proliferation caused by premature differentiation. The short-hair phenotype is consistent with *Hr* promoting differentiation at the expense of proliferation as in other mouse mutants predicted to have increased differentiation (Mann et al., 1993; Miettinen et al., 1995; Threadgill et al., 1995; Prowse et al., 1999; Brakebusch et al., 2000). In addition, the hair cycle is accelerated in K14-rHr transgenic mice, as hair follicles reenter anagen at postnatal day 24 (P24) an age at which wild-type skin is still in telogen. Sebaceous glands were not detected in K14-rHr skin at P10, although sebaceous glands are detected later in development, indicating that overexpression of *Hr* delays sebaceous gland differentiation.

Thus, the K14-rHr transgenic phenotype is generally consistent with the model that HR promotes hair cell fate while suppressing progression into cells that populate the epidermis and sebaceous glands. In addition to providing support for the model of *Hr* function derived from the $Hr^{-/-}$ phenotype, these mice were essential for experiments demonstrating the role of *Hr* in hair regrowth.

7. HR is necessary and sufficient for hair regrowth

The phenotype of *Hr* mutant mice and humans shows that *Hr* is necessary for normal hair cycling but does not specify the role of *Hr* in the hair cycle. The bulge activation hypothesis specifies that reentry into anagen is controlled by factors secreted from the DP, which reactivate the bulge to produce hair bulb progenitor cells (Cotsarelis et al., 1990). In catagen, the hair bulb and lower portion of the follicle are retracted and destroyed, leaving the bulge in close proximity to the DP (Cotsarelis et al., 1990). In *Hr* mutant mice, the catagen phase is aberrant and the hair follicles disintegrate into cell clusters below the dermis. It has been postulated that the defect in hair regrowth in *Hr* mutants is because the DP and bulge remain separated in *Hr* mutant skin (Montagna et al., 1952; Mann, 1971; Paus and Foitzik, 2004). However, analysis showing expression of HR protein in K14-positive keratinocyte progenitor cells at the transition between the rest and growth phases of the hair cycle suggested that HR could have an active role in reinitiating hair growth. To test the hypothesis that expression of HR in these cells can reinitiate hair growth, we generated mice that express HR only in K14-positive cells by crossing the K14-rHr transgenic mice with $Hr^{-/-}$ mice. The resulting mice still lose their hair, but remarkably, just a few days later the "transgenic rescue" mice regenerate hair follicles and produce normal hair (Fig. 7) (Beaudoin et al., 2005). Thus, HR expression in progenitor keratinocytes is sufficient to induce hair regrowth. Expression in these cells did not rescue hair loss, which may be due to a function of HR in K14-negative hair bulb cells or because the overall timing of *Hr* expression in the rescue is the not the same as in wild type. The histology of mutant and rescue mouse skin is initially similar, with epithelial cells present in the dermis of both knockout and rescue skin (Beaudoin et al., 2005).

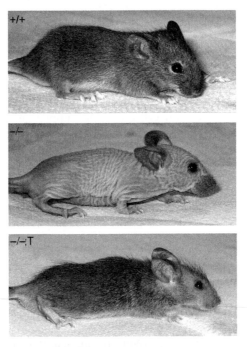

Fig. 7. Expression of HR in keratinocyte progenitor cells rescues hair regrowth in $Hr^{-/-}$ skin. Mice expressing HR in K14-positive cells (–/–; T) regrow hair, demonstrating that HR expression in these cells is sufficient for hair regrowth. Wild type (+/+), $Hr^{-/-}$(−/−). Adapted from Beaudoin et al. (2005). (See Color Insert.)

Since hair regrows in the rescue mouse, contact between the DP and bulge is likely not required for induction of hair growth. Thus, our data suggests that a diffusible signal is exchanged between the DP and bulge and that this signaling is defective in *Hr* mutant mice.

8. Transcriptional regulation by HR *in vivo*

Since HR is a transcriptional corepressor, the defects observed in *Hr* mutant skin presumably arise from altered gene expression caused by a defect in HR corepressor activity. Changes in gene expression in $Hr^{-/-}$ skin were identified using microarray analysis of RNA from $Hr^{-/-}$ skin, at an age at which there are no observable morphological changes (P12). Since HR is a repressor, loss of HR should result in upregulation of gene expression. Microarray analysis of 8000 genes revealed significant (>twofold) upregulation of 14 genes in $Hr^{-/-}$ skin (Zarach et al., 2004). Most of the identified genes either play established roles in epidermal differentiation (filaggrin, loricrin, keratin 10) or have been associated with epidermal differentiation (caspase-14, keratinocyte differentiation associated protein, calmodulin-4) (Eckhart et al., 2000; Lippens et al., 2000; Oomizu et al., 2000; Koshizuka et al., 2001; Chien et al., 2002; Hwang and Morasso, 2003). Although direct regulation of these genes by HR is yet to be demonstrated, three of these genes were shown to be upregulated in cultured $Hr^{-/-}$ keratinocytes (Zarach et al., 2004). *In situ* hybridization analyses revealed that transcriptional changes were localized to the upper part of the hair follicle; thus, the site of altered gene expression becomes a distinct epidermal structure (utricle), suggesting that aberrant transcriptional regulation is a cause rather than a consequence of the $Hr^{-/-}$ phenotype (Zarach et al., 2004).

To identify the genes regulated by HR with a role in reactivation of hair growth, the "transgenic rescue" mice were used to provide additional criteria for identifying genes that are regulated by HR in hair follicles. Based on the phenotypes, genes involved in hair regrowth should be (1) upregulated in $Hr^{-/-}$ skin and (2) downregulated (relative to $Hr^{-/-}$) in transgenic rescue skin, at the time hair follicles regenerate (P24). One of the genes that met these criteria, Wnt modulator in surface ectoderm (Wise) encodes a protein that modulates Wnt signaling (Itasaki et al., 2003). Northern analysis confirmed that Wise mRNA expression is upregulated in $Hr^{-/-}$ skin and reduced in transgenic rescue skin, relative to $Hr^{-/-}$ skin (Beaudoin et al., 2005). Expression was also reduced in K14-rHr skin, as expected for overexpression of a repressor (Beaudoin et al., 2005). Preliminary promoter analysis suggests that HR represses transcription of *Wise* (Beaudoin et al., 2005).

The identification and analysis of Wise, a putative inhibitor of Wnt signaling, suggested that HR might be regulating this important signaling pathway through its function as a corepressor. The Wnt family of secreted glycoproteins act via binding to cell surface receptors, which induces the stabilization of cytoplasmic β-catenin. β-catenin then translocates to the nucleus, where it acts as a coactivator for members of the Lef/TCF family of transcription factors, resulting in activation of target gene expression (He et al., 2004; Logan and Nusse, 2004). A number of studies have shown that the canonical Wnt/β-catenin signaling pathway is critical for hair regrowth. Wnt signaling (like HR function) in progenitor keratinocytes has been directly implicated in hair follicle regeneration (Huelsken et al., 2001; Merrill et al., 2001; Niemann et al., 2002; Van Mater et al., 2003; Lo Celso et al., 2004). In transgenic mice that express a Wnt-responsive reporter gene (TOP-Gal), Wnt signaling is activated transiently in the epithelial stem cells during the telogen to anagen transition (DasGupta and Fuchs, 1999; Merrill et al., 2001). Activation of Wnt signaling is necessary for hair follicles to reenter anagen, as blocking Wnt signaling by either loss of β-catenin expression or expression of a mutant Lef1 blocks anagen reinitiation (Huelsken et al., 2001; Merrill et al., 2001; Niemann et al., 2002). Wnt signaling is sufficient for normal anagen reinitiation, as inducing nuclear translocation of β-catenin in a transgenic mouse model causes telogen stage hair follicles to reenter anagen (Van Mater et al., 2003; Lo Celso et al., 2004). Transient Wnt signaling appears to be critical, as overexpression of a stabilized β-catenin induces skin tumors in mice and humans (Gat et al., 1998; Chan et al., 1999; Van Mater et al., 2003; Lo Celso et al., 2004). Recent evidence indicates that activation of Wnt signaling regulates gene expression in epithelial stem cells (Lowry et al., 2005).

The established role of Wnt signaling in hair regrowth suggested that HR regulates hair follicle regeneration by repressing the expression of one or more Wnt inhibitors, thereby allowing Wnt activation. Consistent with this hypothesis, HR protein and Wise mRNA expression are inversely correlated *in vivo*, suggesting that HR is suppressing Wise expression. Of the 19 known mammalian Wnts, two (Wnt10a and 10b) are expressed at the proper time and location in the skin to regulate hair regrowth (Reddy et al., 2001). Wise inhibits Wnt10b-induced reporter gene expression in cultured cells, suggesting that it does so *in vivo* (Beaudoin et al., 2005). Nuclear HR protein colocalizes with expression of a Wnt-responsive reporter gene in the skin of TOP-Gal transgenic mice (DasGupta and Fuchs, 1999; Beaudoin et al., 2005), and no reporter gene expression is detected in the skin of TOP-Gal/$Hr^{-/-}$ mice (Sisk, J. and Thompson, C.C., unpublished observations). In addition, expression of an endogenous Wnt target gene, Axin2, is correlated with the Hr genotype; Axin2 is expressed in wild-type and transgenic rescue skin and is not detected in Hr knockout skin.

Based on these data, our current model for HR action in the skin is that HR protein promotes Wnt/β-catenin activation, and thus hair growth initiation, by repressing expression of soluble Wnt inhibitors at the proper time in the hair cycle (Fig. 8). This model specifies a molecular mechanism to account for the failure of hair follicles to regenerate in Hr mutant skin: inhibition of Wnt signaling caused by the persistent expression of Wise.

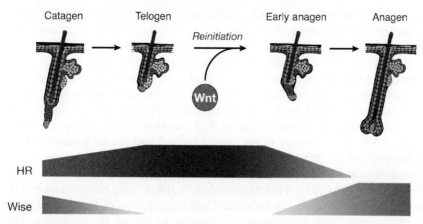

Fig. 8. Model for HR action in hair regrowth. Summary of HR protein expression through the hair cycle in relation to Wise mRNA expression. Catagen is characterized by the upregulation of HR protein and concurrent downregulation of Wise mRNA. HR protein expression during telogen is predicted to repress Wise mRNA expression, allowing activation of Wnt signaling in the bulge. Subsequent to the reactivation of hair growth, HR protein is downregulated while Wise mRNA increases; this predicted increase in Wise will cause a decrease in Wnt signaling. In *Hr* mutants, uncontrolled expression of Wise and possibly other Wnt inhibitors prevents the regrowth of hair by suppressing Wnt signaling. (See Color Insert.)

This work is the first evidence linking *Hr* gene function and Wnt signaling and provides a molecular mechanism for regulating the precise temporal and spatial localization of Wnt signaling required for hair cycle reinitiation (Van Mater et al., 2003; Lo Celso et al., 2004). This model includes a means of maintaining transient Wnt signaling, via the decrease in HR protein expression and concomitant increase in Wise expression subsequent to anagen initiation. The function of HR in the epidermis and sebaceous glands may be through Wnt signaling as well, as HR expression in K14-positive cells also prevents the wrinkling phenotype. This suggests that HR expression also rescues the epidermal and sebaceous phenotypes and is consistent with the idea that HR is regulating epithelial stem cell differentiation via the modulation of Wnt signaling.

The phenotypes of several mouse models support the idea that HR functions in the skin via Wnt signaling. Studies of *Hr^{hr}* mice have noted that sparse hair growth is confined to the tylotrich hair follicles (Mann and Straile, 1961; Mann, 1971). This is strikingly similar to the *Lef1*-null mouse, which lack nontylotrich hair follicles, suggesting these follicles are particularly sensitive to Wnt signaling (van Genderen et al., 1994). K14-rHr transgenic mice have shorter hair, similar to mice in which Wnt signaling is increased by overexpression of *Wnt3* or *dishevelled 2* (Millar et al., 1999). The cysts in *Hr* mutant skin resemble the dermal cysts that form in mice either null for β-catenin or expressing a dominant-negative Lef1 (Huelsken et al., 2001; Merrill et al., 2001). Activation of Wnt signaling also prevents sebocyte differentiation (Niemann et al., 2003), possibly explaining the delayed differentiation of sebaceous glands in K14-rHr transgenic mice.

Regulation of Wnt signaling by HR is unlikely to be solely through Wise. There are a number of established Wnt inhibitors, some of which are expressed in the bulge (Kawano and Kypta, 2003; Fuchs et al., 2004; Morris et al., 2004; Tumbar et al., 2004; Lowry et al., 2005). Although these Wnt inhibitors were not misregulated in *Hr*-null skin at P12, another factor, Soggy (Krupnik et al., 1999), was regulated in the same temporal and spatial pattern as Wise (Thompson et al., 2006). *Soggy* encodes a protein related to Dickkopfs, which inhibit Wnt activation (Krupnik et al., 1999; Logan and Nusse, 2004). Other Wnt inhibitors may play a role independent of HR, or may be regulated by HR with different timing and/or location. Unlike reinitiation of hair growth, hair morphogenesis does not seem to be sensitive to loss of *Hr* expression. HR function may not be required because undifferentiated ectoderm is not subject to the influence of Wnt inhibitors, or an alternative factor may regulate Wnt signaling during morphogenesis.

Although Wnt/β-catenin signaling plays an essential role in hair regrowth, Wnt signaling is not the only pathway necessary for hair regrowth and morphogenesis (Huelsken et al., 2001; Van Mater et al., 2003; Lo Celso et al., 2004; Lowry et al., 2005). Other signaling pathways play important roles in hair follicle morphogenesis and hair cycling, including Shh and BMP signaling (Stenn and Paus, 2001; Millar, 2002). Exactly how these signaling pathways are coordinated *in vivo* is not clear, but HR may regulate other signaling pathway(s) as well. HR may also regulate BMP signaling via Wise, as multiple studies have shown that Wise can also inhibit BMP signaling (Sclerostin-like 1, Balemans and Van Hul, 2002; Ectodin, Laurikkala et al., 2003; USAG-1, Yanagita et al., 2004). In addition, HR may regulate the expression of additional factors that modulate such pathways.

9. *Hr* action in other tissues

Although *Hr* function is most apparent in the skin, *Hr* is also expressed in other tissues. *Hr* mRNA and protein are highly expressed in brain and detected at lower levels in several other tissues, including pituitary, lung, colon, and various tissues of epithelial origin (Thompson, 1996; Cachon-Gonzalez et al., 1999; Potter et al., 2002). Based on HR function in the skin, one may speculate that HR plays a role in epithelial cells in other tissues as well. Notable in this regard are defects observed in the colon of Hr^{rh} mutant mice, which show an increased number of villi attributed to hyperprolifera-tion of the epithelium (Cachon-Gonzalez et al., 1999). The role of HR in regulating epithelial stem cell differentiation may extend to other tissues with regenerative capacity.

In the developing brain, *Hr* expression is induced by TH, significant because of the well-established role of TH in brain development (Thompson, 1996; Thompson and Bottcher, 1997; Thompson and Potter, 2000). Expression of *Hr* is also regulated by TH in other tissues, although notably not in skin or in adult brain (Thompson, 1996). *Hr* mRNA and protein are broadly expressed within the nervous system, with expression detected in cerebellum, somatosensory cortex, hippocampus, dentate gyrus, retina, and inner ear (Thompson, 1996; Cachon-Gonzalez et al., 1999; Potter et al., 2002). Although *Hr* is prominently expressed in brain, its *in vivo* role is not clear. The majority of patients with mutations in *Hr* have not been assessed for neural function; however, there are two cases of patients homozygous for a mutation in *Hr* that suffer from mental retardation (del Castillo et al., 1974; Aita et al., 2000). Although not extensively studied, *Hr* mutant mice show phenotypic changes in the nervous system that include abnormalities in neuronal morphology, inner ear, and retinal cytoarchitecture (Garcia-Atares et al., 1998; Cachon-Gonzalez et al., 1999). Alterations in the cerebellar cortex include changes in the size and shape of Purkinje cells, which may be related to HR interaction with RORα, as both RORα and TH play essential roles in Purkinje cell development and function (Hamilton et al., 1996; Matysiak-Scholze and Nehls, 1997; Dussault et al., 1998; Garcia-Atares et al., 1998; Steinmayr et al., 1998).

10. Conclusions

The biological role of transcriptional activation by nuclear receptors is well established. Far less is known about the role of repression, but the study of corepressors is helping to elucidate the mechanisms and significance of repression by nuclear receptors. The study of HR over the past decade has contributed to this understanding.

HR corepressor function *in vivo* is essential for maintaining skin function. HR appears to be a member of a distinct class of corepressors that are spatially and temporally restricted, as *Hr* mRNA expression is restricted to specific tissues and/or cell types, and HR protein expression is further restricted temporally. The study of the biochemical and biological roles of HR has revealed its mechanism of action in regulating developmental processes in the skin, which will likely apply to other tissues as well.

Acknowledgments

Work cited was supported by a National Institutes of Health grant to C.C. T. (NS41313) and National Research Service Award to G.M.J.B. (NS44744). A special thank you to Jeanne Sisk for helping prepare artwork and figures.

References

Ahmad, W., Faiyaz ul Haque, M., Brancolini, V., Tsou, H.C., ul Haque, S., Lam, H., Aita, V.M., Owen, J., deBlaquiere, M., Frank, J., Cserhalmi-Friedman, P.B., Leask, A., et al. 1998a. Alopecia universalis associated with a mutation in the human hairless gene. Science 279(5351), 720–724.

Ahmad, W., Panteleyev, A.A, Sundberg, J.P., Christiano, A.M. 1998b. Molecular basis for the rhino (hrrh-8J) phenotype: A nonsense mutation in the mouse hairless gene. Genomics 53(3), 383–386.

Aita, V.M., Ahmad, W., Panteleyev, A.A., Kozlowska, U., Kozlowska, A., Gilliam, T.C., Jablonska, S., Christiano, A.M. 2000. A novel missense mutation (C622G) in the zinc-finger domain of the human hairless gene associated with congenital atrichia with papular lesions. Exp. Dermatol. 9(2), 157–162.

Alonso, L., Fuchs, E. 2003. Stem cells in the skin: Waste not, Wnt not. Genes Dev. 17(10), 1189–1200.

Alonso, L.C., Rosenfield, R.L. 2003. Molecular genetic and endocrine mechanisms of hair growth. Horm. Res. 60(1), 1–13.

Balemans, W., Van Hul, W. 2002. Extracellular regulation of BMP signaling in vertebrates: A cocktail of modulators. Dev. Biol. 250(2), 231–250.

Beaudoin, G.M., III, Sisk, J.M., Coulombe, P.A., Thompson, C.C. 2005. Hairless triggers reactivation of hair growth by promoting Wnt signaling. Proc. Natl. Acad. Sci. USA 102 (41), 14653–14658.

Bernerd, F., Schweizer, J., Demarchez, M. 1996. Dermal cysts of the rhino mouse develop into unopened sebaceous glands. Arch. Dermatol. Res. 288(10), 586–595.

Blanpain, C., Lowry, W.E., Geoghegan, A., Polak, L., Fuchs, E. 2004. Self-renewal, multi-potency, and the existence of two cell populations within an epithelial stem cell niche. Cell 118(5), 635–648.

Brakebusch, C., Grose, R., Quondamatteo, F., Ramirez, A., Jorcano, J.L., Pirro, A., Svensson, M., Herken, R., Sasaki, T., Timpl, R., Werner, S., Fassler, R. 2000. Skin and hair follicle integrity is crucially dependent on beta 1 integrin expression on keratinocytes. EMBO J. 19(15), 3990–4003.

Brooke, H.C. 1926. Hairless mice. J. Hered. 17, 173–174.

Burke, L.J., Baniahmad, A. 2000. Co-repressors 2000. FASEB J. 14(13), 1876–1888.

Cachon-Gonzalez, M.B., Fenner, S., Coffin, J.M., Moran, C., Best, S., Stoye, J.P. 1994. Structure and expression of the hairless gene of mice. Proc. Natl. Acad. Sci. USA 91(16), 7717–7721.

Cachon-Gonzalez, M.B., San-Jose, I., Cano, A., Vega, J.A., Garcia, N., Freeman, T., Schimmang, T., Stoye, J.P. 1999. The hairless gene of the mouse: Relationship of phenotypic effects with expression profile and genotype. Dev. Dyn. 216(2), 113–126.

Cavailles, V., Dauvois, S., L'Horset, F., Lopez, G., Hoare, S., Kushner, P.J., Parker, M.G. 1995. Nuclear factor RIP140 modulates transcriptional activation by the estrogen receptor. EMBO J. 14(15), 3741–3751.

Chan, E.F., Gat, U., McNiff, J.M., Fuchs, E. 1999. A common human skin tumour is caused by activating mutations in beta-catenin. Nat. Genet. 21(4), 410–413.

Chen, D., Ma, H., Hong, H., Koh, S.S., Huang, S.M., Schurter, B.T., Aswad, D.W., Stallcup, M.R. 1999. Regulation of transcription by a protein methyltransferase. Science 284(5423), 2174–2177.

Chen, J.D., Evans, R.M. 1995. A transcriptional co-repressor that interacts with nuclear hormone receptors. Nature 377(6548), 454–457.

Chien, A.J., Presland, R.B., Kuechle, M.K. 2002. Processing of native caspase-14 occurs at an atypical cleavage site in normal epidermal differentiation. Biochem. Biophys. Res. Commun. 296(4), 911–917.

Cichon, S., Anker, M., Vogt, I.R., Rohleder, H., Putzstuck, M., Hillmer, A., Farooq, S.A., Al-Dhafri, K.S., Ahmad, M., Haque, S., Rietschel, M., Propping, P., et al. 1998. Cloning, genomic organization, alternative transcripts and mutational analysis of the gene responsible for autosomal recessive universal congenital alopecia. Hum. Mol. Genet. 7(11), 1671–1679.

Clissold, P.M., Ponting, C.P. 2001. JmjC: Cupin metalloenzyme-like domains in jumonji, hairless and phospholipase A2beta. Trends Biochem. Sci. 26(1), 7–9.

Cohen, R.N., Brzostek, S., Kim, B., Chorev, M., Wondisford, F.E., Hollenberg, A.N. 2001. The specificity of interactions between nuclear hormone receptors and corepressors is mediated by distinct amino acid sequences within the interacting domains. Mol. Endocrinol. 15(7), 1049–1061.

Cotsarelis, G., Sun, T.T., Lavker, R.M. 1990. Label-retaining cells reside in the bulge area of pilosebaceous unit: Implications for follicular stem cells, hair cycle, and skin carcinogenesis. Cell 61(7), 1329–1337.

DasGupta, R., Fuchs, E. 1999. Multiple roles for activated LEF/TCF transcription complexes during hair follicle development and differentiation. Development 126(20), 4557–4568.

del Castillo, V., Ruiz-Maldonado, R., Carnevale, A. 1974. Atrichia with papular lesions and mental retardation in two sisters. Int. J. Dermatol. 13(5), 261–265.

Downes, M., Ordentlich, P., Kao, H.Y., Alvarez, J.G., Evans, R.M. 2000. Identification of a nuclear domain with deacetylase activity. Proc. Natl. Acad. Sci. USA 97(19), 10330–10335.

Dussault, I., Fawcett, D., Matthyssen, A., Bader, J.A., Giguere, V. 1998. Orphan nuclear receptor ROR alpha-deficient mice display the cerebellar defects of staggerer. Mech. Dev. 70(1–2), 147–153.

Eckhart, L., Declercq, W., Ban, J., Rendl, M., Lengauer, B., Mayer, C., Lippens, S., Vandenabeele, P., Tschachler, E. 2000. Terminal differentiation of human keratinocytes and stratum corneum formation is associated with caspase-14 activation. J. Invest. Dermatol. 115(6), 1148–1151.

Fernandes, I., Bastien, Y., Wai, T., Nygard, K., Lin, R., Cormier, O., Lee, H.S., Eng, F., Bertos, N.R., Pelletier, N., Mader, S., Han, V.K., et al. 2003. Ligand-dependent nuclear receptor corepressor (LCoR) functions by histone deacetylase-dependent and -independent mechanisms. Mol. Cell. 11(1), 139–150.

Fuchs, E., Tumbar, T., Guasch, G. 2004. Socializing with the neighbors: Stem cells and their niche. Cell 116(6), 769–778.

Garcia-Atares, N., San Jose, I., Cabo, R., Vega, J.A., Represa, J. 1998. Changes in the cerebellar cortex of hairless Rhino-J mice (hr-rh-j). Neurosci. Lett. 256(1), 13–16.

Gat, U., DasGupta, R., Degenstein, L., Fuchs, E. 1998. De novo hair follicle morphogenesis and hair tumors in mice expressing a truncated beta-catenin in skin. Cell 95(5), 605–614.

Glass, C.K., Rosenfeld, M.G. 2000. The coregulator exchange in transcriptional functions of nuclear receptors. Genes Dev. 14(2), 121–141.

Guenther, M.G., Lane, W.S., Fischle, W., Verdin, E., Lazar, M.A., Shiekhattar, R. 2000. A core SMRT corepressor complex containing HDAC3 and TBL1, a WD40-repeat protein linked to deafness. Genes Dev. 14(9), 1048–1057.

Guenther, M.G., Barak, O., Lazar, M.A. 2001. The SMRT and N-CoR corepressors are activating cofactors for histone deacetylase 3. Mol. Cell. Biol. 21(18), 6091–6101.

Hamilton, B.A., Frankel, W.N., Kerrebrock, A.W., Hawkins, T.L., FitzHugh, W., Kusumi, K., Russell, L.B., Mueller, K.L., van Berkel, V., Birren, B.W., Kruglyak, L., Lander, E.S. 1996. Disruption of the nuclear hormone receptor RORalpha in staggerer mice. Nature 379(6567), 736–739.

Hardy, M.H. 1992. The secret life of the hair follicle. Trends Genet. 8(2), 55–61.

He, X., Semenov, M., Tamai, K., Zeng, X. 2004. LDL receptor-related proteins 5 and 6 in Wnt/beta-catenin signaling: Arrows point the way. Development 131(8), 1663–1677.

Heery, D.M., Kalkhoven, E., Hoare, S., Parker, M.G. 1997. A signature motif in transcriptional co-activators mediates binding to nuclear receptors. Nature 387(6634), 733–736.

Horlein, A.J., Naar, A.M., Heinzel, T., Torchia, J., Gloss, B., Kurokawa, R., Ryan, A., Kamei, Y., Soderstrom, M., Glass, C.K., Rosenfeld, M.G. 1995. Ligand-independent repression by the thyroid hormone receptor mediated by a nuclear receptor co-repressor. Nature 377(6548), 397–404.

Hsieh, J.C., Sisk, J.M., Jurutka, P.W., Haussler, C.A., Slater, S.A., Haussler, M.R., Thompson, C.C. 2003. Physical and functional interaction between the vitamin D receptor and hairless corepressor, two proteins required for hair cycling. J. Biol. Chem. 278(40), 38665–38674.

Hu, X., Lazar, M.A. 1999. The CoRNR motif controls the recruitment of corepressors by nuclear hormone receptors. Nature 402(6757), 93–96.

Huang, E.Y., Zhang, J., Miska, E.A., Guenther, M.G., Kouzarides, T., Lazar, M.A. 2000. Nuclear receptor corepressors partner with class II histone deacetylases in a Sin3-independent repression pathway. Genes Dev. 14(1), 45–54.

Huelsken, J., Vogel, R., Erdmann, B., Cotsarelis, G., Birchmeier, W. 2001. Beta-catenin controls hair follicle morphogenesis and stem cell differentiation in the skin. Cell 105(4), 533–545.

Hwang, M., Morasso, M.I. 2003. The novel murine Ca++-binding protein scarf, is differentially expressed during epidermal differentiation. J. Biol. Chem. 278(48), 47827–47833.

Itasaki, N., Jones, C.M., Mercurio, S., Rowe, A., Domingos, P.M., Smith, J.C., Krumlauf, R. 2003. Wise, a context-dependent activator and inhibitor of Wnt signalling. Development 130 (18), 4295–4305.

Jahoda, C.A., Horne, K.A., Oliver, R.F. 1984. Induction of hair growth by implantation of cultured dermal papilla cells. Nature 311(5986), 560–562.

Jepsen, K., Rosenfeld, M.G. 2002. Biological roles and mechanistic actions of co-repressor complexes. J. Cell Sci. 115(Pt. 4), 689–698.

Jepsen, K., Hermanson, O., Onami, T.M., Gleiberman, A.S., Lunyak, V., McEvilly, R.J., Kurokawa, R., Kumar, V., Liu, F., Seto, E., Hedrick, S.M., Mandel, G., et al. 2000. Combinatorial roles of the nuclear receptor corepressor in transcription and development. Cell 102(6), 753–763.

Kao, H.Y., Downes, M., Ordentlich, P., Evans, R.M. 2000. Isolation of a novel histone deacetylase reveals that class I and class II deacetylases promote SMRT-mediated repression. Genes Dev. 14(1), 55–66.

Kawano, Y., Kypta, R. 2003. Secreted antagonists of the Wnt signalling pathway. J. Cell Sci. 116(Pt. 13), 2627–2634.

Kim, T.G., Kraus, J.C., Chen, J., Lee, Y. 2003. JUMONJI, a critical factor for cardiac development, functions as a transcriptional repressor. J. Biol. Chem. 278(43), 42247–42255.

Koshizuka, Y., Ikegawa, S., Sano, M., Nakamura, K., Nakamura, Y. 2001. Isolation of novel mouse genes associated with ectopic ossification by differential display method using ttw, a mouse model for ectopic ossification. Cytogenet. Cell. Genet. 94(3–4), 163–168.

Krupnik, V.E., Sharp, J.D., Jiang, C., Robison, K., Chickering, T.W., Amaravadi, L., Brown, D.E., Guyot, D., Mays, G., Leiby, K., Chang, B., Duong, T., et al. 1999. Functional and structural diversity of the human Dickkopf gene family. Gene 238(2), 301–313.

Laurikkala, J., Kassai, Y., Pakkasjarvi, L., Thesleff, I., Itoh, N. 2003. Identification of a secreted BMP antagonist, ectodin, integrating BMP, FGF, and SHH signals from the tooth enamel knot. Dev. Biol. 264(1), 91–105.

Li, H., Leo, C., Schroen, D.J., Chen, J.D. 1997a. Characterization of receptor interaction and transcriptional repression by the corepressor SMRT. Mol. Endocrinol. 11(13), 2025–2037.

Li, J., Wang, J., Nawaz, Z., Liu, J.M., Qin, J., Wong, J. 2000a. Both corepressor proteins SMRT and N-CoR exist in large protein complexes containing HDAC3. EMBO J. 19(16), 4342–4350.

Li, M., Indra, A.K., Warot, X., Brocard, J., Messaddeq, N., Kato, S., Metzger, D., Chambon, P. 2000b. Skin abnormalities generated by temporally controlled RXRalpha mutations in mouse epidermis. Nature 407(6804), 633–636.

Li, Y.C., Pirro, A.E., Amling, M., Delling, G., Baron, R., Bronson, R., Demay, M.B. 1997b. Targeted ablation of the vitamin D receptor: An animal model of vitamin D-dependent rickets type II with alopecia. Proc. Natl. Acad. Sci. USA 94(18), 9831–9835.

Lindner, G., Botchkarev, V.A., Botchkareva, N.V., Ling, G., van der Veen, C., Paus, R. 1997. Analysis of apoptosis during hair follicle regression (catagen). Am. J. Pathol. 151(6), 1601–1617.

Lippens, S., Kockx, M., Knaapen, M., Mortier, L., Polakowska, R., Verheyen, A., Garmyn, M., Zwijsen, A., Formstecher, P., Huylebroeck, D., Vandenabeeleand, P., Declercq, W. 2000. Epidermal differentiation does not involve the pro-apoptotic executioner caspases, but is associated with caspase-14 induction and processing. Cell Death Differ. 7(12), 1218–1224.

Lo Celso, C., Prowse, D.M., Watt, F.M. 2004. Transient activation of beta-catenin signalling in adult mouse epidermis is sufficient to induce new hair follicles but continuous activation is required to maintain hair follicle tumours. Development 131(8), 1787–1799.

Logan, C.Y., Nusse, R. 2004. The Wnt signaling pathway in development and disease. Annu. Rev. Cell. Dev. Biol. 20, 781–810.

Lowry, W.E., Blanpain, C., Nowak, J.A., Guasch, G., Lewis, L., Fuchs, E. 2005. Defining the impact of {beta}-catenin/Tcf transactivation on epithelial stem cells. Genes Dev. 19(13), 1596–1611.

Mangelsdorf, D.J., Evans, R.M. 1995. The RXR heterodimers and orphan receptors. Cell 83(6), 841–850.

Mann, G.B., Fowler, K.J., Gabriel, A., Nice, E.C., Williams, R.L., Dunn, A.R. 1993. Mice with a null mutation of the TGF alpha gene have abnormal skin architecture, wavy hair, and curly whiskers and often develop corneal inflammation. Cell 73(2), 249–261.

Mann, S.J. 1971. Hair loss and cyst formation in hairless and rhino mutant mice. Anat. Rec. 170 (4), 485–499.

Mann, S.J., Straile, W.E. 1961. New observations in hair loss in the hairless mouse. Anat. Rec. 140, 97–102.

Matysiak-Scholze, U., Nehls, M. 1997. The structural integrity of ROR alpha isoforms is mutated in staggerer mice: Cerebellar coexpression of ROR alpha1 and ROR alpha4. Genomics 43(1), 78–84.

McInerney, E.M., Rose, D.W., Flynn, S.E., Westin, S., Mullen, T.M., Krones, A., Inostroza, J., Torchia, J., Nolte, R.T., Assa-Munt, N., Milburn, M.V., Glass, C.K., et al. 1998. Determinants of coactivator LXXLL motif specificity in nuclear receptor transcriptional activation. Genes Dev. 12(21), 3357–3368.

McKenna, N.J., O'Malley, B.W. 2002. Combinatorial control of gene expression by nuclear receptors and coregulators. Cell 108(4), 465–474.

Merrill, B.J., Gat, U., DasGupta, R., Fuchs, E. 2001. Tcf3 and Lef1 regulate lineage differentiation of multipotent stem cells in skin. Genes Dev. 15(13), 1688–1705.

Miettinen, P.J., Berger, J.E., Meneses, J., Phung, Y., Pedersen, R.A., Werb, Z., Derynck, R. 1995. Epithelial immaturity and multiorgan failure in mice lacking epidermal growth factor receptor. Nature 376(6538), 337–341.

Millar, S.E. 2002. Molecular mechanisms regulating hair follicle development. J. Invest. Dermatol. 118(2), 216–225.

Millar, S.E., Willert, K., Salinas, P.C., Roelink, H., Nusse, R., Sussman, D.J., Barsh, G.S. 1999. WNT signaling in the control of hair growth and structure. Dev. Biol. 207(1), 133–149.

Misiti, S., Schomburg, L., Yen, P.M., Chin, W.W. 1998. Expression and hormonal regulation of coactivator and corepressor genes. Endocrinology 139(5), 2493–2500.

Montagna, W., Chase, H.B., Melaragno, H.P. 1952. The skin of hairless mice. I. The formation of cysts and the distribution of lipids. J. Invest. Dermatol. 19, 83–94.

Moraitis, A.N., Giguere, V. 2003. The co-repressor hairless protects RORalpha orphan nuclear receptor from proteasome-mediated degradation. J. Biol. Chem. 278(52), 5258–52511.

Moraitis, A.N., Giguere, V., Thompson, C.C. 2002. Novel mechanism of nuclear receptor corepressor interaction dictated by activation function 2 helix determinants. Mol. Cell. Biol. 22(19), 6831–6841.

Morris, R.J., Liu, Y., Marles, L., Yang, Z., Trempus, C., Li, S., Lin, J.S., Sawicki, J.A., Cotsarelis, G. 2004. Capturing and profiling adult hair follicle stem cells. Nat. Biotechnol. 22(4), 411–417.

Nagy, L., Kao, H.Y., Chakravarti, D., Lin, R.J., Hassig, C.A., Ayer, D.E., Schreiber, S.L., Evans, R.M. 1997. Nuclear receptor repression mediated by a complex containing SMRT, mSin3A, and histone deacetylase. Cell 89(3), 373–380.

Nagy, L., Kao, H.Y., Love, J.D., Li, C., Banayo, E., Gooch, J.T., Krishna, V., Chatterjee, K., Evans, R.M., Schwabe, J.W. 1999. Mechanism of corepressor binding and release from nuclear hormone receptors. Genes Dev. 13(24), 3209–32016.

Nakamura, M., Sundberg, J.P., Paus, R. 2001. Mutant laboratory mice with abnormalities in hair follicle morphogenesis, cycling, and/or structure: Annotated tables. Exp. Dermatol. 10(6), 369–390.

Niemann, C., Owens, D.M., Hulsken, J., Birchmeier, W., Watt, F.M. 2002. Expression of DeltaNLef1 in mouse epidermis results in differentiation of hair follicles into squamous epidermal cysts and formation of skin tumours. Development 129(1), 95–109.

Niemann, C., Unden, A.B., Lyle, S., Zouboulis Ch, C., Toftgard, R., Watt, F.M. 2003. Indian hedgehog and beta-catenin signaling: Role in the sebaceous lineage of normal and neoplastic mammalian epidermis. Proc. Natl. Acad. Sci. USA 100(Suppl. 1), 11873–11880.

Oliver, R.F., Jahoda, C.A. 1988. Dermal-epidermal interactions. Clin. Dermatol. 6(4), 74–82.

Oomizu, S., Sahuc, F., Asahina, K., Inamatsu, M., Matsuzaki, T., Sasaki, M., Obara, M., Yoshizato, K. 2000. Kdap, a novel gene associated with the stratification of the epithelium. Gene 256(1–2), 19–27.

Ordentlich, P., Downes, M., Xie, W., Genin, A., Spinner, N.B., Evans, R.M. 1999. Unique forms of human and mouse nuclear receptor corepressor SMRT. Proc. Natl. Acad. Sci. USA 96(6), 2639–2644.

Orwin, D.F., Chase, H.B., Silver, A.F. 1967. Catagen in the hairless house mouse. Am. J. Anat. 121(3), 489–507.

Oshima, H., Rochat, A., Kedzia, C., Kobayashi, K., Barrandon, Y. 2001. Morphogenesis and renewal of hair follicles from adult multipotent stem cells. Cell 104(2), 233–245.

Panteleyev, A.A., Ahmad, W., Malashenko, A.M., Ignatieva, E.L., Paus, R., Sundberg, J.P., Christiano, A.M. 1998a. Molecular basis for the rhino Yurlovo (hr(rhY)) phenotype: Severe skin abnormalities and female reproductive defects associated with an insertion in the hairless gene. Exp. Dermatol. 7(5), 281–288.

Panteleyev, A.A., Paus, R., Ahmad, W., Sundberg, J.P., Christiano, A.M. 1998b. Molecular and functional aspects of the hairless (hr) gene in laboratory rodents and humans. Exp. Dermatol. 7(5), 249–267.

Panteleyev, A.A., Botchkareva, N.V., Sundberg, J.P., Christiano, A.M., Paus, R. 1999. The role of the hairless (hr) gene in the regulation of hair follicle catagen transformation. Am. J. Pathol. 155(1), 159–171.

Panteleyev, A.A., Paus, R., Christiano, A.M. 2000. Patterns of hairless (hr) gene expression in mouse hair follicle morphogenesis and cycling. Am. J. Pathol. 157(4), 1071–1079.

Paus, R., Foitzik, K. 2004. In search of the "hair cycle clock": A guided tour. Differentiation 72(9–10), 489–511.

Pazin, M.J., Kadonaga, J.T. 1997. What's up and down with histone deacetylation and transcription? Cell 89(3), 325–328.

Perissi, V., Rosenfeld, M.G. 2005. Controlling nuclear receptors: The circular logic of cofactor cycles. Nat. Rev. Mol. Cell. Biol. 6(7), 542–554.

Perissi, V., Staszewski, L.M., McInerney, E.M., Kurokawa, R., Krones, A., Rose, D.W., Lambert, M.H., Milburn, M.V., Glass, C.K., Rosenfeld, M.G. 1999. Molecular determinants of nuclear receptor-corepressor interaction. Genes Dev. 13(24), 3198–3208.

Potter, G.B., Beaudoin, G.M., III, DeRenzo, C.L., Zarach, J.M., Chen, S.H., Thompson, C.C. 2001. The hairless gene mutated in congenital hair loss disorders encodes a novel nuclear receptor corepressor. Genes Dev. 15(20), 2687–2701.

Potter, G.B., Zarach, J.M., Sisk, J.M., Thompson, C.C. 2002. The thyroid hormone-regulated corepressor hairless associates with histone deacetylases in neonatal rat brain. Mol. Endocrinol. 16(11), 2547–2560.

Privalsky, M.L. 2004. The role of corepressors in transcriptional regulation by nuclear hormone receptors. Annu. Rev. Physiol. 66, 315–360.

Prowse, D.M., Lee, D., Weiner, L., Jiang, N., Magro, C.M., Baden, H.P., Brissette, J.L. 1999. Ectopic expression of the nude gene induces hyperproliferation and defects in differentiation: Implications for the self-renewal of cutaneous epithelia. Dev. Biol. 212(1), 54–67.

Reddy, S., Andl, T., Bagasra, A., Lu, M.M., Epstein, D.J., Morrisey, E.E., Millar, S.E. 2001. Characterization of Wnt gene expression in developing and postnatal hair follicles and identification of Wnt5a as a target of Sonic hedgehog in hair follicle morphogenesis. Mech. Dev. 107(1–2), 69–82.

Sakai, Y., Kishimoto, J., Demay, M.B. 2001. Metabolic and cellular analysis of alopecia in vitamin D receptor knockout mice. J. Clin. Invest. 107(8), 961–966.

Sande, S., Privalsky, M.L. 1996. Identification of TRACs (T3 receptor-associating cofactors), a family of cofactors that associate with, and modulate the activity of, nuclear hormone receptors. Mol. Endocrinol. 10(7), 813–825.

Skorija, K., Cox, M., Sisk, J.M., Dowd, D.R., MacDonald, P.N., Thompson, C.C., Demay, M.B. 2005. Ligand-independent actions of the vitamin D receptor maintain hair follicle homeostasis. Mol. Endocrinol. 19(4), 855–862.

Smith, C.L., O'Malley, B.W. 2004. Coregulator function: A key to understanding tissue specificity of selective receptor modulators. Endocr. Rev. 25(1), 45–71.

Sprecher, E., Bergman, R., Szargel, R., Raz, T., Labay, V., Ramon, M., Baruch-Gershoni, R., Friedman-Birnbaum, R., Cohen, N. 1998. Atrichia with papular lesions maps to 8p in the region containing the human hairless gene. Am. J. Med. Genet. 80(5), 546–550.

Sprecher, E., Bergman, R., Szargel, R., Friedman-Birnbaum, R., Cohen, N. 1999a. Identification of a genetic defect in the hairless gene in atrichia with papular lesions: Evidence for phenotypic heterogeneity among inherited atrichias. Am. J. Hum. Genet. 64(5), 1323–1329.

Sprecher, E., Lestringant, G.G., Szargel, R., Bergman, R., Labay, V., Frossard, P.M., Friedman-Birnbaum, R., Cohen, N. 1999b. Atrichia with papular lesions resulting from a nonsense mutation within the human hairless gene. J. Invest. Dermatol. 113(4), 687–690.

Steel, J.H., White, R., Parker, M.G. 2005. Role of the RIP140 corepressor in ovulation and adipose biology. J. Endocrinol. 185(1), 1–9.

Steinmayr, M., Andre, E., Conquet, F., Rondi-Reig, L., Delhaye-Bouchaud, N., Auclair, N., Daniel, H., Crepel, F., Mariani, J., Sotelo, C., Becker-Andre, M. 1998. Staggerer phenotype in retinoid-related orphan receptor alpha-deficient mice. Proc. Natl. Acad. Sci. USA 95(7), 3960–3965.

Stenn, K.S., Paus, R. 2001. Controls of hair follicle cycling. Physiol. Rev. 81(1), 449–494.

Stoye, J.P., Fenner, S., Greenoak, G.E., Moran, C., Coffin, J.M. 1988. Role of endogenous retroviruses as mutagens: The hairless mutation of mice. Cell 54(3), 383–391.

Sumner, F. 1924. Hairless mice. J. Hered. 15(12), 475–481.

Takeuchi, T., Yamazaki, Y., Katoh-Fukui, Y., Tsuchiya, R., Kondo, S., Motoyama, J., Higashinakagawa, T. 1995. Gene trap capture of a novel mouse gene, jumonji, required for neural tube formation. Genes Dev. 9(10), 1211–1222.

Takeuchi, T., Kojima, M., Nakajima, K., Kondo, S. 1999. jumonji gene is essential for the neurulation and cardiac development of mouse embryos with a C3H/He background. Mech. Dev. 86(1–2), 29–38.

Taylor, G., Lehrer, M.S., Jensen, P.J., Sun, T.T., Lavker, R.M. 2000. Involvement of follicular stem cells in forming not only the follicle but also the epidermis. Cell 102(4), 451–461.

Thompson, C.C. 1996. Thyroid hormone-responsive genes in developing cerebellum include a novel synaptotagmin and a hairless homolog. J. Neurosci. 16(24), 7832–7840.

Thompson, C.C., Bottcher, M.C. 1997. The product of a thyroid hormone-responsive gene interacts with thyroid hormone receptors. Proc. Natl. Acad. Sci. USA 94(16), 8527–8532.

Thompson, C.C., Potter, G.B. 2000. Thyroid hormone action in neural development. Cereb. Cortex 10(10), 939–945.

Thompson, C.C., Sisk, J.M., Beaudoin, G.M.J., III. 2006. Hairless and Wnt signaling: Allies in epithelial stem cell differentiation. Cell Cycle 5(17), 1913–1917.

Threadgill, D.W., Dlugosz, A.A., Hansen, L.A., Tennenbaum, T., Lichti, U., Yee, D., LaMantia, C., Mourton, T., Herrup, K., Harris, R.C., Barnard, J.A., Yuspa, S.H., et al. 1995. Targeted disruption of mouse EGF receptor: Effect of genetic background on mutant phenotype. Science 269(5221), 230–234.

Toyoda, M., Shirato, H., Nakajima, K., Kojima, M., Takahashi, M., Kubota, M., Suzuki-Migishima, R., Motegi, Y., Yokoyama, M., Takeuchi, T. 2003. jumonji downregulates cardiac cell proliferation by repressing cyclin D1 expression. Dev. Cell. 5(1), 85–97.

Tsukada, Y., Fang, J., Erdjument-Bromage, H., Warren, M.E., Borchers, C.H., Tempst, P., Zhang, Y. 2006. Histone demethylation by family of JmjC domain-containing proteins. Nature 439(7078), 811–816.

Tumbar, T., Guasch, G., Greco, V., Blanpain, C., Lowry, W.E., Rendl, M., Fuchs, E. 2004. Defining the epithelial stem cell niche in skin. Science 303(5656), 359–363.

van Genderen, C., Okamura, R.M., Farinas, I., Quo, R.G., Parslow, T.G., Bruhn, L., Grosschedl, R. 1994. Development of several organs that require inductive epithelial-mesenchymal interactions is impaired in LEF-1-deficient mice. Genes Dev. 8(22), 2691–2703.

Van Mater, D., Kolligs, F.T., Dlugosz, A.A., Fearon, E.R. 2003. Transient activation of beta-catenin signaling in cutaneous keratinocytes is sufficient to trigger the active growth phase of the hair cycle in mice. Genes Dev. 17(10), 1219–1224.

Wang, H., Huang, Z.Q., Xia, L., Feng, Q., Erdjument-Bromage, H., Strahl, B.D., Briggs, S.D., Allis, C.D., Wong, J., Tempst, P., Zhang, Y. 2001. Methylation of histone H4 at arginine 3 facilitating transcriptional activation by nuclear hormone receptor. Science 293(5531), 853–857.

Webb, P., Anderson, C.M., Valentine, C., Nguyen, P., Marimuthu, A., West, B.L., Baxter, J.D., Kushner, P.J. 2000. The nuclear receptor corepressor (N-CoR) contains three isoleucine motifs (I/LXXII) that serve as receptor interaction domains (IDs). Mol. Endocrinol. 14(12), 1976–1985.

Wen, Y.D., Perissi, V., Staszewski, L.M., Yang, W.M., Krones, A., Glass, C.K., Rosenfeld, M. G., Seto, E. 2000. The histone deacetylase-3 complex contains nuclear receptor corepressors. Proc. Natl. Acad. Sci. USA 97(13), 7202–7207.

Xu, H.E., Stanley, T.B., Montana, V.G., Lambert, M.H., Shearer, B.G., Cobb, J.E., McKee, D.D., Galardi, C.M., Plunket, K.D., Nolte, R.T., Parks, D.J., Moore, J.T., et al. 2002. Structural basis for antagonist-mediated recruitment of nuclear co-repressors by PPARalpha. Nature 415(6873), 813–817.

Yanagita, M., Oka, M., Watabe, T., Iguchi, H., Niida, A., Takahashi, S., Akiyama, T., Miyazono, K., Yanagisawa, M., Sakurai, T. 2004. USAG-1: A bone morphogenetic protein antagonist abundantly expressed in the kidney. Biochem. Biophys. Res. Commun. 316(2), 490–500.

Yoon, H.G., Chan, D.W., Huang, Z.Q., Li, J., Fondell, J.D., Qin, J., Wong, J. 2003. Purification and functional characterization of the human N-CoR complex: The roles of HDAC3, TBL1 and TBLR1. EMBO J. 22(6), 1336–1346.

Yoshizawa, T., Handa, Y., Uematsu, Y., Takeda, S., Sekine, K., Yoshihara, Y., Kawakami, T., Arioka, K., Sato, H., Uchiyama, Y., Masushige, S., Fukamizu, A., et al. 1997. Mice lacking the vitamin D receptor exhibit impaired bone formation, uterine hypoplasia and growth retardation after weaning. Nat. Genet. 16(4), 391–396.

Zarach, J.M., Beaudoin, G.M.J., III, Coulombe, P.A., Thompson, C.C. 2004. The co-repressor hairless has a role in epithelial cell differentiation in the skin. Development 131(17), 4189–4200.

Nuclear receptor transcriptional coactivators in development and metabolism

Janardan K. Reddy, Dongsheng Guo, Yuzhi Jia, Songtao Yu and M. Sambasiva Rao

Department of Pathology, Feinberg School of Medicine,
Northwestern University, Chicago, Illinois

Contents

Advances in Developmental Biology
Volume 16 ISSN 1574-3349
DOI: 10.1016/S1574-3349(06)16012-3

1. Introduction

The nuclear receptor superfamily in the human comprises 48 members that are ligand-regulated transcriptional factors and includes receptors for steroid and thyroid hormones, vitamin D_3, retinoic acids, peroxisome proliferators, and others (Mangelsdorf et al., 1995). Ligand-activated nuclear receptors regulate diverse biological processes, such as development, differentiation, neoplastic conversion, and several metabolic pathways, by controlling gene expression patterns in a cell- and gene-specific manner after binding to specific DNA sequences called hormone responsive elements (*cis*-elements) located in the promoter regions of target genes (Mangelsdorf et al., 1995; Qi et al., 2000; McKenna and O'Malley, 2002; Spiegelman and Heinrich, 2004). All nuclear receptors share a common structure with a highly conserved DNA-binding domain consisting of two zinc fingers, a C-terminal ligand (hormone)-binding domain, and two transcriptional activation function (AF) domains termed AF-1 located in the N-terminal domain and AF-2 in the ligand-binding domain (Danielian et al., 1992; Mangelsdorf et al., 1995). While AF-1 functions in a ligand-independent manner, the transactivation of AF-2 domain is generally ligand dependent (Danielian et al., 1992). Many of the unliganded nuclear receptors, including those for retinoid and thyroid hormone, are maintained mostly in the nucleus bound to DNA in a repressed state by nuclear receptor corepressors such as SMRT and N-CoR (Chen and Evans, 1995; Horlein et al., 1995). Upon ligand binding, the corepressors dissociate from the nuclear receptor to commence transcription of target genes. Certain other nuclear receptors, such as glucocorticoid receptor (GR), require ligand binding to facilitate their translocation from the cytoplasm to the nucleus where the liganded receptor binds hormone responsive elements in DNA. Binding of ligand to a receptor not only results in the dissociation of corepressor proteins but also initiates an orchestrated recruitment of several transcriptional coactivators that form multisubunit protein complexes to facilitate nucleosome remodeling and the nuclear receptor linking to the basal transcription machinery (Hermanson et al., 2002; McKenna and O'Malley, 2002; Smith and O'Malley, 2004; Roeder, 2005).

2. Nuclear receptor coactivators

During the past decade, nearly 100 nuclear receptor coregulators (coactivators, corepressors, and coactivator-associated proteins) have been identified in an attempt to understand the molecular mechanisms by which nuclear receptors achieve transcriptional activation of specific genes in a tissue/cell-specific manner (Hermanson et al., 2002; McKenna and O'Malley, 2002; Roeder, 2005; Yu et al., 2006). Coactivators and coactivator-associated proteins enhance the transcriptional activity of many nuclear receptors and function as positive regulators of transcription, whereas corepressors repress transcription by disrupting interactions between receptors and basal transcription apparatus (Halachmi et al., 1994; Hermanson et al., 2002). Some corepressors are also known to recruit enzymes that inhibit transcription. The overall transcriptional activity of a particular gene at any time during embryogenesis, development, and maintenance of differentiated function depends on the dynamic interplay between these coregulatory molecules (Hermanson et al., 2002; Smith and O'Malley, 2004).

The cloning of steroid receptor coactivator (SRC)-1 by O'Malley's group (Onate et al., 1995) and the demonstration that it interacts with estrogen and progesterone receptors to enhance their transcriptional activity paved the way for the identification and characterization of additional nuclear receptor coactivators. Coactivators, which interact directly with nuclear receptors, contain one or more highly conserved LXXLL (L, leucine; X, any amino acid) motif(s), termed NR-boxes that mediate recognition of, and binding to, AF-2 region in nuclear receptors (Heery et al., 1997; Torchia et al., 1997; Coulthard et al., 2003). The most well-characterized nuclear receptor coactivators to date are thought to form at least two major multisubunit protein complexes to provide linkage between DNA bound transcription factors and basal transcription machinery. This linkage facilitates the formation and function of preinitiation complex (Roeder, 2005). The first multiprotein coactivator complex, anchored by the universal coactivators, CBP and p300 (collectively called CBP/p300) (Goodman and Smolik, 2000), comprises several proteins including the three members of SRC-1/p160 family that possess histone acetyltransferase activity necessary to modify histones and chromatin structure to open up DNA for transcription (Taatjes and Tjian, 2004; Roeder, 2005). All three members of

SRC-1/p160 family (SRC-1/NCoA-1; SRC-2/TIF2/GRIP-1/; SRC-3/ACTR/ pCIP/RAC3/TRAM-1/AIB1/NCoA3) (Onate et al., 1995; Kamei et al., 1996; Voegel et al., 1996; Zhu et al., 1996; Anzick et al., 1997; Chen et al., 1997; Hong et al., 1997; Li et al., 1997; Torchia et al., 1997; Zhang et al., 2004) bind to the A/B domain of nuclear receptors in a manner similar to that of CBP/p300 (Kobayashi et al., 2000). The p160 coactivators possess histone acetyltransferase activity and complex with CBP/p300 on DNA bound-liganded nuclear receptors. The main function of the CBP/p300 anchored coactivator complex is the regulation of transcription through remodeling of chromatin by acetylating histones (Blanco et al., 1998). CBP/p300 proteins are potent histone acetyltransferases and these interact with nuclear receptors as well as a variety of other of transcription factors (Goodman and Smolik, 2000; Hermanson et al., 2002). Data suggest that the coactivator-associated enhancement of transcription by nuclear receptors involves the recruitment of several coactivator-associated proteins, such as coactivator-associated arginine methyltransferase (CARM1), a member of S-adenosyl-L-methionine-dependent protein arginine methyl transferase (PRMT) family (Chen et al., 1999). The PRMT family consists of five members (PRMT1 through PRMT5), of which PRMT4/CARM1 is known to enhance the function of the p160/ SCR-1 coactivator family (Chen et al., 1999). PRMT4/CARM1 catalyzes the methylation of arginine residues particularly in histone 3 (Chen et al., 1999). CARM1 binds to and methylates the KIX domain of CBP/p300 in addition to its initially described role in histone 3 methylation (Chen et al., 1999). The KIX domain-methylated CBP/p300 fails to activate CREB-dependent gene transcription but is able to enhance nuclear receptor-mediated gene transcription (Xu et al., 2001).

The second multisubunit coactivator complex, variously called TRAP/ DRIP/ARC/Mediator complex, consists of approximately 15–20 proteins and serves as an adaptor for facilitating the signal transduction between enhancer-bound factors and the core transcriptional machinery (Fondell et al., 1996; Naar et al., 1999; Rachez et al., 1999; Hermanson et al., 2002; Taatjes and Tjian, 2004; Roeder, 2005). The TRAP/DRIP/ARC complex has been shown to be necessary for several nuclear receptor-mediated transcription (Fondell et al., 1996; Naar et al., 1999; Rachez et al., 1999; Hermanson et al., 2002). This complex is anchored by PBP(*PPARBP*)/TRAP220/DRIP205/Med1 (Zhu et al., 1997; Yuan et al., 1998; Rachez et al., 1999) and as far as is known peroxisome proliferator-activated receptor-binding protein (PBP) and other Mediator complex proteins lack histone acetyltransferase activity (Roeder, 2005). Other coactivators, such as RIP-140 (L'Horset et al., 1996), PGC-1 (Puigserver et al., 1998), PRIP/ASC2/AIB3/RAP250/NRC/TRBP/NCoA6 (Lee et al., 1999; Caira et al., 2000; Ko et al., 2000; Mahajan and Samuels, 2000; Zhu et al., 2000a), and GAC63 (Chen et al., 2005), are not known to be part of the CBP/ p300 and Mediator complexes. Nevertheless, protein–protein interaction data suggest that these may function as linkers between the first and second multisubunit coactivator complexes to form one large complex albeit transiently (Ko et al., 2000; Zhu et al., 2001; Mahajan et al, 2002; Surapureddi et al., 2002; Misra et al., 2002a). PRIP and PRIP-binding protein PIMT appear to serve as linkers between two major coactivator complexes involved in the multistep model of transcription (Zhu et al., 2001; Hermanson et al., 2002; Misra et al., 2002b; Spiegelman and Heinrich, 2004; Roeder, 2005). Consistent with this view is that liganded peroxisome proliferator-activated receptor (PPAR) α identifies a cofactor protein complex designated PRIC complex that comprises some 25 polypeptides and includes some of the components of both CBP/p300 and Mediator complexes and some coactivator-binding proteins (Surapureddi et al., 2002). Also included in this PRIC complex are other proteins that have not previously been described in association with transcription complexes (CHD5, TOG, MORF) and a few novel polypeptides containing LXXLL signature motifs (Surapureddi et al., 2002). This approach of analyzing proteins bound to PPARα or to PPARα ligand, ciprofibrate, resulted in the identification of two novel coactivators designated PRIC285 (Surapureddi et al., 2002) and PRIC320 (Surapureddi et al., 2006). PRIC285 contains five LXXLL motifs and interacts with several nuclear receptors, including PPARα, PPARγ, RXR, ERα, and TRβ1, implying like many other coactivators that PRIC285 might function as a coactivator for a variety of nuclear receptors (Surapureddi et al., 2002). PRIC285 contains a Uvr helicase-like domain (Surapureddi et al., 2002) and shown to bind to the DNA-binding domain of PPARγ (Tomaru et al., 2006). PRIC320 also contains five LXXLL motifs and interacts with nuclear receptors PPARα, ERα, RXR, and constitutive androstane receptor (CAR) but less avidly with PPARγ (Surapureddi et al., 2006). PRIC320 has two isoforms (PRIC320-1 and PRIC320-2) and reveals the presence of two chromodomains, a helicase/ATPase domain with DEAD-like helicase superfamily motif (Shur and Benayahu, 2005; Surapureddi et al., 2006). Thus, the recognition of this chromatin remodeling function and nuclear receptor coactivator function in PRIC320 and PRIC285 is suggestive of the multiple roles played by these coactivators.

3. Transcriptional coactivator knockout mouse models

The discovery of nuclear receptor cofactors, in particular, a plethora of coactivators and coactivator-binding proteins, points to the complexity in the receptor-regulated transcription of target genes. While the changes in the nuclear receptor content are critical in transcriptional control, it is increasingly likely that the relative content of coactivator proteins also influences gene- and cell-specific transcriptional regulation (Jain et al., 1998; Cook et al., 2000; Hermanson et al., 2002; Lim et al., 2004; Spiegelman and Heinrich, 2004). Emerging evidence also suggests that coactivators may influence the function of many nuclear receptors and other transcription factors. To fully appreciate the *in vivo* biological functions of these coactivators, molecular genetic approaches are being increasingly utilized and the approach of targeted gene knockout mice is beginning to provide valuable clues about their role in development and in the regulation of nuclear receptor-specific gene transcription (Xu et al., 1998; Qi et al., 1999; Ito et al., 2000; Wang et al., 2000; Xu et al., 2000; Zhu et al., 2000b, 2003; Gehin et al., 2002; Jia et al., 2004, 2005a). The following chapter focuses on progress in understanding the role of nuclear receptor coactivators in development, differentiation, and certain metabolic functions based on gene-disrupted mouse models. Some of the coactivators and coactivator-associated proteins appear indispensable for embryonic development while others are considered dispensable. Accordingly, these proteins are designated *essential*, if the gene deletion leads to embryonic lethality and *redundant*, if the gene-deleted mice are viable and grossly normal (Table 1). Nevertheless, mice mutant for a redundant or dispensable coactivator protein appear to manifest specific functional and growth disturbances that may result from resistance to certain nuclear receptor signaling (Xu et al., 1998, 2000; Wang et al., 2000) but such a coactivator may appear redundant for the function of other nuclear receptors (Xu et al., 1998; Qi et al., 1999; Jia et al., 2004, 2005a).

Table 1
Coactivators and coactivator-associated proteins in embryonic development

Essential (embryonic lethality)	
CBP	Tanaka et al., 1997; Yao et al., 1998; Oike et al., 1999; Tanaka et al., 2000
p300	Yao et al., 1998
PBP(*PPARBP*)/TRAP220/MED1	Ito et al., 2000; Zhu et al., 2000b; Landles et al., 2003
PRIP/ASC2/RAP250/NRC/TRBP	Kuang et al., 2002; Antonson, 2003; Zhu et al., 2003; Mahajan et al., 2004
TRAP100	Ito et al., 2002
Redundant (viable phenotype)	
SRC-1 (NCoA1)	Xu et al., 1998; Qi et al., 1999
TIF2 (SRC-2)	Gehin et al., 2002
p/CIP (SRC-3)	Wang et al., 2000; Xu et al., 2000
CARM1 (PRMT4)	Yadav et al., 2003
PGC-1α	Lin et al., 2004; Leone et al., 2005

4. SRC-1/p160 family of coactivators in development and energy metabolism

4.1. SRC-1 null mice

During embryogenesis, SRC-1 expression at embryonic day 9.5 (E9.5) and E13.5 mouse embryos occurs mostly in the somites, cartilage, olfactory epithelium, and neuroepithelium (Jain et al., 1998). In the adult, SRC-1 is expressed to a variable extent in many tissues, including mammary epithelium of nonpregnant, pregnant, and lactating mice (Jain et al., 1998). SRC-1 gene knockout mice were generated to delineate the role of this coactivator in ER, PR, and PPARα signaling (Xu et al., 1998; Qi et al., 1999). SRC-1$^{-/-}$ mice are viable and fertile (Xu et al., 1998; Qi et al., 1999) but exhibit partial resistance to several hormones, including estrogen, progestin, thyroid hormone, and androgen (Xu et al., 1998; Weiss et al., 1999). These mice also show defects in ductal branching of breast epithelium, a finding consistent with SRC-1 expression in the breast (Jain et al., 1998; Xu et al., 1998). Although the intact SRC-1 null mice were fertile, they showed subdued responses to sex hormonal stimuli after orchiectomy or ovariectomy (Xu et al., 1998). SRC-1 null mice, when challenged with a PPARα ligand, such as Wy-14,643 or ciprofibrate, display characteristic pleiotropic responses, including hepatomegaly, peroxisome proliferation in hepatocytes, and PPARα target gene activation, similar to that noted in SRC$^{+/+}$ littermates (Qi et al., 1999). These observations suggest that SRC-1 is not essential for PPARα-mediated transcriptional activation *in vivo*, implying that coactivators are likely to function more efficiently in certain cell types and influence the transcriptional activity of a subset of nuclear receptors (Qi et al., 1999). Whether the SRC-1 redundancy in PPARα function is due to the compensatory presence of SRC-2 and SRC-3 remains to be examined (Qi et al., 1999).

SRC-1 deficient mice exhibit moderate motor dysfunction and delayed development of cerebellar Purkinje cells (Nishihara et al., 2003). These mice also exhibit altered hypothalamic–pituitary–adrenal axis function under conditions of stress and a state of glucocorticoid resistance (Winnay et al., 2006). In the adult mouse brain and during embryonic development, SRC-1 is highly expressed in olfactory epithelium, neuroepithelium, and cartilage suggesting that SRC-1 gene disruption affects nervous system development and leading to moderate motor dysfunction (Jain et al., 1998; Nishihara et al., 2003).

4.2. TIF2/SRC-2 null mice

Mice lacking TIF2/SRC-2 are also viable, but the fertility of both sexes is impaired (Gehin et al., 2002). Age-dependent testicular degeneration and defects in spermiogenesis (teratozoospermia) accounted for male hypofertility, whereas a placental hypoplasia appears to be responsible for female hypofertility in SRC-2 gene-deleted mice (Gehin et al., 2002). Defects in TIF2$^{-/-}$ testis include accumulation of lipid droplets in Sertoli cells and degeneration of germinal epithelium in seminiferous tubules (Gehin et al., 2002). In older TIF2 null males, testicular degeneration is manifested by vacuolation of the seminiferous epithelium, detachment of spermatocytes, and spermatids and formation of multinucleate giant cells (i.e., symplasts). Lipid metabolism in Sertoli cells is altered in SRC-2 null possibly due to changes in the function of many nuclear receptors (Gehin et al., 2002).

4.3. p/CIP/SRC-3 null mice

Ablation of p/CIP/AIB1/RAC3/ACTR/TRAM-1(SRC-3) gene revealed that it is required for normal somatic growth from E13.5 through maturity (Wang et al., 2000; Xu et al., 2000). The short stature or dwarfism seen in SRC-3$^{-/-}$ mice is attributed to altered regulation of insulin-like growth factor-1 (IGF-1) gene expression in specific tissues (Wang et al., 2000). SRC-3 gene deletion also leads to a delayed puberty, reduced female reproductive function, and somewhat blunted mammary development (Xu et al., 2000). Although SRC-1$^{-/-}$ and SRC-3$^{-/-}$ mutants were fertile, it appears that these two coactivators do not compensate for the loss of fertility in SRC-2$^{-/-}$ mutants and these three members of p160 family play distinct roles *in vivo* (Gehin et al., 2002).

SRC-3 overexpression on the other hand stimulates cell growth and modulates Akt signaling pathways in a steroid-independent manner, implying that SRC-3 also functions as a coactivator for other transcription factors (Zhou et al., 2003). Likewise, in SRC-3 null mice Akt signaling is downregulated in normally SRC-3 expressing tissues (Wang et al., 2000; Zhou et al., 2003). Consistent with these observations is the finding that SRC-3 is amplified or overexpressed in many cancers, including breast and prostate cancers (Anzick et al., 1997; Torres-Arzayus et al., 2004; Yan et al., 2006).

4.4. SRC-1/SRC-2 and SRC-1/SRC-3 double nulls

Mice deficient in both SRC-1 and TIF2/SRC-2 (Picard et al., 2002; Mark et al., 2004), or deficient in SRC-1 and p/CIP/SRC-3) (Wang et al., 2006) were generated to elucidate the physiological roles of these coactivators. A majority of the SRC-1$^{-/-}$/SRC-2$^{-/-}$ and SRC-1$^{-/-}$/SRC-3$^{-/-}$ double-knockout mice died neonatally generally before 3 weeks of age (Mark et al., 2004; Wang et al., 2006). At birth, the lungs of SRC-1$^{-/-}$/SRC-2$^{-/-}$ mutant mice were immature and failed to inflate properly resulting in cyanosis (Mark et al., 2004). Both SRC-1$^{-/-}$/SRC-2$^{-/-}$ and SRC-1$^{+/-}$/SRC-2$^{-/-}$ (i.e., inactivation of SRC-1 alleles in the TIF2 null genetic background) male mice exhibit complete sterility due to severe oligo- and teratozoospermia, and early and severe testis degeneration more profound than that found in mice with only SRC-2 gene deletion (Gehin et al., 2002; Mark et al., 2004). The testicular degenerative changes are reminiscent of changes detected in testes of elderly men, suggesting a possible role for p/160/SRC-1 family members in testicular senescence (Mark et al., 2004). Thus, it would appear that SRC-1 partially compensates for the loss of SRC-2 in SRC-2 null males and that these two coactivators appear to encompass partially redundant functions involving postnatal development and male reproduction (Gehin et al., 2002; Mark et al., 2004).

4.5. Energy balance in p160 mutant mice

Mice lacking both SRC-1 and p/CIP/SRC-3 (SRC-1$^{-/-}$/SRC-3$^{-/-}$ double nulls) are lean and resistant to high-fat diet induced obesity (Wang et al., 2006). These mice show a developmental arrest in interscapular brown fat due to a failure in the induction of selective PPARγ target genes involved in adipogenesis and mitochondrial uncoupling (Wang et al., 2006). Also of significance is that these double null mice consume more food but do not become obese, in large part, due to increased basal metabolic rate and induction of heightened levels of physical activity (Wang et al., 2006). It appears that significantly decreased SCD1 in white adipose tissue and liver leads to increased basal metabolic burning of energy and reduced lipogenesis (Wang et al., 2006).

It is of interest to note that SRC-1 and SRC-2 single-knockout mice yielded interesting information regarding energy storage and metabolism (Picard et al., 2002). While loss of SRC-1 enhances the propensity for the development of obesity, SRC-2 gene deletion results in increased adaptive thermogenesis and protects against obesity (Picard et al., 2002). It appears that these two coactivators play opposing roles in energy utilization (SRC-1 in energy burning and SRC-2 in energy conservation) and that the relative level of SRC-1/SRC-2 molecules can influence energy balance. SRC-1 null mice on high-fat diet develop obesity but not SRC-1/SRC-3 double nulls, suggesting that the members of this p160 family of coactivators exert different effects both singly and in combination. Since SRC-1/SRC-3 double nulls do not become obese despite the excessive consumption of high-energy high-fat diets, compounds capable of disrupting the function of both SRC-1 and p/CIP might have therapeutic value in combating obesity in those with excessive energy consumption (Wang et al., 2006). In summary, the combined loss of SRC-1 and SRC-3 appears to exert high energy burning effects in the organism, possibly by inducing physical overactivity, despite the finding that loss of SRC-1 alone enhances the propensity for obesity. SRC-2 appears to influence energy storage in that lack of TIF2 decreases PPARγ activity in white adipose tissue and reduces fat accumulation, whereas SRC-1 plays a role in energy burning (Picard et al., 2002).

In an effort to explain the molecular events responsible for differential energy regulation by SRC-1/p160 family members, a genomic approach using microarray analysis was used to identify the subsets of genes that are altered in the livers of SRC-1$^{-/-}$, SRC-2$^{-/-}$, and SRC-3$^{-/-}$ mice (Jeong et al., 2006). Liver exerts a critical role in the regulation of whole body energy homeostasis, including lipogenesis and lipid catabolism (Browning and Horton, 2004; Reddy and Rao, 2006). Liver gene expression profiles of SRC-1/p160 mouse mutants demonstrated that these coactivators exert variable changes in the expression levels of genes essential for energy homeostasis (Jeong et al., 2006). In the livers of SRC-2$^{-/-}$ mice, the expression of energy expenditure genes is increased, while there is a decrease in the expression of energy storage genes (Jeong et al., 2006). In contrast, in SRC-1 null mouse liver, there is an increase in the expression of genes involved in glycolysis and in glycogen and fatty acid synthesis. These studies suggest that SRC-2 influences the overexpression of fatty acid catabolizing enzymes, while downregulating fatty acid, cholesterol, and steroid biosynthetic pathways in liver (Jeong et al., 2006).

5. PBP/TRAP220/DRIP205/MED1 gene deletion results in embryonic lethality

PBP/TRAP220/ARC/DRIP205/MED1 is the pivotal subunit of the TRAP/ Mediator complex, an evolutionarily conserved multisubunit protein complex (Fondell et al., 1996; Zhu et al., 1997; Naar et al., 1999; Rachez et al., 1999; Hermanson et al., 2002; Roeder, 2005). PBP/TRAP220 binds to a variety of nuclear receptors, such as PPAR, TR, ER, VDR, RXR, RAR, FXR, via two conserved LXXLL motifs and this *PBP/PPARBP* gene is amplified or over-expressed in human breast cancers (Zhu et al., 1997; Yuan et al., 1998; Zhu et al., 1999; Torra et al., 2004; Udayakumar et al., 2006). PBP/MED1 also interacts with tumor suppressor p53 gene and five GATA family members, including GATA-2, and possibly other transcription factors (Crawford et al., 2002; Lottin-Divoux et al., 2005; Gordon et al., 2006). PBP/TRAP220 also interacts directly with p300 and PGC-1 (Roeder, 2005). These interactions imply that PBP might exert a more global function in homeostatic regulation. As pointed out earlier, PBP/TRAP220, in its capacity as an anchor for TRAP/ Mediator complex, facilitates the linkage between histone acetyltransferase containing CBP/p160 protein complex and the basal transcription machinery in regulating transcription by RNA polymerase II (Roeder, 2005).

Ablation of the *PBP/PPARBP* gene in mice is embryonic lethal around E11.5, further attesting to the essentiality of this coactivator in ontogenesis (Ito et al., 2000; Zhu et al., 2000b; Misra et al., 2002a; Landles et al., 2003). The embryonic lethality, in PBP gene-deleted mice, is attributed, in part, to defects in development of placental vasculature, similar to those encountered in PPARγ mutants (Barak et al., 1999; Hemberger and Zechner, 2004). Around midgestation (E10) in the mouse development, the labyrinth layer of the placenta develops to exhibit a complex capillary network to provide transport mechanism for nutrient needs of the growing fetus (Hemberger and Zechner, 2004). The PBP$^{-/-}$ placentas lack this complex placental capillary network (Zhu et al., 2000b; Crawford et al., 2002 [see Fig. 3A,B in Crawford et al., 2002]) suggesting that placental insufficiency may result in fetal death at E11.5 (Zhu et al., 2000b; Landles et al., 2003) in that tetraploid aggregation assays partially reversed embryonic development until E13.5 (Landles et al., 2003). PBP null embryos manifest defects in heart, eye, and megakaryocytes and die of heart failure because of noncompaction of the ventricular myocardium (Ito et al., 2000; Zhu et al., 2000b; Crawford et al., 2002; Landles et al., 2003). Absence of PBP results in the paucity of retinal pigment, defective lens formation, excessive systemic angiogenesis, and a deficiency in the number of megakaryocytes and an arrest in erythrocytic and myelomonocytic differentiation (Crawford et al., 2002; Urahama et al., 2005). Yolk sac hematopoietic progenitor cells derived from PBP/TRAP220 null embryos were resistant to 1,25-dihydroxyvitamin D3-stimulated monocyte/macrophage differentiation (Urahama et al., 2005). Some of these changes observed in PBP null embryos overlap with those occurring in mice deficient in members of GATA, a family of transcription factors (Molkentin, 2000). These gene knockout studies point to the absolute requirement of PBP/TRAP220 for the normal ontogeny of the cardiac hepatic and hematopoietic systems and in the continuation of embryogenesis (Crawford et al., 2002; Landles et al., 2003; Urahama et al., 2005).

5.1. PBP is needed for adipogenesis

Primary embryonic fibroblasts derived from PBP/TRAP220 null mutants manifested an impaired cell cycle regulation and decrease in thyroid hormone responsiveness, suggesting an abnormal TR function in the absence of PBP/TRAP220 (Ito et al., 2000). PBP/TRAP220 null fibroblasts exhibit refractoriness to PPARγ-stimulated adipogenesis and that exogenously expressed PBP restored these defects (Ge et al., 2002). Of note is that PBP null fibroblasts responded to MyoD-stimulated myogenesis implying that this coactivator functions differently with different transcription factors. It is also of interest to note that osteoclast precursors derived from Paget's disease of bone showed hyperresponsivity to 1,25-dihydroxyvitamin D3 which is attributed to increased levels of nuclear receptor coactivators, such as CBP/p300 and PBP/TRAP220, but not due to increased vitamin D receptor (Kurihara et al., 2004).

6. PBP/TRAP220/MED1 conditional gene deletion

6.1. PBP is required for PPARα function

The embryonic lethality of PBP/TRAP220 null mutants necessitated the creation of conditional null mice using the *Cre-loxP* strategy for elucidating the cell- and gene-specific roles of this coactivator (Jia et al., 2004). The role of PBP in PPARα signaling was evaluated using targeted deletion of this coactivator in mouse liver parenchymal cells (Jia et al., 2004). The livers in PBP$^{Liv-/-}$ mice are generally smaller with diminished liver cell size. In these livers, an occasional liver cell that escapes albumin-Cre–directed targeted deletion of PBP appears somewhat prominent in comparison to smaller PBP$^{-/-}$ liver cells (Fig. 1). The smaller size of PBP$^{-/-}$ hepatocytes may be due to the possible global effects of this coactivator in liver cell function (Jia et al., 2004, 2005a). PBP null hepatocytes fail to respond to peroxisome proliferators in that there is near abrogation of PPARα ligand-induced peroxisome proliferation and liver cell proliferation (Jia et al., 2004). Few hepatocytes, that escape targeted deletion of PBP gene, respond to the PPARα ligand-induced transcriptional activity and exhibit peroxisome proliferation (Fig. 1), providing striking contrast between PBP$^{-/-}$ and PBP$^{+/+}$ hepatocytes due to differential PPARα activation (Jia et al., 2004). While, PBP null livers fail to reveal hepatocellular proliferative activity in response to PPARα ligands, a rare PBP-positive hepatocyte in these livers exhibits DNA synthesis as evaluated by bromodeoxyuridine immunohistochemistry (Jia et al., 2004). Chromatin immunoprecipitation results showed that PBP deficiency leads to a reduced recruitment of cofactors, especially CBP, TRAP150, PRIP, and SRC-1, to the peroxisomal L-bifunctional protein gene promoter (Jia et al., 2004). In PBP null livers, the formation of a PIMT-associated coactivator subcomplex was also affected in that PRIP and CBP were not detected in PIMT-binding protein complex (Jia et al., 2004). These observations establish that PBP is an essential coactivator for *in vivo* PPARα target gene transcription in response to PPARα ligands such as ciprofibrate and Wy-14,643 (Jia et al., 2004). The responses of PBP$^{-/-}$ liver cells appear essentially similar to those exhibited by PPARα$^{-/-}$ hepatocytes (Lee et al., 1995). These observations clearly establish that neither PPARα alone nor PBP/TRAP220 alone is sufficient to elicit transcriptional activation of genes induced by PPARα ligands and that absence of PBP in hepatocytes *in vivo* mimics the absence of PPARα (Jia et al., 2004).

6.2. PBP is required for CAR-mediated acetaminophen hepatotoxicity

The nuclear receptor CAR mediates the response evoked by a class of xenobiotics known as phenobarbital (PB)-like inducers (Swales and Negishi, 2004). Targeted deletion of PBP in liver parenchymal cells

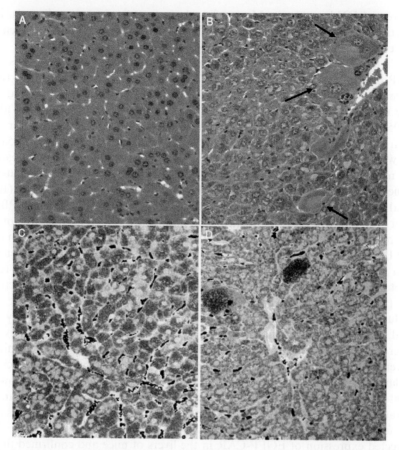

Fig. 1. Abrogation of PPARα ligand-induced peroxisome proliferation in PBP$^{Liv-/-}$ mouse liver. (A) and (B) represent H&E stained, while (C) and (D) represent diaminobenzidine-stained 0.5 μm thick sections from wild-type (A and C) and PBP$^{Liv-/-}$ (B and D) mice treated with Wy-14,643 (0.125% in diet) for 1 week. Note eosinophilic cytoplasmic staining of all liver cells in wild-type and an occasional liver cell that is PBP$^{+/+}$ (arrows) in PBP$^{Liv-/-}$ mouse liver. Diaminobenzidine cytochemistry reveals peroxisome proliferation (dark brown granules in the cytoplasms of all hepatocytes in wild-type (C) and two PBP$^{+/+}$cells (D, arrows) in PBP$^{Liv-/-}$ mouse treated with PPARα ligand. (See Color Insert.)

(PBP$^{Liv-/-}$) results in the abrogation of hypertrophic and hyperplastic influences in liver mediated by CAR ligands, phenobarbital and 1,4-bis-2 [-(3,5-dichloropyridyloxy)]benzene (TCPOBOP) and of acetaminophen-induced hepatotoxicity (Jia et al., 2005a). Short-term exposure (3 days) to either phenobarbital or TCPOBOP increases liver mass in wild-type PBP$^{+/+}$ but not in PBP$^{Liv-/-}$ mice. Both phenobarbital and TCPOBOP induce hepatocellular proliferation in wild-type livers but these CAR-ligands had no effect on hepatocytes lacking PBP (Jia et al., 2005a). As was noted with PPARα ligands, some liver cells that escaped PBP gene deletion showed increased DNA synthesis. Many of the known nuclear receptor ligands, in particular those activating PPARα, TR, and CAR exert direct mitogenic effect in liver (Columbano and Ledda-Columbano, 2003).

CAR interacts with two nuclear receptor-interacting LXXLL motifs in PBP in a ligand-dependent fashion. The interaction between PBP and the C-terminal AF-2 containing portion of CAR suggests that PBP is involved in the regulation of CAR transcriptional activity as a coactivator (Jia et al., 2005a). Mice with liver-specific PBP deletion revealed subdued gene expression changes in liver when exposed to CAR ligands (Jia et al., 2005a). In wild-type mice, treatment with either phenobarbital or TCPOBOP leads to marked induction of CYP2B10, CYP1A2, and other CAR regulated genes, but this response was markedly diminished in PBP null livers (Jia et al., 2005a). Although there was a reduction in CAR mRNA levels in PBP null livers, the lack of response is mostly due to lack of PBP in that adenoviral reconstitution of PBP in PBP$^{Liv-/-}$ mouse livers restored the induction of CYP1A2, CYP2B10, CYP3A11, and other phenobarbital-inducible genes (Jia et al., 2005a). CAR-inducible CYP gene expression is responsible for acetaminophen metabolism-derived hepatotoxic metabolites and liver necrosis (Swales and Negishi, 2004; Qatanani and Moore, 2005). The lack of induction of CYP enzymes accounts for the zoxazolamine-induced paralysis and death in PBP$^{Liv-/-}$ mice (Jia et al., 2005a). Marked centrilobular liver necrosis occurs in phenobarbital or TCPOBOP pretreated wild-type mice given acetaminophen but PBP deficiency abrogated acetaminophen toxicity in large part due to the failure to induce CAR target genes (Fig. 2). It is known that CAR-ligands, such as phenobarbital and TCPOBOP, facilitate the translocation of CAR into the hepatocyte nucleus and this translocation fails to occur in PBP-deficient hepatocytes (Jia et al., 2005a). We used adenoviral-directed expression of EGFP-CAR in the livers of mice and confirmed that even if there is overexpression of CAR in liver cells it does not translocate to the nucleus in the absence of PBP but readily occurs in livers of wild type in response to phenobarbital (Figs. 3 and 4). These observations clearly show that coactivator PBP is not only an important component of CAR-regulated gene transcription but also necessary for the translocation of this receptor from cytoplasm to the nucleus (Jia et al., 2005a; Guo et al., unpublished data).

6.3. Defective liver regeneration in PBP null livers

Studies with PBP$^{Liv-/-}$ mice revealed that PBP is important coactivator in the maintenance of hepatocyte size and hepatocellular regeneration possibly by influencing many transcriptional pathways (Jia et al., 2004, 2005a; Matsumoto et al., 2006). As pointed out earlier, the absence of PBP in liver cells abrogates the liver cell proliferative function of xenobiotic sensing nuclear receptors, PPARα and CAR, but SRC-1 coactivator deficiency has

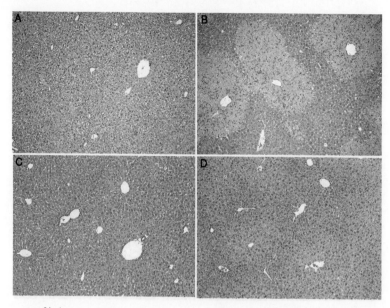

Fig. 2. PBP$^{Liv-/-}$ mice exhibit resistance to acetaminophen-induced liver necrosis. Wild-type (PBP$^{+/+}$) (A and B) and PBP$^{Liv-/-}$ (C and D) mice treated for 3 days with saline (A and C) or with phenobarbital (B and D). Centrizonal liver necrosis occurs in phenobarbital-treated PBP$^{+/+}$ mice given a single dose of acetaminophen (B) but not in PBP$^{Liv-/-}$ mouse (D). (See Color Insert.)

no influence on PPARα ligand-induced liver cell hyperplasia (Qi et al., 1999; Jia et al., 2004, 2005a). The regeneration of liver was also markedly impaired in the PBP$^{Liv-/-}$ mouse following partial hepatectomy (Matsumoto et al., 2006). Expression of several cell cycle regulatory genes was also reduced in PBP$^{Liv-/-}$ mouse liver after partial hepatectomy as compared to partial hepatectomized to wild-type controls (Matsumoto et al., 2006). Increase in IGFBP-1 mRNA in liver was modest at 4 h after partial hepatectomy in PBP$^{Liv-/-}$ mice. It would appear that PBP/MED1 plays a significant role in the maintenance of liver phenotype in the adult and loss of this coactivator leads to hypoproliferative response of hepatocytes.

6.4. PBP is needed for breast development

PBP/TRAP220 interacts with estrogen receptor α (ERα) in the absence of estrogen but this interaction becomes exaggerated in the presence of estrogen (Zhu et al., 1999). Consistent with this observation is that transfection of PBP in CV-1 cells results in the enhancement of estrogen-dependent transcription, indicating that PBP/TRAP220 serves as a coactivator of ER signaling (Zhu et al., 1999; Llopis et al., 2000; Kang et al., 2002). PBP/PPARBP gene was found to be amplified and overexpressed in some breast cancers, suggesting that PBP coactivator might play a role in breast cancer development

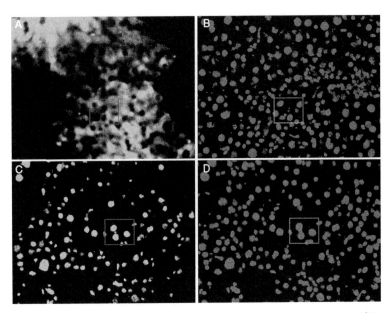

Fig. 3. Phenobarbital-dependent nuclear translocation of CAR in wild-type (PBP$^{+/+}$) mice. Mice were injected with adeno-EGFP-CAR and were analyzed for phenobarbital-mediated nuclear translocation of CAR. Green fluorescence is predominantly cytoplasmic in mice not given phenobarbital untreated (A) and mostly nuclear in mice treated with phenobarbital (C). Nuclei are visualized by DAPI (B and D). [Panels A and B reproduced from Guo et al. (2006) with permission from Elsevier.] (See Color Insert.)

and/or progression (Zhu et al., 1999). PBP is expressed in normal mammary ductal epithelium (Jain et al., 1998), suggesting that PBP might play a role in the mammary gland development. Mice conditionally deficient in PBP in mammary epithelium revealed retarded ductal elongation and branching, impaired lobuloalveolar development, incapacity to produce milk to foster pups even with relatively intact expression of milk proteins (Jia et al., 2005b). The PBP null mutation in breast epithelium imparts an attenuated ductal elongation response to estrogen and reduced mammary lobuloalveolar pro-liferative response when stimulated by combined estrogen and progesterone treatment (Jia et al., 2005b). PBP-deficient mammary epithelial cells fail to form mammospheres, presumably derived from mammary progenitor/stem cells. These studies demonstrate the importance of this coactivator in breast development. Further studies are necessary to ascertain the role of PBP in breast cancer development using this PBP conditional null mouse lineage.

6.5. *PBP deficiency reduces dexamethasone-induced hepatic steatosis*

Liver-specific disruption of *PBP/PPARBP* gene results in near abrogation of PPARα and CAR nuclear receptor function (Jia et al., 2004, 2005a).

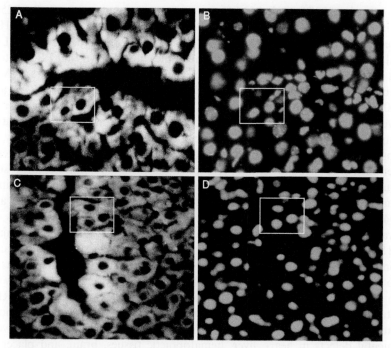

Fig. 4. Phenobarbital-mediated translocation of CAR to liver nuclei does not occur in PBP$^{Liv-/-}$ mouse. PBP$^{Liv-/-}$ were injected with adeno-EGFP-CAR and analyzed for expression. In these PBP liver null mice, CAR expression is confined almost exclusively to the hepatocyte cytoplasm (A and B) and did not translocate to the nucleus in response to phenobarbital (C and D). (See Color Insert.)

Since CAR is regulated by GR, the possibility exists that the lower levels of CAR expression in PBP null livers might be due to GR target gene dysfunction in the absence of PBP (Jia et al., 2005a). Dexamethasone induces fatty liver through GR-regulated pathway and that targeted deletion of PBP in liver cells (PBP$^{Liv-/-}$) results in the abrogation of dexamethasone-induced hepatic steatosis (Jia et al., 2006). These observations suggest that PBP is required for nuclear receptor GR and CAR function and that the absence of PBP prevents GR ligand-induced effects in liver, including steatosis.

7. PRIP/ASC2/RAP250/NRC/TRBP gene deletion is embryonic lethal

Transcription coactivator, peroxisome proliferator-activated receptor interacting protein (PRIP), is also known as nuclear hormone coregulator (NRC) (Mahajan and Samuels, 2000), activating signal cointegrator-2 (ASC-2, AIB3) (Lee et al., 1999), thyroid hormone receptor-binding protein (TRBP)

(Ko et al., 2000), and 250-kDa receptor-associated protein (RAP250) (Caira et al., 2000). Like *PBP/PPARBP* gene (Zhu et al., 1999), PRIP/AIB3 gene is also amplified and overexpressed in breast and other types of cancer (Lee et al., 1999). Protein–protein interaction studies have shown that PRIP interacts with other transcriptional cofactors, including CBP/p300, SRC-1, PIMT (Zhu et al., 2001), TRAP/DRIP130 (Lee et al., 1999; Caira et al., 2000; Ko et al., 2000; Misra et al., 2002b). PRIP and PRIP-interacting protein PIMT appear to serve as linkers between CBP/p300-anchored and PBP-anchored coactivator complexes and thus enhance the transcriptional activation of a wide variety of transcription factors. Studies dealing with PRIP/RAP250/NRC/AIB3 gene deletion in mice have shown that PRIP null mutation is embryonic lethal mostly during E11.5 and E12.5 (postcoitum), a relatively narrow window of time (Kuang et al., 2002; Antonson et al., 2003; Zhu et al., 2003; Mahajan et al., 2004). PRIP deficiency results in the formation of fewer blood vessels in extraembryonic membrane covering the embryo (Zhu et al., 2003; Mahajan and Samuels, 2005). In contrast, wild-type yolk sac presents well-developed and prominent vessels. In addition, superficial vasculature in PRIP null mutants is also not well developed. The liver of wild-type embryos is easily visualized by its rich vasculature through the skin, whereas the liver development in PRIP mutant embryos appears markedly retarded (Fig. 5). PRIP null embryos show clear evidence of growth retardation (Fig. 5) resulting from widespread developmental defects in heart, liver, hematopoietic tissues, brain, and placenta (Kuang et al., 2002; Antonson et al., 2003; Zhu et al., 2003; Mahajan et al.,

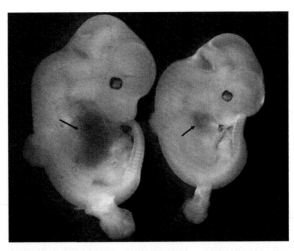

Fig. 5. Transcription coactivator PRIP null mutation in the mouse. Growth retardation and embryonic lethality at E12.5. PRIP[−/−] embryo (right) is smaller than the PRIP[+/+] (left) littermate. Liver in PRIP[−/−] mice is smaller in size as compared to PRIP[+/+] liver (arrow). (See Color Insert.)

2004). Of interest is that heterozygous PRIP/NRC$^{+/-}$ mice exhibit a spontaneous wound healing defect, indicating that this coactivator plays an important role in maintaining the integrity and healing of wounds (Mahajan et al., 2004). Placentas from the PRIP$^{-/-}$ embryos show dramatically reduced spongiotrophoblast layer with islands of spongiotrophoblast-like cells dissociated from trophoblast giant cell layer and migrating into the labyrinth zone (Zhu et al., 2003). The placental developmental defects resulting in dramatic reduction in vasculature observed in PRIP deficiency are reminiscent of some of the placental defects noted in PBP and PPARγ null mutants (Barak et al., 1999; Crawford et al., 2002). The failure of placentation may be one of the pivotal factors in embryonic lethality due to insufficient delivery of oxygen and nutrients to the developing fetus from this defective maternal fetal vasculature.

7.1. PRIP is required for PPARγ-mediated adipogenesis

Mouse embryonic fibroblasts (MEFs), derived from E12.5 embryos with PRIP/NRC/ASC-2 null mutation, revealed a high rate of apoptosis as compared to PRIP$^{+/+}$ wild-type MEFs suggesting that PRIP is involved in growth by preventing apoptosis (Mahajan et al., 2004). Coactivators PBP and PRIP were initially identified as PPARγ coactivators (Zhu et al., 1997, 2000a) implying that these two coactivators might influence PPARγ-mediated adipogenesis. PBP deficiency has been shown to affect adipogenesis (Ge et al., 2002), suggesting that PRIP might also play a crucial role in adipogenic conversion. PRIP$^{-/-}$ MEFs are also refractory to PPARγ-stimulated adipogenesis and they also fail to express adipogenic marker aP2, a PPARγ-responsive gene (Qi et al., 2003). The ChIP assays showed reduced association of PIMT (PRIP-binding protein) and PBP with aP2 gene promoter, suggesting that PRIP is required for maintaining the integrity of cofactor subunit complex(es) that are involved in transcription (Qi et al., 2003) similar to that of PBP/TRAP220 in PPARγ-dependent adipogenesis (Ge et al., 2002).

7.2. PRIP conditional liver null mice respond normally to PPARα and CAR ligands

Disruption of coactivator PBP gene in liver has resulted in a dramatic abrogation of PPARα ligand-induced transcriptional activation in hepatocytes (Jia et al., 2004). In contrast, liver-specific conditional deficiency of PRIP had no effect on the inducibility of PPARα target genes and peroxisome proliferation in mouse liver parenchymal cells (Guo et al., 2006; Sarkar et al., 2006). PRIP null livers revealed hepatomagaly, hepatocellular proliferation, and hepatic peroxisome proliferation, in addition to upregulation of PPARα target genes when treated with PPAR ligands, ciprofibrate for Wy-14,643 (Sarkar et al., 2006). It should be noted that absence of SRC-1 also had no effect on PPARα activation in liver indicating that both SRC-1 and PRIP are redundant for PPARα action, while PBP is required (Qi et al., 1999; Jia et al., 2004; Sarkar et al., 2006). PRIP/ASC-2 dominant negative transgenic mice were shown to be resistant to acetaminophen-induced hepatotoxicity (Choi et al., 2005). The dominant negative fragment of PRIP/ASC-2 covered the first LXXLL motif of PRIP and apparently suppressed no other nuclear receptor function (Choi et al., 2005). In contrast, the liver-specific disruption of PRIP gene had no significant effect on nuclear receptor CAR-regulated transcriptional upregulation of cytochrome P450 genes, CYP1A2, CYP2B10, CYP3A11, and CYP2E1 (Sarkar et al., 2006). CAR ligands phenobarbital and TCPOBOP induced the expression of these genes in PRIP conditional livers. PRIP deficiency in liver did not protect from acetaminophen-induced hepatic necrosis (Fig. 6), unlike that seen in PBP null liver further attesting to the fact that coactivators appear to exert different

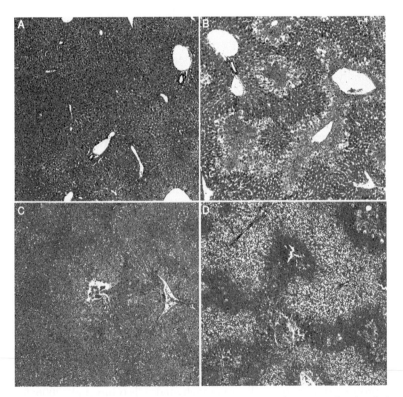

Fig. 6. Liver-specific coactivator PRIP deficiency does not alter acetaminophen-induced hepatotoxicity. PRIP$^{+/+}$ and PRIP$^{Liv-/-}$ mice were given saline (A and C) or phenobarbital (B and D) for 3 days. Mice pretreated with phenobarbital received a single dose of acetaminophen (B and D). Centrizonal liver necrosis occurred in both wild-type and PRIP liver null mice. (See Color Insert.)

effects on a given nuclear receptor transcription. Although PRIP appears to function as a coactivator for CAR, it appears that PRIP conditional null mutation in liver does not abrogate CAR function and acetaminophen-induced centrilobular hepatic necrosis, implying the redundancy of this coactivator.

7.3. PRIP conditional null mutation in mammary glands causes defective mammopoiesis

The role of PRIP in mammary gland development has been investigated using conditional null mutation (Qi et al., 2004). In PRIP-deficient mammary glands, the elongation of ducts during puberty was not affected, whereas the number of ductal branches was markedly decreased (Qi et al., 2004). The failure of ductal branching persisted long after puberty, resulting in decreased alveolar density in pregnancy (Qi et al., 2004). As a consequence, the lactating PRIP deficient mammary glands revealed markedly reduced density of lobuloalveoli and increased numbers of fat cells, while the number of fat cells in lactating wild-type breast is generally minimal. The reduction in lactiferous lobuloalveoli in PRIP deficient breast accounted for reduced milk production and inability to nurse all the pups (Qi et al., 2004). The ductal branching in response to estrogen was also reduced in PRIP deficient mammary epithelium suggesting refractoriness to hormone stimuli in the absence of coactivator PRIP, which is somewhat analogous to that noted in SRC-1 deficient mice (Xu et al., 1998).

8. Coactivator PGC-1α in adaptive thermogenesis and energy metabolism

PPARγ-coactivator-1 (PGC-1) family of inducible coactivators consists of three members: PGC-1α, a cold-inducible coactivator cloned from the brown adipose tissue, which coactivates PPARγ and other nuclear receptors that regulate genes involved in energy metabolism and thermogenesis (Puigserver et al., 1998); PGC-1β, which coactivates the sterol responsive element-binding protein (SREBP) transcription factor family to augment lipogenesis in liver (Lin et al., 2005), and a recently identified third member PGC-1–related coactivator (PRC), whose function is largely unknown (for reviews see Spiegelman and Heinrich, 2004; Finck and Kelly, 2006). PGC-1α and PGC-1β are preferentially expressed in tissues with high oxidative metabolism, such as brown adipose tissue, heart, and slow-twitch skeletal muscle (Spiegelman and Heinrich, 2004; Finck and Kelly, 2006). PGC-1α and PGC-1β stimulate mitochondrial biogenesis and modulate biological programs associated with increased oxidative metabolism in muscle and liver (Lin et al., 2004; Finck and Kelly, 2006). PGC-1α is cold inducible in brown adipose tissue and fasting inducible in liver (Spiegelman and Heinrich, 2004). In the liver, the induction of PGC-1α under conditions of fasting stimulates gluconeogenesis, increases mitochondrial electron transport, and enhances fatty acid β-oxidative capacity and ketogenesis (Spiegelman and Heinrich, 2004). Overexpression of PGC-1α and PGC-1β in cells *in vitro* activates genes that increase energy generation. PGC-1β appears to influence SREBP-mediated lipogenic gene expression but unlike SREBP, PGC-1β has been shown to reduce hepatic steatosis by increasing lipoprotein transport due to activation of liver X receptor (LXRα) which plays a role in hepatic lipid transport (Lin et al., 2005).

In order to elucidate the consequences of loss of function *in vivo*, PGC-1α knockout mice have been generated by Lin et al. (2004) and Leone et al. (2005) and these null mice exhibited some significant dissimilarities possibly due to differences in strain background. PGC-1α null mice generated by Lin et al. (2004) exhibited partial lethality, whereas the knockout mice generated by Leone et al. (2005) revealed a viable phenotype with expected genotype ratios. The PGC-1α null mice are cold sensitive due to defective brown fat function (Lin et al., 2004). The PGC-1α knockout mice have reduced mitochondrial function in the heart in that they showed diminished expression of genes involved in mitochondrial fatty acid oxidation, and oxidative phosphorylation (Lin et al., 2004), but these effects were not prominent in PGC-1α knockout mice generated by a second group (Leone et al., 2005). PGC-1α gene deletion had no appreciable effect on PPARα-regulated fatty acid β-oxidation system genes in liver (Lin et al., 2004; Leone et al., 2005). PGC-1α knockout mice have a lean phenotype and show resistance to diet-induced obesity largely due to their hyperactivity (Lin et al., 2004). This hyperactivity is attributed to spongiform lesions in the striatal region of the brain that controls movement (Lin et al., 2004; Leone et al., 2005).

8.1. Liver-specific PGC-1α null mice

Liver-specific PGC-1α null mice have been generated to investigate whether PGC-1α is a key regulator of the 5-aminolevulinate synthase (ALAS-1) to avoid the confounding systemic effects of whole-body PGC-1α knockout of this coactivator (Handschin et al., 2005). PGC-1α induced in liver, under conditions of fasting, activates ALAS-1 promoter by coactivating transcription factors NRF-1 and FOXO1 (Handschin et al., 2005). Liver-specific PGC-1 knockout did not affect the expression levels of PGC-1β in liver, heart, skeletal muscle, and brown fat (Handschin et al., 2005). In liver-specific PGC-1α–deleted mice, the basal levels of ALAS-1 were significantly lower as compared to total body knockout (Handschin et al., 2005). The induction of ALAS-1 and genes involved in gluconeogenesis after fasting was markedly blunted in the liver-specific knockout mice implying a PGC-1α–mediated link between nutritionally regulated heme biosynthesis and acute porphyria (Handschin et al., 2005). These observations were extended to show that elevated expression of PGC-1α exacerbates the effects of porphyrinogenic chemicals, whereas liver-specific PGC-1 deficiency protected from chemical porphyria (Handschin et al., 2005). In this regard, it should be noted that agents capable of increasing or decreasing PGC-1α levels in liver will influence heme biosynthesis and modulate porphyria. The PGC-1α-floxed mice should prove to be highly useful in generating mice with conditional deletion of this coactivator in heart and other tissues for the analysis of tissue functions of this coactivator.

9. CBP/p300 gene deletion is embryonic lethal

The general transcriptional coactivators CREB-binding protein (CBP) and p300 share 61% overall sequence identity and are evolutionarily conserved (for review see Goodman and Smolik, 2000). These are ubiquitously expressed and bind to the activation domains of transcription factors via well-defined domains (Goodman and Smolik, 2000). Both CBP and p300 possess histone acetyltrans-ferase activity, recruit other coactivators from multisubunit protein complexes, and activate the transcriptional function of a wide variety of transcription factors (Goodman and Smolik, 2000; Kasper et al., 2006). Accordingly, the absence of CBP/p300 is expected to affect the functions of a majority of transcription factors. Consistent with this view is that both these factors are important for development and lack of expression of these general coactivators results in embryonic lethality (Tanaka et al., 1997; Yao et al, 1998; Kung et al., 2000). Mice nullizygous for CBP or p300 die between E9.5 and E11.5 (Yao et al., 1998; Oike et al., 1999; Kung et al., 2000). Mice mutant for p300 exhibit defects in neurulation, cell proliferation, and heart development (Yao et al., 1998). Likewise, mice null for CBP reveal severe open neural tube defects similar to those observed in p300$^{-/-}$ embryos (Yao et al., 1998; Tanaka et al., 2000). The functional roles of CBP and p300 in hematopoietic develop-ment are of interest in view of the interaction of these coactivators with large numbers of transcription factors that participate in T and B cell development (Rebel et al., 2002; Kasper et al., 2006). Chromosomal translocations involving CBP or p300 have been known to provide gain-function support in some types of human leukemia (Blobel, 2002). Furthermore, it is important to note that p300 heterozygotes also manifest considerable embryonic mortality (Yao et al., 1998) and that embryonic lethality was predominant in CBP/p300 double heterozygotes (Yao et al., 1998).

In humans, monoallelic mutation of the CBP locus gives rise to Rubenstein–Taybi syndrome, a disease characterized by craniofacial, skeletal, and cardiac defects (Petrij et al., 1995; Giles et al., 1998). Individuals with this syndrome display broad thumbs and broad big toes and are mentally retarded. Mice haploin sufficient for CBP exhibit developmental defects, including skeletal abnormalities (Tanaka et al., 1997). The relative roles of CBP and Gli-3 in these skeletal developmental defects are not clear.

9.1. CBP/p300 conditional nulls

Mice with conditional knockout alleles for CBP and p300 have been developed (Kasper et al., 2006). These mice were used for delineating the role of these coactivators in thymocyte development and their roles in T cell and macrophage gene expression (Kasper et al., 2006). Loss of either p300 or CBP resulted in a decrease in $CD4^+CD8^+$ double-positive thymocytes and an increase in the percentage of $CD8^+$ single-positive thymocytes. Also of interest is that T cells devoid of both CBP and p300 failed to develop normally. Detailed analysis of these conditional null mice revealed that CBP and p300 exhibit considerable redundancy in T cell and macrophage functions. The generation of $CBP/^{flox}$ and $p300/^{flox}$ mice will allow cell/tissue-specific deletions of these genes using the Cre-recombination approach for detailed investigations to delineate the functions of these coactivators singly and in combination in different cell types.

10. Coactivator-associated proteins

While there is considerable interest in elucidating the roles of various coactivators, increasing attention is also devoted to the identification of proteins that interact with coactivators but are not known to interact directly with nuclear receptors (Chen et al., 1999; Zhu et al., 2001; Mahajan et al., 2002). A protein designated as coactivator-associated arginine methyltransferase (CARM1/PRMT4), which associates with members of p160 family, has been known to possess methyltransferase activity and it can methylate histones (Chen et al., 1999; Koh et al., 2001). To assess the importance of CARM1 in transcriptional regulation, mice with CARM1 gene knockout were generated (Yadav et al., 2003). While $CARM1^{+/-}$ mice were normal and fertile, all $CARM1^{-/-}$ mice died before weaning (Yadav et al., 2003). In general $CARM1^{-/-}$ embryos were smaller than the $CARM1^{+/+}$ and $CARM1^{+/-}$ littermates and their lungs failed to inflate adequately and displayed smaller air spaces in the lung. $CARM1^{-/-}$ MEFs from E12.5 embryos revealed reduced estrogen responsive gene expression emphasizing the criticality of this coactivator interacting protein in hormone action (Yadav et al., 2003).

The multisubunit TRAP/MED coactivator complex contains several$^{-/-}$ peptides, including PBP/TRAP220 and TRAP100 among others (Fondell et al., 1996; Ito et al., 2002). Deletion of TRAP100 gene leads to early embryonic lethality in the mouse and reveal attenuation of the function of many transcriptional activators (Ito et al., 2002). It is also pertinent to note that TRAP100 deletion results in an incomplete TRAP complex that lacks TRAP95 and TRAP150β, implying that TRAP100 might anchor a subcomplex in the larger TRAP/MED complex (Ito et al., 2002).

Two other coactivator interacting proteins, namely PIMT and NIF-1, have been identified that interact with PRIP/NRC/TRBP (Zhu et al., 2001; Mahajan et al., 2002). PIMT is evolutionarily conserved and has a conserved methyltransferase domain that appears to hypermethylate the cap structure of yeast snRNAs and snoRNAs (Mouaikel et al., 2002). In the *Drosophila*, PIMT homologue plays an important role in development (Komonyi et al., 2005). Since PIMT interacts with PRIP, CBP, PBP, and possibly other coactivators (Misra et al., 2002a,b), it is likely that PIMT may prove to be an essential coactivator-associated protein in transcription.

11. Perspective

It is becoming increasingly clear that transcriptional regulation involves the participation not only of DNA-binding transcription factors but also a plethora of proteins that are known variously as cofactors, coregulators, and coactivators. A majority of cofactors identified to date have been shown to interact directly with transcription factors to enhance the transcriptional activity and hence are known as coactivators. Others, in particular, the proteins that bind coactivators may exert a bridging role in assembling various subcomplexes that possibly influence gene- and cell-specific transcription. Gene knockout mouse models are beginning to provide valuable insights into the complex role of these proteins in development, differentiation, and metabolic functions. Because of embryonic lethality that results from the gene deletion, some of the coactivators are considered essential and to investigate the functional roles of such molecules, it becomes necessary to generate mice with floxed alleles for conditional targeting.

Acknowledgments

This work was supported by NIH Grants R01 CA104578, and GM23750, and Joseph L. Mayberry, Sr., Endowment Fund, Northwestern University.

References

Antonson, P., Schuster, G.U., Wang, L., Rozell, B., Holter, E., Flodby, P., Treuter, E., Holmgren, L., Gustafsson, J.-A. 2003. Inactivation of the nuclear receptor coactivator RAP250 in mice results in placental vascular dysfunction. Mol. Cell. Biol. 23, 1260–1268.

Anzick, S.L., Kononen, J., Walker, R.L., Azorsa, D.O., Tanner, M.M., Guan, X.Y., Sauter, G., Kallioniemi, O.P., Trent, J.M., Meltzer, P.S. 1997. AIB1, a steroid receptor coactivator amplified in breast and ovarian cancer. Science 277, 965–968.

Barak, Y., Nelson, M.C., Ong, E.S, Jones, Y.Z., Ruiz-Lozano, P., Chien, K.R., Koder, A., Evans, R.M. 1999. PPARγ is required for placental, cardiac, and adipose tissue development. Mol. Cell 4, 585–595.

Blanco, J.C., Minucci, S., Lu, J., Yang, X.-J., Walker, K., Evans, R.M., Chen, H., Nakatani, Y., Ozato, K. 1998. The histone acetylase PCAF is a nuclear receptor coactivator. Dev. Biol. 12, 1638–1651.

Blobel, G.A. 2002. CBP and p300: Versatile coregulators with important roles in hematopoietic gene expression. J. Leukoc. Biol. 71, 545–556.

Browning, J.D., Horton, J.D. 2004. Molecular mediators of hepatic steatosis and liver injury. J. Clin. Invest. 114, 147–152.

Caira, F., Antonson, P., Pelto-Huikko, M., Treuter, E., Gustafsson, J.A. 2000. Cloning and characterization RAP250, a novel nuclear receptor coactivator. J. Biol. Chem. 275, 5308–5317.

Chen, J.D., Evans, R.M. 1995. A transcriptional co-repressor that interacts with nuclear hormone receptors. Nature 377, 454–457.

Chen, H., Lin, R.J., Schlitz, R.L., Chakrabarti, D., Nash, A., Nagy, L., Privalsky, M.L., Nakatani, Y., Evans, R.M. 1997. Nuclear receptor coactivator ACTR is a novel histone acetyltransferase and forms a multimeric activation complex with p/CAF and CBP/p300. Cell 90, 569–580.

Chen, D., Ma, H., Hong, H., Koh, S.S., Huang, S.-M., Schurter, B.T., Aswad, D.W., Stallcup, M. R. 1999. Regulation of transcription by a protein methyltransferase. Science 284, 2174–2177.

Chen, Y.-H., Kim, J.H., Stallcup, M.R. 2005. GAC63, a GRIP1-dependent nuclear receptor coactivator. Mol. Cell. Biol. 25, 5965–5972.

Choi, E., Lee, S., Yeom, S.-Y., Kim, G.H., Lee, J.W., Kim, S.-W. 2005. Characterization of activating signal cointegrator-2 as a novel transcriptional coactivator of the xenobiotic nuclear receptor constitutive androstane receptor. Mol. Endocrinol. 19, 1711–1719.

Columbano, A., Ledda-Columbano, G.M. 2003. Mitogenesis by ligands of nuclear receptors: An attractive model for the study of the molecular mechanisms implicated in liver growth. Cell Death Differ. 10, S19–S21.

Cook, W.S., Yeldandi, A.V., Rao, M.S., Hashimoto, T., Reddy, J.K. 2000. Less extrahepatic induction of fatty acid β-oxidation enzymes by PPARα. Biochem. Biophys. Res. Commun. 278, 250–257.

Coulthard, V.H., Matsuda, S., Heery, D.M. 2003. An extended LXXLL motif sequence determines the nuclear receptor binding specificity of TRAP220. J. Biol. Chem. 278, 10942–10951.

Crawford, S.E., Qi, C., Misra, P., Stellmach, V., Rao, M.S., Engel, J.D., Zhu, Y., Reddy, J.K. 2002. Defects of the heart, eye, and megakaryocytes, in peroxisome proliferator activator receptor-binding protein (PBP) null embryos implicate GATA family of transcription factors. J. Biol. Chem. 277, 3585–3592.

Danielian, P.S., White, R., Lees, J.A., Parker, M.G. 1992. Identification of a conserved region required for hormone-dependent transcriptional activation by steroid hormone receptors. EMBO J. 11, 1025–1033.

Finck, B.N., Kelly, D.P. 2006. PGC-1 coactivators: Inducible regulators of energy metabolism in health and disease. J. Clin. Invest. 116, 615–622.

Fondell, J.D., Ge, H., Roeder, R.G. 1996. Ligand induction of a transcriptionally active thyroid hormone receptor coactivator complex. Proc. Natl. Acad. Sci. USA 93, 8329–8333.

Ge, J., Guermah, M., Yuan, C.-X., Ito, M., Wallberg, A., Spiegelman, B.M., Roeder, R.G. 2002. Transcription coactivators TRAP220 is required for PPAR gamma 2-stimulated adipogenesis. Nature 417, 563–567.

Gehin, M., Martk, M., Dennefeld, C., Dierich, A., Gronemeyer, H., Chambon, P. 2002. The function of TIF2/GRIP1 in mouse reproduction is distinct from those of SRC-1 and p/CIP. Mol. Cell. Biol. 22, 5923–5937.

Giles, R.H., Peters, D.J., Breuning, M.H. 1998. Conjunction dysfunction: CbP/p300 in human disease. Trends Genet. 14, 178–183.

Goodman, R.H., Smolik, S. 2000. CBP/p300 in cell growth, transformation, and development. Genes Dev. 14, 1553–1577.

Gordon, D.F., Tucker, E.A., Tundwal, K., Hall, H., Wood, W.M., Ridgway, E.C. 2006. MED220/TRAP220 functions as a transcriptional coactivator with Pit-1 and GATA-2 on the TSH promoter in thyrotropes. Mol. Endocrinol. 20, 1073–1089.

Guo, D., Sarkar, J., Ahmed, M.R., Viswakarma, N., Jia, Y., Yu, S., Rao, M.S., Reddy, J.K. 2006. Peroxisome proliferator-activated receptor (PPAR)-binding protein (PBP) but not PPAR-interacting protein (PRIP) is required for nuclear translocation of constitutive androstane receptor in mouse liver. Biochem. Biophys. Res. Comm. 347, 485–495.

Halachmi, S., Marden, E., Martin, G., Mackay, H., Abhondanza, C., Brown, M. 1994. Estrogen receptor-associated proteins: Possible mediators of hormone-induced transcription. Science 264, 1455–1458.

Handschin, C., Lin, J., Rhee, J., Peyer, A.-K., Chin, S., Wu, P.-H., Meyer, U.A., Spiegelman, B.M. 2005. Nutritional regulation of hepatic heme biosynthesis and Porphyria through PGC-1α. Cell 122, 505–515.

Heery, D.M., Kalkhoven, E., Hoare, S., Parker, M.G. 1997. A signature motif in transcriptional co-activators mediates binding to nuclear receptors. Nature 387, 733–736.

Hemberger, M., Zechner, U. 2004. Genetic and genomic approaches to study placental development. Cytogenet. Genome Res. 105, 257–269.

Hermanson, O., Glass, C.K., Rosenfeld, M.G. 2002. Nuclear receptor coregulators: Multiple modes of modification. Trends Endocrin. Metab. 13, 55–60.

Hong, H., Kohli, K., Garabedian, M.J., Stallcup, M.R. 1997. GRIP1, a transcriptional coactivator for the AF-2 transactivation domain of steroid, thyroid, retinoid, and vitamin D receptors. Mol. Cell. Biol. 17, 2735–2744.

Horlein, A.J., Naar, A.M., Heinzel, T., Torchia, J., Gloss, B., Kurokawa, R., Ryan, A., Kamei, Y., Soderstrom, M., Glass, C.K., Rosenfeld, M.G. 1995. Ligand-independent repression by the thyroid hormone receptor mediated by a nuclear receptor co-repressor. Nature 377, 397–404.

Ito, M., Yuan, C.-X., Okano, H.J., Darnell, R.B., Roeder, R.G. 2000. Involvement of the TRAP220 component of the TRAP/SMCC coactivator complex in embryonic development and thyroid hormone action. Mol. Cell 5, 683–693.

Ito, M., Okano, H.J., Darnell, R.B., Roeder, R.G. 2002. The TRAP100 component of the TRAP/mediator complex is essential in broad transcriptional events and development. The EMBO J. 21, 3464–3475.

Jain, S., Pulikuri, S., Zhu, Y., Qi, C., Kanwar, Y.S., Yeldandi, A.V., Rao, M.S., Reddy, J.K. 1998. Differential expression of the peroxisome proliferator-activated receptor γ (PPARγ) and its coactivators steroid receptor coactivator-1 and PPAR-binding protein PBP in the brown fat, urinary bladder, colon, and breast of the mouse. Am. J. Pathol. 153, 349–354.

Jeong, J.-W., Kwak, I., Lee, K.Y., White, L.D., Wang, X.-P., Brunicardi, F.C., O'Malley, B.W., DeMayo, F.J. 2006. The genomic analysis of the impact of steroid receptor coactivators (SRCs) ablation on hepatic metabolism. Mol. Endocrinol. 20, 1138–1152.

Jia, Y., Qi, C., Kashireddi, P., Surapureddi, S., Zhu, Y.J., Rao, M.S., Le Roith, D., Chambon, P., Gonzalez, F.J., Reddy, J.K. 2004. Transcription coactivator PBP, the peroxisome proliferators-activated receptor (PPAR)-binding protein, is required for PPARα-regulated gene expression in liver. J. Biol. Chem. 279, 24427–24434.

Jia, Y., Guo, G.L., Surapureddi, S., Sarkar, J., Qi, C., Guo, D., Xia, J., Kashireddi, P., Yu, S., Cho, Y.W., Rao, M.S., Kemper, B., et al. 2005a. Transcription coactivator peroxisome proliferators-activated receptor binding protein/mediator 1 deficiency abrogates acetaminophen hepatotoxicity. Proc. Natl. Acad. Sci. USA 102, 12531–12536.

Jia, Y., Qi, C., Zhang, Z., Zhu, Y.T., Rao, S.M., Zhu, Y.-J. 2005b. Peroxisome proliferator-activated receptor-binding protein null mutation results in defective mammary gland development. J. Biol. Chem. 280, 10766–10773.

Jia, Y., Yu, S., Rao, M.S., Reddy, J.K. 2006. Transcription coactivators PBP/MED1 deficiency prevents dexamethasone-induced fatty liver. FASEB J. 20, A226.

Kamei, Y., Xu, L., Heinzel, T., Torchia, J., Kurokawa, R., Gloss, B., Olin, S.-C., Heyman, R.A., Rose, D.W., Glass, C.K., Rosenfeld, M.G. 1996. A CBP integrator complex mediates transcriptional activation and AP-1 inhibition by nuclear receptors. Cell 85, 403–414.

Kang, Y.K., Guermah, M., Yuan, C.-X., Roeder, R.G. 2002. The TRAP/Mediator coactivator complex interacts directly with estrogen receptor and through the TRAOP220 subunit and directly enhances estrogen receptor function *in vitro*. Proc. Natl. Acad. Sci. USA 99, 2642–2647.

Kasper, L.H., Fukuyama, T., Biesen, M.A., Boussouar, F., Tong, C., de Pauw, A., Murray, P.J., van Deursen, J.M.A., Brindle, P.K. 2006. Conditional knockout mice reveal distinct functions for the global transcriptional coactivators CBP and p300 in T-cell development. Mol. Cell. Biol. 26, 789–809.

Ko, L., Cardona, G.R., Chin, W.W. 2000. Thyroid hormone receptor binding protein an LXXLL motif-containing protein, functions as a general coactivator. Proc. Natl. Acad. Sci. USA 97, 6212–6217.

Kobayashi, Y., Kitamoto, T., Masuhiro, Y., Watanabe, M., Kase, T., Metzger, D., Yanagisawa, J., Kato, S. 2000. p300 mediates functional synergism between AF-1 and AF-2 of estrogen receptor and by interacting directly with the N-terminal A/B domains. J. Biol. Chem. 275, 15645–15651.

Koh, S.S., Chen, D., Lee, Y.-H., Stallcup, M.R. 2001. Synergistic enhancement of nuclear receptor function by p160 coactivators and two coactivators with protein methyltransferase activities. J. Biol. Chem. 276, 1089–1098.

Komonyi, O., Pápai, G., Enunlu, I., Muratoglu, S., Pankotai, T., Kopitova, D., Mary, P., Udvardy, A., Boros, I. 2005. DTL, the *Drosophila* homolog of PIMT/Tgs1 nuclear receptor coactivator-interacting protein/TNA methyltransferase, has an essential role in development. J. Biol. Chem. 280, 12397–12404.

Kuang, S.-Q., Liao, L., Zhang, H., Pereira, F.A., Yuan, Y., DeMayo, F.J., Ko, L., Xu, J. 2002. Deletion of the cancer-amplified coactivator AIB3 results in defective placentation and embryonic lethality. J. Biol. Chem. 277, 45356–45360.

Kung, A.L., Rebel, V.I., Bronson, T., Ch'ng, L.E., Sieff, C.A., Livingston, D.M., Yao, T.-P. 2000. Gene dose-dependent control of hematopoiesis and hematologic tumor suppression by CPB. Genes Dev. 14, 272–277.

Kurihara, N., Ishizuka, S., Demulder, A., Menaa, C., Roodman, G.D. 2004. Paget's disease: A VDR coactivator disease? J. Steroid Biochem. Mol. Biol. 89–90, 321–325.

Landles, C., Chalk, S., Steel, J.H., Rosewell, I., Spencer-Dene, B., Lalani, E.-N., Parker, M.G. 2003. The thyroid hormone receptor-associated protein TRAP220 is required at distinct embryonic stages in placental, cardiac, and hepatic development. Mol. Endocrinol. 17, 2418–2435.

Lee, S.S., Pineau, T., Drago, J., Lee, E.J., Owens, J.W., Kroetz, D.L., Fernandez-Salguero, P.M., Westphal, H., Gonzalez, F.J. 1995. Targeted disruption of the alpha isoform of the peroxisome proliferator-activated receptor gene in mice results in abolishment of the pleiotropic effects of peroxisome proliferators. Mol. Cell. Biol. 15, 3012–3022.

Lee, S.K., Anzick, S.L., Choi, J.E., Bubendorf, L., Guan, X.Y., Jung, Y.R., Kallioniemi, O.P., Kononen, J., Trent, J.M., Azorsa, D., Jhun, B.H., Cheong, J.H., et al. 1999. A nuclear factor, ASC-2, is a cancer amplified transcriptional coactivator essential for ligand-dependent transactivation by nuclear receptors *in vivo*. J. Biol. Chem. 274, 34283–34293.

Leone, T.C., Lehman, J.J., Finck, B.N., Schaeffer, P.J., Wende, A.R., Boudina, S., Courtois, M., Wozniak, D.F., Sambandam, N., Bernal-Mizrachi, C., Chen, Z., Holloszy, J.O., et al. 2005.

PGC-1α deficiency causes multi-system energy metabolic derangements: Muscle dysfunction, abnormal weight control and hepatic steatosis. PLoS Biol. 3, e101.

L'Horset, F., Dauvois, S., Heery, D.M., Cavaills, V., Parker, M.G. 1996. RIP-140 interacts with multiple nuclear receptors by means of two distinct sites. Mol. Cell. Biol. 16, 6029–6030.

Li, H., Gomes, P.J., Chen, J.D. 1997. RAC3, a steroid/nuclear receptor-associated coactivator that is related to SRC-1 and TIF2. Proc. Natl. Acad. Sci. USA 94, 8479–8484.

Lim, H.J., Moon, I., Han, K. 2004. Transcriptional cofactors exhibit differential preference toward peroxisome proliferator-activated receptors α and δ in uterine cells. Endocrinology 145, 2886–2895.

Lin, J., Wu, P.-H., Tarr, P.T., Lindenberg, K.S., St-Pierre, J., Zhang, C.-Y., Mootha, V.K., Jäger, S., Vianna, C.R., Reznick, R.M., Cui, L., Manieri, M., et al. 2004. Defects in adaptive energy metabolism with CNS-linked hyperactivity in PGC-1α null mice. Cell 119, 121–135.

Lin, J., Yang, R., Tarr, P.T., Wu, P.-H., Handschin, C., Li, S., Yang, W., Pei, L., Uldry, M., Tontonoz, P., Newgard, C.B., Spiegelman, B.M. 2005. Hyperlipidemic effects of dietary saturated fats mediated through PGC-1β coactivation of SREBP. Cell 120, 261–273.

Llopis, J., Westin, S., Ricote, M., Wang, J., Cho, C.Y., Kurokawa, R., Mullen, T.-M., Rose, D.W., Rosenfeld, M.G., Tsien, R.Y., Glass, C.K. 2000. Ligand-dependent interactions of coactivators steroid receptor coactivator-1 and peroxisome proliferator-activated receptor binding protein with nuclear hormone receptors can be imaged in live cells and are required for transcription. Proc. Natl. Acad. Sci. USA 97, 4363–4368.

Lottin-Divoux, S., Barel, M., Frade, R. 2005. RB18A enhances expression of mutant p53 protein in human cells. FEBS Lett. 579, 2323–2326.

Mahajan, M.A., Samuels, H.H. 2000. A new family of nuclear receptor coregulators that integrate nuclear receptor signaling through CREB-binding protein. Mol. Cell. Biol. 20, 5048–5063.

Mahajan, M.A., Samuels, H.H. 2005. Nuclear hormone receptor coregulator: Role in hormone action, metabolism, growth and development. Endocrine. Rev. 26, 583–597.

Mahajan, M.A., Murray, A., Samuels, H.H. 2002. NRC-interacting factor 1 is a novel cotransducer that interacts with and regulates the activity of the nuclear hormone receptor coactivator NRC. Mol. Cell. Biol. 22, 6883–6894.

Mahajan, M.A., Das, S., Zhu, H., Tomic-Canic, M., Samuels, H.H. 2004. The nuclear hormone receptor coactivator NRC is a pleiotropic modulator affecting growth, development, apoptosis, reproduction, and wound repair. Mol. Cell. Biol. 24, 4994–5004.

Mangelsdorf, D.J., Thummel, C., Beato, M., Herrlich, P., Schutz, G., Umesono, K., Blumberg, B., Kastner, P., Mark, M., Chambon, P., Evans, R.M. 1995. The nuclear receptor superfamily: The second decade. Cell 83, 835–839.

Mark, M., Yoshida-Komiya, H., Gehin, M., Liao, L., Tsai, M.-J., O'Malley, B.W., Chambon, P., Xu, J. 2004. Partially redundant functions of SRC-1 and TIF2 in postnatal survival and male reproduction. Proc. Natl. Acad. Sci. USA 101, 4453–4458.

Matsumoto, K., Jia, Y., Yu, S., Rao, M.S., Reddy, J.K. 2006. Transcription coactivators PBP/MED1 regulates hepatocyte size and hepatocellular regeneration. FASEB J. 20, A2380.

McKenna, N.J., O'Malley, B.W. 2002. Combinatorial control of gene expression by nuclear receptors and coregulators. Cell 108, 465–474.

Misra, P., Owuor, E.D., Li, W., Yu, S., Qi, C., Meyer, K., Zhu, Y.J., Rao, M.S., Kong, A.N., Reddy, J.K. 2002. Phosphorylation of transcriptional coactivator peroxisome proliferator activator receptor-binding protein (PBP): Stimulation of transcriptional regulation by mitogen-activated protein kinase. J. Biol. Chem. 277, 48745–48754.

Misra, P., Qi, C., Yu, S., Shah, S.H., Cao, W.-Q., Rao, M.S., Thimmapaya, B., Zhu, Y., Reddy, J.K. 2002. Interaction of PIMT with transcription coactivators CBP, p300, and PBP differential role in transcriptional regulation. J. Biol. Chem. 277, 20011–20019.

Molkentin, J.D. 2000. The zinc finger-containing transcription factors GATA-4, -5, and -6: Ubiquitously expressed regulators of tissue-specific gene expression. J. Biol. Chem. 275, 38949–38952.

Mouaikel, J., Verheggen, C., Bertrand, E., Tazi, J., Bordonne, R. 2002. Hypermethylation of the cap structure of both yeast snRNAs and snoRNAs requires a conserved methyltransferase that is localized to the nucleolus. Mol. Cell 9, 891–901.

Naar, A.M., Beaurang, P.A., Zhou, S., Abraham, S., Solomon, W., Tjian, R. 1999. Composite coactivator ARC mediates chromatin-directed transcriptional activation. Nature 398, 828–832.

Nishihara, E., Yoshida-Komiya, H., Chan, C.-S., Liao, L., Davis, R.L., O'Malley, B.W., Xu, J. 2003. SRC-1 null mice exhibit moderate motor dysfunction and delayed development of cerebellar Purkinje cells. J. Neurosci. 23, 213–222.

Oike, Y., Takakura, N., Hata, A., Kaname, T., Akizuki, M., Yamaguchi, Y., Yasue, H., Araki, K., Yamamura, K., Suda, T. 1999. Mice homozygous for a truncated form of CREB-binding protein exhibit defects in hematopoiesis and vaculo-angiogenesis. Blood 93, 2771–2779.

Onate, S.A., Tsai, S.Y., Tsai, M.-J., O'Malley, B.W. 1995. Sequence and characterization of coactivator of the steroid hormone receptor superfamily. Science 270, 1354–1357.

Petrij, F., Giles, R.H., Dawerse, H.G., Saris, J.J., Hennekam, R.C., Masuno, M., Tommerup, N., van Ommen, O.G., Goodman, R.H., Peters, D.J. 1995. Rubenstein-Taybi syndrome caused by mutations in the transcriptional coactivator CBP. Nature 376, 348–351.

Picard, F., Gehin, M., Annicotte, J., Rocchi, S., Champy, M.F., O'Malley, B.W., Chambon, P., Aurex, J. 2002. SRC-1 and TIF-2 control energy balance between white and brown adipose tissues. Cell 111, 931–941.

Puigserver, P., Wu, Z., Park, C.W., Graves, R., Wright, M., Spiegelman, B.M. 1998. A cold inducible coactivator of nuclear receptors linked adaptive thermogenesis. Cell 92, 829–839.

Qatanani, M., Moore, D.D. 2005. CAR, the continuously advancing receptor, in drug metabolism and disease. Curr. Drug Metab. 6, 329–339.

Qi, C., Zhu, Y., Pan, J., Yeldandi, A.V., Rao, M.S., Maeda, N., Subbarao, V., Pulikuri, S., Hashimoto, T., Reddy, J.K. 1999. Mouse steroid receptor coactivator-1 is not essential for peroxisome proliferator-activated receptor α-regulated gene expression. Proc. Natl. Acad. Sci. USA 96, 1585–1590.

Qi, C., Zhu, Y., Reddy, J.K. 2000. Peroxisome proliferator-activated receptors, coactivators, and downstream targets. Cell Biochem. Biophys. 32, 187–204.

Qi, C., Surapureddi, S., Zhu, Y.J., Yu, S., Kashireddy, P., Rao, M.S., Reddy, J.K. 2003. Transcriptional coactivator PRIP, the peroxisome proliferator-activated receptor γ (PPARγ)-interacting protein, is required for PPARγ-mediated adipogenesis. J. Biol. Chem. 278, 25281–25284.

Qi, C., Kashireddy, P., Zhu, Y.T., Rao, S.M., Zhu, Y.-J. 2004. Null mutation of peroxisome proliferator-activated receptor-interacting protein in mammary glands causes defective mammopoiesis. J. Biol. Chem. 279, 33696–33701.

Rachez, C., Lemon, B.D., Suldan, Z., Bromleigh, V., Gamble, M., Naar, A.M., Erdjument-Bromage, H., Tempst, P., Freedman, L.P. 1999. Ligand-dependent transcription activation by nuclear receptors requires the DRIP complex. Nature 398, 824–828.

Rebel, V.I., Kung, A.L., Tanner, E.A., Yang, H., Bronson, R.T., Livingston, D.M. 2002. Distinct roles for CREB-binding protein and p300 in hematopoietic stem cell self-renewal. Proc. Natl. Acad. Sci. USA 99, 14789–14794.

Reddy, J.K., Rao, M.S. 2006. Lipid metabolism and liver inflammation II. Fatty liver disease and fatty acid oxidation. Am. J. Physiol. 290, G852–G858.

Roeder, R.G. 2005. Transcriptional regulation and the role of diverse coactivators in animal cells. FEBS Lett. 579, 909–915.

Sarkar, J., Jia, Y., Qi, C., Rao, M.S., Reddy, J.K. 2006. Transcription coactivators PRIP is not essential for the function of nuclear receptors CAR and PPARα in liver. FASEB J. 20, A225.

Shur, I., Benayahu, D. 2005. Characterization and functional analysis of CreMM, a novel chromodomain helicase DNA binding protein. J. Mol. Biol. 352, 646–655.

Smith, C.L., O'Malley, B.W. 2004. Coregulator function: A key to understanding tissue specificity of selective receptor modulators. Endocrine. Rev. 25, 45–71.

Spiegelman, B.M., Heinrich, R. 2004. Biological control through regulated transcriptional coactivators. Cell 119, 157–167.

Surapureddi, S., Yu, S., Bu, H., Hashimoto, T., Yeldandi, A.V., Kashireddy, P., Cherkaoui-Malki, M., Qi, C., Zhu, Y.J., Rao, M.S., Reddy, J.K. 2002. Identification of a transcriptionally active peroxisome proliferator-activated receptor α-interacting cofactor complex in rat liver and characterization of PRIC285 as a coactivator. Proc. Natl. Acad. Sci. USA 99, 11836–11841.

Surapureddi, S., Viswakarma, N., Yu, S., Guo, D., Rao, M.S., Reddy, J.K. 2006. PRIC320, a transcription coactivator, isolated from peroxisome proliferator-binding protein complex. Biochem. Biophys. Res. Commun. 343, 535–543.

Swales, K., Negishi, M. 2004. CAR, driving into the future. Mol. Endocrinol. 18, 1589–1598.

Taatjes, D.J., Tjian, R. 2004. Structure and function of CRSP/Med2: A promoter-selective transcriptional coactivator complex. Mol. Cell 14, 675–683.

Tanaka, Y., Naruse, I., Maekawa, T., Masuya, H., Shiroishi, T., Ishii, S. 1997. Abnormal skeletal patterning in embryos lacking a single Cbp allele: A partial similarity with Rubinstein-Taybi syndrome. Proc. Natl. Acad. Sci. USA 94, 10215–10220.

Tanaka, Y., Naruse, I., Hong, T., Xu, M., Nakahata, T., Maekawa, T., Ishii, S. 2000. Extensive brain hemorrhage and embryonic lethality in a mouse null mutant of CREB-binding protein. Mech. Dev. 95, 133–145.

Tomaru, T., Satoh, T., Yoshino, S., Ishizuka, T., Hashimoto, K., Monden, T., Yamada, M., Mori, M. 2006. Isolation and characterization of a transcriptional cofactor and its novel isoform that bind the deoxyribonucleic acid-binding domain of peroxisome proliferator-activated receptor-γ. Endocrinology 147, 377–388.

Torchia, J., Rose, D.W., Inostroza, J., Kamei, Y., Westin, S., Glass, C.K., Rosenfeld, M.G. 1997. The transcriptional coactivator p/CIP binds CBP and mediates nuclear receptor function. Nature 387, 677–684.

Torra, I.P., Freedman, L.P., Garabedian, M.J. 2004. Identification of DRIP205 as a coactivators from the farnesoaid X receptor. J. Biol. Chem. 279, 36184–36191.

Torres-Arzayus, M.I., Font de Mora, J., Yuan, J., Vazquez, F., Bronson, R., Rue, M., Sellers, W.R., Brown, M. 2004. High tumor incidence and activation of the PI3K/AKT pathway in transgenic mice define AIB1 as an oncogene. Cancer Cell 6, 263–274.

Udayakumar, T.S., Belakvadi, M., Choi, K.-H., Pandey, P.K., Fondell, J.D. 2006. Regulation of aurora-A kinase gene expression via GABP recruitment of TRAP220/MED1. J. Biol. Chem. 281, 14691–14699.

Urahama, N., Ito, M., Sada, A., Yakushijin, K., Yamamoto, K., Okamura, A., Minagawa, K., Hato, A., Chihara, K., Roeder, R.G., Matsui, T. 2005. The role of transcriptional coactivator TRAP220 in myelomonocytic differentiation. Genes Cells 10, 1127–1137.

Voegel, J.J., Heine, M.J., Zechel, C., Chambon, P., Gronemeyer, H. 1996. TIF2, a 160 kDa transcriptional mediator for the ligand-dependent activation function AF-2 of nuclear receptors. EMBO J. 15, 3667–3675.

Wang, Z., Rose, D.W., Hermanson, O., Liu, F., Herman, T., Wu, W., Szeto, D., Gleiberman, A., Krones, A., Pratt, K., Rosenfeld, R., Glass, C.K., et al. 2000. Regulation of somatic growth by the p160 coactivator p/CIP. Proc. Natl. Acad. Sci. USA 97, 13549–13554.

Wang, Z., Qi, C., Krones, A., Woodring, P., Zhu, X., Reddy, J.K., Evans, R.M., Rosenfeld, M.G., Hunter, T. 2006. Critical roles of the p160 transcriptional coactivators p/CIP and SRC-1 in energy balance. Cell Metab. 3, 111–122.

Weiss, R.E., Xu, J., Ning, G., Pholenz, J., O'Malley, B.W., Refetoff, S. 1999. Mice deficient in the steroid receptor coactivator 1 (SRC-1) are resistant to thyroid hormone. EMBO J. 18, 1900–1904.

Winnay, J.N., Xu, J., O'Malley, B.W., Hammer, G.D. 2006. Steroid receptor coactivator-1-deficient mice exhibit altered hypothalamic-pituitary-adrenal axis function. Endocrinology 147, 1322–1332.

Xu, J., Qiu, Y., DeMayo, F.J., Tsai, S.Y., Tsai, M.J., O'Malley, B.W. 1998. Partial hormone resistance in mice with disruption of the steroid receptor coactivator-1 (SRC-1) gene. Science 279, 1922–1925.

Xu, J., Liao, L., Ning, G., Yoshida-Komiya, H., Deng, C., O'Malley, B.W. 2000. The steroid receptor coactivator SRC-3 (p/CIP/RAC3/AIB1/ACTR/TRAM-1) is required for normal growth, puberty, female reproductive function, and mammary gland development. Proc. Natl. Acad. Sci. USA 97, 6379–6384.

Xu, W., Chen, H., Du, K., Asahara, H., Tini, M., Emerson, B.M., Montminy, M., Evans, R.M. 2001. A transcriptional switch mediated by cofactor methylation. Science 294, 2507–2511.

Yadav, N., Lee, J., Kim, J., Shen, J., Hu, M.C.-T., Aldaz, C.M., Bedford, M.T. 2003. Specific protein methylation defects and gene expression perturbations in coactivator-associated arginine methyltransferase 1-deficient mice. Proc. Natl. Acad. Sci. USA 100, 6464–6468.

Yan, J., Tsai, S.Y., Tsai, M.-J. 2006. SRC-3/AIB1: Transcriptional coactivator in oncogenesis. Acta Pharmacol. Sin. 27, 387–394.

Yao, T.-P., Oh, S.P., Fuchs, M., Zhou, N.D., Ch'ng, L.-E., Newsome, D., Bronson, R.T., Li, E., Livingston, D.M., Eckner, R. 1998. Gene dosage-dependent embryonic development and proliferation defects in mice lacking the transcriptional integrator p300. Cell 93, 361–372.

Yu, S., Sarkar, J., Reddy, J.K. 2006. Differential expression of nuclear receptor coregulators in mouse tissues. FASEB J. 20, A227.

Yuan, C.X., Ito, M., Fondell, J.D., Fu, Z.Y., Roeder, R.G. 1998. The TRAP220 component of a thyroid hormone receptor-associated protein (TRAP) coactivator complex interacts directly with nuclear receptors in a ligand-dependent fashion. Proc. Natl. Acad. Sci. USA 95, 7939–7944.

Zhang, H., Yi, X., Sun, X., Yin, N., Shi, B., Wu, H., Wang, D., Wu, G., Shang, Y. 2004. Differential gene regulation by the SRC family of coactivators. Genes Dev. 18, 1753–1765.

Zhou, G., Hashimoto, Y., Kwak, I., Tsai, S., Tsai, M.-J. 2003. Role of the steroid receptor coactivator SRC-3 in cell growth. Mol. Cell. Biol. 23, 7742–7755.

Zhu, Y., Qi, C., Calandra, C., Rao, M.S., Reddy, J.K. 1996. Cloning and identification of mouse steroid receptor coactivator-1 (mSRC-1), as a coactivator of peroxisome proliferator-activated receptor γ. Gene Expr. 6, 185–195.

Zhu, Y., Qi, C., Jain, S., Rao, M.S., Reddy, J.K. 1997. Isolation and characterization of PBP, a protein that interacts with peroxisome proliferator-activated receptor. J. Biol. Chem. 272, 25500–25506.

Zhu, Y., Qi, C., Jain, S., Le Beau, M.M., Espinosa, R., III, Atkins, G.B., Lazar, M.A., Yeldandi, A.V., Rao, M.S., Reddy, J.K. 1999. Amplification and overexpression of peroxisome proliferator-activated receptor binding protein (PBP/PPARBP) gene in breast cancer. Proc. Natl. Acad. Sci. USA 96, 10848–10853.

Zhu, Y., Kan, L., Qi, C., Kanwar, Y.S., Yeldandi, A.V., Rao, M.S., Reddy, J.K. 2000a. Isolation and characterization of peroxisome proliferator-activated receptors (PPAR) interacting protein (PRIP) as a coactivator for PPAR. J. Biol. Chem. 275, 13510–13516.

Zhu, Y., Qi, C., Jia, Y., Nye, J.S., Rao, M.S., Reddy, J.K. 2000b. Deletion of PBP/PPARBP, the gene for nuclear receptor coactivator peroxisome proliferator-activated receptor-binding protein, results in embryonic lethality. J. Biol. Chem. 275, 14779–14782.

Zhu, Y., Qi, C., Cao, W.-Q., Yeldandi, A.V., Rao, M.S., Reddy, J.K. 2001. Cloning and characterization of PIMT, a protein with methyltransferase domain, which interacts with and enhances nuclear receptor coactivator PRIP function. Proc. Natl. Acad. Sci. USA 98, 10380–10385.

Zhu, Y.J., Crawford, S.E., Stellmach, V., Dwivedi, R.S., Rao, M.S., Gonzalez, F.J., Qi, C., Reddy, J.K. 2003. Coactivator PRIP, the peroxisome proliferator-activated receptor-interacting protein, is a modulator of placental, cardiac, hepatic and embryonic development. J. Biol. Chem. 278, 1986–1990.

Index

Nicolas Rotman, Liliane Michalik, Béatrice Desvergne and Walter Wahli, Chapter 2, Table 1. Timing of expression of the three PPAR isotypes during fetal and early postnatal development in the rodent

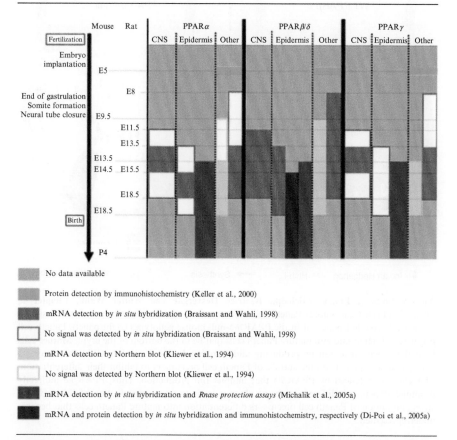

Expression data within the placenta are not summarized in the table. CNS, central nervous system.

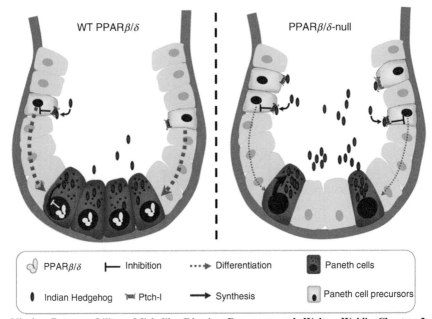

Nicolas Rotman, Liliane Michalik, Béatrice Desvergne and Walter Wahli, Chapter 2, Figure 5. PPARβ/δ promotes Paneth cell differentiation. An intestinal crypt scheme is shown in the wild type (left panel) or in the PPARβ/δ-null mouse genotypes (right panel). Paneth cell precursors differentiate into mature Paneth cells that lie in the bottom of the crypt. Mature cells limit their own formation by producing and secreting Indian Hedgehog (Ihh) that interacts with Patched-1 (Ptch-1) on the surface of Paneth cell precursors to inhibit their differentiation. This effect is balanced by PPARβ/δ that inhibits Ihh production. Thus, PPARβ/δ indirectly promotes Paneth cell differentiation; its absence leads to an overproduction of Ihh and to a strong inhibition of Paneth cell precursors differentiation into mature Paneth cells that become underrepresented.

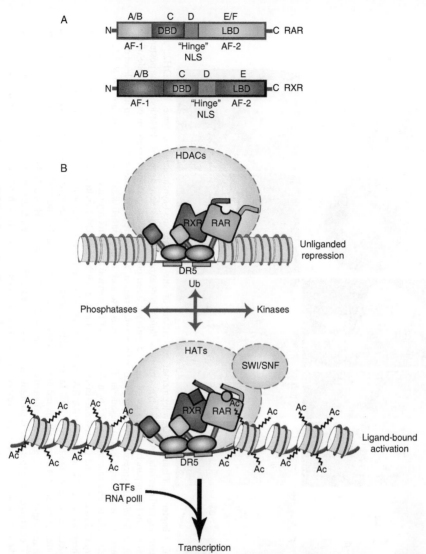

Tracie Pennimpede, Don Cameron and Martin Petkovich, Chapter 3, Figure 1. Retinoid receptor structure and regulation of RA-responsive genes. (A) The modular A–F domain structure of RARs and RXRs. The A/B region contains the AF-1 domain, the C region contains the DNA-binding domain (DBD), D is the "hinge" region containing a nuclear localization signal (NLS), the E region contains the ligand-binding domain (LBD) and AF-2. The F region is not found in RXRs. (B) A schematic representation of the current model of retinoid-regulated gene activation showing RAR/RXR heterodimer bound to a DR5 response element. Gene repression via chromatin compaction results from aporeceptor association with a large protein complex containing various corepressors with HDAC activity. Ligand binding then induces transcriptional activation by the recruitment of various coactivator proteins with HAT activity and ATP-dependent chromatin remodelers like SWI/SNF. This leads to chromatin decompaction and the recruitment of the basal transcription machinery. Various phosphorylation events and the ubiquitin-proteasome pathway have been suggested to play a role in transcriptional regulation, although the *in vivo* sequence of events is still unclear. Ub, ubiquitin; Ac, acetyl groups; GTFs, general transcription factors, HATs, histone acetyltransferases; HDACs, histone deacetylases.

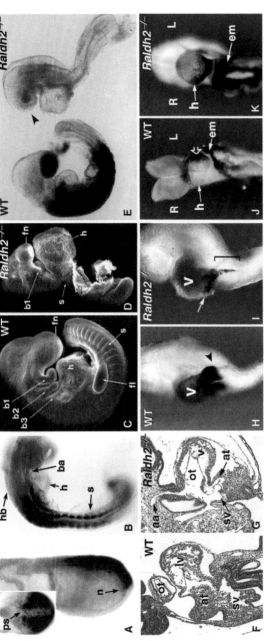

Karen Niederreither and Pascal Dollé, Chapter 4, Figure 2. Developmental expression of the retinoic acid-synthesizing enzyme RALDH2, and early embryonic abnormalities in corresponding null mutants. (A) In wild-type embryos, *Raldh2* gene expression is first seen during gastrulation in the posterior embryonic mesoderm up to the node (n) region (main panel: profile view of an E7.5 embryo; inset: posterior view showing expression on each side of the primitive streak, ps). (B) Eventually at E8.5, *Raldh2* is expressed at highest levels in the cervical region, with sharp boundaries of expression at the level of the posterior hindbrain (hb), branchial arches (ba), and heart tube (h) mesoderm. Expression is also seen in the epithelialized somites (s). (C and D) *Raldh2*−/− null mutants exhibit severe morphogenetic defects, characterized at E9.5 by a shortening of the cervical and trunk region with truncated somites (s; Fig. 3), a markedly dilated heart (h) cavity, a truncated forebrain and frontonasal (fn) region and a lack of visible branchial arches (b2, b3), except for the first arch (b1). (E) *Raldh2*−/− embryos at E8.5 show no detectable activity of the RA-sensitive RARE-*lacZ* transgene, except in the facial region which expresses *Raldh3* (arrowhead). (F and G) Histological analysis of the heart at E9.5 shows highly hypoplastic inflow cavities (atrium and sinus venosus, at and sv, respectively), defective trabeculation of the ventricular (v) myocardium and lack of morphogenesis of ventricular and outflow tract (ot) chambers, resulting in an abnormal juxtaposition of inflow and outflow (aortic arches, aa) circulations in the *Raldh2*−/− mutants. (H and I) Abnormal distribution of *Tbx5* transcripts in the heart of E9.5 *Raldh2*−/− embryos. While this gene is normally strongly expressed in the posterior heart chambers of wild-type embryos (H), its expression in mutants is limited to a narrow band of cells (white arrow), with an ill-defined posterior boundary (bracket). (J and K) An abnormal posterior location of *Hand1* transcripts, which normally become lateralized along the

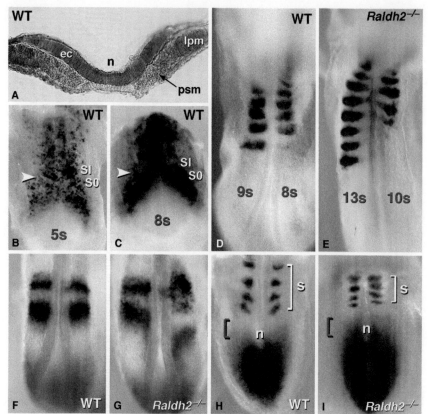

Karen Niederreither and Pascal Dollé, Chapter 4, Figure 3. Retinoic acid deficiency in $Raldh2^{-/-}$ mutants affects the bilateral symmetry of somite development. (A–C) In E8 (two somite stage) wild-type mouse embryos, RA signaling as detected by the activity of the RARE-*lacZ* reporter transgene occurs symmetrically in the left and right presomitic mesoderm, and is seen in the ectodermal layer of the node. During the first cycles of somitogenesis, the RA signal progresses caudally along the rostral presomitic mesoderm, as RA-responding cells are consistently seen at a comparable distance from the last formed intersomitic boundary (arrowheads) and epithelialized somite (SI, S0 designating the somite undergoing epithelialization; somite stages are indicated in red). (D and E) Asymmetric L–R somitic development in $Raldh2^{-/-}$ embryos. Two examples are shown, with the presence of one (D) or three (E) additional somites formed on the left side (somite numbers are indicated in red). Somites are marked by the expression of *Uncx4.1*. (F and G) Desynchronized waves of expression of an oscillating gene (*Lunatic fringe*) along the left and right presomitic mesoderm of $Raldh2^{-/-}$ embryos. (H and I) Combined detection of *Uncx4.1* and *Fgf8* transcripts in six somite stage embryos reveals an abnormal compaction of somites (s), as well as anteriorization of the *Fgf8* mRNA gradient along both sides of the node (n; see red brackets), in $Raldh2^{-/-}$ embryos. Panel A from Sirbu and Duester (2006), Nat. Cell Biol. 8, 271; B–I from Vermot and Pourquié (2005), Science 308, 563, with permission.

---

prospective left wall of the left ventricle during left–right (L–R) heart looping (arrow in J), indicates an impaired looping morphogenesis in $Raldh2^{-/-}$ embryos (K). *Hand1* expression in extraembryonic membranes (em) is unaltered. Panels A and B from Niederreither et al. (1997), Mech. Dev. 62, 67; C–E and J and K from Niederreither et al. (1999), Nat. Genet. 21, 444; F–I from Niederreither et al. (2001), Development 128, 1019, with permission.

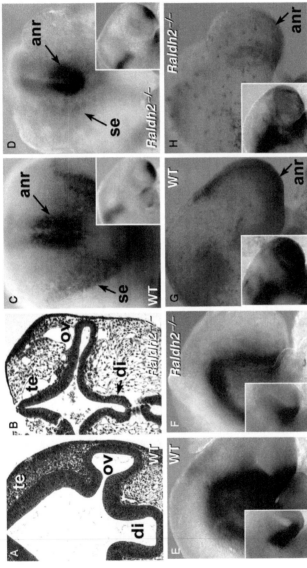

Karen Niederreither and Pascal Dollé, Chapter 4, Figure 4. Abnormal development and patterning of the embryonic forebrain in *Raldh2⁻/⁻* mutants. (A and B) Histological analysis of E9.5 embryos shows an abnormally thin forebrain neuroepithelium, with a lack of formation of the telencephalic (te) vesicles and of optic vesicle (ov) morphogenesis. di, diencephalon. (C and D) *Fgf8* expression is not affected in the anterior neural ridge (or in other embryonic areas, for example, at the midhindbrain boundary: compare insets), but is defective in the facial surface ectodem, of *Raldh2⁻/⁻* embryos (main panels: E9.5; insets: E8.5). (E and F) Decreased mRNA expression of *Gli1* along the ventral forebrain neuroepithelium (main panels) and of *NKx2.1* in the ventral diencephalon and telencephalon (insets) indicates defective Shh signaling in *Raldh2⁻/⁻* embryos. (G and H) Likewise, analysis of phosphorylated ERK1/2 proteins (main panels) and *Mkp3* mRNA (insets) indicate a decrease in response to FGF signaling in the rostral forebrain of the RA-deficient embryos. From Ribes et al. (2006), Development 133, 351, with permission.

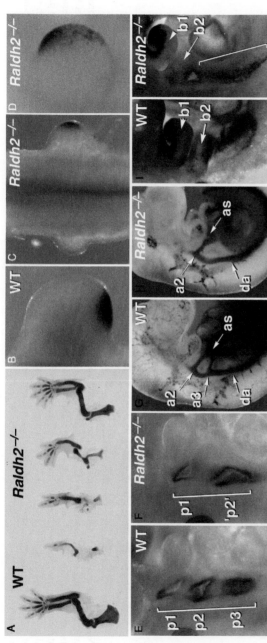

Karen Niederreither and Pascal Dollé, Chapter 4, Figure 5. Rescue of the lethality of $Raldh2^{-/-}$ embryos unveils novel developmental defects. (A) $Raldh2^{-/-}$ fetuses at E14.5 after a minimal RA-rescue from E7.5 to E8.5 have highly hypoplastic forelimbs (left-most sample, compare to the adjacent wild-type forelimb). By extending the duration and the doses of RA supplied, growth and patterning of the $Raldh2^{-/-}$ forelimbs could be improved, although the best rescued limbs displayed abnormal anterior–posterior patterning (notice a six-digit polydactyly in the right-most sample). Cartilaginous skeletal patterns were revealed by alcian blue staining. (B–D) At E10.5, the rescued $Raldh2^{-/-}$ embryos fail to activate *Sonic Hedgehog* (*Shh*), or express it in a inappropriate distal/anterior location within the forelimb buds (compare with the posteriorly restricted *Shh* domain in the wild-type limb bud). (E and F) *Pax9* specific expression in pharyngeal endoderm reveals abnormal development of branchial pouches (p1–p3) in rescued E9.5 $Raldh2^{-/-}$ mutants. The mutants exhibit a normal first pouch (p1) between the first and second branchial arches, and posteriorly have a single pouch-like structure (p2). (G and H) Impaired development of posterior branchial arches is also seen by the lack of third aortic arches (a3) following intracardiac ink injections (as, aortic sac; da, dorsal aorta). (I and J) *Fg f8* is expressed at low levels in a patchy domain along the pharyngeal endoderm of $Raldh2^{-/-}$ embryos (bracket), whereas it is normally expressed at high levels in the branchial pouch endoderm of wild-type embryos. In contrast, *Fgf8* expression levels are normal in the first arch (b1) ectoderm in mutants. A–D from Niederreither et al. (2002), Development 129, 3563; E–J from Niederreither et al., Development 130, 2525, with permission.

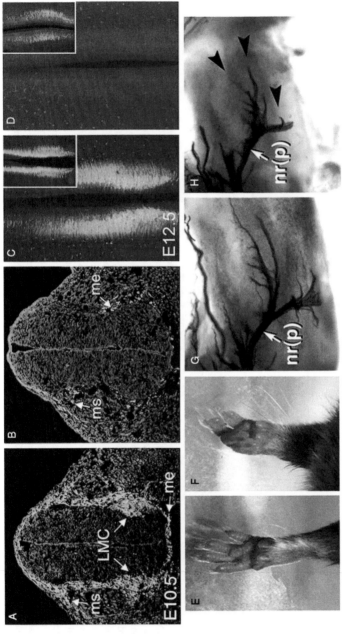

Karen Niederreither and Pascal Dollé, Chapter 4, Figure 6. Conditional mutagenesis of *Raldh2* leads to viable mutants with a loss of function in forelimb-innervating developing motor neurons. (A–D) Immunodetection of RALDH2 protein on transverse sections at brachial levels of the spinal cord of E10.5 wild-type and *RARβ-Cre;Raldh2* embryos (A and B) showing lack of protein in the lateral motor columns of the mutant, but persistent expression in adjacent meningeal (me) and mesenchymal (ms) cells. (C and D) are flat-mounts of E12.5 whole spinal cords analyzed by confocal microscopy (main panels: brachial levels; insets: lumbar levels). (E and F) Abnormal forelimb phenotype of adult *RARβ-Cre;Raldh2* mice. (G and H) Deficient outgrowth of the extensor branch (ramus profundus, p) of the nerve radialis (nr) in E13.5 *RARβ-Cre;Raldh2* embryos (arrowheads), as seen by whole-mount antineurofilament staining. From Vermot et al. (2005b), Development 132, 1611, with permission.

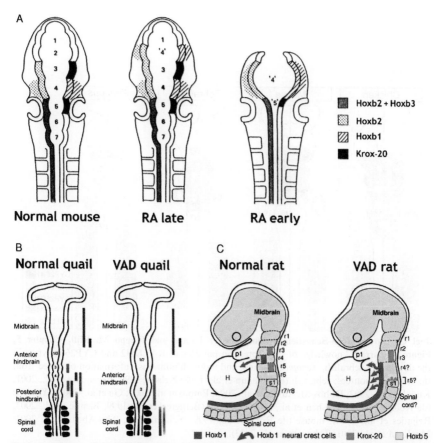

Joel C. Glover, Jean-Sébastien Renaud, Xavier Lampe and Filippo M. Rijli, Chapter 5, Figure 1. The teratogenic effects of retinoid excess and deficiency on early patterning of the hindbrain. (A) Stage-dependent teratogenesis in the mouse embryo caused by exposure to exogenous retinoic acid (modified from Wood et al., 1994). (B) The effects of retinoid deficiency in the VAD quail embryo (modified from Gale et al., 1999). (C) The effects of retinoid deficiency in the VAD rat embryo (modified from White et al., 2000).

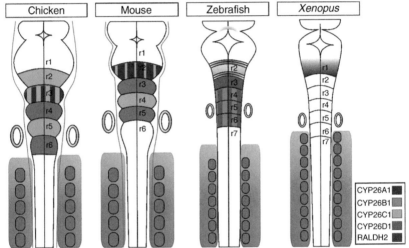

Joel C. Glover, Jean-Sébastien Renaud, Xavier Lampe and Filippo M. Rijli, Chapter 5, Figure 3. Current knowledge of the expression patterns of RALDH2 and CYP26 enzymes in the developing zebrafish, *Xenopus*, chicken, and mouse hindbrains. Information obtained from the following sources: zebrafish (Hollemann et al., 1998; Nelson, 1999; Begemann et al., 2001; Kudoh et al., 2002; Dobbs-McAuliffe et al., 2004; Emoto et al., 2005; Gu et al., 2005); *Xenopus* (de Roos et al., 1999; Chen et al., 2001); chicken (Berggren et al., 1999; Blentic et al., 2003; Reijntjes et al., 2004); mouse (de Roos et al., 1999; Nelson, 1999; Abu-Abed et al., 2001; MacLean et al., 2001; Tahayato et al., 2003).

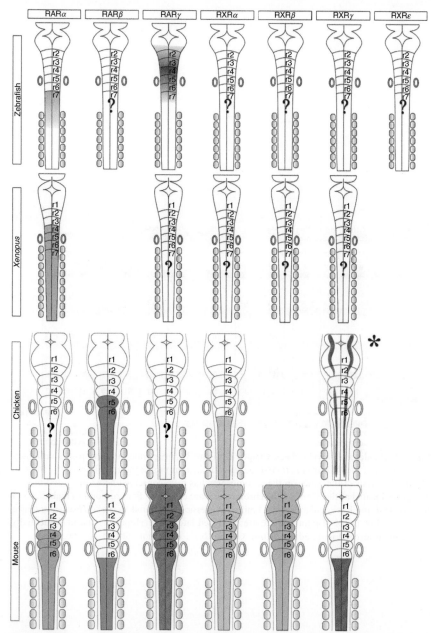

Joel C. Glover, Jean-Sébastien Renaud, Xavier Lampe and Filippo M. Rijli, Chapter 5, Figure 4. Current knowledge of the expression patterns of retinoid receptors in the developing hindbrains of zebrafish, *Xenopus*, chicken, and mouse embryos. Information obtained from the following sources: zebrafish (Joore et al., 1994; Jones et al., 1995; Sharma et al., 2003); *Xenopus* (Sharpe, 1992; Dekker et al., 1994; Ho et al., 1994; Marklew et al., 1994; Sharpe, 1994; Crawford et al., 1995); chicken (Maden et al., 1991; Smith and Eichele, 1991; Smith et al., 1994; Hoover and Glover, 1998; *note that RXRγ is expressed in this pattern in the chicken only at later stages); mouse (Ruberte et al., 1990, 1991, 1992, 1993; Dolle et al., 1994; Mollard et al., 2000).

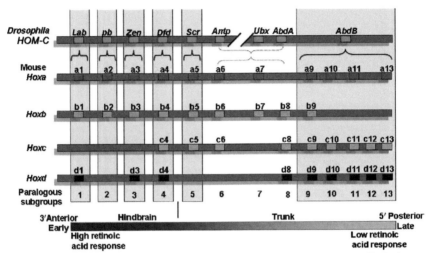

Charlotte Rhodes and David Lohnes, Chapter 6, Figure 3. *Hox* paralogous groups. Evolutionary conservation of the *Drosophila HOM-C* genes with the four mouse paralogous clusters, *Hoxa* to *Hoxd*. The order of the genes on the chromosome has been conserved between fly and mouse. The genes located at the 3′ end of the chromosome are expressed first, respond to high doses of retinoic acid and have a more rostral limit of expression compared to the 5′ *Hox* genes which respond to low doses of retinoic acid, are expressed later in development, and at sequentially more caudal levels. Figure adapted from Gilbert, S.F., Dev. Biol., 6th ed., p. 366.

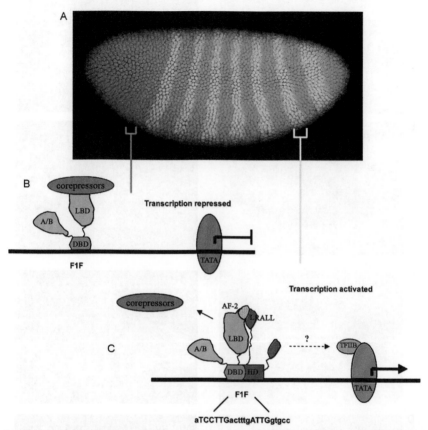

Leslie Pick, W. Ray Anderson, Jeffrey Shultz and Criag T. Woodard, Chapter 8, Figure 2. Ftz mimics an NR coactivator for Ftz-F1. (A) Ftz and Ftz-F1 expression overlaps in the seven Ftz stripes. Ftz-F1 (green) is expressed and nuclear localized in all somatic cells of the embryo, shown here at the blastoderm stage. The embryo at this stage consists of approximately 5000 cells located around the circumference of the egg. Ftz is expressed in seven stripes, each 3–4 cells wide, in the primordia of the even-numbered parasegments that are missing in *ftz* mutants. Ftz protein was detected with a rhodamine-coupled secondary antibody (red); because expression overlaps precisely with Ftz-F1, stripes appear yellow. (B) Ftz-F1 is inactive in cells that do not express Ftz. When Ftz-F1 binds DNA in Ftz-nonexpressing (green) cells, corepressors are recruited to inhibit transcription. (C) Ftz-F1 and Ftz cooperate to activate target gene expression. Ftz and Ftz-F1 bind cooperatively to composite Ftz-F1/Ftz-binding sites, exemplified by F1F. Interaction of the Ftz LRALL sequence with the Ftz-F1 AF-2 domain displaces corepressors, allowing productive transcription complexes to form.

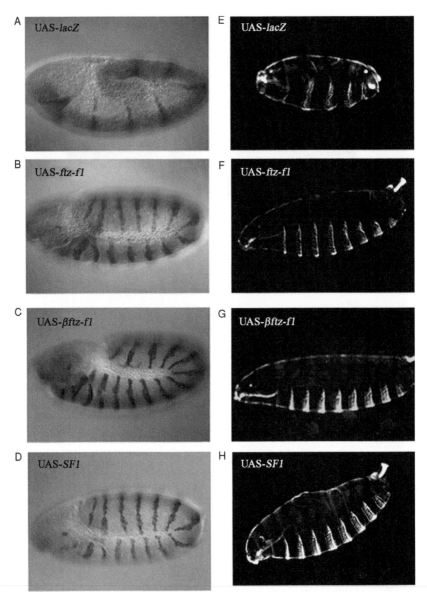

Leslie Pick, W. Ray Anderson, Jeffrey Shultz and Criag T. Woodard, Chapter 8, Figure 3.
Evolutionarily conserved protein domains are necessary for Ftz-F1 function *in vivo*. (A–D)
Expression of En and (E–H) pair-rule defects in maternal-specific *ftz-f1²⁰⁹* mutants were
unaffected by expression of *lacZ* but were rescued by expression of *ftz-f1*, *βftz-f1*, or the murine
orthologue SF-1, using the GAL4-UAS system with an NGT40 driver (Tracey et al., 2000).

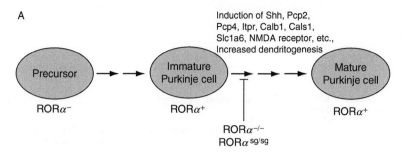

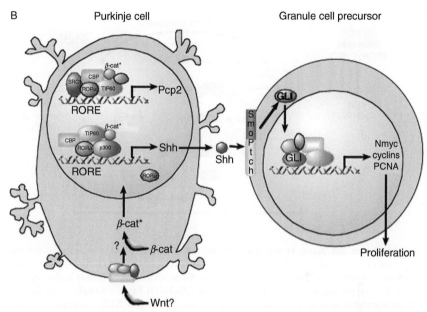

Anton M. Jetten and Joung Hyuck Joo, Chapter 10, Figure 1. Functions of RORα in the development of the cerebellum. (A) RORα plays a critical role in the maturation of Purkinje cells. Immature Purkinje cells arise from RORα⁻ precursor cells. RORα becomes highly expressed in postmitotic Purkinje cells at E12.5 of mouse embryonic development. RORα is required for dendritogenesis to proceed and for the induction of a number of genes normally expressed in mature Purkinje cells. Dendritogenesis and the expression of several genes normally expressed in mature Purkinje cells are inhibited in RORα-deficient mice. (B) RORα regulates expression of several genes, including *Pcp2* and *Shh*, directly by binding to ROREs in their promoter regions. The composition of the coactivator complexes assembled by RORα appears to be distinct for each gene, suggesting that the promoter context plays a critical role. These coactivators include β-catenin (β-cat), SRC1, p300, CBP, and Tip60. Activation of β-catenin by Wnt signaling might influence the transcriptional activity of RORα. RORα may act downstream of Wnt and because *Shh* is an RORα target gene, RORα may function as a mediator between the Wnt and Shh signaling pathways. Shh released by Purkinje cells interacts with Patched (Ptch) receptors present on granule cell precursors and lead to the activation of GLI transcription factors. Subsequently, this results in the transcriptional activation of several growth regulatory genes, including PCNA and cyclins, and an induction of the proliferation of these cells. The reduced Shh expression in Purkinje cells from RORα-deficient mice is a major cause of the cerebellar atrophy observed in these mice (Landis and Sidman, 1978; Herrup and Mullen, 1979; Hamilton et al., 1996; Goldowitz and Hamre, 1998; Steinmayr et al., 1998; Wallace, 1999; Gold et al., 2003, 2006; Kenney et al., 2003; Boukhtouche et al., 2006b).

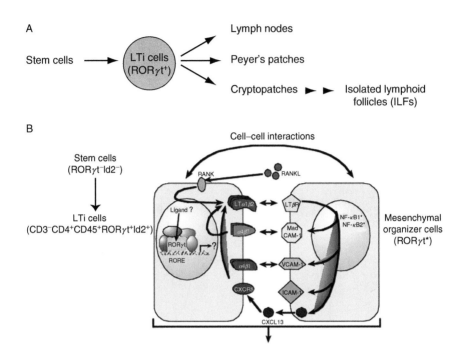

Anton M. Jetten and Joung Hyuck Joo, Chapter 10, Figure 2. RORγt is essential for the development of secondary lymphoid tissues. (A) Lymphoid tissue inducer (LTi) cells are derived from hematopoietic stem cells and are essential in the development of lymph nodes, Peyer's patches, and cryptopatches. The isolated lymphoid follicles (ILFs) are thought to be derived from cryptopatches after the colonization of the intestine by bacteria. RORγt is required for the generation and/or survival of LTi cells. The absence of these lymphoid tissues in RORγ-deficient mice is due to a deficiency of LTi cells in these mice. (B) Role of RORγt and LTi cells in lymph node development. The transcription factors Id2 and RORγt are both essential for the generation of LTi cells (CD4$^+$CD3$^-$ CD45$^+$IL-7Rα$^+$RORγt$^+$Id2$^+$). LTi cells are recruited to lymph node anlagen through their interaction with mesenchymal organizer cells. Interactions between these two cell types are key in the recruitment of LTi cell from the circulation to the lymph node anlagen. Activation of the RANK signaling pathway enhances the expression of lymphotoxins (LT) in LTi cells and promotes lymph node development. Binding of LTα1β2 on LTi cells to the LTβR on mesenchymal organizer cells plays a key role in lymph node development. The latter results in the activation of NF-κB pathways and induction of VCAM-1, ICAM-1, and MAdCAM-1 in mesenchymal organizer cells and their subsequent binding to integrins on LTi cells further promote their interaction and the recruitment of additional LTi cells. In addition, induction of various chemokines in mesenchymal organizer cells, including CXCL13 and CCL19, interact with their corresponding receptors on LTi cells thereby promoting their interaction and allow the recruitment of monocytes, T and B lymphocytes. (Yokota et al., 1999; Kurebayashi et al., 2000; Sun et al., 2000; Jetten et al., 2001; Cupedo et al., 2002; Eberl and Littman, 2003; Eberl et al., 2004; Jetten, 2004; Lipp and Muller, 2004; Eberl, 2005).

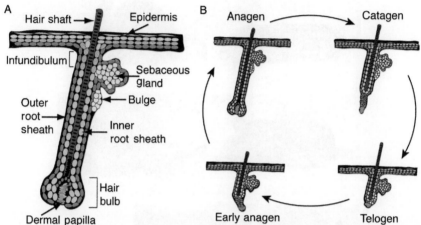

Catherine C. Thompson and Gerard M. J. Beaudoin, III, Chapter 11, Figure 4. Schematic diagram of the skin and hair cycle. (A) Schematic section through skin showing three major epithelial components, the epidermis, sebaceous gland, and hair follicle. The undifferentiated, progenitor keratinocytes (green) are found in the outer root sheath (ORS) and the basal layer of the epidermis. These progenitor cells are generated by the stem cells in the bulge (yellow) and migrate along the ORS to the hair bulb, to produce the hair shaft and inner root sheath, or to the sebaceous gland or through the infundibulum to the epidermis. The interaction of the mesenchymal cells (dermal papilla) and bulge stem cells is thought to regulate hair morphogenesis and cycling. (B) Schematic representation of the hair cycle. The hair follicle cycles from periods of hair growth (anagen) to an apoptosis-driven regression of the hair bulb and lower follicle (catagen) followed by a period of rest (telogen). The follicle exits telogen and reenters anagen to regenerate the hair bulb. Adapted from Beaudoin et al. (2005).

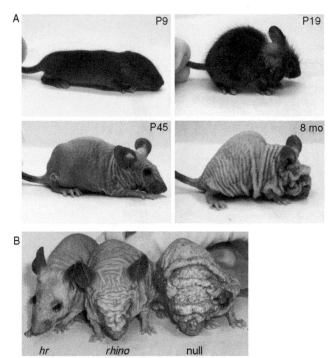

Catherine C. Thompson and Gerard M. J. Beaudoin, III, Chapter 11, Figure 5. Phenotype of *Hr* mutant mice. (A) *Hr*$^{-/-}$ mice show progressive hair loss. Initial hair growth is normal (P9), but after shedding the hair does not grow back (P21, P45). The skin becomes progressively more wrinkled with age (8 months). (B) Different *Hr* alleles show varying degrees of skin wrinkling. Adult mice homozygous for *hairless* (hr, *Hr*^{hr}), *rhino* (*Hr*^{rh}), and null (*Hr*$^{-}$) alleles; note that wrinkling is most severe in null.

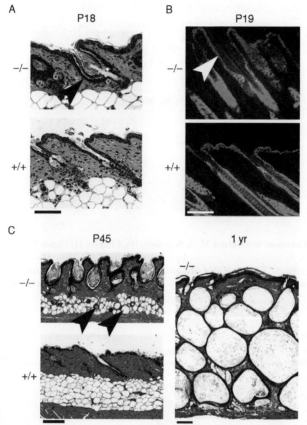

Catherine C. Thompson and Gerard M. J. Beaudoin, III, Chapter 11, Figure 6. Histological changes in skin of *Hr* mutant mice. (A) Hematoxylin–eosin stained sections from wild-type (+/+) and $Hr^{-/-}$ skin at P18. Arrowhead indicates widening of the top of the hair follicle adjacent to the epidermis (utricle) in $Hr^{-/-}$(−/−) skin. Scale bar: 50 µm. (B) Utricles are composed of epidermal cells. Immunofluorescence for keratin expression demonstrates cells in the utricle wall express K10 (arrowhead, red), an epidermal marker, instead of K17 (green). Scale bar: 50 µm. (C) (Left) Dermal cysts develop in $Hr^{-/-}$(−/−) skin. Hematoxylin–eosin stained sections of wild-type (+/+) and $Hr^{-/-}$ skin at P45. Arrowheads indicate cysts found in the dermis. (Right) Dermal cysts increase in size in $Hr^{-/-}$ skin. Hematoxylin–eosin stained section of skin from an $Hr^{-/-}$(−/−) mouse at 1-year old. Scale bars: 20 µm. Adapted from Zarach et al. (2004).

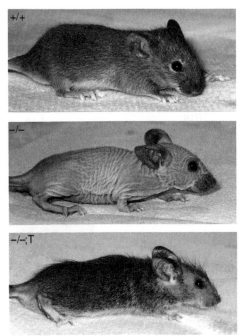

Catherine C. Thompson and Gerard M. J. Beaudoin, III, Chapter 11, Figure 7. Expression of HR in keratinocyte progenitor cells rescues hair regrowth in $Hr^{-/-}$ skin. Mice expressing HR in K14-positive cells (−/−; T) regrow hair, demonstrating that HR expression in these cells is sufficient for hair regrowth. Wild type (+/+), $Hr^{-/-}$ (−/−). Adapted from Beaudoin et al. (2005).

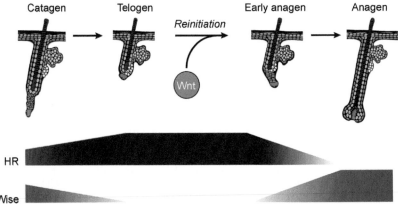

Catherine C. Thompson and Gerard M. J. Beaudoin, III, Chapter 11, Figure 8. Model for HR action in hair regrowth. Summary of HR protein expression through the hair cycle in relation to Wise mRNA expression. Catagen is characterized by the upregulation of HR protein and concurrent downregulation of Wise mRNA. HR protein expression during telogen is predicted to repress Wise mRNA expression, allowing activation of Wnt signaling in the bulge. Subsequent to the reactivation of hair growth, HR protein is downregulated while Wise mRNA increases; this predicted increase in Wise will cause a decrease in Wnt signaling. In *Hr* mutants, uncontrolled expression of Wise and possibly other Wnt inhibitors prevents the regrowth of hair by suppressing Wnt signaling.

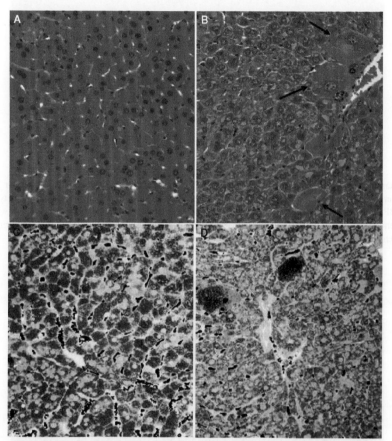

**Janardan K. Reddy, Dongsheng Guo, Yuzhi Jia, Songtao Yu and M. Sambasiva Rao, Chapter 12,
Figure 1**. Abrogation of PPARα ligand-induced peroxisome proliferation in PBP$^{Liv-/-}$ mouse
liver. (A) and (B) represent H&E stained, while (C) and (D) represent diaminobenzidine-stained
0.5 μm thick sections from wild-type (A and C) and PBP$^{Liv-/-}$ (B and D) mice treated with
Wy-14,643 (0.125% in diet) for 1 week. Note eosinophilic cytoplasmic staining of all liver cells in
wild-type and an occasional liver cell that is PBP$^{+/+}$ (arrows) in PBP$^{Liv-/-}$ mouse liver.
Diaminobenzidine cytochemistry reveals peroxisome proliferation (dark brown granules in the
cytoplasms of all hepatocytes in wild-type (C) and two PBP$^{+/+}$ cells (D, arrows) in PBP$^{Liv-/-}$
mouse treated with PPARα ligand.

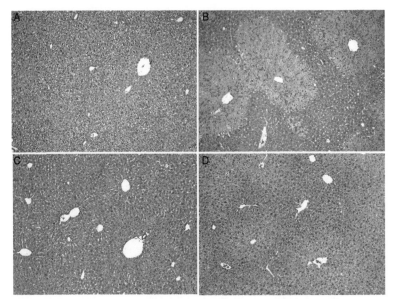

Janardan K. Reddy, Dongsheng Guo, Yuzhi Jia, Songtao Yu and M. Sambasiva Rao, Chapter 12, Figure 2. PBP$^{Liv-/-}$ mice exhibit resistance to acetaminophen-induced liver necrosis. Wild-type (PBP$^{+/+}$) (A and B) and PBP$^{Liv-/-}$ (C and D) mice treated for 3 days with saline (A and C) or with phenobarbital (B and D). Centrizonal liver necrosis occurs in phenobarbital-treated PBP$^{+/+}$ mice given a single dose of acetaminophen (B) but not in PBP$^{Liv-/-}$ mouse (D).

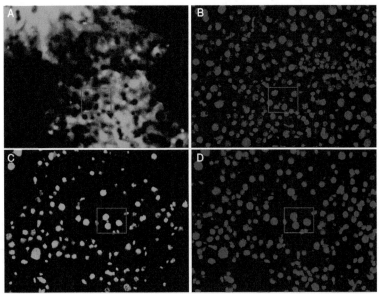

Janardan K. Reddy, Dongsheng Guo, Yuzhi Jia, Songtao Yu and M. Sambasiva Rao, Chapter 12, Figure 3. Phenobarbital-dependent nuclear translocation of CAR in wild-type (PBP$^{+/+}$) mice. Mice were injected with adeno-EGFP-CAR and were analyzed for phenobarbital-mediated nuclear translocation of CAR. Green fluorescence is predominantly cytoplasmic in mice not given phenobarbital untreated (A) and mostly nuclear in mice treated with phenobarbital (C). Nuclei are visualized by DAPI (B and D). [Panels A and B reproduced from Guo et al. (2006) with permission from Elsevier.]

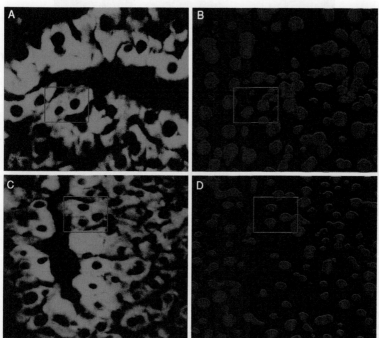

**Janardan K. Reddy, Dongsheng Guo, Yuzhi Jia, Songtao Yu and M. Sambasiva Rao, Chapter 12,
Figure 4**. Phenobarbital-mediated translocation of CAR to liver nuclei does not occur in PBP$^{Liv-/-}$
mouse. PBP$^{Liv-/-}$ were injected with adeno-EGFP-CAR and analyzed for expression. In these PBP
liver null mice, CAR expression is confined almost exclusively to the hepatocyte cytoplasm (A and
B) and did not translocate to the nucleus in response to phenobarbital (C and D).

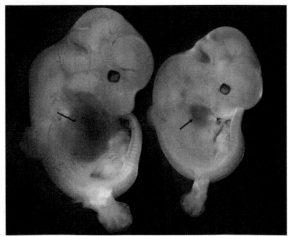

Janardan K. Reddy, Dongsheng Guo, Yuzhi Jia, Songtao Yu and M. Sambasiva Rao, Chapter 12, Figure 5. Transcription coactivator PRIP null mutation in the mouse. Growth retardation and embryonic lethality at E12.5. PRIP$^{-/-}$ embryo (right) is smaller than the PRIP$^{+/+}$ (left) littermate. Liver in PRIP$^{-/-}$ mice is smaller in size as compared to PRIP$^{+/+}$ liver (arrow).

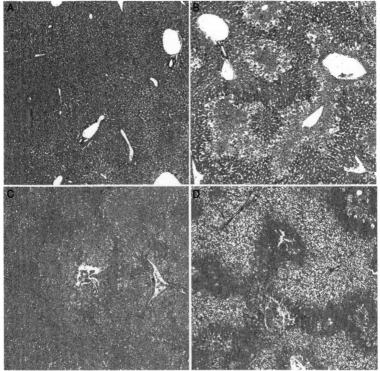

Janardan K. Reddy, Dongsheng Guo, Yuzhi Jia, Songtao Yu and M. Sambasiva Rao, Chapter 12, Figure 6. Liver-specific coactivator PRIP deficiency does not alter acetaminophen-induced hepatotoxicity. PRIP$^{+/+}$ and PRIP$^{Liv-/-}$ mice were given saline (A and C) or phenobarbital (B and D) for 3 days. Mice pretreated with phenobarbital received a single dose of acetaminophen (B and D). Centrizonal liver necrosis occurred in both wild-type and PRIP liver null mice.

Printed and bound by CPI Group (UK) Ltd, Croydon, CR0 4YY

08/05/2025

01864966-0005